Beleges. Nachdruck
14.6.2005 P.J.B.

Jürgen Renn (Ed.)

Einstein's Annalen Papers

The Complete Collection 1901 – 1922

Jürgen Renn (Ed.)

Einstein's Annalen Papers
The Complete Collection 1901 – 1922

WILEY-VCH Verlag GmbH & Co. KGaA

Editor

Prof. Jürgen Renn
Max Planck Institute
for the History of Science, Berlin

1. Auflage 2005
  1. Nachdruck der 1. Auflage 2005

This book was carefully produced. Nevertheless, editors, authors and publisher do not warrant the information contained therein to be free of errors. Readers are advised to keep in mind that statements, data, illustrations, procedural details or other items inadvertently be inaccurate.

Library of Congress Card-No.: applied for

British Library Cataloging-in-Publication Data:
A catalogue record for this book is available from the British Library

Bibliographic information published by
Die Deutsche Bibliothek
Die Deutsche Bibliothek lists this publication in the Deutsche Nationalbibliografie; detailed bibliographic data is available in the Internet at <http://dnb.ddb.de>.

© 2005 WILEY-VCH Verlag GmbH & Co. KGaA, Weinheim

All rights reserved (including those of translation into other language). No part of this book may be reproduced in any form – nor transmitted or translated into machine language without written permission from the publishers. Registered names, trademarks, etc. used in this book, even when not specifically marked as such, are not to be considered unprotected by law.

Printed in the Federal Republic of Germany
Printed on acid-free paper

Printing   Druckhaus „Thomas Müntzer" GmbH, Bad Langensalza
Bookbinding   Druckhaus „Thomas Müntzer" GmbH, Bad Langensalza

ISBN-13: 978- 3-527-40564-0
ISBN-10: 3-527-40564-X

Contents

Introduction
    Jürgen Renn 9

* * *

Einstein and the quantum hypothesis
    David C. Cassidy 15

* * *

Einstein's invention of Brownian motion
    Jürgen Renn 23

* * *

The optics and electrodynamics of 'On the Electrodynamics of Moving Bodies'
    Robert Rynasiewicz 38

* * *

Of pots and holes: Einstein's bumpy road to general relativity
    Michel Janssen 58

* * *

Folgerungen aus den Capillaritätserscheinungen [AdP **4**, 513 (1901)]
    A. Einstein 87

* * *

Ueber die thermodynamische Theorie der Potentialdifferenz zwischen Metallen
und vollständig dissociirten Lösungen ihrer Salze und über eine elektrische Methode
zur Erforschung der Molecularkräfte [AdP **8**, 798 (1902)]
    A. Einstein 99

* * *

Kinetische Theorie des Wärmegleichgewichtes und des zweiten Hauptsatzes
der Thermodynamik [AdP **9**, 417 (1902)]
    A. Einstein 117

* * *

Eine Theorie der Grundlagen der Thermodynamik [AdP **11**, 170 (1903)]
    A. Einstein 135

* * *

Zur allgemeinen molekularen Theorie der Wärme [AdP **14**, 354 (1904)]
A. Einstein 154

* * *

Über einen die Erzeugung und Verwandlung des Lichtes betreffenden
heuristischen Gesichtspunkt [AdP **17**, 132 (1905)]
A. Einstein 164

* * *

Über die von der molekularkinetischen Theorie der Wärme geforderte Bewegung
von in ruhenden Flüssigkeiten suspendierten Teilchen [AdP **17**, 549 (1905)]
A. Einstein 182

* * *

Zur Elektrodynamik bewegter Körper [AdP **17**, 891 (1905)]
A. Einstein 194

* * *

Ist die Trägheit eines Körpers von seinem Energieinhalt abhängig? [AdP **18**, 639 (1905)]
A. Einstein 225

* * *

Eine neue Bestimmung der Moleküldimensionen [AdP **19**, 289 (1906)]
A. Einstein 229

* * *

Zur Theorie der Brownschen Bewegung [AdP **19**, 371 (1906)]
A. Einstein 248

* * *

Zur Theorie der Lichterzeugung und Lichtabsorption [AdP **20**, 199 (1906)]
A. Einstein 259

* * *

Das Prinzip von der Erhaltung der Schwerpunktsbewegung und die Trägheit der Energie
[AdP **20**, 627 (1906)]
A. Einstein 268

* * *

Über eine Methode zur Bestimmung des Verhältnisses der transversalen
und longitudinalen Masse des Elektrons [AdP **21**, 583 (1906)]
A. Einstein 275

* * *

© 2005 WILEY-VCH Verlag GmbH & Co. KGaA, Weinheim

Die Plancksche Theorie der Strahlung und die Theorie der spezifischen Wärme
[AdP **22**, 180 (1907)]
    A. Einstein 280

* * *

Über die Gültigkeitsgrenze des Satzes vom thermodynamischen Gleichgewicht und
über die Möglichkeit einer neuen Bestimmung der Elementarquanta [AdP **22**, 569 (1907)]
    A. Einstein 292

* * *

Berichtigung zu meiner Arbeit: „Die Plancksche Theorie der Strahlung etc."
[AdP **22**, 800 (1907)]
    A. Einstein 296

* * *

Über die Möglichkeit einer neuen Prüfung des Relativitätsprinzips [AdP **23**, 197 (1907)]
    A. Einstein 297

* * *

Bemerkungen zu der Notiz von Hrn. Paul Ehrenfest: „Die Translation deformierbarer
Elektronen und der Flächensatz" [AdP **23**, 206 (1907)]
    A. Einstein 300

* * *

Über die vom Relativitätsprinzip geforderte Trägheit der Energie [AdP **23**, 371 (1907)]
    A. Einstein 303

* * *

Über die elektromagnetischen Grundgleichungen für bewegte Körper [AdP **26**, 532 (1908)]
    A. Einstein and J. Laub 317

* * *

Über die im elektromagnetischen Felde auf ruhende Körper ausgeübten
ponderomotorischen Kräfte [AdP **26**, 541 (1908)]
    A. Einstein and J. Laub 327

* * *

Berichtigung zur Abhandlung: „Über die elektromagnetischen Grundgleichungen
für bewegte Körper" [AdP **27**, 232 (1908)]
    A. Einstein and J. Laub 337

* * *

Bemerkungen zu unserer Arbeit: „Über die elektromagnetischen Grundgleichungen
für bewegte Körper" [AdP **28**, 445 (1909)]
    A. Einstein and J. J. Laub 339

* * *

Bemerkung zu der Arbeit von D. Mirimanoff: „Über die Grundgleichungen ..."
[AdP **28**, 885 (1909)]
    A. Einstein 343

* * *

Über einen Satz der Wahrscheinlichkeitsrechnung und seine Anwendung
in der Strahlungstheorie [AdP **33**, 1096 (1910)]
    A. Einstein and L. Hopf 347

* * *

Statistische Untersuchung der Bewegung eines Resonators in einem Strahlungsfeld
[AdP **33**, 1105 (1910)]
    A. Einstein and L. Hopf 357

* * *

Theorie der Opaleszenz von homogenen Flüssigkeiten und Flüssigkeitsgemischen
in der Nähe des kritischen Zustandes [AdP **33**, 1275 (1910)]
    A. Einstein 368

* * *

Bemerkung zu dem Gesetz von Eötvös [AdP **34**, 165 (1911)]
    A. Einstein 392

* * *

Eine Beziehung zwischen dem elastischen Verhalten und der spezifischen Wärme
bei festen Körpern mit einatomigem Molekül [AdP **34**, 170 (1911)]
    A. Einstein 398

* * *

Bemerkungen zu den P. Hertzschen Arbeiten: „Über die mechanischen Grundlagen
der Thermodynamik" [AdP **34**, 175 (1911)]
    A. Einstein 403

* * *

Bemerkung zu meiner Arbeit: „Eine Beziehung zwischen dem elastischen Verhalten ..."
[AdP **34**, 590 (1911)]
    A. Einstein 405

* * *

Berichtigung zu meiner Arbeit: „Eine neue Bestimmung der Moleküldimensionen"
[AdP **34**, 591 (1911)]
    A. Einstein 406

* * *

Elementare Betrachtungen über die thermische Molekularbewegung in festen Körpern
[AdP **35**, 679 (1911)]
    A. Einstein 408

* * *

Über den Einfluß der Schwerkraft auf die Ausbreitung des Lichtes [AdP **35**, 898 (1911)]
    A. Einstein 425

* * *

Thermodynamische Begründung des photochemischen Äquivalentgesetzes
[AdP **37**, 832 (1912)]
    A. Einstein 436

* * *

Lichtgeschwindigkeit und Statik des Gravitationsfeldes [AdP **38**, 355 (1912)]
    A. Einstein 444

* * *

Zur Theorie des statischen Gravitationsfeldes [AdP **38**, 443 (1912)]
    A. Einstein 460

* * *

Nachtrag zu meiner Arbeit: „Thermodynamische Begründung des photochemischen
Äquivalentgesetzes" [AdP **38**, 881 (1912)]
    A. Einstein 476

* * *

Antwort auf eine Bemerkung von J. Stark: „Über eine Anwendung
des Planckschen Elementargesetzes ..." [AdP **38**, 888 (1912)]
    A. Einstein 480

* * *

Relativität und Gravitation. Erwiderung auf eine Bemerkung von M. Abraham
[AdP **38**, 1059 (1912)]
    A. Einstein 481

* * *

Bemerkung zu Abrahams vorangehender Auseinandersetzung
„Nochmals Relativität und Gravitation" [AdP **39**, 704 (1912)]
    A. Einstein 487

* * *

Einige Argumente für die Annahme einer molekularen Agitation beim absoluten Nullpunkt
[AdP **40**, 551 (1913)]
    A. Einstein and O. Stern 489

* * *

Die Nordströmsche Gravitationstheorie vom Standpunkt des absoluten Differentialkalküls
[AdP **44**, 321 (1914)]
    A. Einstein and A. D. Fokker 500

* * *

Antwort auf eine Abhandlung M. v. Laues „Ein Satz der Wahrscheinlichkeitsrechnung und seine Anwendung auf die Strahlungstheorie" [AdP **47**, 879 (1915)]
    A. Einstein 509

* * *

Die Grundlage der allgemeinen Relativitätstheorie [AdP **49**, 769 (1916)]
    A. Einstein 517

* * *

Über Friedrich Kottlers Abhandlung „Über Einsteins Äquivalenzhypothese und die Gravitation"
[AdP **51**, 639 (1916)]
    A. Einstein 572

* * *

Prinzipielles zur allgemeinen Relativitätstheorie [AdP **55**, 241 (1918)]
    A. Einstein 577

* * *

Bemerkung zu der Franz Seletyschen Arbeit „Beiträge zum kosmologischen System"
[AdP **69**, 436 (1922)]
    A. Einstein 582

Introduction

Scope of the present volume

This volume presents Einstein's 49 contributions to *Annalen der Physik*, together with four introductory essays based on recent historical studies. The first three essays, by David Cassidy, Jürgen Renn, and Robert Rynasiewicz, discuss key aspects of the scientific revolution triggered by the pathbreaking papers of Einstein's *annus mirabilis* 1905, which changed our understanding of space, time, matter, and radiation. Various ramifications of these papers are worked out in Einstein's subsequent contributions to the *Annalen*. These papers document Einstein's further exploration of the quantum hypothesis and the triumphs of statistical physics as well as various stages of Einstein's journey from special to general relativity. General relativity is the subject of the fourth historical essay, by Michel Janssen.

The earliest contributions were written just after Einstein graduated with a teacher's diploma from the Swiss Federal Polytechnic School in Zurich; the latest while Einstein was working in Berlin as a member of the Prussian Academy of Sciences and as director of the Kaiser-Wilhelm Institute of Physics. The rise of Nazism in Germany put an end to this glorious period of the history of science. Einstein was forced to emigrate from Germany in 1933 and was never to return again. This volume, published in the centenary of Einstein's *annus mirabilis*, offers the reader a comprehensive overview of the breathtaking scope and depth of the investigations of the towering figure of 20th-century physics, focusing on his most productive years. The dramatically changing historical circumstances under which these papers were written may also serve as a reminder of the fragility of the scientific enterprise and the need both to reflect on its contexts and to strengthen it by civil courage, just as Einstein has taught us.

Foundation and role of the *Annalen*

The *Annalen der Physik,* one of the most influential journals in the history of physics, was founded in 1790 by Friedrich Albert Carl Gren, a professor of physics and chemistry at Halle University. As is described in the masterful account of the rise of theoretical physics by Christa Jungnickel and Russel McCormmach (*Intellectual Mastery of Nature*, University of Chicago Press), the original mission of the *Annalen* was to familiarize its German-speaking readership with the results of investigations pertaining to the mathematical and chemical parts of the theory of nature, including reports from other journals, foreign as well as German. From the outset, the spirit of the journal was international and integrative and continued to be so under the subsequent editors, in particular Ludwig Wilhelm Gilbert and Johann Christian Poggendorff, who succeeded in turning it into a principal point of reference for the German-speaking scientific community in physics and chemistry, which included not only university professors, but also teachers, doctors, and apothecaries.

Original contributions published in the *Annalen* were soon translated or reported in foreign journals. In spite of the rising specialization, the editors paid close attention to the interconnections between the broad variety of subjects treated in the articles. While the emphasis was on experimental work, the rising significance of theoretical contributions was acknowledged as well. The wide distribution of the *Annalen*, which was available not only in university libraries, but also in secondary and technical schools, furthered the formation of a broadly accessible scientific culture. Accordingly, the *Annalen* remained open to contributions not only from established physicists and institute directors, but also to articles submitted by students, assistants, and teachers. Its role as an intellectual reference point was reinforced by the foundation of the *Beiblätter*, which offered brief reports on work not published in the *Annalen*.

The subjects treated in the *Annalen* over the years reflect the development of research in 19th century physics and chemistry. Under the editorship of Gustav Wiedemann, who took office after Poggendorff's death in 1877, the broad perspective of the journal was maintained and occasionally even included articles on the history of science. All in all, the journal was transformed into a means of communication oriented towards the increasingly professionalized community of physicists. Yet the growing hints at the existence and relevance of a microworld of atoms and molecules for the understanding of nature kept alive the promise of unity in the dispersive multitude of results published in the *Annalen*.

The *Annalen* as food for thought

This was roughly the situation when the young Einstein began to avidly study the *Annalen*, which had been edited since 1900 by Paul Drude. Drude's work on an atomistic theory of conduction in metals was of special interest to Einstein and the precocious young student even entered into a controversy with Drude.

Einstein's originality is often attributed to his autodidactic training. But the possibility to learn independently obviously very much depends on the availability of appropriate reading material. Although academically isolated, it was the *Annalen* that offered Einstein an up-to-date overview of contemporary physics, stimulating many of the original ideas he pursued during his student days and his time at the Swiss patent office in Bern. His contemporary correspondence suggests that he often believed he just had to put the pieces of a puzzle together in order to achieve a breakthrough, pieces he often found in papers he read in the *Annalen*.

Apart from Drude's work on the electron theory of metals, which eventually stimulated Einstein's development of statistical mechanics, he also read Max Planck's work on black-body radiation and Philipp Lenard's studies of the photoelectric effect, which triggered his work on the light quantum hypothesis. Also Wien's report on the problematic attempts to detect the translatory motion of the ether offered an important stepping stone towards the rejection of the ether and the formulation of the special theory of relativity.

The *Annalen* as a source of income

The *Annalen* also served as a source of modest additional income for Einstein, who wrote more than twenty reports for its *Beiblätter* – mainly on the theory of heat – thus demonstrating an impressive mastery of the contemporary literature. This activity started in 1905 and probably resulted from his earlier publications in the *Annalen* in this field. Going by his publications between 1900 and early 1905, one would conclude that Einstein's specialty was thermodynamics.

Beginners' papers

The collection begins with what Einstein later designated his two "worthless beginners' papers," one on capillarity published in early 1901 and the other on dilute salt solutions published in 1902. Both are dedicated to an investigation of the nature of molecular forces through the effect of such forces on phenomena in liquids, a subject Einstein also planned to investigate for his dissertation, a plan he then abandoned.

His early exploration of a molecular theory of solutions nevertheless helped shape many of the techniques used in the dissertation he did complete in 1905. It dealt with the determination of molecular dimensions. It was published in the *Annalen* in 1906 and is included in this collection. The investigations documented by Einstein's first papers also provided a motivation for generalizing the methods of the kinetic theory, and for establishing statistical mechanics independently from Josiah Gibbs.

Statistical mechanics

The pivotal role of statistical mechanics in Einstein's early work is clearly visible in this collection. While its development was obviously driven by his early atomistic speculations, the statistical framework he established between 1902 and 1904 provided the backbone for his papers on the light quantum and on

Brownian motion of 1905. It pointed to the crucial role of fluctuations in discerning the non-classical character of heat radiation, and revealed atomic dimensions in his analysis of Brownian motion.

The *annus mirabilis* 1905

Without detracting from the singularity of Einstein's 1905 papers in the history of science, this collection may help to frame these contributions in the context of his intellectual development, as is discussed in the historical essays opening this volume. The 1905 papers deal with subjects as diverse as heat radiation, Brownian motion, and the electrodynamics of moving bodies. How were these topics related in Einstein's mind? In view of his earlier publication record and of insights gained from his contemporary correspondence, it seems plausible to assume that one unifying theme goes back to Einstein's early pursuit of atomistic ideas, which includes both the quest for evidence for the existence of atoms and speculative ideas such as that of a corpuscular constitution of light.

Later these speculations turned into the exploration of the limits of classical physics, as Einstein encountered them when critically reading the *Annalen*. His perception of these limits was sharpened by the philosophical acumen he had developed through his reading of authors such as Hume, Kant, Mach, and Poincaré. All three of the revolutions that Einstein initiated in 1905 originated from problems at the borders between the major conceptual domains of classical physics; mechanics, electrodynamics, and thermodynamics. Special relativity emerged from the electrodynamics of moving bodies, an area at the intersection of electrodynamics and mechanics; the light quantum hypothesis can be seen as an attempt to cope with the problem of heat radiation, a problem at the intersection of electrodynamics and thermodynamics; while Einstein's work on Brownian motion deals with a borderline problem of mechanics and thermodynamics.

The year 1905 was just the beginning of Einstein's career and of the scientific revolution triggered by his pathbreaking contributions. This becomes evident from his own subsequent publications, which show that Einstein's contributions should not be seen as a series of isolated achievements, but as integrated in a lively scientific context, involving collaborative efforts and discussions – polemics even – with his colleagues.

Electrodynamics in moving media

Einstein's 1908 paper with Jakob Laub on the electrodynamics of moving media was, for instance, a direct continuation of his 1905 work on the electrodynamics of moving bodies, which focused on microscopic electron theory, extending it, following prior work by Minkowski, to the macroscopic theory of electromagnetic and optical phenomena in polarizable and magnetizable material media in motion. It was in this context that Einstein was first confronted with the four-dimensional spacetime formalism developed by Minkowski. In their own work, Einstein and Laub avoided this formalism, the value of which Einstein only gradually learned to appreciate.

Specific heats

The present collection also documents Einstein's early efforts to further explore the consequences of his revolutionary interpretation of Planck's formula for black-body radiation as hinting at a non-classical foundation of physics. Such an exploration was needed all the more since Einstein's interpretation – in particular the light quantum hypothesis – met, in contrast to his other 1905 achievements, with little sympathy from his established colleagues.

A first milestone of this exploration was Einstein's 1907 paper on the specific heat of solid bodies, which exploited the insight into the non-classical behavior of atomic oscillators for a new understanding of the thermal properties of solid bodies, in particular at lower temperatures. The experimental confirmation in Nernst's laboratory of the prediction of the decrease of specific heats with temperature turned out to be crucial for Einstein's career and his eventual move to Berlin in 1914.

Elastic behavior of solids

This line of research is continued in a paper of 1911 about the relation between molecular vibrations and optical wavelengths in the infrared region, which exploits the connection that Einstein had established between molecular vibrations and specific heats. He thus succeeded in propagating the quantum discontinuity from its original locus in radiation theory to yet another range of physical phenomena, identifying, very much in the vein of his early atomistic speculations, a link between the thermal and mechanical properties of a solid.

Collaboration with Hopf

Planck and others remained skeptical of Einstein's claim that a new radiation theory was required. Challenged by this skepticism, Einstein in 1910 published two papers together with Ludwig Hopf on the statistical properties of the radiation field. Their main purpose was to provide support for the claim that classical radiation theory leads to unacceptable implications for heat radiation and that Planck's radiation formula does imply a break with classical physics.

Critical opalescence

Einstein's 1910 work on critical opalescence was both a direct continuation of his earlier work on fluctuations and a reaction to a contemporary issue raised by the Polish physicist, Marian von Smoluchowski, who in 1905 had analyzed independently from Einstein the statistical properties of Brownian motion. In 1908 Smoluchowski published a paper on critical opalescence in the *Annalen*, which dealt with the optical effects occurring near the critical point of a gas and near the critical point of a binary mixture of liquids. In his paper, Einstein provided a quantitative derivation of the effect from a treatment of density fluctuations. His key insight was that both critical opalescence and the blue color of the sky can be explained with the help of such density fluctuations, which originate from the atomistic constitution of matter.

Photochemical equivalence law

Another contribution illustrating Einstein's attempts to explore the quantum hypothesis at a time when he had already begun to despair about ever capturing it in a coherent theory is his influential 1912 paper about the photochemical equivalence law, the beginning of a line of research that would lead him in 1916 to his ground-breaking rederivation of Planck's law based on the concepts of spontaneous and induced emission.

Zero-point energy

The 1913 paper by Einstein and Otto Stern also testifies to the early struggle to understand the status of Planck's radiation law and its implications for applying the quantum hypothesis to the atomistic conception of matter. Einstein and Stern attempted to develop a quantum theory of rotating diatomic molecules, which show that the notion of zero-point energy – first introduced by Planck in his "second quantum theory" – could be used to interpret measurements of the specific heat of hydrogen at low temperatures. But Einstein soon became skeptical of some of the arguments in this paper and considered zero-point energy, as he put it in a letter to his friend Paul Ehrenfest, "as dead as a doornail."

Light deflection

While ever more desperate about the quantum, Einstein became increasingly involved with the idea of formulating a relativistic field theory of gravitation, modeled on electromagnetic field theory. As early as

1907, while working on a review of special relativity, he had realized that, if such a theory were to incorporate Galileo's principle that all bodies fall with the same acceleration, it would require yet another fundamental revision of our concepts of space and time. This led Einstein to formulate his famous equivalence principle, by which gravitation and inertia ultimately became as intertwined as the electric and the magnetic field in the first relativity revolution.

This collection contains some of the early papers marking Einstein's path from special to general relativity such as his 1911 paper predicting the deflection of light by the gravitational field of the sun.

Static gravitational fields

The collection also includes a number of papers illustrating some of the heuristic strategies Einstein adopted as well as some of the obstacles he had to overcome in his search for a relativistic field theory of gravitation. As documented by the papers in this volume, he started in 1912 by treating the special case of a static gravitational field with the help of the equivalence principle, which allowed him to use knowledge about acceleration in the absence of gravity to draw conclusions about physical effects in the presence of a gravitational field.

While making impressive advances in this way, such as the prediction of light deflection and his recognition of the need for non-Euclidean geometry, these early successes consolidated a framework of expectations rooted in classical physics, many of which had to be abandoned or seriously modified before general relativity could be established.

Controversy with Abraham

One can argue that, unlike special relativity, general relativity was essentially the achievement of a single man. As a matter of fact, most of Einstein's established colleagues were skeptical about his attempt to build a new theory of gravitation on the idea of curved spacetime described by a ten-component metric tensor rather than the familiar scalar potential of Newton's theory.

It is important to realize, however, that Einstein was not only supported by some friends and collaborators such as his Swiss companions Marcel Grossmann and Michele Besso and by the astronomer Erwin Freundlich, but that he also had to face competitors and opponents who provided his endeavor with a scientific context that was crucial for the emergence of general relativity. It was Max Abraham, for instance, and not Einstein, who first formulated a comprehensive gravitational field theory in 1912, thus challenging Einstein to integrate his own considerations based on the equivalence principle into a coherent theory as well. Our collection contains the papers resulting from these efforts while offering some glimpses of the heated controversy in which this early competition resulted.

Nordström's special relativistic theory of gravitation

While Einstein was initially convinced that the problem of gravitation could not successfully be addressed within the framework of special relativity, Abraham's failed attempt to provide such a theory was followed by a more convincing theory developed by Gunnar Nordström in the years between 1912 and 1913. Nordström's theory was a serious competitor of nascent general relativity. It might well have become the dominating relativistic theory of gravitation for some time had it not been for Einstein's philosophically motivated quest to combine such a theory with the attempt to generalize the principle of relativity. This collection features a paper resulting from a collaboration with Adriaan Fokker, which shows how Nordström's theory can be reformulated in terms of the absolute differential calculus, the mathematical language Einstein had adopted in his own search for a field theory of gravitation. In this way, it became possible to compare the two approaches more directly and to reveal the assumptions underlying Nordström's theory. At the same time it suggested that Nordström's theory, like Einstein's, went beyond special relativity and would likewise involve curved space-time.

Foundations of general relativity

It took Einstein eight years, from 1907 to 1915, to attain his goal of a relativistic field theory of gravitation that preserved both the heritage of mechanics and that of field theory. The drama of this struggle with the conceptual foundations of classical and special relativistic physics is documented by Einstein's research manuscripts, by his correspondence, by several intermediary publications, and in particular by the famous sequence of communications to the Prussian Academy of November 1915.

A comprehensive reconstruction of this drama including key sources appears elsewhere (*The Genesis of General Relativity*, Kluwer Academic Publishers, edited by J. Renn). The present collection features the outcome of this quest – the general theory of relativity – in the form of Einstein's first masterful exposition of the finished theory in his famous 1916 contribution to the *Annalen*. This paper bears clear traces of the gestation period of the theory, as is demonstrated in the historical essay of Michel Janssen.

Cosmology

Einstein's subsequent work on general relativity is no longer extensively documented in the *Annalen*. As a newly minted member of the Prussian Academy in Berlin, his outlet of choice in this period are the Academy's own *Sitzungsberichte*. Both the four celebrated papers of November 1915 documenting the final breakthrough in Einstein's search for a relativistic field theory of gravity and the famous paper on cosmology of 1917 appeared in the *Sitzungsberichte*. This volume, however, does contain a short but important paper of 1918 on the foundations of general relativity, in which Einstein formally introduced what he called "Mach's Principle," the requirement that matter fully determines the metric field. The volume ends with a short paper of 1922 providing at least a hint at the fate of general relativity, which was subsequently turned from a philosophically motivated integration of the classical knowledge about gravitation with the kinematics of relativity into the theoretical foundation of modern cosmology describing an expanding universe. In this 1922 paper, Einstein reacted to a proposal by Franz Selety for resolving Einstein's objections to Newtonian cosmology of 1917 by what he called a "hierarchical molecular world." Einstein rejected this proposal because it did not, in his view, comply with Mach's principle. He also rejected the interpretation of the spiral nebulae as galaxies similar to our own milky way, referring to the evidence of contemporary observations. The cosmological mission of general relativity was yet to be accomplished.

Perspective

The present collection offers a first entry point into Einstein's work, which is being published comprehensively in an annotated documentary edition by the *Collected Papers of Albert Einstein* (Princeton University Press). Here the reader will find more extensive commentaries and annotations that offer insights into the genesis and historical context of Einstein's papers. In line with Einstein's legacy and spirit of broadly sharing scientific knowledge, the Editor-in-Chief of the *Annalen*, Ulrich Eckern, and WILEY-VCH have consented, in agreement with the Albert Einstein Archives at the Hebrew University Jerusalem and the *Collected Papers*, and in collaboration with the Max Planck Society, to make the papers in this collection freely accessible on the Internet.

Acknowledgements I would like to thank Lindy Divarci for her role as editorial assistant in the preparation of this volume.

Jürgen Renn (Berlin)

Einstein and the quantum hypothesis

David C. Cassidy*

Natural Science Program, Hofstra University, Hempstead, NY 11549, USA

Published online 14 February 2005

Key words Einstein, light quantum, quantum hypothesis, photoelectric effect.
PACS 01.65.+g, 05.20.Dd, 42.50.Ct, 44.40.+a

In 1905 Einstein was the first to propose that light behaves in some circumstances as if it consists of localized units, or quanta, of energy–light quanta. He showed that this hypothesis could account for several phenomena, including in particular the photoelectric effect [1]. In 1922 Einstein was awarded the Nobel Prize for Physics for the year 1921, "for his services to Theoretical Physics, and especially for his discovery of the law of the photoelectric effect". [2]

In 1900 Max Planck had derived a law for the frequency distribution of the energy of thermal, or blackbody, radiation contained in a material cavity of perfectly reflecting walls. He did so by assuming that the energy density is divided into discrete energy elements distributed over a large number of assumed charged harmonic oscillators in equilibrium with the thermal radiation [3]. In his subsequent papers, Einstein showed in 1906 that Planck's theory agreed implicitly with the light-quantum hypothesis, since it required the energy quantization of the charged oscillators in amounts corresponding to emitted and absorbed light quanta [4]. In 1907 Einstein demonstrated the necessity of quantizing the harmonic oscillators in a derivation of Planck's law using methods of statistical mechanics, and he showed that the quantization of the atomic oscillators in a solid crystal lattice results in an equation for the specific heat of solids as a function of temperature that deviates from the classical Dulong-Petit law with decreasing temperature [5]. Together, these papers demonstrated that Planck's radiation law, which had been experimentally confirmed, could not be reconciled with either of the two pillars of classical theory, Maxwell's electromagnetic theory or the kinetic-molecular theory of matter. Fundamental revisions of both electromagnetic and mechanical theory were required in order to encompass the quantum nature of both radiation and matter. In a letter to his friend Conrad Habicht in 1905, Einstein called his forthcoming light-quantum paper "very revolutionary, as you will see". [6]

The quantum hypothesis always remained for Einstein a problem and a puzzle, rather than a solution. He called his first proposal of the hypothesis, for the behavior of light, only "a heuristic viewpoint" that enabled natural explanations of certain phenomena. In 1907 he declared that the "new conception of the phenomena of light emission and light absorption... still in no way possesses the character of a complete theory, but it is remarkable in so far as it enables the understanding of a series of regularities." [7] In 1909 Einstein wrote to his first collaborator Jakob Laub of his intense struggle with the problem of light quanta: "I am ceaselessly occupied with the problem of the constitution of radiation... This question of quanta is so extraordinarily important and difficult that everyone should be concerned with it." [8]

Einstein struggled with the quantum hypothesis for the rest of his life. He was never satisfied with quantum mechanics, even though he had contributed greatly to bringing it about. He never regarded it as the fundamental theory that was needed. He pointed to Max Born's later statistical interpretation of the wave function as an indication of this. In 1926 he wrote to Born: "The quantum mechanics is very worthy of

---

* E-mail: david.cassidy@hofstra.edu

regard. But an inner voice tells me that this is not the true Jacob. The theory yields a lot, but it hardly brings us closer to the secrets of the Old One. In any case I am convinced that He does not throw dice." [9]

A great deal of historical literature has been written about Einstein and the early quantum theory. With the publication of *The Collected Papers of Albert Einstein*, a good understanding of that early history is now available. The remainder of this paper explores the background to Einstein's work, his early papers on the quantum hypothesis, and the reception of his work, based upon this extensive literature [10].

Background

During his student years at the ETH in Zurich in the period 1896–1900, Einstein frequently studied on his own the works of the leading physicists at the time. Among the works he studied was Ernst Mach's *Wärmelehre*, which contained a discussion of the frequency distribution of the energy of black-body radiation, including Kirchhoff's work on the problem. The problem was of interest not only to the nascent electric lighting industry, but also to theoretical physicists for its universal character. Among other early developments was the "Wien displacement law," introduced by Wilhelm Wien in 1894. It related the energy E_λ emitted by a black-body at temperature T in the interval of wavelength, λ to $\lambda + d\lambda$:

$$E_\lambda = \lambda^{-5} \phi(\lambda T), \qquad (1)$$

were $\phi(\lambda T)$ is a universal function, independent of the material substance.

Two years later, Wien inferred that the function ϕ, expressed in terms of frequency ν, is an exponential function of ν and T, and that the energy density ρ_ν per unit frequency is:

$$\rho_\nu = \alpha \nu^3 e^{-\beta \nu / T} \qquad (2)$$

where α and β are constants. (The symbols in all equations have been made uniform and in some cases updated to modern notation). Friedrich Paschen and others in Berlin experimentally confirmed this expression for visible radiation up to large values of ν/T [11].

Beginning in 1897 Berlin theorist Max Planck, drawn to the study of black-body radiation in search of an electromagnetic account of irreversibility, attempted to derive Wien's result from electromagnetic and thermodynamic principles. He considered thermal radiation in a cavity in equilibrium at temperature T with charged matter consisting, for simplicity, of charged simple harmonic oscillators (electric dipoles), acting as emitters and absorbers of radiation. (Since the spectral distribution was independent of the material structure of the cavity, he could make this assumption.) Further assuming a completely random distribution of the radiant energy (what he called "natural radiation") at equilibrium, he obtained an important fundamental expression relating the radiation density ρ_ν per unit frequency on the one hand to the average energy E_ν of the material oscillators on the other:

$$\rho_\nu = \frac{8\pi\nu^2}{c^3} E_\nu, \qquad (3)$$

where c is the speed of light. After further assumptions about the thermodynamic entropy of the oscillators, Planck obtained through mathematical manipulation an expression for the average energy E_ν of an oscillator that, when substituted into Eq. (3), yielded Wien's distribution law, Eq. (2) [12].

If instead of calculating the average energy E_ν from the entropy function of the oscillators, Planck had used the kinetic-molecular theory and its well-established equipartition theorem, which gave an average energy of kT for each vibration mode, he would have obtained the famous Rayleigh-Jeans law, which Rayleigh first published in 1900, with a correction by Jeans in 1905 [13]:

$$\rho_\nu = \frac{8\pi\nu^2}{c^3} kT. \qquad (4)$$

As is well-known, this expression yields a catastrophic infinite total energy density when integrated over the infinite frequency spectrum of thermal radiation. Also, for constant T, it fails at higher radiation frequencies, where Wien's formula prevails. But experiments by Rubens and Kurlbaum in Berlin showed that Eq. (4) did in fact agree with the data at low frequencies, where Wien's formula was now found to fail [14].

After Rubens and Kurlbaum privately informed Planck of their findings, Planck set out in 1900 to join the two distribution laws together into a single equation for the energy density ρ_ν of black-body radiation per unit frequency at temperature T, applicable to all radiation frequencies [15]. Again through ingenious mathematical manipulation of the entropy function, Planck was able to obtain a suitable expression:

$$\rho_\nu = \frac{A\nu^3}{e^{\beta\nu/T} - 1}, \tag{5}$$

where A and β are constants. The challenge now was to derive this equation from *physical* considerations.

For this task Planck turned to the molecular-kinetic theory of the assumed oscillators, specifically to Boltzmann's equation (which Einstein later called "Boltzmann's principle") for the entropy of a canonical ensemble of oscillators, $S = k \log W$. Here k, introduced by Planck, is Boltzmann's constant, which is equal to R/N, where R is the gas constant, and N is Avogadro's, or Loschmidt's, number. Planck designated S as the entropy of a system of N identical oscillators of fundamental frequency ν. W is the statistical probability of a state, based upon the number of discrete "complexions" compatible with the total energy of the system. Planck realized that in order to calculate W using combinatorials, the oscillator energies could not be continuous, as far as this calculation was concerned. In order to make the calculation, Planck divided the total energy E into an integral number P of energy elements ϵ: $E = P\epsilon$, where $\epsilon = h\nu$, h being a constant – subsequently called Planck's constant. The distribution of the P energy elements over the N oscillators yielded an entropy function S that provided an expression for the average energy of an oscillator E_ν that, when substituted into Eq. (3), resulted in the now-famous Planck law for the frequency distribution of the energy density of black-body radiation:

$$\rho_\nu = \frac{8\pi\nu^2}{c^3} \frac{h\nu}{e^{h\nu/kT} - 1}. \tag{6}$$

Rubens immediately confirmed Planck's result, but Planck was not satisfied. His division of the energy into discrete elements distributed discontinuously over the oscillators was obviously inconsistent with classical physics. The constant h represented a unit of action, but there was no principle of conservation of action in classical physics. For Planck, the constant h and the discrete energy elements were only artifacts of an attempt to derive the energy distribution formula. He spent the rest of the decade attempting to revise the derivation so as to rid physics of the constant that now bears his name.

Enter Einstein

Einstein, too, was dissatisfied with Planck's derivation. In an undated letter attributed to 1901, he wrote his friend Mileva Marić: "About Max Planck's studies on radiation, misgivings of a fundamental nature have arisen in my mind, so that I am reading his article with mixed feelings" [16]. He did not publish on black-body radiation until the last of his fundamental studies of statistical mechanics in 1904, but he was apparently still thinking about Planck's work [17]. Yet in his 1905 paper on the light-quantum hypothesis, he did not make direct use of Planck's result, even though he mentioned Planck's work and even wrote his equation. Apparently he viewed Planck's theory, as he said a year later: "in a certain sense as an antithesis to my work." [18] Einstein was interested in 1905 in the behavior of the radiant energy, not the oscillators, and he was apparently still uncertain what to make of Planck's derivation. Instead, Einstein focused on the Wien distribution law, Eq. (2), and its consequences "with the help of the Boltzmann principle" [19].

As in his other papers in 1905, Einstein began by observing an asymmetry in physics. In this case it was "a profound formal difference" between Maxwell's electromagnetic theory, involving continuous processes in space, and the kinetic-molecular theory of mechanics, involving discrete particles. Einstein demonstrated how the utilization of both viewpoints in handling a single problem can lead to difficulties in the case of black-body radiation [20]. Here he considered a cavity containing a low-density gas and a large number of charged oscillators in equilibrium with radiation at temperature T. While electromagnetic theory leads to Eq. (3) for the energy density of the radiation, as Planck had derived, dynamical equilibrium requires an average energy of the oscillators corresponding to the vibration mode ν given by the equipartition theorem: $E_\nu = (R/N)T = kT$, which leads directly to the untenable Rayleigh-Jeans formula for all frequencies. (Einstein's use of R/N is in general replaced by k hereafter.)

Planck's experimentally confirmed law could be regarded as independent to a certain extent from Planck's derivation of it. Since it yielded Wien's law at high values of ν/T, where the Rayleigh-Jeans law of classical mechanics failed completely, Einstein turned to Wien's law, Eq. (2), where he expected to find, and did find, a startling new insight regarding the nature of the radiation.

Einstein began by showing that if one slowly changes the volume of a black-body radiation cavity from v_0 to v, while keeping the energy E of the radiation of frequency ν constant, then the change in the entropy of the radiation, $S - S_0$, is:

$$S - S_0 = \frac{E}{\beta\nu} \log\left(\frac{v}{v_0}\right). \tag{7}$$

Einstein then showed that, using Boltzmann's equation for the entropy, the entropy change of an ideal gas of n particles as its volume slowly changes from v_0 to v is:

$$S - S_0 = nk \log\left(\frac{v}{v_0}\right). \tag{8}$$

The similarly between Eqs. (7) and (8) was striking. Einstein examined the similarity more closely. Equation (8) arises from Boltzmann's principle, $S = k \log W$, by assuming a canonical ensemble of n ideal-gas particles distributed randomly throughout the volume at equilibrium and moving completely independently of one another. In that case, the probability of finding all n particles in the subvolume v of v_0 is

$$W = \left(\frac{v}{v_0}\right)^n,$$

which yields Eq. (8). Equation (8) could arise only by assuming a canonical ensemble of completely random and independently moving particles.

From this result and Eq. (7), Einstein inferred that the probability of finding all of the radiation in the cavity's subvolume v of v_0 is:

$$W = \left(\frac{v}{v_0}\right)^{E/\beta\nu k},$$

whereby $E/k\beta\nu$ behaves, in analogy to n, as an integer. From this Einstein made the profound conclusion:

> Monochromatic radiation of low density (within the region of validity of the Wien distribution law) behaves with respect to thermal phenomena as if it consisted of independent energy quanta of magnitude $R\beta\nu/N[= h\nu]$ [21].

Apparently because he was considering only Wien's law, Einstein did not then identify R/N with k and $k\beta$ with h, the constant in Planck's formula, where $\beta = h/k$.

Although Einstein considered the hypothesis of energy quanta only a "heuristic viewpoint," he immediately applied it to an explanation of several observed phenomena, including in particular the then puzzling phenomenon of the photoelectric effect. Originally discovered by Heinrich Hertz in 1886, the photoelectric effect, since studied by Philipp Lenard, had indicated evidence of what we now call a threshold frequency ν_0; an increase with frequency of the kinetic energies of the electrons emitted from the surface of a metal when irradiated by ultraviolet light; and electron kinetic energies that were independent of the intensity of the light [22]. Although these were still tentative and sometimes qualitative results, they could not be reconciled with Maxwellian wave theory, where radiation energy is proportional to the intensity and is spread over an expanding wave. Einstein, however, argued that these observations could be easily explained if light is treated instead as if it consisted of a stream of independent, particle-like light quanta. When a light quantum strikes an electron in the metal, it delivers its entire energy, $h\nu$ (in current notation), to the electron, which would escape the metal with a maximum kinetic energy $h\nu$, less the work P required to escape the metal:

$$KE_{\max} = h\nu - P. \qquad (9)$$

Einstein's predicted result was not confirmed until nearly a decade later, when in 1914 Robert A. Millikan began obtaining results that he finally published in 1916. Still resistant, like many others, to the notion of light quanta, Millikan declared the results a complete confirmation of Einstein's predicted equation, but not a confirmation of Einstein's "bold, not to say reckless, hypothesis" of light quanta. Most physicists still did not accept the existence of light quanta until Compton's publication of the "Compton effect" in 1923 [23].

Einstein's analysis of Planck's formula

If Einstein's 1905 analysis of Wien's distribution law, a limiting case of Planck's law, suggested a possible unification of physics on the basis of the discrete mechanics of particles and light quanta, his analysis of Planck's radiation formula from 1906 to 1909 rendered the matter more complicated. Einstein had been pondering Planck's formula since at least 1901. His work on the foundations of statistical mechanics during the intervening years provided him the tools and the fundamental insights he needed for the task [24].

Equation (3), representing the interaction of radiation with matter, relates the energy density of electromagnetic radiation per unit frequency to the average mechanical energy of the emitting and absorbing oscillators in the same frequency range [25]. Planck's derivation of this equation, said Einstein, is consistent with the Maxwell-Lorentz theory for disordered equilibrium radiation interacting with oscillators, but application of the molecular-kinetic theory to a canonical ensemble of charged oscillators in equilibrium with the radiation yielded the disastrous Rayleigh-Jeans law. "How is it," Einstein asked in 1906, "that Herr Planck did not arrive at the same formula, but rather at the expression [Eq. (6)]?" [26]

The answer lay in contradictions. Examining a canonical ensemble of oscillators in equilibrium with the radiation at temperature T, Einstein first showed that Planck's derivation contradicted Boltzmann's definition of the probability function W in $S = k \log W$. At equilibrium all of the possible "complexions" of the oscillators consistent with the total energy E are assumed to be equally probable in Boltzmann's principle, but Planck's distribution of P discrete energy elements among the N oscillators meant that, in fact, not all complexions are equally probable; only those energy states that contain an integral multiple of energy elements ϵ are allowed. Second, said Einstein, while Planck's derivation of Eq. (3) relied upon the assumption that the oscillator energies vary continuously, the derivation of the average energies E_ν leading to Planck's law relied upon a discontinuous variation of their energies. This led Einstein to the profound conclusion that Planck was right: if Planck's equation is to be derived at all, not only must the energy of the radiation be quantized but, it appears, so too must the energies of the oscillators: "The energy of an elementary resonator can take on only values that are integral multiples of $(R/N)\beta\nu$; the energy of a resonator is changed jump-wise through absorption and emission, that is by whole-number multiples of $(R/N)\beta\nu$ [i.e., $h\nu$]" [27].

So, how was it that Maxwell's continuous wave theory, which no longer seemed applicable to either radiation or resonators, was able to yield the valid equation, Eq. (3), which leads to Planck's law? Einstein's answer still holds today. "Planck's theory thus entails the assumption: Although the Maxwell theory is not applicable to elementary resonators, the *mean* energy of an elementary resonator located in a radiation cavity is equal to that which one calculates by means of the Maxwell theory of electricity" [28].

A year later, in 1907, Einstein returned again to an analysis of the origin of the Planck formula. It was not possible, he now concluded definitively, to obtain the Planck formula without "a modification of the molecular-kinetic theory of heat" [29].

Einstein began by demonstrating once again that a consistent application of the full apparatus of statistical mechanics to a canonical ensemble of oscillators of frequency ν in equilibrium with thermal radiation leads inexorably to the Rayleigh-Jeans law, Eq. (4). In order to arrive at the Planck formula, Einstein once again preserved Eq. (3), derived from matter and radiation interacting in thermal equilibrium. By assuming that Maxwell's theory does yield the correct relationship between radiation and average oscillator energies, he then showed that the only way to obtain Planck's formula from the apparatus of statistical mechanics was to introduce an arbitrary restriction of the oscillator energies to integral multiples of $h\nu$. He did this by introducing a weighting function into the calculation of the Boltzmann probability function W such that the probability of an oscillator state is zero if its energy is not infinitesimally close to 0 or to an integral multiple of the elementary energy $h\nu$, and 1 otherwise. "We must assume," he declared, "that the mechanism of the energy transfer is such that the energy of an elementary body [or combinations of elementary oscillating bodies – *his note*] can take on exclusively only the values 0, $h\nu$, $2h\nu$, etc." [30]. From now on, Einstein began referring more often to the "quantum hypothesis," rather than to the "light-quantum hypothesis."

In the same paper of 1907 Einstein suggested that if Planck's theory "hits the core of the matter," then we should expect that the quantization of mechanical energy will remove other contradictions of theory with experiment. He did just that by presenting a quantum theory of specific heat, where inexplicable deviations from the classical Dulong-Petit law, derived using the equipartition theorem, had already been observed. Einstein treated the atoms as quantized simple harmonic oscillators elastically bound in a solid lattice, much as in radiation theory. This resulted in an expression for the specific heat of the solid as a function of temperature that approaches the classical Dulong-Petit value when $kT/h\nu$ is about 0.9 or greater, but which declines to zero with decreasing temperature (decreasing $kT/h\nu$), as shown in the graph on page 186 of his paper – an obvious deviation from the classical Dulong-Petit value. Three years later Walther Nernst and F. A. Lindemann obtained experimental confirmation of Einstein's prediction for a number of solids. It was the first confirmation of the quantum hypothesis outside the field of radiation studies. In 1911 Nernst declared: "That the observations in their totality provide a brilliant confirmation of the quantum theory of Planck and Einstein is obvious" [31].

The future mechanics

In 1910 Nernst, with the financial support of the Belgian philanthropist, Ernst Solvay, organized an international conference of the world's leading quantum theorists and experimentalists to meet in Brussels the following year. The purpose was to examine current theories of matter, to study the quantum hypothesis, and to begin the search for "a future mechanics." [32] Two years earlier Einstein had already pointed out a direction in which to seek the future mechanics. In January 1909 Einstein submitted a paper to the *Physikalische Zeitschrift* [33] which, he told H. A. Lorentz, "contains several considerations, from which it follows for me that not only the mechanics of molecules, but also the Maxwell-Lorentz electrodynamics cannot be brought into harmony with the [Planck] radiation formula." [34] Planck's formula indicated that both the radiation and the matter exhibited quantum behavior. Although his paper delivered at the meeting was not published in the *Annalen der Physik*, it is an important conclusion to Einstein's work on the light-quantum hypothesis during that decade.

One of the considerations that Einstein presented in his paper involved another statistical calculation: a calculation of the mean square fluctuation $(\Delta E)^2$, of the energy of black-body radiation in a volume V with unit frequency ν and a corresponding energy density ρ given by Planck's law. The calculation yielded an expression consisting of two terms:

$$(\Delta E)^2 = V d\nu \left[h\nu\rho + \left(\frac{c^2}{8\pi\nu^2} \right) \rho^2 \right] . \qquad (10)$$

Einstein was able to identify the first term with the fluctuation expected for a collection of particle-like light quanta of energy $h\nu$; the second term could be identified as arising from the interference of waves of frequency ν. The appearance of these two terms together in one equation was the direct consequence of assuming Planck's law for ρ. They represented two independent and incompatible interpretations of light, each one predominate at opposite extremes. For high frequencies and low temperatures, where Wien's law is valid, the first term dominates and represents the particle interpretation. For low frequencies and higher temperatures, where the classical Rayleigh-Jeans law is valid, the second term dominates and represents the wave interpretation of light. Since both terms appear independently in the fluctuation equation as a linear summation, this result, and its basis in Planck's law, indicated to Einstein that in the future both interpretations – wave and particle – will be needed for a complete account of the behavior of light. It was the first indication of what we now call the wave-particle duality [35]. "It is my opinion," Einstein declared later in 1909, "that the next phase of the development of theoretical physics will bring us a theory of light that can be interpreted as a sort of fusion of the wave and the emission [particle] theory of light." [36] Two years later, French physicist Marcel Brillouin concluded, after hearing all of the reports presented to the subsequent Solvay Congress: "It appears certain that from now on it will be necessary to introduce into our physical and chemical concepts a discontinuity, an element varying by jumps, which we had no idea of a few years ago." [37]

Einstein's work and his fundamental insights into the quantum nature of matter and radiation rendered the quantum hypothesis a permanent, if often puzzling, constituent of physics ever since.

References

The edition *Collected Papers of Albert Einstein* is abbreviated CPAE, followed by the volume and page numbers.

[1] A. Einstein, Ann. Phys. (Leipzig) **17**, 132 (1905), reprinted in this volume (p. 165).
[2] Nobel Prize citation, Nobel Archive online, http://nobelprizes.com/nobel/nobel.html.
[3] M. Planck, Ann. Phys. (Leipzig) **1**, 69 and 719 (1900); Ann. Phys. (Leipzig) **4**, 553 (1901).
[4] A. Einstein, Ann. Phys. (Leipzig) **20**, 199 (1906), reprinted in this volume (p. 260).
[5] A. Einstein, Ann. Phys. (Leipzig) **22**, 180 (1907), reprinted in this volume (p. 281).
[6] Einstein to Habicht, 18 or 25 May 1905, in: CPAE 5, 31.
[7] A. Einstein, Ann. Phys. (Leipzig) **22**, 180 (1907), see p. 180, reprinted in this volume (p. 281).
[8] Einstein to Laub, 17 May 1909, in: CPAE 5, 187.
[9] Einstein to Born, 4 December 1926, in: A. Einstein, H. Born, M. Born, Briefwechsel 1916–1955 (Nymphenburger Verlagshandlung, München, 1969), pp. 129–130.
[10] For example, in alphabetical order:
J. Büttner, J. Renn, and M. Schemmel, Stud. Hist. Philos. Mod. Phys. **34**, 37–59 (2003).
A. Hermann, Frühgeschichte der Quantentheorie (1899–1913) (Physik Verlag, Mosbach/Baden, 1969).
M. Jammer, The Conceptual Development of Quantum Mechanics (McGraw-Hill, New York, 1966).
C. Jungnickel and R. McCormmach, Intellectual Mastery of Nature, vol. 2 (University of Chicago Press, Chicago, 1986), chap. 26.
H. Kangro, Vorgeschichte des Planckschen Strahlungsgesetzes (F. Steiner, Wiesbaden, 1970).
M. Klein, Arch. Hist. Exact Sci. **1**, 459 (1962); The Natural Philosopher **1**, 83 (1963); The Natural Philosopher **2**, 59 (1963); Science **148**, 173 (1965).
M. Klein, in: Einstein, A Centenary Volume, edited by P. French (Harvard University Press, Cambridge, 1979), p. 133.
M. Klein, in: Albert Einstein: Historical and Cultural Perspectives, The Centennial Symposium in Jerusalem,

edited by G. Holton and Y. Elkana (Princeton University Press, Princeton, 1982), p. 39.

M. Klein, in: Some Strangeness in the Proportion: A Centennial Symposium to Celebrate the Achievements of Albert Einstein (Addison-Wesley, Reading, 1988), p. 161.

T. S. Kuhn, Black-Body Theory and the Quantum Discontinuity, 1894–1912 (Oxford University Press, New York, 1978). J. Mehra and H. Rechenberg, The Historical Development of Quantum Theory, vol. 1 (Springer, New York, 1982).

A. Pais, "Subtle is the Lord...", The Science and the Life of Albert Einstein (Oxford University Press, New York, 1982).

J. Stachel, in: Einstein's Miraculous Year: Five Papers That Changed the Face of Physics, edited by J. Stachel (Princeton University Press, Princeton, 1998), p. 165.

J. Stachel et al., in: The Collected Papers of Albert Einstein, vol. 2, edited by J. Stachel et al. (Princeton University Press, Princeton, 1989), p. 134.

[11] F. Paschen, Berl. Ber. 1899, 405.
[12] M. Planck, Verh. Dtsch. Phys. Ges. **2**, 237 (1900).
[13] J. W. S. Rayleigh, Philos. Mag. **49**, 539 (1900).
J. H. Jeans, Philos. Mag. **10**, 91 (1905).
[14] H. Rubens and F. Kurlbaum, Ann. Phys. (Leipzig) **4**, 649 (1901), reporting work in 1900.
[15] M. Planck, Verh. Dtsch. Phys. Ges. **2**, 202 (1900).
[16] Einstein to Marić, 4 April 1901, in: CPAE 1, 284.
[17] A. Einstein, Ann. Phys. (Leipzig) **14**, 354 (1904), see pp. 360–362, reprinted in this volume (p. 161–163).
[18] A. Einstein, Ann. Phys. (Leipzig) **20**, 199 (1906), see p. 199, reprinted in this volume (p. 260).
[19] A. Einstein, Ann. Phys. (Leipzig) **17**, 132 (1905), see p. 141, reprinted in this volume (p. 174).
[20] M. Klein, in: Einstein, A Centenary Volume, edited by P. French (Harvard University Press, Cambridge, 1979), p. 133.
[21] A. Einstein, Ann. Phys. (Leipzig) **17**, 132 (1905), see p. 143, reprinted in this volume (p. 176).
[22] P. Lenard, Ann. Phys. (Leipzig) **1**, 486 (1900); Ann. Phys. (Leipzig) **2**, 359 (1900); Ann. Phys. (Leipzig) **8**, 149 (1902); Ann. Phys. (Leipzig) **12**, 449 (1903).
[23] R. A. Millikan, Phys. Rev. **7**, 18 and 355 (1916).
A. H. Compton, Phys. Rev. **21**, 483 (1923).
[24] See the works by J. Stachel [10].
[25] For a modern derivation, see Jammer [10], Appendix A.
[26] A. Einstein, Ann. Phys. (Leipzig) **20**, 199 (1906), see p. 200, reprinted in this volume (p. 261).
[27] A. Einstein, Ann. Phys. (Leipzig) **20**, 199 (1906), see p. 202, reprinted in this volume (p. 263).
[28] A. Einstein, Ann. Phys. (Leipzig) **20**, 199 (1906), see p. 203, reprinted in this volume (p. 264).
[29] A. Einstein, Ann. Phys. (Leipzig) **22**, 180 (1907), see p. 180, reprinted in this volume (p. 281).
[30] A. Einstein, Ann. Phys. (Leipzig) **22**, 180 (1907), see p. 184, reprinted in this volume (p. 285); here Einstein again wrote $(R/N)\beta\nu$ instead of $h\nu$.
[31] W. Nernst, Sitz.ber. Kgl. Preuss. Akad. Wiss. **306**, 310 (1911).
[32] Discussion of H. A. Lorentz, in: La théorie du rayonnement et les quanta, edited by P. Langevin and M. De Broglie (Gauthier-Villars, Paris, 1912), p. 8.
[33] A. Einstein, Phys. Z. **10**, 185 (1909); in: CPAE 2, 542.
[34] Einstein to Lorentz, 30 March 1909, in: CPAE 5, 166.
[35] See Pais [10], Chap. 21.
M. Klein, The Natural Philosopher **3**, 86 (1964).
[36] A. Einstein, Verh. Dtsch. Phys. Ges. **7**, 482–482 (1909); in: CPAE 2, 482–483.
[37] M. Brillouin, remark, in: La théorie du rayonnement et les quanta, edited by P. Langevin and M. De Broglie (Gauthier-Villars, Paris, 1912), p. 451.

Einstein's invention of Brownian motion

Jürgen Renn*

Max-Planck-Institut für Wissenschaftsgeschichte, Wilhelmstr. 44, 10117 Berlin, Germany

Published online 14 February 2005

Key words Brownian motion, kinetic theory of heat, statistical physics, stochastic processes, thermodynamics of solutions
PACS 05.40.Jc, 05.20.Dd, 05.40.-a, 82.60.Lf

The atomistic revolution[1]

The impact of Einstein's work

Einstein's 1905 paper on Brownian motion was an essential contribution to the foundation of modern atomism [20]. Atomism as understood in science today presupposes, like its predecessor rooted in the theories of nature from Greek antiquity and from early modern times, that matter is constituted by small entities. But it no longer assumes that the properties and the behavior of these entities can simply be inferred from the familiar physical laws governing our macroscopic environment, nor that a description of matter in terms of its atomistic constituents can be exhaustive. Einstein succeeded in interpreting the irregular movements of small particles suspended in a liquid as visible evidence for the molecular motions constituting the heat of a ponderable body according to the kinetic theory of heat. But he did so by radically changing the understanding of these irregular motions which he no longer conceived as being characterized by a velocity in the classical sense but as a stochastic process that can only be described with the help of statistical methods. It is therefore not surprising that Einstein's work on Brownian motion also became one of the pillars of modern statistical thermodynamics and, more generally, of the physics of stochastic processes.

In the sequel to his groundbreaking work, Einstein published several other related articles, extending the subject to Brownian motion in condensers and the fluctuations of heat radiation. His work aroused widespread interest among physicists and chemists, as indicated by Einstein's correspondence with other scientists interested in the subject, in particular Conrad Röntgen, Richard Lorenz, Marian von Smoluchowski, and The Svedberg.[2] In 1906 the Polish physicist von Smoluchowski submitted a paper on the kinetic theory of Brownian motion to the *Annalen* that was stimulated by Einstein's papers but represented results which he had derived independently. While Smoluchowski's argument was different from Einstein's, his results were – apart from a numerical factor – essentially equivalent. Einstein's interpretation of Brownian motion soon also received striking experimental confirmation by Jean Perrin and others. This success furthered the general acceptance of atomism and helped to convert the then still numerous skeptics. Indeed, while in the nineteenth century atomism was widely employed as a working hypothesis in numerous fields of physics and chemistry, it was accepted as a physical reality only after the impressive accumulation

---

* E-mail: renn@mpiwg-berlin.mpg.de
[1] This essay is based on my earlier contributions to the subject, in particular, the editorial note on Brownian motion in [10, 206–222], and [44].
[2] See Wilhelm Röntgen to Einstein, 18 September 1906 [11, Doc. 40]; Richard Lorenz to Einstein, 15 November 1907 [11, Doc. 65]; and Einstein to Marian von Smoluchowski, 11 June 1908 [11, Doc. 105]. For evidence of early correspondence between Svedberg and Einstein, see The Svedberg to Einstein, 8 December 1919 [12, Doc. 202].

of evidence in the early twentieth century to which Einstein's interpretation of Brownian motion was a key contribution.

This must also have been Einstein's own view. He made a point of sending at least one of his papers on Brownian motion to Ernst Mach, one of the skeptics with regard to atomism, emphasizing the direct relationship between Brownian motion and "thermal motion".[3] In a popular account from 1915, he wrote: "Under the microscope one, to some extent, immediately sees a part of thermal energy in the form of mechanical energy of moving particles."[4] In his "Autobiographical Notes", written towards the end of his life, Einstein summarized his view of the influence his work had on Brownian motion [27, p. 49]:

> The agreement of these considerations [on Brownian motion] with experience together with Planck's determination of the true molecular size from the law of radiation (for high temperatures) convinced the sceptics, who were quite numerous at that time (Ostwald, Mach) of the reality of atoms.

Einstein's work as a historical puzzle

When turning to the origins of Einstein's eminently successful work on Brownian motion, one is confronted with a puzzle. He did not mention Brownian motion in the title of his paper, which he evidently wrote "without knowing that observations concerning Brownian motion were already long familiar".[5] In other words, Einstein must have somehow "invented" Brownian motion all by himself. Although he had evidently heard about it, he had no concise empirical information and essentially derived the properties of Brownian motion solely from theoretical considerations. That Einstein was only vaguely familiar with observations of Brownian motion is also suggested in a letter he wrote in May 1905 to his friend and discussion partner Conrad Habicht, a famous letter in which Einstein listed four of the five pathbreaking papers on which he was working during his miracle year.[6] The paper on Brownian motion was, after the paper on the light quantum and the dissertation on the determination of molecular dimensions, the third on Einstein's list, before the relativity paper, which he had only outlined at that time:

> The third proves that, on the assumption of the molecular theory of heat, bodies on the order of magnitude 1/1000 mm, suspended in liquids, must already perform an observable random motion that is produced by thermal motion; in fact, physiologist have observed <unexplained> motions of suspended small, inanimate, bodies, which motions they designate as "Brownian molecular motion."[7]

The puzzle of the origin of Einstein's Brownian motion paper raises a number of questions that the following shall attempt to answer: How could Einstein predict the non-classical properties of a phenomenon about which he had apparently no precise information? How was his study of Brownian motion related to his other concerns in 1905, which ranged from the constitution of radiation to the electrodynamics of moving bodies? And why was an explanation of Brownian motion as being due to the motion of atoms and molecules only achieved at the beginning of the twentieth century although both the atomistic hypothesis and the phenomenon itself were long familiar at that time? We shall begin our discussion by addressing the last question.

---

[3] Einstein to Ernst Mach, 9 August 1909 [11, Doc. 174] and 17 August 1909 [11, Doc. 175].
[4] [26, p. 261].
[5] [27, p. 44; translation, p. 45].) See also Einstein to Michele Besso, 6 January 1948 [1, Call Nr. 7-382.00], and Einstein to Carl Seelig, 15 September 1952 [1, Call Nr. 39-040].
[6] Einstein to Conrad Habicht, 18 or 19 May 1905 [11, Doc. 27].
[7] "Die dritte beweist, daß unter Voraussetzung der molekularen Theorie der Wärme in Flüssigkeiten suspendirte Körper von der Größenordnung 1/1000 mm bereits eine wahrnehmbare ungeordnete Bewegung ausführen müssen, welche durch die Wärmebewegung erzeugt ist; es sind <unerklärte> Bewegungen lebloser kleiner suspendirter Körper in der That beobachtet worden von den Physiologen, welche Bewegungen von ihnen "Brownsche Molekularbewegung" genannt wird." Unless indicated otherwise, translations are taken from the English companion volumes to the "Collected Papers of Albert Einstein."

A marginal problem of classical science

The exclusion of alternative explanations

The first systematic study of the irregular movement of microscopic particles suspended in a liquid goes back to the botanist Robert Brown, who published his careful observations in 1828.[8] He examined a great variety of materials which he suspended in a liquid. These ranged from the pollen of plants to fragments of an Egyptian sphinx. He also explored an equally great variety of possible causes of the irregular motion of the suspended particles, ranging from currents within the liquid via an interaction among the particles to the creation of small air bubbles. In this way Brown was able to exclude a number of potential explanations of the irregular movement of the suspended particles and, in particular, the claim that this is a characteristic of organic materials only and somehow an expression of "life." Nevertheless, Brownian motion failed to become a subject of broad interest to physicists, at least until the middle of the nineteenth century. Meanwhile several papers appeared in which the influence of specific circumstances on the properties of Brownian motion were studied, such as the temperature of the liquid, capillarity, convection currents of the liquid, evaporation, the illumination of the particles, electrical forces, and the role of the environment.[9]

While from today's perspective, these early works may appear to be more or less fruitless and part of a prehistory that may just be disregarded, the knowledge accumulated by these works turned out to be crucial for the later identification of Brownian motion as a special case of the kind of motion that we call heat. First, all the various experiments taken together pointed to the persistence and the ubiquity of the phenomenon. Second, they increasingly excluded *a priori* possible explanations that made use of the specific circumstances under which the phenomenon was produced. If the explanations by Einstein and von Smoluchowski found such a rapid and successful acceptance as evidence in favor of an atomistic constitution of matter, this was not least due to the fact that alternative explanations of Brownian motion had been thoroughly pursued over nearly a century and eventually discarded after careful examination.

Brownian motion as a challenge for the kinetic theory

Since the mid-nineteenth century, several authors considered the kinetic theory of heat as a possible explanation of Brownian motion.[10] This is hardly surprising as the kinetic theory, as developed by Rudolf Clausius, James Clerk Maxwell, Ludwig Boltzmann and others, became an ever more powerful tool to explain thermal phenomena on a mechanical basis. Still there was a variety of factors that could be taken into account when attempting to explain Brownian motion within this framework, such as the temperature and viscosity of the liquid, the differences in specific heats of the particles and the liquid, as well as the magnitude and velocity of the particles. In addition, the role of additional factors outside the scope of the kinetic theory, such as electric interactions, could still not be excluded and gave rise to further investigations. But the potential complexity of the phenomenon was not the only reason it remained in a kind of "epistemic isolation," that is, why it failed to become a key subject of the great works on the kinetic theory of heat.

One of the reasons for the marginal role that Brownian motion continued to play in nineteenth-century physics was a matter of perspective. Indeed, the focus of the kinetic theory was oriented more towards a reconstruction of the laws of phenomenological thermodynamics than towards the discovery of deviations from these laws, even if these were the statistical fluctuations that must occur if the interpretation of heat as a kind of motion is correct.[11] Boltzmann's *Gastheorie*, for instance, explicitly denies that the thermal motion of molecules in a gas leads to observable motions of suspended bodies.[12] Another reason was the intrinsic difficulties of applying the kinetic theory to Brownian motion. Since the 1870s, several scientists had pursued the idea that Brownian motion might be explained as the result of collisions between suspended

---

[8] See [4].
[9] For contemporary reviews of research on Brownian motion, see [50] and [13]. For historical accounts, see [5, 36].
[10] For historical discussion, see [5, § 3].
[11] See the historical discussion in [44].
[12] See [3, pp. 111–112].

particles and the molecules of the liquid, among them Delsaulx, Carbonelle, and Gouy.[13] Gouy supported this explanation by performing further experiments excluding alternative accounts. While the qualitative explanation of Brownian motion with the help of the kinetic theory thus became ever more plausible, serious problems occurred as soon as such an explanation made use of quantitative arguments.

It was this kind of quantitative argument that was used by the cytologist Karl von Nägeli in 1879 against the kinetic explanation of Brownian motion.[14] The argument was based on the equipartition theorem at the center of the kinetic theory. According to this theorem, in thermal equilibrium the energy of a physical system is equally distributed over its internal degrees of freedom, the energy portion of each single degree of freedom being proportional to the absolute temperature. It was therefore possible to calculate the average velocity of the molecules of the liquid, and then use the laws of elastic collision to obtain the velocity of a suspended particle. Nägeli concluded from this argument that the velocity of such a particle, because of its comparatively large mass, would be vanishingly small. This internal contradiction of an explanation of Brownian motion as the motion of a very large molecule in thermal equilibrium with all the smaller molecules of the liquid was confirmed in 1900 by the work of Felix Exner, actually a supporter of the kinetic explanation.[15] He performed extensive measurements of the velocity of Brownian motion and observed that it decreases when larger particles are suspended and increases with rising temperature, as must be the case according to the kinetic theory. When he calculated, however, the kinetic energy of the molecules on the basis of his velocity measurements, he found values that were dramatically smaller than those implied by the kinetic theory of heat.

Consequently, by the turn of the century, Brownian motion had emerged as a veritable challenge to classical physics, even if this challenge was not broadly acknowledged due to the apparent marginality of the phenomenon, at least from the perspective of the majority of the physics community. With practically all other accounts excluded, the kinetic theory had emerged as the most viable option for explaining the phenomenon, yet failed to provide an adequate quantitative understanding.

Preparing for a breakthrough

Einstein's perspective

How did Einstein encounter the problem of Brownian motion and why was he able to provide a solution to this problem that had apparently escaped his predecessors? Einstein's perspective distinguished itself from that of the majority of his contemporaries by his broad orientation beyond the narrow confines of subdisciplinary specialization. Since his early student days, he was interested in atomism as a conceptual tool for identifying hidden links between physical phenomena, which otherwise seemed to have no connection, such as the specific heats of solids and their optical transparency, or the thermal and electrical conductivity of metals. He also speculated about the relation between molecular and gravitational forces and about the possibility of a direct conversion of the kinetic energy of molecules and atoms into light. The search for a conceptual unity of physics, which became the hallmark of Einstein's research, had probably been prepared by his early reading of popular scientific literature in which the heritage of the Romantic idea that science could account for the unity of nature was preserved.[16]

Atomismus was, however, not only one of the principal conceptual tools employed in Einstein's quest for a scientific world view, but was also widely exploited in contemporary professional research within the disciplines and subdisciplines of classical science, which ranged from Lorentz's electron theory of electromagnetism via Boltzmann's kinetic gas theory to chemistry. In view of this spread of atomistic ideas, it is rather surprising that attempts to relate the various usages of the concept to each other, in order to extract a coherent overall picture, were not more prominent. Attempts at a conceptual synthesis had been

---

[13] See the discussion in [13].
[14] See [34].
[15] See [30].
[16] For a more extensive discussion, see [44].

almost obligatory for any scientist in the age of the Scientific Revolution. But in the nineteenth century, the age of specialization, these attempts were no longer part of the ordinary pursuit of science and were rather left to a few philosopher-scientists such as Ernst Mach and Henri Poincaré, whose works the young Einstein ardently devoured. He read, for instance, Poincaré's *Science et hypothèse*, which contains a brief discussion of Gouy's work on Brownian motion, and emphasizes Gouy's argument that Brownian motion violates the second law of thermodynamics, i.e,. the principle of the irreversibility of thermodynamic processes.[17] Einstein was therefore not only familiar with the potential of atomism as a conceptual bond between phenomena studied in isolation from each other because of the specialist outlook of contemporary science. He was also aware of the precarious status of concepts such as atoms and the ether, which were often uncritically presupposed in contemporary scientific arguments, without carefully examining their meaning and their relation to empirical evidence that was not just limited to the special problem on which a particular investigation happened to focus.

During Einstein's student years, the kinetic theory of heat was the subject of a heated controversy between Ernst Mach, Wilhelm Ostwald, Georg Helm, and Ludwig Boltzmann.[18] Mach rejected the existence of entities not directly accessible to sense-experience, and was skeptical, in particular, about the existence of atoms. Although Einstein criticized Boltzmann for a lack of emphasis on the comparison of his theory with observation,[19] he enthusiastically embraced the atomistic principles of Boltzmann's theory.[20] He must have therefore found it challenging when he read in Boltzmann's *Gastheorie*[21], that Boltzmann, presumably reacting to the above-mentioned controversy, suggested that he was isolated in his support of the kinetic theory.[22] Indeed Einstein's interests soon turned from the details of atomistic explanations to the quest for facts, "which would guarantee as much as possible the existence of atoms of definite finite size," as he later remembered.[23]

The theory of solutions

There can be little doubt, however, that Einstein's perspective on the problem of Brownian motion was as much shaped by the specific problems he dealt with in his prior research as by the general, philosophical outlook outlined above. Even in his first two papers, published in 1901 and 1902 and later disqualified as worthless beginner's papers,[24] he familiarized himself with some of the ideas that figured in his later work on Brownian motion, in particular the nature of diffusion processes and the application of thermodynamics to the theory of solutions.[25] In [17], for instance, he suggested replacing semipermeable walls in thermodynamic arguments with external conservative forces, a method he stated to be particularly useful for treating arbitrary mixtures. In his subsequent papers on statistical physics, Einstein generalized the idea of external conservative forces,[26] and, as we shall see in more detail below, noted the significant role of fluctuations in statistical physics.

---

[17] See [43, p. 209].

[18] For Einstein's reading of Mach, see Einstein to Mileva Marić, 10 September 1899 [9, Doc. 54]; for his reading of [37, 1893], see Einstein to Wilhelm Ostwald, 19 March 1901 [9, Doc. 92]; for his reading of [2, 3], see Einstein to Marić, 10 September 1899, 13 September 1900, and 19 September 1900 [9, Docs. 54, 75, and 76].

[19] On 30 April 1901, Einstein wrote to Mileva Marić: "At present I am again studying Boltzmann's theory of gases. Everything is very nice, but there is too little stress on the comparison with reality" ("Ich studiere gegenwärtig wieder Boltzmanns Gastheorie. Alles ist sehr schön, aber zu wenig Wert gelegt auf den Vergleich mit der Wirklichkeit.") [9, Doc. 102].

[20] Einstein to Mileva Marić, 13 September 1900 [9, Doc. 75].

[21] See the preceding note.

[22] See the preface to [3]; for accounts of the dispute, see [6, pp. 96–98;], [14, pp. 416ff.].

[23] See [27, p. 44; translation, p. 45].

[24] See [16, 17]. For their qualification, see Einstein to Johannes Stark, 7 December 1907 [11, Doc. 66]. See also the editorial note "Einstein on the Nature of Molecular Forces" in [10, pp. 3–8].

[25] For a discussion of Einstein's earlier interest in diffusion, see the editorial note "Einstein's Dissertation on the Determination of Molecular Dimensions" in [10, 177–179].

[26] See, in particular, [18, § 10].

Einstein's contemporary correspondence suggests that the theory of solutions, including the issues of semipermeable membranes and osmotic pressure, must have played an even larger role in his thinking than is directly apparent from his published papers. In 1903, he discussed the concepts of semipermeable membrane and osmotic pressure in his correspondence with Michele Besso, and expressed interest in Sutherland's hypothesis on the mechanism of semipermeable membranes.[27] The theory of solutions provided, as elaborated by Jacobus H. van't Hoff and later by Walther Nernst, a concise analogue to the kinetic theory of gases, and therefore offered an important field of exploration to someone like Einstein, who was interested in extending the range of the applicability of atomistic ideas.[28] The theory of solutions must also have been central to the thesis Einstein submitted for his doctoral degree, a thesis he eventually withdrew.[29] He used this again in 1905 as the central subject of the dissertation with which he finally succeeded in obtaining his degree.

Einstein's doctoral thesis provided much of the framework essential to his analysis of Brownian motion.[30] It was also motivated by a goal similar to the one in his paper on Brownian motion, i.e., to offer evidence for the existence of atoms and molecules and to determine their size. The dissertation proposed a new method for measuring atomic dimensions, explaining how Avogadro's number could be found by considering large sugar molecules in solution. The basic procedure was to set up two equations for two unknowns, from which Avogadro's number and the size of the solute molecules could then be calculated. Physically, one of the equations was derived from the change in viscosity due to the addition of sugar molecules to the solution, while the other used a relation between the diffusion coefficient of the sugar molecules and the viscosity of the solution. The first of these equations was derived from rather involved hydrodynamical calculations. The other equation, relating diffusion and viscosity, turned out to be crucial for the analysis of Brownian motion as well. It may therefore be worthwhile to examine its conceptual roots more closely. Its derivation is based ultimately on establishing a bridge between a bulk phenomenon, diffusion, and the motion of an individual particle as affected by viscosity acting as a friction force due to the environment of the particle. How did Einstein manage to build this bridge and how did the very idea to look for such a relation emerge?

The relation between diffusion and viscosity

These questions lead back to Einstein's first two papers, which deal with the theory of solutions and to his continued concern with their thermodynamic properties. In his second paper, dedicated to the properties of electrolytic solutions, he questioned the legitimacy of applying the laws of thermodynamics and concepts such as osmotic pressure to such solutions, even if no semipermeable membranes are available to provide experimental meaning to such pressure. He addressed this question by claiming that such devices can be substituted by conservative forces acting on the substances under consideration. In the derivation of the relation between diffusion and viscosity in his later paper on Brownian motion, he made use of such forces, which now played the role of an intermediate between the motion of individual particles and the bulk process of diffusion.

In this paper, Einstein considered particles suspended in a liquid and analyzed the dynamic equilibrium of these particles under the assumption that the individual particles are subject to the influence of a force depending only on position. This force is hence an example of the kind of fictitious conservative forces that Einstein had introduced earlier to replace unrealizable semipermeable membranes in thermodynamic considerations. It is therefore not surprising that his thermodynamic considerations led him to conclude that the position-dependent force is counterbalanced by a force due to the osmotic pressure. Clearly, this conclusion relates a force acting on the individual particles to a bulk property of the suspended particles –

---

[27] See Michele Besso to Einstein, 7–11 February 1903 [11, Doc. 6], which indicates that there was additional correspondence on this subject. See [51] for Sutherland's hypothesis.

[28] See [55] and [35].

[29] See the extensive discussion in [44].

[30] See the editorial note "Einstein's Dissertation on the Determination of Molecular Dimensions," in [10, pp. 177–179] and [44].

their osmotic pressure. Einstein then considered the dynamic equilibrium of the suspended particles from a second perspective, i.e., as a balance of the motion of the individual particles under the influence of the fictitious force and a process of diffusion. To describe the motion of the particles in the liquid, he relied on Stokes' law between the force exerted, the viscosity of the liquid, and the velocity attained by the particles. To describe the diffusion process, Einstein simply made use of the definition of the diffusion coefficient as relating transport and density of the suspended particles.

On the basis of these results, Einstein was now able to couple the two balance equations, one derived from a thermodynamic argument, and the other from relating diffusion to the motion of the individual particles determined with the help of Stokes' law. He could thus eliminate the fictitious force and directly establish a relation between the diffusion coefficient and osmotic pressure. As the latter involves, according to the kinetic theory of heat, Avogadro's number, Einstein finally arrived at an expression of the diffusion coefficient in terms of atomic sizes.

$$D = \frac{RT}{N} \frac{1}{6\pi \, kP} \tag{1}$$

where R the gas constant, T the temperature, N Avogadro's number, k the viscosity, and P the radius of the solute molecules or the suspended particles.

In his dissertation, he used this equation together with the equation relating atomic sizes to the change of viscosity derived from hydrodynamic considerations to derive values for the atomic dimensions from experimental data on diffusion and viscosity. In his paper on Brownian motion, he rederived the viscosity-diffusion equation, albeit in a more elegant manner. He now made use of the methods of statistical mechanics that he had developed in previous years to describe the irregular motion of suspended particles.

Statistical mechanics and heat radiation

Einstein's establishment, independently of Willard Gibbs, of statistical mechanics between 1902 and 1904 was motivated by the quest mentioned above to extend the methods of the kinetic theory beyond gases to include a wide range of physical systems such as the electron gas in metals and heat radiation, not least in order to provide additional evidence for the atomic hypothesis. The methods he developed imposed only the most general requirements on the systems studied and did not depend on knowledge of the detailed interaction between the constituents of a systems as is the case in kinetic gas theory, where collision dynamics plays a major role. As Einstein wrote in a letter to his friend Marcel Grossmann in 1904, who at that time was a student of mathematics working on non-Euclidean geometry:[31]

> There is a remarkable similarity between us. ... You treat geometry without the parallel axiom, and I treat the atomistic theory of heat without the kinetic hypothesis.

Einstein's generic approach to the statistical properties of physical systems may thus appear to have been well-suited to examine the properties of a phenomenon such as Brownian motion. But by 1904, he had evidently neither heard about it nor could he conceive its existence. In his last paper on statistical mechanics, he did study fluctuation phenomena and even derived an expression for mean square deviations from the average value of the energy of a system, which he interpreted as expressing a condition for the stability of a physical system involving Boltzmann's constant, thus giving a new meaning to this constant.[32] But when it came to the issue of the observability of such fluctuations, Einstein claimed that radiation in thermal equilibrium was the only system for which experience suggested that it exhibits observable fluctuations. His argument was that, for a radiation cavity whose linear dimensions are chosen so as to be comparable to

---

[31] "Es waltet eine merkwürdige Ähnlichkeit zwischen uns. ... Du behandelst die Geometrie ohne das Parallelenaxiom, ich die atomistische Wärmelehre ohne die kinetische Hypothese." Einstein to Marcel Grossmann, 6 April 1904 [11, Doc. 17].

[32] At this time, however, Einstein regarded black-body radiation as the only physical system for which experience suggests the existence of observable energy fluctuations." See [19, p. 361].

the wavelength corresponding to the maximum energy in the black body spectrum, the fluctuations should be of the same order of magnitude as the mean energy.

The numerical values given to illustrate this claim indicate that Einstein must had used Wien's formula to describe the energy spectrum of heat radiation.[33] He may well have considered as early as 1904 what we would call today a photon gas, starting from the speculative assumption that heat radiation may be conceived of as a collection of light quanta whose energy is given by their frequency according to $E = h\nu$. He did, however, then cast his arguments into a form that made them independent from a specific interpretation of heat radiation and only came back to the light quantum hypothesis a year later, now as a means of interpreting Planck's radiation formula, at least in the range in which it can be reasonably approximated by Wien's formula.

Revisiting fluctuations

By 1905, Einstein's views of the observability of fluctuation phenomena had changed. He had now assembled all the tools necessary to construct a model of observable fluctuation phenomena in a material system by integrating the results achieved whilst writing his dissertation, i.e., the study of dissipation and diffusion phenomena, and those accumulated whilst studying fluctuation phenomena in the context of statistical mechanics and its application to heat radiation. First of all, fluctuations in heat radiation could be related directly to a material process if a mirror is exposed to them, which, as a consequence of the radiation impinging on it and the friction force it suffers at the same type should exhibit Brownian motion-like behavior. This thought experiment was discussed at length in Einstein's later publications, but was apparently already conceived by 1905, as is indicated by later recollections. It represents, so to speak, the missing link between Einstein's principal concerns at the time – heat radiation, statistical physics, and the electrodynamics of moving bodies.

Second, and perhaps of even greater consequence to Einstein's invention of Brownian motion, the argument at the core of his dissertation with which he had inferred that the observability of fluctuations in the case of heat radiation can be transferred directly to the model of large molecules in solution. Increasing the linear dimensions of such particles in analogy to considering wavelengths comparable in size to the radiation cavity would not change the character of the particles as partaking in a world governed by the kinetic theory of heat, but might actually make their random motions visible. The crucial function of the dissertation model was thus to provide a framework in which scaling of this kind, first conceived for the case of radiation, made physical sense for a material process. All that was needed was to mentally transform a solvent with large molecules into a suspension with minuscule but observable particles. Such a transition was particularly plausible as the commonly made distinction between suspensions and solutions in nineteenth-century chemistry had, by the turn of the century, gradually lost its absolute character.[34] The absence of any fundamental difference between solutions and suspensions became clear in 1902, when observations using the newly invented ultramicroscope[35] made it possible to resolve many colloidal solutions into their constituents.[36]

In summary, Einstein's invention of Brownian motion was just as much prepared by his quest to identify evidence in favor of the atomic hypothesis as by the specific research problems he had tackled, in particular, in the course of his long-standing interest in the theory of solutions. Combining a model that had assumed a central role in this pursuit – the model of suspended particles undergoing diffusion in a liquid – with his search for observable fluctuation phenomena, he was naturally led to consider the irregular motion that must

---

[33] For this claim, see [46].

[34] For a contemporary discussion of the distinction between solution and suspensions, see the introduction to [56]. For a discussion of colloidal chemistry and its relation to the study of Brownian motion, see [36, pp. 98–102].

[35] The ultramicroscope, developed by Henry Siedentopf and Richard A. Zsgimondy, is based on a new illumination technique that makes it possible to observe the diffraction discs of otherwise invisible objects; it increased the limit of visibility to approximately 5×10^{-1} micron. For a contemporary discussion of ultramicroscopes, see [8, Chap. 3].

[36] See [49].

be exhibited by such particles. The formula he had derived for expressing the diffusion coefficient in terms of atomic dimensions and the size of the particles would now make it possible to extract information about the atomic scale from the irregular motion of the suspended particles to the atomic scale, if this motion of individual particles could be related to the bulk property of diffusion. To bridge this last gap, Einstein needed a crucial conceptual leap in his analysis of Brownian motion, conceiving it as a kind of process hitherto unknown in classical physics. Before coming to this last step, it is helpful to reconsider the problem of Brownian motion from a somewhat larger perspective, comparing it to the other problems for which Einstein achieved equally important conceptual breakthroughs, in particular, the problem of heat radiation, which triggered the quantum revolution, and the problem of the electrodynamics of moving bodies, which gave rise to the relativity revolution.

Brownian Motion as a Borderline Problem

The three partite division of classical physics

The physical problems at the center of Einstein's *annus mirabilis* 1905 have a common feature that becomes evident only when considered from the perspective of the long-term development of physical knowledge. They may be characterized as borderline problems of classical physics because they are all situated at the intersection of two of the three major subdomains of contemporary physics: mechanics, thermodynamics, and electrodynamics. The problem of heat radiation in thermal equilibrium thus constituted a borderline problem between thermodynamics and the theory of radiation, i.e., electrodynamics. The electrodynamics of moving bodies represented a borderline problem between electrodynamics and the theory of motion, i.e., mechanics. And problems of the kinetic theory of heat, such as Brownian motion, can be understood as borderline problems between thermodynamics and mechanics. Borderline problems require the integration of knowledge resources from different domains. They also serve as catalysts that unveil conceptual conflicts between these domains such as the conflict between the constancy of the speed of light and the relativity principle in the case of the electrodynamics of moving bodies.

The hidden bond between Einstein's revolutionary papers of 1905 was their precise concern with such borderline problems. This concern emerged, in Einstein's intellectual biography, in connection with his quest, mentioned above, to establish a conceptual unity of physics by probing the explanatory power of what may be called a speculative or "interdisciplinary" atomism. In this way, the 1905 problems entered his intellectual focus as problems related to more than one conceptual framework of classical physics rather than special issues pertaining to one of its subdomains. Brownian motion could thus be perceived as a borderline problem in the sense that it represented a challenge both for classical phenomenological thermodynamics and its counterpart, the kinetic theory of heat. It constituted a concrete problem allowing the confrontation of their different perspectives. As a consequence, contradictions were engendered that suggested a rethinking and restructuring of the given architecture of knowledge.

Brownian motion as a paradox

Einstein addressed the problem of Brownian motion in his 1905 paper in exactly this way, as a paradox between thermodynamics and the kinetic theory. He argued that, from the perspective of phenomenological thermodynamics, small particles suspended in a liquid should, after a while, simply achieve a thermal equilibrium with the surrounding liquid and certainly not continue to perform irregular motions. According to classical thermodynamics, the second law is not just a statistical assertion but claims absolute validity. The scepticism toward a merely statistical interpretation of the second law precisely represented a serious obstacle to the conception of Brownian motion as the result of collisions between the suspended particles and the molecules of the liquid.

From the perspective of the kinetic theory, on the other hand, nothing distinguishes these particles in principle from atoms and molecules but their size. They should therefore be exposed to the permanent collisions ensuring the thermal equilibrium in the liquid and partake themselves in this thermal motion. But, as mentioned earlier, Brownian motion also confronted the kinetic theory with a serious problem, the lack of agreement between the observed velocities and the velocities calculated theoretically on the basis of the equipartition theorem. It is remarkable that the equipartition theorem, crucial to Einstein's investigations of heat radiation and consolidated in the context of his work on statistical mechanics, plays no role in his work on Brownian motion. He thus may well have been aware of the difficulty of ascribing a velocity in the ordinary sense to the particles suspended in a liquid.

The irregular motion of such particles hence manifests a conflict between two domains of classical physics in a way similar to the conflict between the relativity postulate and the constancy of the speed of light embodied in the electrodynamics of moving bodies, as well as to the conflict between the assumption of a continuum of wavelengths and the assumption of an equipartition of energy in the case of heat radiation. Einstein's reaction to these conflicts was similar in all three cases as well, and distinguished itself in a similar way to that of most of his contemporaries, who looked at such problems from a specialist perspective. While it was plausible to assume that these conflicts were due to a failure somewhere buried in the conceptual foundations of one of the domains involved, Einstein's remarkable overview of the knowledge of classical physics as well as his philosophical acumen caused him to be skeptical with regard to all of these domains and dare to look for new concepts that were capable of overcoming what he saw as a fundamental crisis of classical physics.

Instead of relying on the classical concept of an ether as used in Lorentz's electron theory, abandoning essentials of classical mechanics such as the relativity principle, Einstein audaciously conceived of new concepts of space and time to resolve the problems of the electrodynamics of moving bodies. He was thereby able to preserve insights from both electrodynamics and mechanics, rather than sacrificing one for the sake of the other. In his treatment of the behavior of small suspended particles, Einstein similarly combined insights of the kinetic theory and of thermodynamics, i.e., of micro- and macrophysics, without reducing one to the other. Instead he proposed new laws for the domain of "mesoscopic" physics, which was recognized for the first time as an autonomous level of physical knowledge.

Reinterpreting the results of classical physics

As we have seen above, Einstein inferred from the kinetic theory that a suspension of small particles must possess an osmotic pressure just as in the case for a solution of molecules. If this pressure is distributed in a spatially inhomogeneous way, it gives rise to a compensatory diffusion process whose bulk properties can be calculated, as also discussed above, with help of Stokes' law determining the moveability of the particles in a viscous fluid. In this way, Einstein obtained an equation for the diffusion coefficient figuring in the partial differential equation, determining the relation between spatial and temporal change of the concentration $f(x,t)$ of a substance in solution:

$$\frac{\partial f}{\partial t} = D \frac{\partial^2 f}{\partial x^2}, \qquad (2)$$

where D is the diffusion coefficient. This equation was first established by Adolf Fick, following the work of Fourier for the conduction of heat and that of Ohm for the conduction of electricity [31]. Einstein now reinterpreted this equation in a way analogous to his reinterpretation of the Lorentz transformations for the electrodynamics of moving bodies and to his reinterpretation of Planck's black-body formula. While largely preserving the technical framework of these results in the works of Lorentz and Planck respectively, Einstein had profoundly changed their conceptual meaning, thus creating the new kinematics of the theory of special relativity and introducing the revolutionary idea of light quanta. He did so in a process of reflection that may be described as a "Copernicus process" since Copernicus as well had largely kept the deductive machinery of traditional astronomy when changing its basic conceptual structure [45]. In the context of his work on fluctuation phenomena, Einstein similarly gave a radically new interpretation to the traditional diffusion equation, thus effectively "inventing" Brownian motion as a theoretical concept.

Instead of considering the diffusion equation as describing the overall distribution of a solute substance, Einstein now interpreted it as determining the probability distribution of the irregular displacements of the individual particles. The introduction of such a distribution could draw on his previous experience with such probability distributions in his work on statistical mechanics.[37] In his paper, he wrote with regard to the above equation:[38]

> This is the familiar differential equation for diffusion, and D can be recognized as the diffusion coefficient. Another important consideration can be linked to this development. We assumed that all the individual particles are referred to the same coordinate system. However, this is not necessary since the motions of the individual particles are mutually independent. We will now refer the motion of each particle to a coordinate system whose origin coincides at time $t = 0$ with the position of the center of gravity of the particle in question, with the difference that $f(x,t)dx$ now denotes the number of particles whose X-coordinate has *increased* between the times $t = 0$ and $t = t$ by a quantity lying between x and $x + dx$. Thus, the function f varies according to equation (1) in this case as well.

In this way, Einstein managed to identify the irregular motion of the suspended particles, now described not as a movement in the ordinary sense along a continuous trajectory but as a stochastic process governed by the function $f(x,t)$, as the elementary process corresponding to diffusion as a bulk phenomenon. He assumed the existence of a time interval that was short with respect to the observation time, but long enough to treat the motions of a suspended particle in two successive time intervals independently of each other. The displacement of the suspended particles can then be described by a probability distribution that determines the number of particles displaced by a certain distance in each time interval.

On the basis of this new interpretation, the solution of the diffusion equation, when combined with Einstein's expression for the diffusion coefficient, now results in an expression for the mean square displacement, λ_x as a function of time. Einstein suggested that this expression could be used experimentally to determine Avogadro's number N:

$$\lambda_x = \sqrt{t}\sqrt{\frac{RT}{N}\frac{1}{3\pi kP}}, \qquad (3)$$

where t is the time, and as before R the gas constant, T the temperature, k the viscosity, and P the radius of the suspended particles.

Repercussions

Brownian motion continued to play the role of a borderline problem in Einstein's subsequent publications, in which he related it not only to thermodynamics and the kinetic theory, but to electrodynamics and to radiation theory as well. In his second paper on the subject, he elaborated on the relation between Brownian motion and the foundations of the molecular theory of heat [21, p. 371]. There he took up his earlier result for energy fluctuations derived in 1904,[39] and applied it to a system subject to an external force in order to calculate the probability of deviations from the equilibrium value – due to irregular molecular motions – of a suitable observable parameter of the system. He also derived a formula for the vertical distribution of

---

[37] For Einstein's first use of probability distributions in his papers on statistical physics, see [18, p. 422].
[38] "Dies ist die bekannte Differenzialgleichung der Diffusion, und man erkennt, daß D der Diffusionskoeffizient ist. An diese Entwicklung läßt sich noch eine wichtige Überlegung anknüpfen. Wir haben angenommen, daß die einzelnen Teilchen alle auf dasselbe Koordinatensystem bezogen seien. Dies ist jedoch nicht nötig, da die Bewegungen der einzelnen Teilchen voneinander unabhängig sind. Wir wollen nun die Bewegung jedes Teilchens auf ein Koordinatensystem beziehen, dessen Ursprung mit der Lage des Schwerpunktes des betreffenden Teilchens zur Zeit $t = 0$ zusammenfällt, mit dem Unterschiede, daß jetzt $f(x,t)dx$ die Anzahl der Teilchen bedeutet, deren X-Koordinaten von der Zeit $t = 0$ bis zur Zeit $t = t$ um eine Größe *gewachsen* ist, welche zwischen x und $x + dx$ liegt. Auch in diesem Falle ändert sich also die Funktion f gemäß Gleichung (1)." [20, p. 558.]
[39] See [19, §4].

suspended particles under the influence of gravitation. As a further example, Einstein considered a system involving heat radiation, a charged harmonic oscillator in thermal equilibrium with a gas and heat radiation, thus establishing a bridge between thermodynamics, kinetic theory, and electrodynamics.

In further papers, Einstein analyzed voltage fluctuations in a condenser [22], and he returned to the issue of heat radiation [24]. Following up on work by Smoluchowski, he also dealt with the phenomenon of critical opalescence [25], showing that critical opalescence and the blue color of the sky, while not obviously related to each other, are both due to density fluctuations caused by the molecular constitution of matter. Einstein determined the pressure fluctuations in black-body radiation from the condition that the momentum they convey to a small mirror moving through the radiation precisely compensates for the momentum lost due to the average radiation pressure on the mirror.[40] The application of techniques developed for Brownian motion to the problem of heat radiation supported Einstein's controversial claim that Planck's formula for the energy spectrum of heat radiation is not compatible with the classical understanding of radiation.[41] His results show that the fluctuations due to the radiation field can neither be exclusively explained by interference phenomena of classical radiation nor by statistical fluctuations in a gas of light quanta that is conceived as a collection of classical particles.

While Einstein's work on Brownian motion had repercussions in a broad variety of fields, its central impact was, as pointed out in the beginning, on the acceptance of atomism in the early twentieth century. This impact was made possible above all by the pathbreaking experiments of Jean Perrin, that were publicized in numerous articles and books, and in particular in his best-selling and very readable "Les Atomes," published in 1913 [41]. Perrin began his experiments in 1908 and pursued them very much in line with Einstein's thinking on the subject. He then tested a formula equivalent to Einstein's for the vertical distribution of suspended particles under the influence of gravitation.[42] Perrin had also realized that the analogy established by van't Hoff between an ideal gas and a solution could be extended to colloidal solutions and suspensions, and that this analogy provides an excellent means for obtaining evidence in favor of the atomistic hypothesis.[43]

Most striking was, however, Perrin's detailed, quantitative confirmation of almost all of Einstein's predictions for the stochastic behavior of suspended particles, thus transforming the latter's "invention" of Brownian motion into powerful experimental evidence for the atomistic hypothesis. Earlier experiments were either merely qualitative in nature, e.g., Felix Ehrenhaft's observations of displacements of aerosol particles, Victor Henri's cinematographical measurements of displacements of suspended particles,[44] or Max Seddig's study of the temperature dependence of Brownian motion;[45] or they were quantitative in character, e.g., those by Svedberg [53] but still conceived within the conceptual framework of the kinetic theory, assuming that one could actually measure the velocity of Brownian particles.[46] In 1907 Einstein even wrote a paper dedicated to correcting basic flaws in Svedberg's work [23].[47]

Perrin, on the other hand, was fully aware that the work of Einstein and Smoluchowski had established a new conceptual basis for the analysis of Brownian motion. In his masterpiece "Les Atomes" he wrote [42]:

> Einstein and Smoluchowski have defined the activity of the Brownian motion in the same way. Previously we have been obliged to determine the "mean velocity of agitation" by following as

---

[40] A similar argument is given in more detail in [29].

[41] See [7].

[42] Although Einstein's name is mentioned in [39] in connection with the validity of the equipartition theorem for suspended particles, none of his papers are cited.

[43] This conceptual background is discussed at length in [40, pp. 166ff.].

[44] See [15] and [32]; for a discussion of Henri's work, see [36, p. 126].

[45] See [47, 48]. For contemporary discussions of Seddig's work, see the discussion remarks by Einstein and Seddig in [28], and [40, p. 204]; for a recent account, see [36, pp. 125–126].

[46] See [52, pp. 856–859].

[47] For Svedberg's attempt to defend his experimental analysis, see [54]. For a review of criticisms of Svedberg's work, see [33, pp. 210–212].

nearly as possible the path of a grain. Values so obtained were always a few microns per second for grains of the order of a micron. But such evaluations of the activity are *absolutely wrong*.

In 1909, Einstein gratefully wrote to Perrin:

I would have thought it impossible to investigate Brownian motion with such precision; it is fortunate for this material that you have taken it up.[48]

Ironically, the confirmation of the age-old atomistic hypothesis came in a historical moment when the massive evidence accumulated in its favor hinted at the limits of the classical understanding of atomism as well. This eventually gave way to an understanding of matter on the basis of a new physics just underway – not least due to Einstein's contributions.

Looking back at a revolution

Einstein's exploration of the statistical properties of physical processes, such as Brownian motion, amounted to a reversal of perspective with respect to that of classical physics, as he was well aware himself. In his papers on statistical physics, this becomes particularly clear with regard to his interpretation of Boltzmann's principle, which relates the thermodynamic entropy of a physical system to the statistical probability of its states. In a lecture on Boltzmann's principle from 1910,[49] Einstein argued that this principle can be applied in two different ways. When starting from a complete atomistic picture of a system, one can calculate the probability of its states and then determine, with the help of Boltzmann's principle, the entropy and hence the thermodynamic behavior of the system. This is the perspective of classical kinetic theory.

For Einstein, however, the real significance of Boltzmann's principle was rather its reverse application to the case for which no complete atomistic picture of a system was available, as was the case for those systems that were suspected to exhibit non-classical behavior such as heat radiation. In that case, the most important application of Boltzmann's principle was, in Einstein's view, to infer from the observed thermodynamic behavior of a system the statistical probability of its single states. One would thus be able to judge the extent to which the system deviates from the behavior expected according to classical thermodynamics, for instance, by exhibiting the kind of fluctuation behavior represented by Brownian motion. This reversal of perspective is another instance of the Copernicus process mentioned above.

The profound conceptual implications of the atomistic revolution triggered by Einstein's work on Brownian motion also become apparent from the discussion in his 1910 lecture of the question of whether physical facts are causally connected in a complete way. He argued that – in view of the unpredictable irregularity of Brownian motion – this question has to be definitely answered in the negative. Einstein concluded, however, that the very fact that we are able to obtain the statistical laws for such fluctuation phenomena suggests that, on a theoretical level, we have nevertheless to maintain the presupposition of a complete causal determination of physical events, although we can never hope to receive an immediate confirmation of this conception by ever more refined observations – a remarkable stance at the dawn of quantum mechanics.

References

[1] AEA: Albert Einstein Archives, The Jewish National and University Library, The Hebrew University of Jerusalem, Israel, http://www.alberteinstein.info.
[2] L. Boltzmann, Vorlesungen über Gastheorie, Part 1, Theorie der Gase mit einatomigen Molekülen, deren Dimensionen gegen die mittlere Weglänge verschwinden (Johann Ambrosius Barth, Leipzig, 1896).

---

[48] Einstein to Jean Perrin, 11 November 1909, [11, Doc. 186]. "Ich hätte es für unmöglich gehalten, die Brown'sche Bewegung so präzis zu untersuchen; es ist ein Glück für diese Materie, dass Sie sich ihrer angenommen haben."
[49] "Über das Boltzmann'sche Prinzip und einige unmittelbar aus demselben fliessende Folgerungen." *Vorlesung für die Physikalische Gesellschaft Zürich*, 2 November 1910, Zangger Nachlaß, Zentralbibliothek Zürich.

[3] L. Boltzmann, "Vorschlag zur Restlegung gewisser physikalischer Ausdrücke", in: Verhandlungen der Gesellschaft Deutscher Naturforscher und Ärzte, 70. Versammlung zu Düsseldorf, 19–24 September 1898, pp. 67–68; reprinted in: Wissenschaftliche Abhandlungen, edited by Fritz Hasenöhrl (Vogel, Leipzig, 1909).
[4] R. Brown, Edinb. New Philos. J. **5**, 358–371 (1828).
[5] S. G. Brush, Arch. Hist. Exact Sci. **5**, 1–36 (1968).
[6] S. G. Brush, The Kind of Motion We Call Heat: A History of the Kinetic Theory of Gases in the 19th Century, Book 1, Physics and the Atomists. Book 2, Statistical Physics and Irreversible Processes (North-Holland, Amsterdam, 1976).
[7] J. Büttner, J. Renn, and M. Schemmel, Stud. Hist. Philos. Mod. Phys. **34**, 37–59 (2003).
[8] A. Cotton and H. Mouton, Les ultramicroscopes et les objets ultramicroscopiques (Masson, Paris, 1906).
[9] CPAE 1, edited by J. Stachel, D. C. Cassidy, R. Schulmann, and J. Renn, The Collected Papers of Albert Einstein, Vol. 1, The Early Years, 1879–1902 (Princeton University Press, Princeton, 1987).
[10] CPAE 2, edited by J. Stachel, D. C. Cassidy, J. Renn, and R. Schulmann, The Collected Papers of Albert Einstein, Vol. 2, The Swiss Years: Writings, 1900–1909 (Princeton University Press, Princeton, 1989).
[11] CPAE 5, edited by M. J. Klein, A. J. Kox, and R. Schulmann, The Collected Papers of Albert Einstein, Vol. 5, The Swiss Years: Correspondence, 1902–1914 (Princeton University Press, Princeton, 1993).
[12] CPAE 9, edited by D. Kormos Buchwald, R. Schulmann, J. Illy, D. J. Kennefick, and T. Sauer, The Collected Papers of Albert Einstein, Vol. 9, The Berlin Years: Correspondence, January 1919–April 1920 (Princeton University Press, Princeton, 2004).
[13] G. L. De Haas-Lorentz, Die Brownsche Bewegung und einige verwandte Erscheinungen (Friedrich Vieweg und Sohn, Braunschweig, 1913).
[14] R. J. Deltete, The Energetics Controversy in Late Nineteenth-Century Germany: Helm, Ostwald and Their Critics, 2 vols., Ph.D. dissertation, Yale University (1983).
[15] F. Ehrenhaft, Sitz.ber., Kaiserl. Akad. Wiss. Math.-Nat.wiss. Kl. IIa **116** 1139–1149 (1907).
[16] A. Einstein, Ann. Phys. (Leipzig) **4**, 513–523 (1901), reprinted in this volume (pp. 88–98).
[17] A. Einstein, Ann. Phys. (Leipzig) **8**, 798–814 (1902), reprinted in this volume (pp. 100–116).
[18] A. Einstein, Ann. Phys. (Leipzig) **9**, 417–433 (1902), reprinted in this volume (pp. 118–134).
[19] A. Einstein, Ann. Phys. (Leipzig) **14**, 354–362 (1904), reprinted in this volume (pp. 155–163).
[20] A. Einstein, Ann. Phys. (Leipzig) **17**, 549–560 (1905), reprinted in this volume (pp. 182–193).
[21] A. Einstein, Ann. Phys. (Leipzig) **19**, 371–381 (1906), reprinted in this volume (pp. 248–258).
[22] A. Einstein, Ann. Phys. (Leipzig) **22**, 569–572 (1907), reprinted in this volume (pp. 292–295).
[23] A. Einstein, Zeitschrift für Elektrochemie und angewandte physikalische Chemie **13**, 41–42 (1907), reprinted in [10, Doc. 40].
[24] A. Einstein, Physikalische Zeitschrift **10**, 323–324 (1909).
[25] A. Einstein, Ann. Phys. (Leipzig) **33**, 1275–1298 (1910), reprinted in this volume (pp. 368–391).
[26] A. Einstein, "Theoretische Atomistik", in: Die Kultur der Gegenwart. Ihre Entwicklung und ihre Ziele, Part 3, Sect. 3, Vol. 1, Physik, edited by P. Hinneberg and E. Warburg (B. G. Teubner, Leipzig, 1915), pp. 251–263.
[27] A. Einstein, Autobiographical Notes: A Centennial Edition, translated and edited by P. A. Schilpp (La Salle, Illinois, 1979).
[28] A. Einstein et al., Phys. Z. **10**, 779–780 (1909), reprinted in CPAE 2, Doc. 58.
[29] A. Einstein and L. Hopf, Ann. Phys. (Leipzig) **33**, 1105–1115 (1910), reprinted in this volume (pp. 357–367).
[30] F. M. Exner, Ann. Phys. (Leipzig) **2**, 843–847 (1900).
[31] A. Fick, Ann. Phys. Chem. **4**, 59–86 (1855).
[32] V. Henri, C.R. Acad. Sci. **146**, 1024–1026 (1908).
[33] M. Kerker, ISIS **67**, 190–216 (1976).
[34] K. von Nägeli, Sitz.ber., Kgl. Bayerische Akad. Wiss. München, Math.-phys. Kl. **9**, 389–453 (1879).
[35] W. Nernst, Z. Phys. Chem. Stöchiometrie Verwandtschaftslehre **2**, 613–637 (1888)).
[36] M. J. Nye, Molecular Reality: A Perspective on the Scientific Work of Jean Perrin (Macdonald, London, 1972)
[37] W. Ostwald, Lehrbuch der allgemeinen Chemie, 2nd rev. ed., Vol. 1, Stöchiometrie (Wilhelm Engelmann, Leipzig, 1891).
[38] W. Ostwald, Lehrbuch der allgemeinen Chemie. 2nd rev. ed. Vol. 2, part 1, Chemische Energie (Wilhelm Engelmann, Leipzig, 1893).
[39] J. Perrin, C.R. Acad. Sci. **146**, 967–970 (1908).

[40] J. Perrin, "Les preuves de la réalité moléculaire (Etude spéciale des émulsions)", in: La théorie du rayonnement et les quanta. Rapports et discussions de la réunion tenue à Bruxelles, du 30 octobre au 3 novembre 1911. Sous les auspices de M. E. Solvay, edited by P. Langevin and M. de Broglie, (Gauthier-Villars, Paris).

[41] J. Perrin, Les Atomes (Félix Alcan, Paris, 1913).

[42] J. Perrin, Atoms, translated by D. Ll. Hammick (Ox Bow Press, Woodbridge, Conn., 1990).

[43] H. Poincaré, La science et l'hypothèse (E. Flammarion, Paris, 1902).

[44] J. Renn, "Einstein's Controversy with Drude and the Origin of Statistical Mechanics: A New Glimpse from the 'Love Letters'", in: Einstein: The Formative Years, 1879–1909, Einstein Studies, Vol. 8, edited by D. Howard and J. Stachel (Birkhäuser, Boston, 2000), pp. 107–157.

[45] J. Renn, Phys. J. **3**, 49–55 (2004).

[46] J. Renn and R. Rynasiewicz, "The Puzzle of Einstein's Annus Mirabilis", in: Cambridge Companion to Einstein, edited by M. Janssen and C. Lehner (Cambridge University Press, New York, in press); J. Renn, "Einstein as a Disciple of Galileo: A comparative study of Concept Development in Physics", in: Einstein in Context, edited by M. Beller, R. Cohen, and J. Renn (Cambridge University Press, Cambridge, 1993).

[47] M. Seddig, Sitz.ber., Ges. Beförderung der gesammten Naturwissenschaften, 182–188 (1907).

[48] M. Seddig, Phys. Z. **9**, 465–468 (1908).

[49] H. Siedentopf and R. A. Zsigmondy. Ann. Phys. (Leipzig) **10**, 1–39 (1903).

[50] M. von Smoluchowski, Ann. Phys. (Leipzig) **21**, 756–780 (1906).

[51] W. Sutherland, Philos. Mag. J. Sci. **44**, 493–498 (1897).

[52] T. Svedberg, Z. Elektrochem. angewandte phys. Chem. **12**, 853–860 (1906).

[53] T. Svedberg, Z. Elektrochem. angewandte phys. Chem. **12**, 909–910 (1906).

[54] T. Svedberg, Z. Phys. Chem. Stöchiometrie Verwandtschaftslehre **71**, 571–576 (1910).

[55] J. H. Van 't Hoff, Z. Phys. Chem. Stöchiometrie Verwandtschaftslehre **1**, 481–508 (1887).

[56] R. A. Zsigmondy, Zur Erkenntnis der Kolloide. Über irreversible Hydrosole und Ultramikroskopie (Gustav Fischer, Jena, 1905).

The optics and electrodynamics of 'On the Electrodynamics of Moving Bodies'

Robert Rynasiewicz[*]

Department of Philosophy, Johns Hopkins University, Baltimore, MD 21218, USA

Received 21 December 2004
Published online 14 February 2005

Key words Einstein, special relativity, electrodynamics, optics, history of physics.
PACS 01.65.+g, 03.30.+p, 03.50.De, 03.65.Sq

Introduction

Einstein began using the expression "the electrodynamics of moving bodies" as early as 1899 [1] with reference to the problem whether, when a body moves through the ether, it drags the ether in its interior and immediate surroundings along with it, or whether instead the ether remains completely stationary. At the time, experiments could be cited in support of either alternative. An article by Wien [2], which Einstein read and wrote to Wien about that year [3], contains a review of both optical and purely electrodynamical experiments in connection with how they bear on the issue. For example, Fizeau's measurement of the speed of light in moving water [4] and the phenomenon of stellar aberration present difficulties for a totally dragged ether. In contrast, the null result of the Michelson-Morley interferometer experiment [5] and Röntgen's failure to find a magnetic effect depending on the orientation of a charged condenser with respect to the motion of the earth [6] speak prima facie against a completely stationary ether. Part of Wien's point is that various hypotheses, as to how much ether is dragged in proportion to the mass of a body and to what extent beyond the boundaries of the body, fare differently with respect to different experiments.

The paper 'On the Electrodynamics of Moving Bodies' [7] presents Einstein's ultimate solution to this nexus of problems, a solution that ironically renders the ether, whether stationary or dragged, completely 'superfluous', by proposing a revolutionary new kinematics based on the extension of the Galilean principle of relativity to the domain of optics and electrodynamics, a doctrine that later came to be known as the special theory of relativity. Einstein's ultimate solution, as proposed in that 1905 paper, contains yet a deeper irony.

The paper contains two parts, the 'Kinematical Part' and the 'Electrodynamic Part'. The former derives the Lorentz transformations from first principles and explores their immediate consequences in terms of length contraction, time dilation, and the composition of velocities. The Electrodynamical Part shows how to understand Maxwell's equations in this new setting and how to treat an array of problems in the optics and electrodynamics of moving bodies. Practitioners today view the special theory of relativity as the final perfecting step in the development, beginning with Faraday and Maxwell, of the first field theory in the history of physics. As a matter of formal geometry, Minkowk29i space-time is the natural arena for for Maxwell's theory. Indeed, doubts have often been expressed that the latter has a coherent formulation in the background of the classical space-time of Newtonian mechanics. Therefore, it is commonplace to regard special relativity as well-motivated to the extent that Maxwell's equations are deemed valid and to

---

[*] E-mail: ryno@jhu.edu

view Einstein's goal from this perspective, namely as the endeavor to secure the crowning perfection of Maxwell's theory.

Yet only a few months prior to setting out the theory of relativity, Einstein had sent off to the *Annalen der Physik* a paper challenging the very conception of radiation mandated by Maxwell's theory of light. The thesis put forward in that paper, with the somewhat obscure title, 'On a Heuristic Point of View Concerning the Production and Transformation of Light' [8], is that the energy of light radiating from a source is not distributed continuously over an ever expanding region of space as demanded by the electromagnetic theory, or any wave theory of light, but instead consists of a finite number of energy quanta, localized at spatial points, which move without dividing, and which can be absorbed and produced only as indivisible units. Far from being a believer in the exact validity of Maxwell's theory, as were almost all physicists until well into the 1920's, Einstein had already developed the conviction that it was drastically inadequate in a number of domains to which it could be applied only under appropriate limiting conditions.

The larger irony of 'On the Electrodynamics of Moving Bodies' is thus that Einstein developed the theory of relativity as a consequence of having failed to find a foundational alternative to Maxwell's theory. This casts the 'Electrodynamic Part' of the 1905 paper in a new light. Although it expresses no overt skepticism as to the validity of Maxwell's theory, it nonetheless addresses just those applications Einstein had previously worked through in order to explore the limits of the validity of that theory.

When the paper is read with this understanding, not only do the details stand out in greater relief, the character of paper as a whole becomes clearer as a "transition" document, marking a passage from one set of concerns to another. The problems addressed in the Electrodynamical Part drove Einstein, albeit in round about ways, to the discovery of the self-standing doctrine as set out in the Kinematical Part. This doctrine yielded a secure and independent justification, previously lacking, for the approach he had explored for the problems of the Electrodynamical Part. However, at the time of writing 'On the Electrodynamics of Moving Bodies', it had only just occurred to Einstein that the new kinematics held consequences beyond this group of optical and electrodynamical problems. Only after several months did the consequences for the relation between mass and energy dawn on him. Over the course of the next two years he continued to explore this new terrain of general dynamical implications.

This perspective on the emergence of special relativity will dominate the exposition to follow. The first task is to prepare the way by clarifying the content of the postulates of the theory, their experimental support, and Einstein's principal motivation for developing the theory. Then the structure and content of the 1905 relativity paper will be analyzed from Einstein's perspective on the microstructure of radiation and the limited adequacy of Maxwell's theory in other domains of application. This will lead to questions about the precise timing of Einstein's discovery of various ingredients of the theory. The character of the 1905 paper as a transitional document is then spelled more fully. Finally, we will see how our conception of special relativity as a foundational doctrine is just as out of step with Einstein's conception of the theory as is the idea that he created special relativity in order to secure a prized place for Maxwell's theory.

The postulates and their support

When Einstein visited Japan in 1922, he gave an impromptu lecture, 'How I Created the Theory of Relativity', at the bequest of his hosts at Kyoto University. As Einstein spoke in German, the physicist Jan Ishiwara took notes in Japanese. These were published the following year in the Japanese periodical *Kaizo* and reprinted in 1971 as a chapter of Ishiwara's *Einstein Kōen-Roku* [9]. Although the account is rather sketchy, a number of English translations have subsequently appeared in print [10] and been subjected to intense scrutiny [11] in connection with what they say, or do not say, about Einstein's awareness, prior to 1905, of the Michelson and Morley experiment to measure the velocity of the earth through the ether.

The special theory of relativity is founded on two postulates, the principle of (special) relativity and a postulate Einstein called 'the principle of the constancy of the velocity of light'. The latter has been repeatedly misstated in textbooks and popularizations as the assertion that the speed of light is the same in all inertial frames, and typically the Michelson-Morley experiment has been adduced as the principal experimental support for it. This has engendered the myth that Einstein created the special theory of relativity in direct response to the null result of that experiment.

The *light postulate*, as historians have come refer to it [12], states only that the velocity of light is independent of the motion of its source in *some* given frame. The official enunciation of the postulate in the 1905 paper occurs in §2: 'Each light ray travels in the "stationary" coordinate system with the definite velocity [c], independently of whether this light ray is emitted from a stationary or a moving body', where the "stationary" coordinate system is an inertial frame, which the contemporary reader is naturally invited to identify with the rest frame of the ether. The light postulate is in fact a trivial consequence of Maxwell's theory, or of any wave theory of light based on the assumption of a stationary ether, and thus a completely uncontroversial assumption for Einstein's contemporary reader. Rather, it is the principle of relativity, with which Einstein notes the light postulate is only apparently incompatible, that stands in need of justification. In later expositions, Einstein repeatedly invoked the Michelson-Morely experiment specifically in support for the principle of relativity [12]. If Einstein meant to include it among the 'failed attempts to confirm the motion of the earth relative to the "light-medium" ' mentioned in the second paragraph of the 1905 paper, then there too the Michelson-Morley experiment is invoked exclusively in support of the principle of relativity.

In order to see that the experiment provides no support for the light postulate, or for the so-called invariance of c, consider an alternative to the light postulate historically associated with corpuscular theories of light, namely, the *emission hypothesis*, which asserts that the velocity of light (in at least one frame) is the vector sum of the velocity of the source with the velocity of emission relative to the source. Assuming the classical velocity addition rule, it follows that this holds in all frames, as required by the principle of relativity. Since the light source used in the Michelson-Morley experiment is mounted on the interferometer, and hence stationary with respect to it, the null result of the experiment is in full accord with the emission hypothesis. From this it is clear that that experiment can support only the principle of relativity.

The primary motivation given for the principle of relativity in the 1905 paper, however, is a consideration of a very different sort. The opening paragraph points to an asymmetry in the standard way Maxwell's theory was applied to a commonplace phenomenon which itself doesn't display the asymmetry. If a magnet and a conductor move relative to one another, an electric current is generated in the conductor. According to the customary understanding of Maxwell's electrodynamics, the cause of the current is different depending on which of the two is regarded to be in absolute motion. If the magnet moves and the conductor is at rest, there arises an electric field with a definite associated energy, which then produces the current. In contrast, if the magnet is at rest and the conductor moves, there is no electric field. Instead the current is thought to be generated by an "electromotive force" corresponding to which there is no associated energy. Nevertheless, the intensity and direction of the electric current is the same in the two cases, assuming the same relative motion of the magnet and conductor.

Why should Einstein have introduced such an example as the primary motivation for the principle of relativity? In the manuscript draft of an article, 'Fundamental Ideas and Methods of the Theory of Relativity, Presented in their Development' [13], intended for, but never published in *Nature*, Einstein mentions that 'in developing the special theory of relativity, the following... thought concerning Faraday electromagnetic induction played a leading role for me.' [14] He rehearses the example of the relatively moving magnet and conductor, explaining specifically how the Maxwell-Lorentz theory treats the two cases differently. In the first case, the electric field $\vec{E}$ arises from the temporal variation of the magnetic field $\vec{H}$ according to the Maxwell equation

$$\frac{1}{c}\frac{\partial \vec{H}}{\partial t} = -\vec{\nabla} \times \vec{E}. \tag{1}$$

In the second case, the electrons in the conductor have a velocity v through the ether and thus suffer an electromotive force corresponding to the $\vec{v} \times \vec{H}$ term of the Lorentz force equation:

$$\vec{F} = q \left(\vec{E} + \frac{1}{c} \left(\vec{v} \times \vec{H} \right) \right). \tag{2}$$

Einstein continues:

> The thought that here two essentially different cases are involved was for me intolerable. The difference between the two could be, according to my conviction, only a difference in the choice of viewpoint, and not a real difference. Judged from the magnet, there was certainly *no* electric field present; judged from the current-loop there certainly was one such present. The existence of the electric field was thus a relative one, according to the state of motion of the coordinate system used, and, leaving out of consideration the state of motion of the observer, [i.e.] coordinate system, only the electric and magnetic field *together* could be granted a kind of objective reality. The phenomenon of electromagnetic induction forced me to postulate the (special) Principle of Relativity. [15]

We now take it for granted that the electromagnetic field corresponds to a single mathematical object, the Maxwell tensor, which at a each space-time point can be decomposed into electric and magnetic components depending on the choice of a time-like unit vector and its associated Minkowski-orthogonal space-like hypersurface. Even at a level of minimal mathematical sophistication, the idea of the electric and magnetic fields as frame relative quantities is endemic to the conception of the electromagnetic field as a self-subsistent entity, rather than as a state of some quasi-material continuous medium. The source of Einstein's vexation cannot be appreciated without recapturing the pre-relativistic point of view that the electric and magnetic field quantities at each point pertain to the intrinsic state of the ether. The idea that these quantities could somehow be frame independent was as unthinkable as the idea that the temperature at a given point of a continuous body should somehow depend on one's state of motion with respect to it. Thus, the very possibility of a principle of relativity for electrodynamics required first the fashioning of an entirely new conception of the electric and magnetic field quantities.

Appended to the above passage is a revealing footnote:

> The difficulty to be overcome lay then in the constancy of the vacuum-velocity of light, which initially I believed I would have to give up. Only after years of fumbling did I notice that the difficulty rested on the arbitrariness of fundamental kinematical concepts. [16]

In other words, Einstein initially thought that in order to implement the principle of relativity, not only must the electric and magnetic fields be thought of as frame relative quantities, but also the velocity of light must depend on the motion of its source. In other words, the principle of relativity mandated not just a re-conception of Maxwell's theory, but a wholesale rejection of it and the establishment an alternative based on the emission hypothesis. A variety of independent sources confirm that Einstein struggled in vain to find such a theory for years prior to 1905.

After the introduction of the special theory of relativity, a number of physicists, beginning with Ritz in 1908, explored alternatives of this sort in order to avoid the new conception space and time while retaining the principle of relativity [17]. The first experimental evidence directly supporting the light postulate appeared only in 1913 with De Sitter's observations on binary stars [18]. Without having any direct experimental evidence for or against it, Einstein gave up on the emission hypothesis prior to 1905 primarily because it requires the introduction of additional arbitrary assumptions even in the simplest of cases, such as the velocity of light reflected from a moving mirror. [19] It is unlikely, though, that he conceived of the light postulate as an alternative principle, clearly separable from the remainder of Maxwell's theory, until May of 1905. Rather, his abandonment of the emission hypothesis would have been a protracted process of comparing the treatment of various problems on a case by case basis from various points of view. With a growing disenchantment with the emission hypothesis, the major problems areas remained threefold:

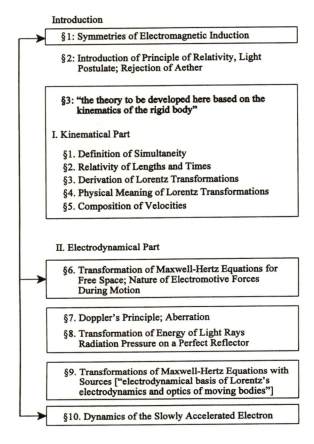

Fig. 1 Thematic structure of "On the Electrodynamics of Moving Bodies".

1. the ponderomotive forces exerted by external fields on electrically or magnetically charged bodies,
2. the behavior of radiation in free space, and
3. the electromagnetic properties of bulk matter, including their influence on optical phenomena.

The first of these Einstein would have approached with the relativistic conception of the electric and magnetic fields in hand. The second he would have addressed, eventually, if not initially, with the heuristic of the light quantum hypothesis. For the third, he could consider bulk matter phenomenologically as a continuum, but would ultimately want and need an account based on an atomic-molecular theory of matter.

Structure and contents of the 1905 relativity paper

These concerns are evident in the structure and contents of the 1905 paper. Fig. 1 groups the various sections of the paper as they relate to these. That the theme of the relativistic conception of the field is chosen as a binding thread – from the very first paragraph, introducing the puzzling symmetric behavior of electromagnetic induction, to the interpretation of the field transformation equations in §6 which resolves that puzzle, to the final section (§10) on the dynamics of the slowly accelerated electron, which is the one direct and immediate application of the idea in the entire paper– should not come as a surprise. These correspond to the first problem area above. The two sections, §7 and §8, clearly fit together insofar as they treat optical problems for radiation in free space, thus corresponding to the second problem area. §9 addresses Lorentz's electron theory of matter and comprises Einstein's resolution of the third. The 'Kinematical Part', presented as a self-standing doctrine based on the two postulates and the recognition of the conventional character of

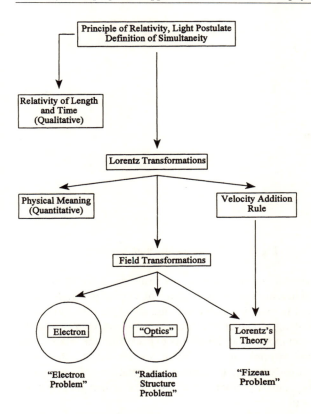

Fig. 2 Logical structure of "On the Electrodynamics of Moving Bodies"

distant simultaneity, represents a point of view Einstein achieved only in late May of 1905, some five weeks before the completion of the paper.

Fig. 2 sketches the logical structure of the paper. The primary thrust of the 'Kinematical Part', apart from securing a derivation of the Lorentz transformation from the postulates and the definition of simultaneity alone, is to explore their physical interpretation. Even the derivation of the velocity addition rule, which interestingly has but one application in the 'Electrodynamical Part', proceeds as an exploration of the limiting character of the velocity of light and the group character of the Lorentz transformations. The three leaves at the bottom of Fig. 2 represent the end applications in the 'Electrodynamical Part' corresponding to the three problem areas mentioned above. It is to these that we now turn.

The radiation structure problem

Einstein's light quantum paper [8] was the first of the five works that Einstein produced in his annus mirabilis of 1905. In a paper the following year [20], Einstein remarked that at the time he wrote the 1905 light quantum paper, he conceived of that earlier work as standing in opposition to Planck's theory. It offered an explanation of the Wien portion of the blackbody spectrum, as well as of photoluminescence, the photo-electric effect, and the ionization of gases, in terms of an elementary process of the direct conversion of light energy into energy of motion, and vice-versa, without the intermediary of electron resonators. This idea of "direct conversion" had occurred to Einstein' [21] in early 1901 in critical response [22] to Planck's efforts to derive Wien's spectral law, which at the time Planck thought to be strictly valid, from the statistical behavior of resonators [23]. Thus, it would appear that the idea of light quanta played a background role for many years leading to 1905.

The two sections of the 1905 relativity paper that address the properties of radiation in free space, §7 and §8, do not address, as might be thought, the interpretation of null results of optical experiments to

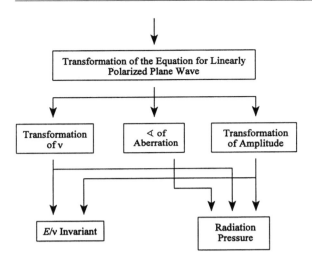

Fig. 3 "Radiation–Structure" problem

detect the motion of the earth. Those are accounted for immediately by the principle of relativity. Nor are these sections just a potpourri of results in relativistic optics. The chain of derivations falls into a distinctive pattern, as sketched in Fig. 3, the terminal points of which constitute concerns related directly to the conception of radiation as the propagation of light quanta: the transformation of energy and the calculation of radiation pressure. For both, it was incumbent that the results expected on the basis of the light quantum hypothesis concur with those obtained by applying the principle of relativity to Maxwell's theory of light. Conversely, one might expect that, already having the principle of relativity as a necessary constraint, Einstein approached them first as light quantum problems, the results of which subsequently guided his approach to them as problems in wave optics.

At the beginning of §7 Einstein introduces the equations for a linearly polarized plane wave in terms of coordinates for the "stationary" frame. The inverse Lorentz transformations are used to transform these to equations expressed in terms of coordinates and fields adapted to the moving frame. The relativistic formulas for stellar aberration and Doppler's principle fall out immediately by regrouping terms. The derivations of these formulae involve only the coordinate transformations. An indication that the material in the two sections underwent editorial shuffling is evident when Einstein turns to the transformation of wave amplitude, which involves using the field transformations. Instead of stating the transformation equation for wave amplitude, he gives the formula for transformation of the *square* of wave amplitude. Only later, in §8, after discussing of the transformation of radiation energy, does he state the equation for wave amplitude

$$A' = A \frac{1 - (v/c)\cos\varphi}{\sqrt{1 - v^2/c^2}} \tag{3}$$

in order to treat the problem of radiation pressure. Of the two, the latter is conceptually the simpler.

Radiation pressure

In 1873 Maxwell had predicted on the basis of his electromagnetic theory of light that radiation should exert a pressure on a non-conducting surface [24]. Shortly thereafter, Bartoli [25] and Boltzmann [26] argued, on purely thermodynamic grounds independently of Maxwell's theory, that light must exert a pressure on a completely reflecting surface equal to double the energy density of the radiation. Radiation pressure continued to play an important role in the development of the theory of blackbody radiation. For example, in the course of the derivation of his celebrated displacement law, Wien calculated the modification of thermal radiation density as a result of reflection from a moving mirror [27].

Einstein's task in the 1905 paper is roughly the inverse: to calculate the pressure exerted by monochromatic radiation from the change in its energy upon reflection from a moving mirror. If v is the velocity of the mirror and P the radiation pressure, then the energy difference represents the work Pv done on, or by, the mirror. As a problem in wave optics, the change in radiation density is computed from the change in wave amplitude squared together with the angles of incidence and reflection as given by the formula for aberration.

As a light quantum problem, all that is required is the resultant Doppler shift of the reflected radiation. Let $\mathcal{N}$ be the number of light quanta per unit volume in the stationary frame and φ the angle between the x-axis and the direction of propagation. Then the number N of light quanta striking the mirror in unit time is $\mathcal{N}(c\cos\varphi - v)$ and the incident energy is $Nh\nu$. Since the numbers of incident and reflected light-quanta are the same, the reflected energy is $Nh\nu_r$, where ν_r is the frequency after reflection. The radiation pressure is thus given simply by the equation

$$Pv = Nh(\nu - \nu_r). \tag{4}$$

To solve for P, all that is left to be done is to determine ν_r. This involves only two applications of the relativistic Doppler transformation and one application of the formula for aberration, corresponding to Einstein's derivation to the expression for ν''', which is the same quantity as our ν_r. Substituting and solving:

$$P = 2\mathcal{N}h\nu \frac{(\cos\varphi - v/c)^2}{1 - v^2/c^2}, \tag{5}$$

which agrees with the result presented in the paper if one identifies $\mathcal{N}h\nu$ as the energy per unit volume in place of the wave-optic expression $A^2/8\pi$.

Given the prominent role that radiation pressure played in the theory of blackbody radiation, it's hard to imagine that Einstein did not consider the problem in these terms, especially considering how far more complicated the calculation becomes as a problem cast strictly terms of wave optics. As a light quantum problem, the case of $\varphi = 0$ for radiation striking the mirror orthogonally is no more than a back-of-the-envelope calculation, requiring, for a first order approximation, only the classical Doppler formula. Moreover, the relativistic correction in this case is particularly easy to guess. The principle of relativity requires that if ν transforms to ν' by a velocity boost v and if ν' transforms to ν'' by the inverse velocity boost by $-v$, then $\nu = \nu''$. Using the classical Doppler shift,

$$\begin{aligned}\nu'' &= \nu'(1 + v/c) \\ &= \nu(1 - v/c)(1 + v/c) \\ &= \nu(1 - v^2/c^2),\end{aligned}$$

suggesting that the classical expression needs only to be corrected by dividing by the square root of $1 - v^2/c^2$ in order to be brought into conformity with the principle of relativity. A similar technique, reviewed below, permits a direct derivation of the relativistic field transformation equations, and thus acquisition of the relativistic transformation of wave amplitude. This suffices to set up a pressing puzzle for the transformation of radiation energy.

The transformation of radiation energy

If monochromatic radiation has frequency ν in the stationary frame and there are N quanta in a given region of space, the energy E of the radiation contained in that region is $Nh\nu$. If light quanta have an objective reality, then each and every one of these N quanta must exist for the moving observer as well, although the energy of each quantum will be $h\nu'$ and hence the total energy E' of those quanta is $Nh\nu'$. By the relativistic Doppler formula, the ratio of the energies as measured in the respective frames is then

$$\frac{E'}{E} = \frac{1 - (v/c)\cos\varphi}{\sqrt{1 - v^2/c^2}}. \tag{6}$$

From the point of view of wave optics, as Einstein indicates at the beginning of §8, the energy of radiation per unit volume is directly proportional to the square of the wave amplitude. If A is the amplitude as measured in the stationary frame, the energy per unit volume is $A^2/8\pi$. Similarly, if A' is the wave amplitude as measured in the moving frame, then $A'^2/8\pi$ must be regarded as the energy of radiation per unit volume in the moving frame. According to the transformation of wave amplitude (3),

$$\frac{A'^2}{A^2} = \frac{(1-(v/c)\cos\varphi)^2}{1-v^2/c^2}. \tag{7}$$

At first sight, this appears to stand in conflict with (6), at least if one presupposes that the light quanta occupying a unit volume in one frame occupy a unit volume in the other.

However, it is completely evident from considerations of radiation pressure that the *density* of light quanta is not even first order invariant if the light postulate is assumed. Suppose a beam of light quanta propagating in the positive x direction strikes a mirror moving with velocity v in the same direction. The number of quanta striking the mirror per unit time is $\mathcal{N}(c-v)$, where $\mathcal{N}$ is the density of quanta per unit volume in the stationary frame. As judged from the rest frame of the mirror, the number of quanta incident per unit time is $\mathcal{N}'c$, where $\mathcal{N}'$ is the density of quanta per unit volume in the mirror frame. Assuming that the measure of unit time is the same in the two frames, at least to first order, these two numbers must be the same. Hence, to first order,

$$\mathcal{N}' = \mathcal{N}(1-v/c). \tag{8}$$

At first sight, this of course would be taken to be a reductio of the compatibility the light quantum hypothesis with the light postulate, and thus so much the worse for the latter. Alternatively, if the light postulate is given as fixed, as required by Maxwell's theory, it provides an insight on how to square (6) with (7).

The exact relation between $\mathcal{N}$ and $\mathcal{N}'$ can be discovered by equating the quantum-theoretic expression for energy per unit volume with the wave-optical expression. Then

$$\mathcal{N}'h\nu' = \frac{A'^2}{8\pi} = \frac{A^2}{8\pi}\frac{(1-v/c)^2}{1-v^2/c^2} = \mathcal{N}h\nu\frac{1-v/c}{\sqrt{1-v^2/c^2}}.$$

Thus,

$$\mathcal{N}' = \mathcal{N}\frac{1-v/c}{\sqrt{1-v^2/c^2}}.$$

The ratio of the volumes containing numerically the same light quanta is then the inverse of the factor on the right hand side. This, however, falls short of a justification, since it implicitly assumes what needs to be shown, viz., the invariance of the number of light quanta.

Einstein's method of justification is to fix a region across the boundary of which there is no energy flux, so that that the region always contains, as Einstein puts it, the same "light-complex." Formally, this is the same idea as that of riding a light wave, something Einstein described in 1949 in his *Autobiographical Notes* [28] as a "paradox" he had first thought about at the age of sixteen:

> If I pursue a beam of light with the velocity c ..., I should observe such a beam of light as an electromagnetic field at rest though spatially oscillating. There seems to be no such thing, however, neither on the basis of experience nor according to Maxwell's equations. From the very beginning it appeared to me intuitively clear that, judged from the standpoint of such an observer, everything would have to happen according to the same laws as for an observer who, relative to the earth, was at rest. For how should the first observer know, or be able to determine, that he is in a state of fast uniform motion? [29]

There, Einstein tells us that it was renewed reflection upon this "paradox" that led to the final insight necessary for the special theory of relativity.

In order to illustrate, it is convenient to specialize to the case, $\varphi = 0$, of light propagating in the direction of the positive x-axis. Then, instead of a sphere of radius R that encompass the same "light-complex" as Einstein uses in the 1905 paper, we can use the surface of a cube of volume R^3 moving with a velocity c in the positive x-direction. So, consider the surface given by the intersection of six planes specified, using stationary coordinates, by the equations $y_0 = 0 = z_0$, $y_1 = R = z_1$, $x_0 = ct$, and $x_1 = ct + R$. The enclosed volume V is

$$V = (x_1 - x_0)(y_1 - y_0)(z_1 - z_0) = R^3.$$

Since the x_0 and the x_1 planes move in the positive x-direction with velocity c, and since the other planes are all orthogonal to the direction of light propagation, there is no energy flux through the surface. The same surface is described in terms of the coordinates (ξ, η, ζ, τ) of the moving system by the equations $\eta_0 = 0 = \zeta_0$, $\eta_1 = R = \zeta_1$, and

$$\frac{\xi_0 + v\tau}{\sqrt{1 - v^2/c^2}} = \frac{c(\tau + v\xi_0/c^2)}{\sqrt{1 - v^2/c^2}}, \tag{9}$$

$$\frac{\xi_1 + v\tau}{\sqrt{1 - v^2/c^2}} = \frac{c(\tau + v\xi_1/c^2)}{\sqrt{1 - v^2/c^2}} + R. \tag{10}$$

Subtracting (9) from (10) yields

$$\xi_1 - \xi_0 = R \frac{\sqrt{1 - v^2/c^2}}{1 - v/c}.$$

Thus, the volume V' as measured in the moving system is:

$$V' = (\xi_1 - \xi_0)(\eta_1 - \eta_0)(\zeta_1 - \zeta_0) = R^3 \frac{\sqrt{1 - v^2/c^2}}{1 - v/c}.$$

If it is not immediately clear that the difference in volume of the given "light complex" in the two frames is due to the fact that the surfaces orthogonal to the direction of motion are judged at different times in the different frames, reflecting a difference in the criterion for of simultaneity as given by their respective coordinates, then the point is reinforced that physical consequences of the Lorentz transformations can be obtained without a clear understanding of the interpretation of those transformations.

Finally, since the energy in the region is the energy per unit volume multiplied times the volume:

$$\frac{E'}{E} = \frac{(A'^2/8\pi)V'}{(A^2/8\pi)V} = \frac{1 - v/c}{\sqrt{1 - v^2/c^2}} = \frac{\nu'}{\nu}. \tag{11}$$

In the 1905 paper, Einstein wraps up his derivation with what is clearly an allusion to the light quantum hypothesis: 'It is noteworthy that the energy and the frequency of a light-complex vary with the state of motion of the observer according to the same law.' What he does not draw attention to there is the intimate relation of this result to the relative character of simultaneity.

The electron-problem

The conception of the electron at the turn of the twentieth century was hardly that of today's or even of the 1920's. The term 'electron' had been introduced specifically in connection with the recognition of a minimum unit of charge, negative or positive. Whether such fundamental charges had to be attached to anything material was an open question. Einstein himself toyed with the idea of electrodynamics as the science of disembodied charges in 1899:

> The introduction of the term "ether" into theories of electricity has led to the conception of a medium whose motion can be described, without, I believe, being able to ascribe physical meaning to it. I think that electrical forces can be directly defined only for empty space, something also emphasized by Hertz. Further, electrical currents will have to be thought of not as "the disappearance of electrical polarization over time," but as the motion of true electrical charges whose physical reality appears to be confirmed by electrochemical equivalents. ... Electrodynamics would then be the theory of the movements of moving electricities and magnetisms in empty space. [1]

It was widely known that the small apparent mass of the constituents of cathode- and beta-rays could be ascribed to a resistance to change in motion arising from the self-inductance of the charges. Drude adopted this view in his electron theory of metals [30]. Einstein was of the same view in 1901:

> I have gotten a hold of a study on electron theory by Paul Drude with which I am in heartfelt agreement ... There's no doubt that Drude is a brilliant man. He too assumes that it is predominantly the negative electrical charges without ponderable mass that determine thermal and electrical phenomena in metals, as occurred to me just before leaving Zurich. [31]

If all elementary constituents of matter are positive or negative electrons and the apparent masses of these are electromagnetic in origin, then the idea of mass as "quantity of matter" plays no foundational role. Wien had in fact proposed replacing mechanics with an "electromagnetic world-view" as an alternative foundations for all of physics [32]. Not only the notion of inertial mass, but also the laws of mechanics, would be derived from electrodynamics, thereby removing the threat of a dualism with mechanics and electrodynamics playing equally fundamental roles. As Einstein retrospectively described this stage of development of physics:

> Then only field energy would be left, and the particle would be merely a domain containing an especially high density of field energy. In that case one could hope to deduce the concept of the mass point together with the equations of motion of the particles from the field equations – the disturbing dualism would have been removed. [33]

The success of this program depended on the outcomes of experiments to measure the apparent mass of the electron mass at extremely high velocities. The expected electrodynamical mass, however, depended on how one modeled the electron in terms of its shape and charge distribution at high velocities. Lorentz proposed a contractile [34], Abraham a perfect rigid electron [35], yielding different predictions.

A potential worry was how to account for the stability of the electron given the electrostatic self-repulsion of its charge density. Abraham, for example, just opted to take the rigidity of the electron as a surd, non-dynamical fact [35]. Einstein preferred a dynamical explanation and thus inferred that Maxwell's equations fail at the scale of the electron. As he later explained:

> Maxwell's equations, however, did not permit the derivation of the equilibrium of the electricity that constitutes a particle. Only different, nonlinear field equations could possibly accomplish such a thing. But no method existed for discovering such field equations without deteriorating into adventurous arbitrariness. [33]

Such adventurous arbitrariness was also implicit in the contemporary approaches to the dynamics of the electron insofar as they needed to introduce hypotheses as to the detailed geometry of the electron as a finite charge density in order to calculate its reaction to its self-field, which electrostatically is a billion or so times stronger in the immediate vicinity of the electron than any experimentally produceable applied field. For a charge at rest, only the applied field needs to be taken into consideration in considering the force exerted on it. For a charge density in motion, the self-inductance goes to infinity as the linear dimensions shrink to zero. Since on the received view, the true electric and magnetic field strengths, and thus the net force on the electron, are given in the ether frame, assumptions as to the geometry of the electron's geometry charge density become crucial.

Einstein was able to avoid entirely any such assumptions in his treatment of the dynamics of the slowly accelerating electron in §10 of the 1905 paper, and he undoubtedly saw this as a major virtue. There he introduces the electron as 'a point-like particle provided with an electric charge ε.' This stands in stark contrast to the description of the electron in §9 – '[an] electric charge [density ϱ] permanently attached to a small rigid body' – which is the conception at the basis of Lorentz's optics and electrodynamics of moving bodies. Einstein could treat the electron as a point charge since, on the relativistic conception of the electromagnetic field, the force exerted on the electron can be evaluated in the instantaneous rest frame of the electron. Thus the reaction to its self-field need not be considered, and the net force due to applied fields is just the charge times the electric field it sees in its rest frame. If the electron has a rest mass μ, then the acceleration in the instantaneous rest frame is given immediately by Newton's second law. The apparent mass is then obtained by translating the acceleration back to the stationary frame. Given the asymmetry in the transformation equations between the x-direction (the instantaneous direction of motion of the electron) and the y and z directions orthogonal to it, Einstein gets expressions, using Abraham's terminology [35], for the "longitudinal" mass and the "transverse" mass of the electron as functions of the velocity of the electron relative to the stationary frame. These expressions diverge from those obtained by Abraham on the assumption of a perfectly rigid electron. Although it is unlikely that Einstein had by then encountered Lorentz's expressions for these quantities, he was likely aware through indirect sources [36] that the ratio of the two agreed with Lorentz's predictions.

In a letter to Michele Besso from January 1903, Einstein indicated that he planned to undertake a comprehensive study of the electron theory in the near future [36]. If it is true that he had spent "years" in fruitless pursuit of emission theories in response to the asymmetries of electromagnetic induction, by the time he did so, he would have been in possession of the relativistic conception of the electromagnetic field. From there it is a few short steps, based on symmetry conditions alone, to the field transformation equations.

Using Einstein's notation Y for the y component of the electric "force" and N for the z component of the magnetic "force," it is clear from any number of treatments of electromagnetic induction that Y must transform, at least to first order accuracy, as

$$Y' = Y - \frac{v}{c}N.$$

The analogy of magneto-motive forces demands that N transforms, again to first order, as

$$N' = N - \frac{v}{c}Y.$$

The inverse transformations are given by substituting $-v$ for v:

$$Y = Y' + \frac{v}{c}N', \tag{12}$$

$$N = N' + \frac{v}{c}Y'. \tag{13}$$

Composing the first transformations with their inverses yields:

$$Y = \left(1 - \frac{v^2}{c^2}\right)Y, \tag{14}$$

$$N = \left(1 - \frac{v^2}{c^2}\right)N, \tag{15}$$

from which it is easy to see that conformity with the principle of relativity can be achieved by correcting the first-order transformations by the inverse of the square root of $(1 - v^2/c^2)$.

A more formal derivation might start with the assumption that the sought after transformations should be given by

$$Y' = \varphi(v/c)\left(Y - \frac{v}{c}N\right), \tag{16}$$

$$N' = \varphi(v/c)\left(N - \frac{v}{c}Y\right), \tag{17}$$

where φ is an arbitrary function of v/c to be solved for. By composing the transformations with their inverses and noting that symmetry requires $\varphi(v/c) = \varphi(-v/c)$, the required expression follows immediately.

Since the velocity c plays a distinguished role as a limiting velocity in these transformation equations, they are obviously not at all suitable for an emission hypothesis. But to the extent that doubts as to the plausibility of the emission hypothesis emerge, it becomes natural to explore the behavior of the Maxwell-Hertz equations for free space under these transformations. The principle of relativity demands that these equations co-vary under the field transformation. The Lorentz transformations then fall out following the inverse line of reasoning in §6 in which Einstein derives the field transformations from the Lorentz transformations. This is all that is needed, formally, to begin to investigation the dynamics of the slowly accelerated electron from a new perspective.

Lorentz's theory

The final problem area treated in the Electrodynamical Part of the 1905 relativity paper is the electromagnetic behavior of material media. The gist of §9 is to show that Lorentz's theory in this domain is in accord with the principle of relativity, and hence to the extent that this theory is successful in this regard for stationary media, it remains successful in the general case. Although Einstein had been acquainting himself with the results of the latest research shortly after they appeared in the *Annalen der Physik*, as for example with Planck [23] on thermal radiation and Drude [30] on the electron theory of metals, he became familiar with Lorentz's approach to electromagnetic theory only at a relatively late date.

Einstein had learned of the electromagnetic theory of light, and in particular Hertz's discovery of electromagnetic waves, while still a youth. However, his conception of the ether was then still that of a mechanical, elastic medium. In an essay written in 1895 for his uncle Caesar Koch, he suggests probing the state of dynamical stress corresponding to magnetic fields by looking for variations the velocity of light should suffer from a change in the modulus of elasticity [38].

Although Maxwell's theory was not taught in his courses in electricity and magnetism at the ETH, Einstein had, by 1899, familiarized himself [1] with Hertz's formulation of Maxwell's theory, which treated ether and matter on a par as fundamentally electro-magnetic continua differing only in regard to inductive capacity, magnetic permeability, and mobility [39]. This macroscopic, phenomenological approach, although having the virtue of presenting a concise formulation of Maxwell's theory explicitly in the form of field equations as we would recognize them, could offer no underlying explanation for the various electro-optical properties of different substances, the phenomenon of dispersion, or even the difference between conductors and insulators. It did, however, offer a symmetric account of electromagnetic induction insofar as all induced currents result from the relative motion of lines of electromagnetic force. Einstein had also studied Drude's 1894 *Physik des Aethers*, which engaged to some extent in hypothetical micro-structural explanations, for example, of dispersion, by constructing Hertzian continuum models on the molecular scale [40].

Lorentz pioneered a fundamentally different picture. One crucial move was to separate the electromagnetic field completely from matter, which in the context of the period meant to separate ether from matter completely. Lorentz assumed the ether is completely stationary, in the sense that its parts do not move with respect to one another, and also that is omnipresent, in the sense that it pervades the interior of matter even at the atomic scale. A second crucial move was to introduce indivisible charge densities of atomic or subatomic size, which he first referred to as 'ions' but subsequently as 'electrons'. On the one hand they served as sources for the electric and magnetic forces that propagate through the ether. On the other, by being linked to matter, they provided the mechanism for the electromagnetic field to exert ponderomotive forces on bodies, as expressed quantitatively by the Lorentz force equation. Electrical currents then consisted of the convection of free electrons. Bound electrons could serve as emitters and absorbers of radiation by oscillating or, alternatively, as sources of polarization under static displacement. This electron-theoretic picture proved to be enormously fecund for explaining not only the electro-magneto and optical properties of material media at rest, but also for phenomena involving moving media.

In his canonical formulation as presented in the *Versuch* [41] of 1895, Lorentz also established that the motion of the earth through the ether cannot be measured by experiments accurate to first order in v/c. He demonstrated this for strictly electric and magnetic phenomena by considering in turn the general cases of electrostatics, stationary currents, and induction. For optical phenomena, he developed an ingenious formal tool, the theorem of corresponding states, for generating solutions for a moving system of bodies – lenses, apertures, prisms, or whatever – from solutions for a corresponding system of stationary bodies. The theorem works by exploiting a first-order formal covariance that results from substituting into Maxwell's equations, expressed in moving coordinates, auxiliary variables for the field quantities, which for the case of free space reduce to

$$\vec{E}' = \vec{E} + \frac{1}{c}\left(\vec{v} \times \vec{H}\right), \tag{18}$$

$$\vec{H}' = \vec{H} - \frac{1}{c}\left(\vec{v} \times \vec{E}\right), \tag{19}$$

and the variable, called "local time,"

$$t' = t - \frac{1}{c^2}\left(v_x \xi + v_y \eta + v_z \zeta\right), \tag{20}$$

where t is the true time, (v_x, v_y, v_z) the velocity components of the moving system, and (ξ, η, ζ) the spatial coordinates relative to the moving system. It needs to be emphasized that Lorentz did not interpret $\vec{E}'$ and $\vec{H}'$ as the fields as measured in the moving system or the local time t' as time measured in the moving system. Since in general the solutions for moving systems are not solutions for stationary systems, he again had to proceed category by category to establish null results to first order. In each application, after finding $\vec{E}'$ and $\vec{H}'$ as functions of $(\xi, \eta.\zeta, t')$, eqs. (18)–(20) needed to be consulted in order to obtain the true fields $\vec{E}$ and $\vec{H}$ as functions of $(\xi, \eta.\zeta, t)$ in order to describe the physical situation in terms of coordinates adapted to the moving system of bodies.

But the purpose of the corresponding states theorem was not confined to predicting null results. A particularly salient example is the derivation of Fizeau's result [4], confirmed to greater quantitative accuracy by Michelson and Morley [42], that the speed of light in a transparent substance moving with a velocity v relative to the source is

$$V = c/n + v(1 - 1/n^2), \tag{21}$$

where n is the index of refraction.

Although Lorentz required no assumptions beyond the fundamental theory for systems at rest in order to show that all experiments accurate to first order in v/c to measure the ether drift should be null, he was forced to introduce a special hypothesis to explain the second-order null result of the Michelson-Morley experiment [5]. This was the famous Lorentz-Fitzgerald contraction hypothesis that a rigid body experiences an attenuation of its length by a factor of $\sqrt{1 - v^2/c^2}$ longitudinally to its direction of motion through the ether. Lorentz gave a plausibility argument for the hypothesis. The equilibrium state of moving system of bodies interacting via electrodynamical forces must be contracted by just that factor in comparison to a corresponding stationary system of bodies. Thus, to the extent that the intermolecular forces responsible for the equilibrium state of macroscopic rigid bodies are ultimately electromagnetic in origin, such a contraction factor is to be expected.

Although Wien's 1898 article on the ether problem [2], which, as we have seen, Einstein read in 1899, mentions how Lorentz treated a number of outstanding problems in the *Versuch*, in particular, both Fizeau's result and the null-result of Michelson and Morley, Einstein had not familiarized himself with the *Versuch* even by the end of 1901. All evidence indicates, however, that when he did get his hands on that work, it had a deep impact. Einstein later recollected:

> the physicist of the present generation regards the point of view achieved by Lorentz as the only possible one; at the time, however, it was a surprising and audacious step, without which the later development would not have been possible. [43]

Although there are many ways in which the *Versuch* would have impressed him, Einstein mentions in his Kyoto lecture having attempted to 'discuss the Fizeau experiment on the assumption that the Lorentz equations for electrons should hold in the frame of reference of the vacuum as originally discussed by Lorentz. ...I spent almost a year trying to modify the idea of Lorentz in the hope of resolving this problem'. [9] Fizeau's experiment, indeed, appears to present a quandary for the emission hypothesis. If a light source and a medium with index of refraction n are relatively stationary, the velocity of light in the medium is c/n. If the light source and the medium move away from another with a velocity v, then, according to the principle of relativity, the resultant velocity of light in the medium should not depend on whether the medium moves with velocity v or the light source moves with velocity $-v$. Suppose the light source moves with velocity $-v$. On the emission hypothesis, the velocity of light in free space, relative to the rest frame of the medium, is $c - v$. The hypothesis that the resultant velocity c_m in the medium is $c/n - v$ can be ruled out on the grounds that a Galilean transformation to the rest frame of the source yields the velocity $c_m' = c/n$ for light through the moving medium, in conflict with Fizeau's result (21). Hence, it is natural to assume that the index of refraction modifies the relative velocity $c - v$ of light in free space, i.e., that $c_m = (c - v)/n$. But transforming back to the rest frame of the light source yields

$$c_m' = c/n + v(1 - 1/n),$$

which is tantalizingly close to (21), but still inadequate. How could one modify Lorentz's theory to restore the missing power of two without do violence to the principle of relativity? In later writings, Einstein repeatedly invoked the general success of Lorentz's theory as grounds for accepting the light postulate [44]. Some, especially among the earlier, cite specifically Fizeau's result [45]. One of these, a manuscript from 1912, discusses the experiment specifically in connection with the emission hypothesis [19].

It is now well known, as shown by Laue [46], that Fizeau's result can be derived directly from the relativistic velocity addition rule. Although the only application that Einstein makes of this rule in the 1905 paper is in §9, specifically, to the convection current term in Lorentz's equations, there is no indication that Einstein was then aware that Fizeau's result can be seen as a purely kinematical consequence of special relativity.

The May 1905 discovery

To appreciate fully Einstein's annus mirabilis of 1905, it helps to keep in mind the swift pace at which he completed the first four works that spring:

- March 17: 'On a Heuristic Point of View Concerning the Production and Transformation of Light' (dated March 17, received March 18). [8]
- April 30: 'A New Determination of Molecular Dimensions' (dated April 30, submitted as a doctoral dissertation to the University of Zurich on July 20). [47]
- May 11: 'On the Movement of Small Particles Suspended in Stationary Liquids Required by the Molecular-Kinetic Theory of Heat' (dated May, received May 11). [48]

- June 30: 'On the Electrodynamics of Moving Bodies' (dated June, received June 30). [7]

A number of independent sources allow us to fix in this chronology the realization of the pivotal insight necessary for laying the groundwork of the kinematics of special relativity. According to the biography by Einstein's son-in-law, Rudolph Kayser:

> Through many long years of hope and disappointment Albert carried on his experiments till he reached the final solution to his problem. When he held the key, with which he was to open the closed door, he despaired and said to his friend [Michele Besso] "I am going to give it up."
>
> But the next day it was in the greatest excitement that he took up his duties at the [patent] office. ... Feverishly he whispered to his friend that now at last he was on the right track. He had made the revolutionary discovery that the traditional conception of the absolute character of simultaneity was a mere prejudice, and that the velocity of light was independent of the motion of the coordinate systems. Only five weeks elapsed between this discovery and the first formulation of the special theory of relativity in the treatise entitled "Towards the Electrodynamics of Moving Bodies." [49]

Similarly, the notes from the Kyoto lecture report:

> By chance a friend of mine in Bern helped me out. It was a beautiful day when I visited him with this problem. I started the conversation with him in the following way: "Recently I have been working on a difficult problem. Today I come here to battle against that problem with you." We discussed every aspect of this problem. Then suddenly I understood where the key to this problem lay. Next day I came back to him again and said to him, without even saying hello, "Thank you. I've completely solved the problem." An analysis of the concept of time was my solution. Time cannot be absolutely defined, and there is an inseparable relation between time and signal velocity. With this new concept, I could resolve all the difficulties completely for the first time.
>
> Within five weeks the special theory of relativity was completed. [9]

Einstein's *Autobiographical Notes* gives us some clue as to the substance of the deliberations. For years, since shortly after 1900, he says, it had been clear to him

> ... that neither mechanics nor electrodynamics could (except in limiting cases) claim exact validity. Gradually I despaired of the possibility of discovering the true laws by means of constructive efforts based on known facts. The longer and the more desperately I tried, the more I came to the conviction that only the discovery of a universal formal principle could lead us to assured results. [50]

Such a principle, Einstein informs us, resulted from thinking about the "paradox" of chasing a light beam he had first thought of at the age of sixteen, which, as we have seen above, represents itself from the point of view of the light quantum hypothesis as the problem examining the same "light complex" in different frames of reference. There we saw how the difference in photon density from frame to frame is an artifact of the use of different standards of simultaneity in different frames. [51] Einstein continues in the *Autobiographical Notes*:

> One sees that in this paradox the germ of the special relativity theory is already contained. Today everyone know, of course, that all attempts to clarify this paradox satisfactorily were condemned to failure as long as the axiom of the absolute character of time, or of simultaneity, was rooted unrecognized in the unconscious. [51]

This poses a fundamental historical question. To what extent did Einstein have the quantitative solution to this problem before having the interpretive solution based on the analysis of clock synchronization and signal speed? In other words, was he in possession of the Lorentz transformations prior to recognizing the conventional character of simultaneity, or did he only know that the Maxwell-Hertz equations for free space are covariant under the first-order transformations (18), (19), and Lorentz's local time (20)? A number of considerations need to be weighed in the balance.

First, it is clear from the light quantum paper that the affair with the emission hypothesis had been broken off for some time. In the introduction of that paper he hazards that Maxwell's theory of light, to the extent that it is valid for time-average quantities, will never be replaced by another. Also, in deriving the relation between entropy density and energy density of blackbody radiation in §3, he begins with the premise that the observable properties of radiation are completely determined once energy density as a function of frequency is given. A footnote alludes to his earlier pursuit and subsequent rejection of an emission theory:

> This assumption is arbitrary. One will naturally stick with this simplest assumption so long as experiment does not force one to give it up.

As he put it to Ehrenfest in 1912:

> ... I considered what is more probable, the principle of the constancy of c as required by Maxwell's theory, or the constancy of c exclusively for an observer who sits with the light source. I decided in favor of the former because I was of the conviction that all light is defined by frequency and intensity alone entirely independently of whether it comes from a moving or from a stationary light source. [52]

Thus, Einstein's ruminations in May were not directed at the plausibility of the emission hypothesis as an alternative to Maxwell's theory. Although he knew experiment could still prove him wrong, he had already set that aside as a remote possibility at least many months earlier.

Second, the available accounts of the May discovery portray it as one in which the discovery of the conventional character of simultaneity immediately resolves Einstein's perplexity to such a extent that few, if any doubts remain as to whether that insight suffices as a solution. This would involve having ruled out such obvious candidates as clock transport for establishing absolute simultaneity. It would also involve seeing that a definition of simultaneity using signal velocity is *all* that is needed (in addition to ideal rods and clocks) to erect a new kinematics. Part of that is to recognize length contraction as a simultaneity effect.

Third, Einstein knew full well just how unorthodox the suggestion to resurrect a corpuscular theory of light would appear, even as a heuristic. Late May, in a letter to his friend Conrad Habicht, he described one of the four works of that spring as 'very revolutionary, as you will see'. [53] It was not the relativity paper, but the light quantum paper. Einstein had ambitions still to launch an academic career [54], and the decision to publish the light quantum paper was surely grave and calculated. What made him so confident?

Given that Einstein presented the light quantum hypothesis on the assumption that Maxwell's theory, rather than an emission theory, is valid for time average field quantities, it would be surprising if he had not investigated the relation of the light quantum hypothesis to wave optics in free space under this assumption. In light of the contraction hypothesis, first-order covariance falls short of full assurance of consistency. The discovery that energy and frequency co-vary exactly, however, would appear as an unexpected confirmation of the plausibility of the light quantum hypothesis from independent considerations.

As for the problem of how to interpret the Lorentz transformations, the role of Lorentz's local time in first-order covariance presents as much of an interpretive challenge. The principle of relativity demands that the phenomena behave *as if* the effective coordinates are given by transformations other than the Galilean. It might be hoped that a fundamental theory would yield an explanation why this is so. But even without such, a conviction in the instrumental adequacy of a Lorentz covariant electrodynamics of moving bodies can suffice.

Broader implications

The 1905 paper has a Janus faced character. On the one hand its professed goal is to present, as the title indicates, an electrodynamics of moving bodies. On the other, the kinematical doctrine introduced to that end, exceeds that electrodynamics in generality, since it presupposed nothing from it other than the validity of the light postulate. It thus promises to have consequences beyond the electrodynamics.

The 1905 paper is an awkward transitional document in this regard. The Kinematical Part strives to lay out testable experimental consequences, not dependent on electrodynamics, where it can. Not yet expunged from the Electrodynamical Part, however, is the suggestion, in §6, that the relativistic field transformation equations indicate that the standard expression for the Lorentz force, even as an "auxiliary" concept, is correct only to first order. Traces of this remain in §10, leading up to the system of equations (A). It can be debated whether the discussion thereafter, which treats the force term in Newton's second law as frame invariant, is an artifact of preserving for the reader's sake the "usual way of looking at things" in terms of "longitudinal" and "transverse" mass, or whether it is because Einstein did not yet know how to cast Newton's second law for point masses in relativistic form, as Planck showed the following year [55]. However, it is clear that Einstein appreciated at the time there should be general dynamical consequences beyond electrodynamics proper. He asserts that the expressions for mass, as well as the expression for the kinetic energy of the electron, should also hold for ponderable material points. A ponderable material point, he explains, can be made into an electron ('in our sense') by the addition of 'an *arbitrarily small* electric charge' – an allusion to the fact that Maxwell's theory provides no explanation of the quantization of charge, and thus is far from a fundamental theory.

Several months passed before Einstein realized how deep the general dynamical consequences might be. To Conrad Habicht he confided:

> A consequence of the electrodynamical work has just occurred to me. The principle of relativity together with Maxwell's equations require namely, that mass of a body is a direct measure for the energy contained in it; light carries mass. A noticeable decrease of the mass of radium must result. The thought is amusing and captivating; but whether the Lord is not laughing and leading me around by the nose, I cannot know. [56]

A short paper that September, 'Does the Inertial of a Body Depend on Its Energy Content?' [57], argues that if a body emits light of energy E, its mass will be reduced by E/c^2, neglecting fourth and higher powers in v/c. Einstein claims that it is inessential that the energy is given up in the form of radiation, but does not give an argument there. The problem was addressed in two more papers [58] before he arrived at the canonical formulation of the equivalence of mass and energy in his exhaustive 1907 review article [59] for the *Jahrbuch der Radioaktivität*. By then, he had hit on the principle of equivalence and turned his sights on gravity.

The heuristic status of special relativity

In closing, there is yet one additional facet in which our present conception of special relativity stands in marked contrast to Einstein's. We tend to view of it as a foundational theory subject to only to the qualifications mandated by general relativity. Even in the context of the latter, it serves as foundational keystone by determining the structure of the tangent space at each point of a relativistic manifold.

Einstein, though, saw it as an interim tool to guide the search for a fundamental theory. In a short remark [59] published in 1907, Einstein corrected Ehrenfest's characterization of relativistic electrodynamics [60] as a "closed system" which would yield an answer, as a matter of pure deduction, to questions such as whether force-free uniform motion is possible for an electron whose rest shape is neither spherical or ellipsoidal. The principle of relativity in conjunction with the light postulate, he clarified, is only a "heuristic principle" that allows one make connections between otherwise seemingly independent regularities, a principle which allows certain laws to be traced back to others, similar to the second law of thermodynamics. The comparison with the second law is revealing. Einstein had shown in his Brownian motion paper [48] of 1905, that, because of fluctuation phenomena, it holds only as a constraining principle commanding only course grained validity. Special relativity regarded by itself, as Einstein explained in response to Ehrenfest, contains only statements about rigid rods, clocks and light signals. The notion of a rigid rod stands in only as an idealization. The notion of a truly rigid body, because it would permit super-luminal propagation, is impossible according to the the theory [61].

As Einstein emphasized in 1907, 'at the present we do not have a complete world view in accord with the principle of relativity' [61]. Do we now?

References

[1] CPAE 1, edited by J. Stachel, D. C. Cassidy, R. Schulmann, and J. Renn, The Collected Papers of Albert Einstein, Vol. 1, The Early Years, 1879–1902 (Princeton University Press, Princeton, 1987), Doc. 52, p. 225. For further historical discussion, see M. Janssen and J. Stachel, "The Optics and Electrodynamics of Moving Bodies", Preprint 265, Max Planck Institute for the History of Science, Berlin; J. Norton, "Einstein's Special Theory of Relativity and the Problems in the Electrodynamics of Moving Bodies that led him to it", in: Cambridge Companion to Einstein, edited by M. Janssen and C. Lehner (Cambridge University Press, Cambridge, in press); J. Stachel, Editorial Note, in: CPAE 2, 3–8; J. Stachel, "The Young Einstein: Poetry and Truth", in: Einstein from 'B' to 'Z', (Birkhäuser, Boston, 2002), pp. 21–38.

[2] CPAE 1, edited by J. Stachel, D. C. Cassidy, R. Schulmann, and J. Renn, The Collected Papers of Albert Einstein, Vol. 1, The Early Years, 1879–1902 (Princeton University Press, Princeton, 1987), Doc. 52, p. 233.

[3] A. Einstein, Doc. 57, Collected Papers Vol. 1 (Princeton University Press, Princeton, 1987), p. 233.

[4] H. Fizeau, C. R. **33**, 349–355 (1851).

[5] A. Michelson and E. Morley, Am. J. Sci. **34**, 333–345 (1887).

[6] W. Röntgen, Ann. Phys. (Leipzig) **35**, 264 (1888).

[7] A. Einstein, Ann. Phys. (Leipzig) **17**, 891–921 (1905), reprinted in this volume (pp. 194–224).

[8] A. Einstein, Ann. Phys. (Leipzig) **17**, 132–148 (1905), reprinted in this volume (pp. 165–181).

[9] J. Ishiwara, Einstein Kóen-Roku (Tokyo-Tosho, Tokyo, 1971), pp. 78–88.

[10] T. Ogawa, Jpn. Stud. Hist. Sci. **18**, 73–81; A. Ono, Phys. Today, 45–47 (August 1982).

[11] R. Itagaki, Science **283**, 1457–1458 (1999).

[12] J. Stachel, Astron. Nachr. (Germany) **303**, 47–53 (1982).

[13] CPAE 7, edited by M. Janssen, R. Schulmann, J. Illy, C. Lehner, and D. Buchwald, The Collected Papers of Albert Einstein, Vol. 7, The Berlin Years: Writings, 1918–1921 (Princeton University Press, Princeton, 2002), Doc. 31, p. 245–281.

[14] CPAE 7, edited by M. Janssen, R. Schulmann, J. Illy, C. Lehner, and D. Buchwald, The Collected Papers of Albert Einstein, Vol. 7, The Berlin Years: Writings, 1918–1921 (Princeton University Press, Princeton, 2002), Doc. 31, p. 264.

[15] CPAE 7, edited by M. Janssen, R. Schulmann, J. Illy, C. Lehner, and D. Buchwald, The Collected Papers of Albert Einstein, Vol. 7, The Berlin Years: Writings, 1918–1921 (Princeton University Press, Princeton, 2002), Doc. 31, pp. 264–265.

[16] CPAE 7, edited by M. Janssen, R. Schulmann, J. Illy, C. Lehner, and D. Buchwald, The Collected Papers of Albert Einstein, Vol. 7, The Berlin Years: Writings, 1918–1921 (Princeton University Press, Princeton, 2002), Doc. 31, Note 34, p. 280.

[17] For references, see W. Pauli, Theory of Relativity (Dover, New York, 1958), pp. 5–9.

[18] W. de Sitter, Phys. Z. **14**, 429 (1913).

[19] CPAE4, edited by M. J. Klein, A. J. Kox, J. Renn, and R. Schulmann, The Collected Papers of Albert Einstein, Vol. 4, The Swiss Years: Writings, 1912–1914 (Princeton University Press, Princeton, 1995), Doc. 1, p. 35.

[20] A. Einstein, Ann. Phys. (Leipzig) **20**, 199–206 (1906), reprinted in this volume (pp. 260–267).

[21] CPAE 1, edited by J. Stachel, D. C. Cassidy, R. Schulmann, and J. Renn, The Collected Papers of Albert Einstein, Vol. 1, The Early Years, 1879–1902 (Princeton University Press, Princeton, 1987), Doc. 102, p. 295.

[22] CPAE 1, edited by J. Stachel, D. C. Cassidy, R. Schulmann, and J. Renn, The Collected Papers of Albert Einstein, Vol. 1, The Early Years, 1879–1902 (Princeton University Press, Princeton, 1987), Doc. 97, p. 286.

[23] M. Planck, Ann. Phys. (Leipzig) **1**, 69–122 (1900).

[24] J. C. Maxwell, A Treatise On Electricity and Magnetism (Clarendon Press, Oxford 1873), § 792.

[25] A. Bartoli, Nuovo Cimento **15**, 195 (1883).

[26] L. Boltzmann, Wied. Ann. **22**, 31 (1884).

[27] W. Wien, Sitzungsber. der Berl. Akad., 1893, pp. 55–62; Ann. Phys (Leipzig) **52**, 132–165 (1894).
[28] A. Einstein, Autobiographical Notes, translated and edited by P. A. Schilpp (Open Court, La Salle, 1979).
[29] A. Einstein, Autobiographical Notes, translated and edited by P. A. Schilpp (Open Court, La Salle, 1979), pp. 48–51.
[30] P. Drude, Ann. Phys. (Leipzig) **1**, 566–613 (1900).
[31] CPAE 1, edited by J. Stachel, D. C. Cassidy, R. Schulmann, and J. Renn, The Collected Papers of Albert Einstein, Vol. 1, The Early Years, 1879–1902 (Princeton University Press, Princeton, 1987), Doc. 96, p. 284–285.
[32] W. Wien, Ann. Phys. (Leipzig) **5**, 501–513 (1900).
[33] A. Einstein, Autobiographical Notes, translated and edited by P. A. Schilpp (Open Court, La Salle, 1979), pp. 34–35.
[34] H. A. Lorentz, K. Ned. Akad. Wet. **7**, 507 (1899).
[35] M. Abraham, Ann. Phys. (Leipzig) , **10**, 105–179 (1903).
[36] M. Abraham, Ann. Phys. (Leipzig) , **14**, 236–287 (1904).
[37] CPAE5, edited by M. J. Klein, A. J. Kox, and R. Schulmann, The Collected Papers of Albert Einstein, Vol. 5, The Swiss Years: Correspondence, 1902–1914 (Princeton University Press, Princeton, 1993), Doc. 5, p. 11.
[38] CPAE 1, edited by J. Stachel, D. C. Cassidy, R. Schulmann, and J. Renn, The Collected Papers of Albert Einstein, Vol. 1, The Early Years, 1879–1902 (Princeton University Press, Princeton, 1987), Doc. 5, p. 6–9.
[39] H. Hertz, Wied. Ann. **40**, 577–624 (1890); **41**, 369–399 (1890).
[40] P. Drude, Physik des Aethers (Ferdinand Enke, Stuttgart, 1894), p. 518.
[41] H. A. Lorentz, Versuch einer Theorie der electrischen und optischen Erscheinungen in bewegten Körpern (E. J. Brill, Leiden, 1895).
[42] A. Michelson and E. Morley, Am. J. Sci. **31**, 377 (1886).
[43] A. Einstein, Autobiographical Notes, translated and edited by P. A. Schilpp (Open Court, La Salle, 1979), pp. 32–33.
[44] A. Einstein, Über die spezielle und die allgemeine Relativitätstheorie (Vieweg, Braunsweig, 1917), §7; The Meaning of Relativity (Princeton University Press, Princeton, 1922), p. 27.
[45] A. Einstein, Arch. Sci. Phys. Nat. **29**, 125–144 (1910).
[46] M. Laue, Ann. Phys. (Leipzig) **23**, 989 (1907).
[47] A. Einstein, Eine neue Bestimmung der Moleküldimensionen (K. J. Wyss, Bern, 1906).
[48] A. Einstein, Ann. Phys. (Leipzig) **17**, 549–560 (1905), reprinted in this volume (pp. 182–193).
[49] A. Reiser and R. Kayser, Albert Einstein: A Biographical Portrait (Albert and Charles Boni, New York, 1930), pp. 68–69.
[50] A. Einstein, Autobiographical Notes, translated and edited by P. A. Schilpp (Open Court, La Salle, 1979), pp. 48–49.
[51] A. Einstein, Autobiographical Notes, translated and edited by P. A. Schilpp (Open Court, La Salle, 1979), pp. 50–51.
[52] For a variant interpretation, see J. Norton, Arch. Hist. Exact Sci. **59**, 45–105 (2004).
[53] CPAE5, edited by M. J. Klein, A. J. Kox, and R. Schulmann, The Collected Papers of Albert Einstein, Vol. 5, The Swiss Years: Correspondence, 1902–1914 (Princeton University Press, Princeton, 1993), Doc. 409, p. 485.
[54] CPAE5, edited by M. J. Klein, A. J. Kox, and R. Schulmann, The Collected Papers of Albert Einstein, Vol. 5, The Swiss Years: Correspondence, 1902–1914 (Princeton University Press, Princeton, 1993), Doc. 27, p. 31.
[55] R. Schulmann, Einstein der Pragmatiker, Lecture at Max-Planck-Institu für Wissenschaftsgeschichte, 18 November 2004.
[56] M. Planck, Verh. Dtsch. Phys. Ges. **8**, 136–141 (1906).
[57] CPAE5, edited by M. J. Klein, A. J. Kox, and R. Schulmann, The Collected Papers of Albert Einstein, Vol. 5, The Swiss Years: Correspondence, 1902–1914 (Princeton University Press, Princeton, 1993), Doc. 28, p. 33.
[58] A. Einstein, Ann. Phys. (Leipzig) **18**, 639–641 (1905), reprinted in this volume (pp. 226–228).
[59] A. Einstein, Ann. Phys. (Leipzig) **20**, 627–633 (1906), reprinted in this volume (pp. 268–274); **23**, 371–384 (1907), reprinted in this volume (pp. 303–316).
[60] A. Einstein, Jahrb. Radioakt. Elektron. **4**, 411–462 (1907).
[61] A. Einstein, Ann. Phys. (Leipzig) **23**, 206–208 (1907), reprinted in this volume (pp. 300–302).
[62] P. Ehrenfest, Ann. Phys. (Leipzig) **23**, 204–205 (1907).
[63] A. Einstein, Ann. Phys. (Leipzig) **23**, 371–384 (1907), reprinted in this volume (pp. 303–316).

Of pots and holes: Einstein's bumpy road to general relativity

Michel Janssen*

Program in History of Science and Technology, University of Minnesota, Tate Laboratory of Physics, 116 Church St. SE, Minneapolis, MN 55455, USA

Published online 14 February 2005

Key words General covariance, equivalence principle, hole argument, Einstein field equations
PACS 04.20.Cv, 01.65.+g, 01.70.+w

General relativity in the *Annalen* and elsewhere

Readers of this volume will notice that it contains only a few papers on general relativity. This is because most papers documenting the genesis and early development of general relativity were not published in *Annalen der Physik*. After Einstein took up his new prestigious position at the Prussian Academy of Sciences in the spring of 1914, the *Sitzungsberichte* of the Berlin academy almost by default became the main outlet for his scientific production. Two of the more important papers on general relativity, however, did find their way into the pages of the *Annalen* [35,41]. Although I shall discuss both papers in this essay, the main focus will be on [35], the first systematic exposition of general relativity, submitted in March 1916 and published in May of that year.

Einstein's first paper on a metric theory of gravity, co-authored with his mathematician friend Marcel Grossmann, was published as a separatum in early 1913 and was reprinted the following year in *Zeitschrift für Mathematik und Physik* [50,51]. Their second (and last) joint paper on the theory also appeared in this journal [52]. Most of the formalism of general relativity as we know it today was already in place in this Einstein-Grossmann theory. Still missing were the generally-covariant Einstein field equations.

As is clear from research notes on gravitation from the winter of 1912–1913 preserved in the so-called "Zurich Notebook,"[1] Einstein had considered candidate field equations of broad if not general covariance, but had found all such candidates wanting on physical grounds. In the end he had settled on equations constructed specifically to be compatible with energy-momentum conservation and with Newtonian theory in the limit of weak static fields, even though it remained unclear whether these equations would be invariant under any non-linear transformations. In view of this uncertainty, Einstein and Grossmann chose a fairly modest title for their paper: "Outline ("Entwurf") of a Generalized Theory of Relativity and of a Theory of Gravitation." The Einstein-Grossmann theory and its fields equations are therefore also known as the "Entwurf" theory and the "Entwurf" field equations.

Much of Einstein's subsequent work on the "Entwurf" theory went into clarifying the covariance properties of its field equations. By the following year he had convinced himself of three things. First, generally-covariant field equations are physically inadmissible since they cannot determine the metric field uniquely. This was the upshot of the so-called "hole argument" ("Lochbetrachtung") first published in an appendix to [51].[2] Second, the class of transformations leaving the "Entwurf" field equations invariant was as broad

---

* E-mail: janss011@tc.umn.edu
[1] An annotated transcription of the gravitational portion of the "Zurich Notebook" is published as Doc. 10 in [11]. For facsimile reproductions of these pages, a new transcription, and a running commentary, see [89].
[2] See Sect. 2 for further discussion of the hole argument.

as it could possibly be without running afoul of the kind of indeterminism lurking in the hole argument and, more importantly, without violating energy-momentum conservation. Third, this class contains transformations, albeit it of a peculiar kind, to arbitrarily moving frames of reference. This, at least for the time being, removed Einstein's doubts about the "Entwurf" theory and he set out to write a lengthy self-contained exposition of it, including elementary derivations of various standard results he needed from differential geometry. The title of this article reflects Einstein's increased confidence in his theory: "The Formal Foundation of the General Theory of Relativity" [30]. As a newly minted member of the Prussian Academy of Sciences, he dutifully submitted his work to its *Sitzungsberichte*, where the article appeared in November 1914. This was the first of many papers on general relativity in the *Sitzungsberichte*, including such gems as [37] on the relation between invariance of the action integral and energy-momentum conservation, [36,40] on gravitational waves, [39], which launched relativistic cosmology and introduced the cosmological constant, and [43] on the thorny issue of gravitational energy-momentum.

In the fall of 1915, Einstein came to the painful realization that the "Entwurf" field equations are untenable.[3] Casting about for new field equations, he fortuitously found his way back to equations of broad covariance that he had reluctantly abandoned three years earlier. He had learned enough in the meantime to see that they were physically viable after all. He silently dropped the hole argument, which had supposedly shown that such equations were not to be had, and on November 4, 1915, presented the rediscovered old equations to the Berlin Academy [31]. He returned a week later with an important modification, and two weeks after that with a further modification [32,34]. In between these two appearances before his learned colleagues, he presented yet another paper showing that his new theory explains the anomalous advance of the perihelion of Mercury [33].[4] Fortunately, this result was not affected by the final modification of the field equations presented the following week.

When it was all over, Einstein commented with typical self-deprecation: "unfortunately I have immortalized my final errors in the academy-papers";[5] and, referring to [30]: "it's convenient with that fellow Einstein, every year he retracts what he wrote the year before."[6] What excused Einstein's rushing into print was that he knew that the formidable Göttingen mathematician David Hilbert was hot on his trail.[7] Nevertheless, these hastily written communications to the Berlin Academy proved hard to follow even for Einstein's staunchest supporters, such as the Leyden theorists H. A. Lorentz and Paul Ehrenfest.[8]

Ehrenfest took Einstein to task for his confusing treatment of energy-momentum conservation and his sudden silence about the hole argument. Ehrenfest's queries undoubtedly helped Einstein organize the material of November 1915 for an authoritative exposition of the new theory. A new treatment was badly needed, since the developments of November 1915 had rendered much of the premature review article of November 1914 obsolete.

In March 1916, Einstein sent his new review article, with a title almost identical to that of the one it replaced, to Wilhelm Wien, editor of the *Annalen*.[9] This is why [35], unlike the papers mentioned so far, can be found in the volume before you.[10] Many elements of Einstein's responses to Ehrenfest's queries ended up in this article. Even though there is no mention of the hole argument, for instance, Einstein does present

---

[3] Einstein stated his reasons for abandoning the "Entwurf" field equations and recounted the subsequent developments in Einstein to Arnold Sommerfeld, 28 November 1915 [15, Doc. 153].

[4] See [21] for an analysis of this paper. That Einstein could pull this off so fast was because he had already done the calculation of the perihelion advance of Mercury on the basis of the "Entwurf" theory two years earlier (see the headnote, "The Einstein-Besso Manuscript on the Motion of the Perihelion of Mercury," in [11, pp. 344–359]).

[5] "Die letzten Irrtümer in diesem Kampfe habe ich leider in den Akademie-Arbeiten [...] verewigt." This comment comes from the letter to Sommerfeld cited in note 3.

[6] "Es ist bequem mit dem Einstein. Jedes Jahr widerruft er, was er das vorige Jahr geschrieben hat." Einstein to Paul Ehrenfest, 26 December 1915 [15, Doc. 173].

[7] See [8,91,94] for comparisons of the work of Einstein and Hilbert toward the field equations of general relativity.

[8] See [68] for discussion of the correspondence between Einstein, Ehrenfest, and Lorentz of late 1915 and early 1916.

[9] Einstein to Wilhelm Wien, 18 March 1916 [15, Doc. 196].

[10] The article is still readily available in English translation in the anthology *The Principle of Relativity* [73]. Unfortunately, this reprint omits the one-page introduction to the paper in which Einstein makes a number of interesting points. He emphasizes the importance of Minkowski's geometric formulation of special relativity, which he had originally dismissed as "superfluous

the so-called "point-coincidence argument", which he had premiered in letters to Ehrenfest and Michele Besso explaining where the hole argument went wrong.[11] The introduction of the field equations and the discussion of energy-momentum conservation in the crucial Part C of the paper – which is very different from the corresponding Part D of [30] – closely follows another letter to Ehrenfest, in which Einstein gave a self-contained statement of the energy-momentum considerations leading to the final version of the field equations.[12] Initially, his readers had been forced to piece this argument together from his papers of November 1914 and 1915. As Einstein announced at the beginning of his letter to Ehrenfest: "I shall not rely on the papers at all but show you all the calculations."[13] He closed the letter asking his friend: "Could you do me a favor and send these sheets back to me as I do not have this material so neatly in one place anywhere else."[14] Einstein may very well have had this letter in front of him as he was writing the relevant sections of [35].

This paper presents a happy interlude in Einstein's ultimately only partially successful quest to banish absolute motion and absolute space and time from physics and establish a truly general theory of relativity.[15] When he wrote his review article, Einstein still thought that general covariance automatically meant relativity of arbitrary motion. The astronomer Willem de Sitter, a colleague of Lorentz and Ehrenfest in Leyden, disabused him of that illusion during a visit to Leyden in the fall of 1916. A lengthy debate ensued between Einstein and De Sitter in the course of which Einstein introduced the cosmological constant in the hope of establishing general relativity in a new way, involving what he dubbed "Mach's principle" in [41].[16] In this paper he proposed a new foundation for general relativity, replacing parts of the foundation laid in [35]. This may well be why he published [41], like [35], in the *Annalen*. Despite its brevity, this then is the other major paper on general relativity contained in this volume.

Einstein had another stab at an authoritative exposition of general relativity in the early twenties, when he agreed to publish a series of lectures he gave in Princeton in May 1921. They appeared two years later in heavily revised form [46].[17] The Princeton lectures superseded the 1916 review article as Einstein's authoritative exposition of the theory, but the review article remains worth reading and is of great historical interest.

In [35] the field equations and energy-momentum conservation are not developed in generally-covariant form but only in special coordinates. Einstein had found the Einstein field equation in terms of these coordinates in November 1915. As explained above, this part of [35] is basically a sanitized version of the argument that had led Einstein to these equations in the first place. The manuscript for an unpublished appendix [13, Doc. 31] to [35] makes it clear that as he was writing his review article, he was already considering redoing the discussion of the field equations and energy-momentum conservation in arbitrary coordinates. In November 1916, he published such a generally-covariant account in the Berlin *Sitzungsberichte* [37]. This paper is undoubtedly much more satisfactory mathematically than the corresponding part of [35] but it does not offer any insight into how Einstein actually found his theory. Reading [37], without

---

erudition" ("überflüssige Gelehrsamkeit"; [88, p. 151]), and the differential geometry of Riemann and others for the development of general relativity. He also acknowledges the help of Grossmann in the mathematical formulation of the theory.

[11] See Sect. 2 for further discussion of the point-coincidence argument.

[12] Einstein to Paul Ehrenfest, 24 January 1916 or later [15, Doc. 185].

[13] "Ich stütze mich gar nicht auf die Arbeiten, sondern rechne Dir alles vor."

[14] "Es wäre mir lieb, wenn Du mir diese Blätter [...] wieder zurückgäbest, weil ich die Sachen sonst nirgends so hübsch beisammen habe."

[15] There are (at least) two separate issues here [20, p. 12–15]. The first issue is whether all motion is relative or whether some motion is absolute. Put differently, is space-time structure something over and above the contents of space-time or is it just a way of talking about spatio-temporal relations? The second issue concerns the ontological status of space-time. Is space-time structure supported by a space-time substance, some sort of container, or is it a set of relational properties? The two views thus loosely characterized go by the names of 'substantivalism' and 'relationism', respectively. Newton is associated with substantivalism as well as with absolutism about motion, Leibniz with relationism as well as with relativism about motion (see, e. g., [2, introduction]; [61, Chap. 8]). It is possible, however, to be an absolutist about motion and a relationist about the ontology of space-time. Although the jury is still out on the ontological question, I shall argue that, while non-uniform motion remains absolute in general relativity, the ontology of space-time in Einstein's theory is best understood in relational rather than substantival terms.

[16] See Sect. 2 for further discussion of Mach's principle.

[17] The Princeton lectures are still readily available in English translation as *The Meaning of Relativity* [47].

having read the November 1915 papers and the 1916 review article, one easily comes away with the impression that Einstein hit upon the Einstein field equations simply by picking the mathematically most obvious candidate for the gravitational part of the Lagrangian for the metric field, namely the Riemann curvature scalar. This is essentially how Einstein himself came to remember his discovery of general relativity. He routinely trotted out this version of events to justify the purely mathematical speculation he resorted to in his work on unified field theory.[18] The 1916 review article preserves the physical considerations, especially concerning energy-momentum conservation, that originally led him to the Einstein field equations, arguably the crowning achievement of his scientific career.

The balance of this essay is organized as follows. Einstein's review article is divided into five parts. The two most important and interesting parts are part A, "Fundamental Considerations on the Postulate of Relativity" (Sects. 1–4) and part C, "Theory of the Gravitational Field" (Sects. 13–18). These two parts are covered in Sects. 2 and 3, respectively. These two sections can be read independently of one another.

The disk, the bucket, the hole, the pots, and the globes[19]

Part A of [35] brings together some of the main considerations that motivated and sustained its author in his attempt to generalize the principle of relativity for uniform motion to arbitrary motion. On the face of it, the arguments look straightforward and compelling, but looking just below the surface one recognizes that they are more complex and, in several cases, quite problematic.

Einstein [35, p. 770] begins with a formulation of the principle of relativity for uniform motion that nicely prepares the ground for the generalization he is after. Both in Newtonian mechanics and in special relativity there is a class of reference frames in which the laws of nature take on a particularly simple form. These inertial frames all move at constant velocity with respect to one another. In the presence of a gravitational field the laws of nature will in general not be particularly simple in any one frame or in any one class of frames. The simplest formulation is a generally-covariant one, a formulation that is the same in all frames, including frames in arbitrary motion with respect to one another. In this sense of relativity, general covariance guarantees general relativity (ibid., 776). This does not mean that observers in arbitrary motion with respect to one another are physically equivalent the way observers in uniform relative motion are. In that more natural sense of relativity, general relativity does not extend special relativity at all.

Einstein's equating of general relativity with general covariance comes in part from a conflation of two different approaches to geometry, a "subtractive" or "top-down" approach associated with the Erlangen program of Felix Klein, and an "additive" or "bottom-up" approach associated with modern differential geometry, which goes back to Bernhard Riemann [86]. In Klein's "subtractive approach" one starts with a description of the space-time geometry with all bells and whistles and then strips away all elements deemed to be descriptive fluff only. Only those elements are retained that are invariant under some group of transformations. Such groups thus characterize the essential part of the geometry. The geometrization of special relativity by Hermann Minkowski (1909) is a picture-perfect example of Klein's "subtractive" approach. Consider Minkowski space-time described in terms of some Lorentz frame, i. e., coordinatized with the help of four orthogonal axes (orthogonal with respect to the standard non-positive-definite Minkowski inner product). Which Lorentz frame is chosen does not matter. The decomposition of space-time into space and time that comes with this choice is not an essential part of the space-time geometry and neither is the state of rest it picks out. These elements are not invariant under transformations of the Lorentz group characterizing the geometry of Minkowski space-time. For instance, a Lorentz boost will map a worldline of a particle at rest onto a worldline of a particle in uniform motion. Lorentz invariance in special relativity is thus directly related to the relativity of uniform motion. The privileged nature of the whole class of uniform motions

---

[18] For further discussion of Einstein's distorted memory of how he found his field equations and the role it played in his propaganda for his unified field theory program, see [66, Sect. 10] and [102], respectively. John Norton [87], however, accepts that Einstein actually did find the Einstein field equations the way he later claimed he did.

[19] I am indebted to Christoph Lehner for his incisive criticism of earlier versions of many of the arguments presented in this section (cf. [64]). For his own take on some of the issues discussed here, see [72].

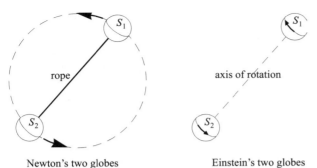

Fig. 1   Absolute rotation.

is an essential part of the geometry. Lorentz transformations will map the set of all possible worldlines of particles at rest in some Lorentz frame onto itself.

In the Riemannian "additive approach" one starts from a bare manifold, a set of points with only a topological and a differential structure defined on it, and adds further structure to turn it into a space-time. Such further structure will typically include an affine connection and a metric so that it becomes possible to tell straight lines from crooked ones and talk about distances. To make sure that no superfluous elements enter into the description of the space-time geometry everything is done in a coordinate-independent manner, if not coordinate-free (i. e., without ever introducing coordinates at all) then at least in a generally-covariant way (i. e., in a way that is exactly the same no matter what coordinates are chosen). Such generally-covariant descriptions can be of space-times with or without preferred states of motion. This already makes it clear that general covariance *per se* has nothing to do with relativity of motion.[20]

Einstein used general covariance in two different ways in his 1916 review article. In Sect. 3, he used it in the spirit of Klein's "subtractive approach" to isolate the essential elements of general relativistic space-times [84, 86]. In Sect. 2, he used it for the implementation of a peculiar principle of relativity distinctly his own.

In Sect. 2 Einstein explained his objection against preferred frames of reference and argued for the need of general covariance using a variant of a thought experiment Newton had used to illustrate that rotation is absolute. These two thought experiments are illustrated in fig. 1, Newton's on the left, Einstein's on the right.

Consider Newton's experiment first [6, p. 12], [7, p. 414]. Two globes, S_1 and S_2, connected by a rope are rotating around their center of gravity far away from any other gravitating matter. Can this situation be distinguished from a situation in which the two globes are not rotating but moving at constant speed in a straight line at a fixed distance from each other? The answer is yes: the tension in the rope will be greater when the globes are rotating.

Einstein asks us to consider the two globes in relative rotation around the line connecting their centers. He has no use for the rope. Newtonian theory tells us that it makes a difference whether S_1 or S_2 is truly rotating. A rotating globe bulges out at its equator. This, Einstein argues, violates Leibniz's principle of sufficient reason. The situation looks perfectly symmetric: S_2 rotates with respect to S_1 and S_1 rotates with respect to S_2. Yet, unless the two globes both happen to rotate with half their relative angular velocity, they behave differently. There is no observable cause to explain this difference in behavior. The Newtonian explanation – that the globe's rotation with respect to (the set of inertial frames of) Newton's absolute space rather than its rotation relative to the other globe is what causes it to bulge out – is unsatisfactory, because the purported cause is not an "observable empirical fact" ("beobachtbare Tatsache"; [35, p. 771]). Special relativity, Einstein claims, inherits this "epistemological defect" ("erkenntnistheoretische[n] Mangel"; ibid.), to which he had been sensitized by Ernst Mach. Situations with two objects in relative motion, such as the globes S_1 and S_2 always look symmetric, regardless of whether the motion is uniform or not, but when the motion is non-uniform the two objects will in general behave differently.

---

[20] See [85] for a review of the (philosophy of) physics literature on the status of general covariance in general relativity.

Fig. 2 Absolute space seems to violate the principle of sufficient reason.

The British philosopher of science Jon Dorling [19] was the first to put his finger on the fallacy in Einstein's reasoning. Imagine attaching ideal clocks to both globes somewhere on their equators. Use these clocks to measure how long one revolution of the other globe takes. According to Newtonian kinematics, the two clocks will record the same time for one revolution. According to special-relativistic kinematics, they will in general record different times because of the phenomenon of time dilation. The difference will be greatest when one of the two globes is at rest with respect to some inertial Lorentz frame in Minkowski space-time. Focus on this special case. The clock on the inertially moving globe measures a longer period of revolution than the clock on the non-inertially moving globe. This is just a variant of the famous twin-paradox scenario. The point of introducing these clocks is to show that the situation of the two globes in relative rotation to one another is *not* symmetric, not even at the purely kinematical level. It therefore need not surprise us that it is not symmetric at the dynamical level either. In the special case in which one globe is moving inertially only the other globe, the one with the lower clock reading, bulges out at its equator.

In Chap. 21 of his popular book on relativity, Einstein [38, p. 49], [48, p. 72] used a charming analogy to get his point across. It can also be used to illustrate Dorling's rejoinder. Consider two identical pots sitting on a stove, only one of which is giving off steam (see Fig. 2). One naturally assumes that this is because only the burner under that one is on. It would be strange indeed to discover that the burners under both pots are turned on (or, for that matter, that both are turned off). That would be a blatant violation of the principle of sufficient reason. Einstein's example of the two globes is meant to convince us that both Newtonian theory and special relativity lead to similar violations of this principle. As with the two pots on the stove, there is no observable difference between the two globes, yet they behave differently. The analogy works for Newtonian theory but not for special relativity. With the kinematics of special relativity, the analogy breaks down immediately. The situation with the two globes does not *look* the same to observers on the two globes, so there is no reason to expect the two globes to *behave* the same.

If we take Einstein at his word in 1916 – that preferred frames of reference are objectionable because they lead to violations of the principle of sufficient reason – we must conclude that Einstein was worried about a problem he had already solved with special relativity by making temporal distances between events, like spatial distances between points, dependent on the path connecting them. Einstein's underestimation of what he had achieved with special relativity compensates for his overestimation of what he had achieved with general relativity. Contrary to what Einstein believed when he wrote his review article in 1916, general covariance does not eliminate absolute motion.

For the further development of physics it was a good thing that Einstein did not fully appreciate what he had accomplished with special relativity. In trying to solve a problem that, unbeknownst to himself, he had already solved, Einstein produced a spectacular new theory of gravity.

The fundamental insight that Einstein would base his new theory on came to him while he was working on a review article on special relativity [24]. Sitting at his desk in the Swiss patent office in Berne one day it suddenly hit him that someone falling from the roof would not feel his own weight.[21] He later called it "the

---

[21] Einstein related this story in a lecture in Kyoto on December 14, 1922 [1, p. 15].

best idea of my life."[22] It told Einstein that there was an intimate connection between acceleration – the kind of motion he wanted to relativize – and gravity. In [26, pp. 360, 366], he introduced the term "equivalence principle" for this connection. Einstein wanted to use this principle to extend the relativity principle from uniform to non-uniform motion.

The equivalence principle explains a striking coincidence in Newton's theory. To account for Galileo's principle that all bodies fall with the same acceleration in a given gravitational field, Newton had to assign the same value to two conceptually clearly distinct quantities, namely inertial mass, the measure of a body's resistance to acceleration, and gravitational mass, the measure of a body's susceptibility to gravity. The equivalence principle removes the mystery of the equality of inertial and gravitational mass by making inertia and gravity two sides of the same coin.

Einstein only formulated the equivalence principle along these lines in his second paper on the foundations of general relativity [41]. In its mature form, the equivalence principle says that inertial effects (i. e., effects of acceleration) and gravitational effects are manifestations of one and the same structure, nowadays called the inertio-gravitational field. How some inertio-gravitational effect breaks down into an inertial component and a gravitational component is not unique but depends on the state of motion of the observer making the call, just as it depends on the state of motion of the observer how an electromagnetic field breaks down into an electric field and a magnetic field [63, pp. 507–509], [64]. In other words, what is relative according to the mature equivalence principle is not motion but the split of the inertio-gravitational field into an inertial and a gravitational component.

Einstein initially did not distinguish these two notions carefully and instead of unifying acceleration and gravity, thereby implementing what I shall call the relativity of the gravitational field, he tried to reduce acceleration to gravity, thereby hoping to extend the relativity principle to accelerated motion. Invoking the equivalence principle, one can reduce a state of acceleration in a gravitational field (i. e., free fall) to a state of rest with no gravitational field present. The man falling from the roof of the Berne patent office and a modern astronaut orbiting the earth in a space shuttle provide examples of this type of situation. One can similarly reduce a state of acceleration in the absence of a gravitational field to a state of rest in the presence of one. An astronaut firing up the engines of her rocket ship somewhere in outer space far from the nearest gravitating matter provides an example of this type of situation. This then is the general principle of relativity that Einstein was able to establish on the basis of the equivalence principle: two observers in non-uniform relative motion can both claim to be at rest if they agree to disagree on whether or not there is a gravitational field present.

This principle is very different from the principle of relativity for uniform motion. Two observers in uniform relative motion are physically equivalent. Two observers in non-uniform relative motion obviously are not. Sitting at one's desk in the patent office does not feel the same as falling from the roof of the building, even though the man falling from the roof can, if he were so inclined, claim that he is at rest and that the disheveled patent clerk whose eyes he meets on the way down is accelerating upward in a space with no gravitational field at all. Likewise, the astronaut accelerating in her rocket in outer space can claim that she is at rest in a gravitational field that suddenly came into being when she fired up her engines and that her hapless colleague, who was hovering in space next to the rocket at that point, is now in free fall in that gravitational field. Despite this nominal relativity of acceleration, the two astronauts will experience this situation very differently.[23]

The physical equivalence in the paradigmatic examples examined above is not between the observers in relative motion with respect to one another, it is between the man at his desk in the first example and the

---

[22] "der glücklichste Gedanke meines Lebens" [14, Doc. 31, p. 21]. For discussion of this oft-quoted passage, see, e. g., [88, p. 178], [63, pp. 507–509], [64].

[23] As late as November 1918 – more than half a year after clarifying the foundations of general relativity [41] – Einstein saw fit to publish an account of the twin paradox along these lines [44]. This 1918 paper not only offered a solution for a problem that had already been solved, it also raised suspicion about the earlier solution by suggesting that the problem called for general relativity. Einstein thus bears some responsibility for the endless confusion over the twin paradox, which is nothing but a vivid example of the path dependence of temporal distances in special as well as in general relativity.

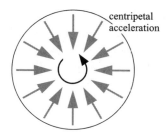 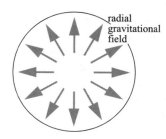

Fig. 3 Rotating disk.

astronaut inside the rocket in the second, and between the man falling from the roof in the first example and the astronaut outside the rocket in the second. Resisting the pull of gravity and accelerating in the absence of gravity feel the same. Likewise, free fall in a gravitational field and being at rest or in uniform motion in the absence of a gravitational field feel the same. These are examples of inertial and gravitational effects that are physically indistinguishable and that get lumped together in the new taxonomy for such effects suggested by the mature equivalence principle. This is arguably one of Einstein's greatest contributions to modern physics. The peculiar general relativity principle for which Einstein originally tried to use the equivalence principle did not make it into the canons of modern physics. It was nonetheless extremely important as a heuristic principle guiding Einstein on his path to general relativity.

The equivalence principle, understood as a heuristic principle, allowed Einstein to infer effects of gravity from effects of acceleration in Minkowski space-time. The most fruitful example of this kind was that of the rotating disk, which is discussed in Sect. 3 of the review article [35, pp. 774–775] and which played a pivotal role in the development of general relativity [96].

Consider a circular disk serving as a merry-go-round in Minkowski space-time (see Fig. 3). Let one observer stand on the merry-go-round and let another stand next to it. The person next to the disk will say that he is at rest and that the person on the disk is subject to centrifugal forces due to the disk's rotation (see the drawing on the left in Fig. 3). Invoking the equivalence principle, the person on the disk will say that she is at rest in a radial gravitational field and that the person next to the disk is in free fall in this field (see the drawing on the right in Fig. 3).[24]

Now have both observers measure the ratio of the circumference and the diameter of the disk. The person next to the disk will find the Euclidean value π. The person on the disk will find a ratio greater than π. After all, according to special relativity, the rods she uses to measure the circumference are subject to the Lorentz contraction, whereas the rods she uses to measure the diameter are not.[25] The spatial geometry for the rotating observer is therefore non-Euclidean. Invoking the equivalence principle, Einstein concluded that this will be true for an observer in a gravitational field as well. This then is what first suggested to Einstein that gravity should be represented by curved space-time.

To describe curved space-time Einstein turned to Gauss's theory of curved surfaces, a subject he vaguely remembered from his student days at the *Eidgenössische Technische Hochschule (ETH)* in Zurich. He had learned it from the notes of his classmate Marcel Grossmann. Upon his return to their *alma mater* as a full professor of physics in 1912, Einstein learned from Grossmann, now a colleague in the mathematics department of the *ETH*, about the extension of Gauss's theory to spaces of higher dimension by Riemann and others.[26] Riemann's theory provided Einstein with the mathematical object with which he could unify the effects of gravity and acceleration: the metric field. The metric makes it possible to identify lines of

---

[24] For the person on the disk, the person standing next to it is rotating and subject to both centrifugal and Coriolis forces. If this person has mass M and moves with angular frequency ω on a circle of radius R, the centrifugal force is $M\omega^2 R$. The Coriolis force provides a centripetal force twice that size, both compensating the centrifugal force and keeping the person in orbit.

[25] This simple argument has been the source of endless confusion. Einstein's clearest exposition can be found in two letters written in response to a particularly muddled discussion of the situation (see Einstein to Joseph Petzoldt, 19 and 23 August 1919 [16, Docs. 93 and 94]).

[26] For accounts of how Einstein's collaboration with Grossmann began, see the Kyoto lecture [1, p. 16] as well as [88, p. 213] and [54, pp. 355–356].

extremal length in curved space-time, so-called metric geodesics. In Riemannian geometry these are also the straightest possible lines, so-called affine geodesics.[27] Free fall in a gravitational field and being at rest or in uniform motion in the absence of a gravitational field are represented by motion along geodesics. Resisting the pull of gravity and accelerating in the absence of gravity are represented by motion that is not along geodesics. In the scenario envisioned in the twin paradox, the stay-at-home moves on a geodesic, whereas the traveller does not. In the example of the rotating disk, the person next to the disk is moving on a geodesic, whereas the person on the disk is not.

No coordinate transformation can turn a geodesic into a non-geodesic or vice versa. Observers moving on geodesics and observers moving on non-geodesics are physically not equivalent to one another. This is just another way of saying that general relativity does not extend the principle of relativity for uniform motion to arbitrary motion. Both in special and in general relativity there are preferred states of motion, namely motion along geodesics.[28]

The relativity principle that Einstein had established on the basis of the equivalence principle, however, will be satisfied if all laws of this new metric theory of gravity, including the field equations for the metric field, are generally covariant. In that case one can choose any wordline, a geodesic or a non-geodesic, as the time axis of one's coordinate system. An observer travelling on that worldline will be at rest in that coordinate system. If her worldline is not a geodesic, she will attribute the inertial forces she experiences to a gravitational field, which will satisfy the generally-covariant field equations. A geodesic and a non-geodesic observer in an arbitrarily curved space-time can thus both claim to be at rest if they agree to disagree about the presence of a gravitational field. As I mentioned above, however, it seems more natural to call this relativity of the gravitational field than relativity of motion.

Either way it was a serious setback for Einstein when his search for field equations in the winter of 1912–1913, undertaken with the help of Grossmann and recorded in the "Zurich Notebook", did not turn up any physically acceptable generally-covariant candidates. The problem continued to bother him in the months following the publication of the "Entwurf" field equations. Eventually (but not ultimately), Einstein made his peace with the limited covariance of these equations. In August 1913, in a vintage Einstein maneuver, he convinced himself that he had not been able to find generally-covariant field equations simply because there were none to be found. Einstein produced two arguments to show that generally-covariant field equations are physically unacceptable. Both arguments are fallacious, but both significantly deepened Einstein's understanding of his own theory.

The first argument, which can be dated with unusual precision to August 15, 1913, was that energy-momentum conservation restricts the covariance of acceptable field equations to linear transformations.[29] Einstein soon realized that the argument turns on the unwarranted assumption that gravitational energy-momentum can be represented by a generally-covariant tensor. The argument was retracted in [52, p. 218, note]. The general insight, however, that there is an intimate connection between energy-momentum conservation and the covariance of the field equations survived the demise of this specific argument and was a key element in Einstein's return to generally-covariant field equations in the fall of 1915 (see Sect. 3 below).

---

[27] The affine connection, which was not introduced until after the formulation of general relativity, is better suited to Einstein's purposes than the metric [100].

[28] A German *Gymnasium* teacher, Erich Kretschmann [70], clearly formulated what it takes for a space-time theory to satisfy a genuine relativity principle. Kretschmann first pointed out that a theory does not satisfy a relativity principle simply by virtue of being cast in a form that is covariant under the group of transformations associated with that principle. With a little ingenuity one can cast just about any theory in such a form. Einstein [41] granted this point, but did not address Kretschmann's proposal in the spirit of Klein's Erlangen program to characterize relativity principles in terms of symmetry groups of the set of geodesics of all space-times allowed by the theory. In special relativity, this would be the group of Lorentz transformations that map the set of all geodesics of Minkowski space-time, the only space-time allowed by the theory, back onto itself. The set of geodesics of all space-times allowed by general relativity has no non-trivial symmetries, so the theory fails to satisfy any relativity principle in Kretschmann's sense. Einstein still did not comment on Kretschmann's proposal after he was reminded of it in correspondence (see Gustav Mie to Einstein, 17–19 February 1918 [15, Doc. 465]). For further discussion of Kretschmann's proposal, see [84, Sect. 8], [85, Sect. 5] and [93].

[29] The date can be inferred from Einstein to H. A. Lorentz, 16 August 1913 [12, Doc. 470], in which Einstein mentions that he had found the argument the day before. For discussion of the argument and its flaws, see [82, Sect. 5].

By the time Einstein retracted this first argument against generally-covariant field equations, it had already been eclipsed by a second one, the infamous hole argument mentioned in the introduction.[30] A memo dated August 28, 1913, in the hand of Einstein's lifelong friend Michele Besso and found among the latter's papers in 1998, sheds some light on the origin of this argument.[31] Shortly after the publication of [50], Einstein and Besso had done extensive calculations to see whether the "Entwurf" theory can account for the anomalous advance of the perihelion of Mercury (see note 4 above). In this context Besso had raised the question whether the field equations uniquely determine the field of the sun [11, Doc. 14, p. 16]. This query may well have been the seed for the following argument recorded in the Besso memo:

> The requirement of covariance of the gravitational equations under arbitrary transformations cannot be imposed: if all matter were contained in one part of space and for this part of space a coordinate system [is given], then outside of it the coordinate system could, except for boundary conditions, still be chosen arbitrarily, so that a unique determinability of the g's cannot occur.[32]

This argument, presumably communicated to Besso by Einstein, turns into the hole argument when space is replaced by space-time and the regions with and without matter are interchanged. In the published version of the argument the point is that the metric field in some small matter-free region of space-time – the "hole" from which the argument derives it name – is not uniquely determined by the matter distribution and the metric field outside the hole.

The hole argument works as follows. Suppose we have generally-covariant field equations that at every point set the result of some differential operator acting on the metric field $g_{\mu\nu}$ equal to the energy-momentum tensor $T_{\mu\nu}$ of matter at that point. Consider a matter distribution such that $T_{\mu\nu}(x) = 0$ for all points inside the hole. Suppose $g_{\mu\nu}(x)$ is a solution for this particular matter distribution. Now consider a coordinate transformation $x \to x'$ that only differs from the identity inside the hole and express the energy-momentum tensor and the metric field in terms of the new primed coordinates. Because of the general covariance we assumed, the field equations in primed coordinates will have the exact same form as the field equations in unprimed coordinates and $\{g'_{\mu\nu}(x'), T'_{\mu\nu}(x')\}$ will be a solution of them. This will still be true – and this is the key observation – if we read x for x' everywhere. Since the energy-momentum tensor vanishes inside the hole and since the coordinate transformation $x \to x'$ is the identity outside the hole, $T'_{\mu\nu}(x) = T_{\mu\nu}(x)$ everywhere. That means that both $g_{\mu\nu}(x)$ and $g'_{\mu\nu}(x)$ are solutions of the field equations in unprimed coordinates for one and the same matter distribution $T_{\mu\nu}(x)$. These two solutions are identical outside the hole but differ inside. The matter distribution (along with boundary conditions for $g_{\mu\nu}$) thus fails to determine the metric field inside the hole uniquely. The only way to avoid this kind of indeterminism, the argument concludes, is to rule out field equations that retain their form under transformations $x \to x'$ such as the one that was used in the construction of the alternative solution $g'_{\mu\nu}(x)$ from the original solution $g_{\mu\nu}(x)$.[33]

---

[30] See [82] and [97] for the classic historical discussions of the hole argument. The argument has also spawned a huge philosophical literature following the publication in [23] of an argument inspired by and named after Einstein's. See, e. g., [5, 83], [20, Chap. 9], [76], and [98, 99].

[31] For detailed analysis of this memo, see [65].

[32] "Die Anforderung der Covarianz der Gravitationsgleichungen für beliebige Transformationen kann nicht aufgestellt werden: wenn in einem Teile des Raumes alle Materie enthalten wäre und für diesen Teil ein Coordinatensystem, so könnte doch ausserhalb desselben das Coordinatensystem noch, abgesehen von den Grenzbedingungen, beliebig gewählt werden, so dass eine eindeutige Bestimmbarkeit der gs nicht eintreten könne." See Fig. 2 of [65] for a facsimile of the page of the Besso memo with this passage.

[33] Another passage in the Besso memo quoted above makes it clear that, even in the embryonic version of the hole argument, Einstein saw the inequality $g_{\mu\nu}(x) \neq g'_{\mu\nu}(x)$ as expressing indeterminism, *not*, as older commentators have suggested (see, e. g., [88, p. 222]), the inequality $g_{\mu\nu}(x) \neq g'_{\mu\nu}(x')$, which merely expresses the non-uniqueness of the coordinate representation of the metric field. Besso wrote: "If in coordinate system 1 [with coordinates x], there is a solution K_1 [i. e., $g_{\mu\nu}(x)$], then this same construct [modulo a coordinate transformation] is also a solution in [coordinate system] 2 [with coordinates x'], K_2 [i. e., $g'_{\mu\nu}(x')$]; K_2, however, [is] also a solution in 1 [i. e., $g'_{\mu\nu}(x)$]" ("Ist im Coordinatensystem 1 eine Lösung K_1, so ist dieses selbe Gebilde auch eine Lösung in 2, K_2; K_2 aber auch eine Lösung in 1"). See [65, Sect. 4] for further discussion.

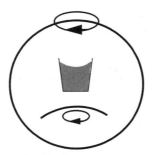

bucket & water rotating shell at rest

bucket & water at rest shell rotating

Fig. 4 Mach's response to Newton's bucket experiment.

Einstein never explicitly retracted the hole argument in print and it was only after he had returned to generally-covariant field equations in November 1915 that he at least addressed the issue in correspondence. Before we turn to this denouement of the hole story, however, we need to examine another strand in Einstein's quest for a general relativity of motion that made it into the 1916 review article.

During the period that he accepted that there could not be generally-covariant field equations, Einstein explored another strategy for eliminating absolute motion. This strategy was directly inspired by his reading of Ernst Mach's response to Newton's famous bucket experiment [6, pp. 10–11]; [7, pp. 412–413].[34] When a bucket filled with water in the gravitational field of the earth is set spinning, the water will climb up the wall of the bucket as it catches up with the bucket's rotation. Newton famously argued that it cannot be the relative rotation of the water with respect to the bucket that is causing this effect.[35] After all, the effect increases as the relative rotation between water and bucket decreases and is maximal when both are rotating with the same angular velocity. The effect, according to Newton, was due to the rotation of the water with respect to absolute space. Mach argued that Newton had overlooked a third possibility: the effect could be due to the relative rotation of the water with respect to other matter in the universe. "Try to fix Newton's bucket," he challenged those taken in by Newton's argument, "and rotate the heaven of fixed stars and then prove the absence of centrifugal forces" [74, p. 279].[36] Mach implied that it should make no difference whether the bucket or the heavens are rotating: in both cases the water surface should become concave. Mach's idea is illustrated in Fig. 4, depicting the bucket, the water, and the earth sitting at the center of a giant spherical shell representing all other matter in the universe.[37] According to Mach, the water surface should be concave no matter whether the bucket and the water or the earth and the shell are rotating. According to Newtonian theory, however, the rotation of the shell will have no effect whatsoever on the water in the bucket.

For most of the reign of the "Entwurf" theory and beyond, Einstein was convinced that this was a problem not for Mach's analysis but for Newton's theory and that his own theory vindicated Mach's account of the bucket experiment. In the spirit of the equivalence principle, Einstein [30, p. 1031] argued that the centrifugal forces responsible for the concave surface of the water in the rotating bucket might just as well be looked upon as gravitational forces due to distant rotating masses acting on the water in a bucket at rest.

---

[34] Looking back on this period in late 1916, Einstein wrote about these Machian ideas: "Psychologically, this conception played an important role for me, since it gave me the courage to continue to work on the problem when I absolutely could not find covariant field equations" ("Psychologisch hat diese Auffassung bei mir eine bedeutende Rolle gespielt; denn sie gab mir den Mut, an dem Problem weiterzuarbeiten, als es mir absolut nicht gelingen wollte, kovariante Feldgleichungen zu erlangen." Einstein to Willem de Sitter, 4 November 1916 [15, Doc. 273]).

[35] For Newton this argument was not so much an argument for absolute space or absolute acceleration as an argument against the Cartesian concept of motion [71], [61, Chap. 7].

[36] "Man versuche das Newtonsche Wassergefäß festzuhalten, den Fixsternhimmel dagegen zu rotieren und das Fehlen der Fliehkräfte nun nachzuweisen" [75, p. 222]. For extensive discussion of Mach's response to Newton's bucket experiment and Einstein's reading and use of it, see [3].

[37] The sad faces that one can discern in these drawings may serve as warning signs that the arguments of Mach and Einstein do not hold up under scrutiny.

To guarantee that Einstein's theory predicts that we get the same concave water surface in both cases in Fig. 4, the field equations need to satisfy two requirements. First, the Minkowski metric expressed in terms of the coordinates of a rotating frame of reference has to be a solution of the vacuum field equations. This was the kind of requirement that Einstein retreated to when he accepted that general covariance could not be had. He hoped that the "Entwurf" field equations would at least allow the Minkowski metric expressed in the coordinates of arbitrarily moving frames as vacuum solutions. Second, the metric field produced by the shell near its center has to be the Minkowski metric in rotating coordinates.

The "Entwurf" field equations satisfy neither of these two requirements. Einstein went back and forth for two years on whether or not the Minkowski metric in rotating coordinates is a vacuum solution. A sloppy calculation preserved in the Einstein-Besso manuscript reassured him in 1913 that it is [11, Doc. 14, pp. 41–42]. Later in 1913 Besso told him it is not.[38] Einstein appears to have accepted that verdict for a few months, but in early 1914 convinced himself on general grounds that it had to be.[39] That the "Entwurf" theory thus seems to account for the bucket experiment along Machian lines is hailed as a great triumph in the systematic exposition of the theory of late 1914 [30, p. 1031]. In September 1915, possibly at the instigation of Besso, Einstein redid the calculation of the Einstein-Besso manuscript and discovered to his dismay that his friend had been right two years earlier.[40] A month later Einstein replaced the "Entwurf" field equations by equations of much broader and ultimately general covariance. The Minkowski metric in its standard diagonal form is a solution of these equations. Their covariance guarantees that it is a solution in rotating coordinates as well.

The second requirement is satisfied neither by the "Entwurf" theory nor by general relativity, although it took a long time for Einstein to recognize this and even longer to accept it. When he calculated the metric field of a rotating shell in 1913 using the "Entwurf" field equations, he chose Minkowskian boundary conditions at infinity and determined how the rotating shell would perturb the metric field of Minkowski space-time.[41] This perturbation does indeed have the form of the Minkowski metric in rotating coordinates near the center of the shell but is much too small to make a dent in the water surface. More importantly, treating the effect of the rotating shell as a perturbation of the metric field of Minkowski space-time defeats the purpose of vindicating Mach's account of the bucket experiment. In this way, after all, the leading term in the perturbative expansion of the field acting on the bucket will not come from distant matter at all but from absolute space, albeit of the Minkowskian rather than the Newtonian variety. This problem will arise for any non-degenerate physically plausible boundary conditions [101, p. 38]. Einstein seems to have had a blind spot for the role of boundary conditions in this problem.

It is important to note that even if both requirements were satisfied, so that the water surfaces have the same shape in the two situations shown in Fig. 4, we would still not have reduced these two situations to one and the same situation looked at from two different perspectives. Consider the shell in the two cases. Its particles are assumed to move on geodesics in both cases, but while the case with the rotating shell requires cohesive forces preventing them from flying apart,[42] the case with the shell at rest does not. This in and of itself shows that the rotation of the water and bucket with respect to the shell and the earth is not relative in the "Entwurf" theory or in general relativity. Einstein added the metric field to the shell, the earth, the bucket, and the water – the material components that for Mach exhausted the system – and the relation between field and matter is very different in the two situations shown in Fig. 4.

Provided that the first of the two requirements distinguished above is satisfied, however, Einstein's own peculiar principle of relativity is satisfied in this case. We can start from the situation on the left with the bucket rotating in Minkowski space-time and transform to a rotating frame in which the bucket is at rest.

---

[38] This can be inferred from the Besso memo discussed above [65, Sect. 3].
[39] See Einstein to H. A. Lorentz, 23 March 1915 [15, Doc. 47].
[40] See Einstein to Erwin Freundlich, 30 September 1930 [15, Doc. 123].
[41] This calculation can be found in the Einstein-Besso manuscript [11, Doc. 14, 36–37]. For further analysis of this calculation, see [65, Sect. 3].
[42] Without such cohesive forces, to put it differently, the shell, as the source of the inertio-gravitational field, will not satisfy the law of energy-momentum conservation as it must both in the "Entwurf" theory and in general relativity (see Sect. 3).

We would have to accept the resulting unphysical degenerate values of the metric at infinity,[43] but we can if we want. Invoking the equivalence principle, an observer at rest in this frame can claim to be at rest in a gravitational field. This observer will claim that the centrifugal forces on the water come from this gravitational field and that the particles that make up the shell are in free fall in this field, which exerts centrifugal as well as Coriolis forces on them.[44] The shell, however, is not the source of this field.

Einstein conflated the situation on the left in Fig. 3, redescribed in a coordinate system in which the bucket is at rest, with the very different situation on the right. He thus believed that meeting the first of the two requirements distinguished above sufficed for the implementation of a Machian account of the bucket experiment. This is clear from a letter he wrote in July 1916. Explaining to Besso how to calculate the field of a rotating ring, a case very similar to that of the rotating shell which Einstein himself had considered in the Einstein-Besso manuscript (see note 41), he wrote

> In *first* approximation, the field is obtained easily by direct integration of the field equations.[45] The second approximation is obtained from the vacuum field equations as the next approximation. The first approximation gives the Coriolis forces, the second the centrifugal forces. That the latter come out correctly is obvious given the general covariance of the equations, so that it is of no further interest whatsoever to actually do the calculation. This is of interest only if one does not know whether rotation-transformations are among the "allowed" ones, i. e., if one is not clear about the transformation properties of the equations, a stage which, thank God, has definitively been overcome.[46]

The general covariance of the Einstein field equations does guarantee that the Minkowski metric in rotating coordinates is a vacuum solution. But it does not follow that this metric field is the same as the metric field near the center of a rotating shell. This would follow if the two situations in Fig. 4 were related to one another simply by a transformation to rotating coordinates. But, notwithstanding Einstein's suggestion to the contrary, they are not.

Einstein's correspondence with Hans Thirring in 1917 shows that this misunderstanding persisted for at least another year and a half.[47] When Thirring first calculated the metric field inside a rotating shell, he was puzzled that he did not simply find the Minkowski metric in rotating coordinates as he expected on the basis of remarks in the introduction of [30]. He asked Einstein about this and Einstein's responses indicate that he shared Thirring's puzzlement and expected there to be an error in Thirring's calculations. When he published his final results, Thirring [101, pp. 33, 38] explained that the metric field inside a rotating shell is not identical to the Minkowski metric in rotating coordinates because of the role of boundary conditions. He cited Einstein and De Sitter [39, 17] for the discussion of the role of boundary conditions. But although they were at the focus of his discussions with De Sitter, Einstein did not breathe a word about boundary conditions in his letters to Thirring.

---

[43] The components $g_{14} = g_{41} = \omega y$, $g_{24} = g_{42} = -\omega x$, and $g_{44} = 1 - \omega^2 r^2$ of the Minkowski metric in a coordinate system rotating with angular velocity ω around the z-axis go to infinity as $r \equiv \sqrt{x^2 + y^2}$ goes to infinity.

[44] For a particle of mass m rotating with angular frequency ω at a distance r from the axis of rotation the centrifugal force and the Coriolis force add up to a centripetal force of size $m\omega^2 r$ needed to keep the particle in its circular orbit (cf. note 24). This is explained in Einstein to Hans Thirring, 7 December 1918 [15, Doc. 405].

[45] Note that no mention is made of the Minkowskian boundary conditions that Einstein had used in his calculation for the case of a rotating shell.

[46] "Das Feld in erster Näherung ergibt sich leicht durch unmittelbare Integration der Feldgleichungen. Die zweite Näherung ergibt sich aus den Vakuumfeldgleichungen als nächste Näherung. Die erste Näherung liefert die Korioliskräfte, die zweite die Zentrifugalkräfte. Dass letztere richtig heraus kommen, ist bei der allgemeinen Kovarianz der Gleichungen selbstverständlich, sodass ein wirkliches Durchrechnen keinerlei Interesse mehr hat. Dies Interesse ist nur dann vorhanden, wenn man nicht weiss ob Rotations-transformationen zu den 'erlaubten' gehören, d. h. wenn man sich über die Transformationseigenschaften der Gleichungen nicht im Klaren ist, welches Stadium gottlob endgültig überwunden ist." Einstein to Michele Besso, 31 July 1916 [15, Doc. 245]. For further discussion, see [62, Sect. 11], [65, Sect. 3].

[47] See Hans Thirring to Einstein, 11–17 July 1917 [15, Doc. 361], Einstein to Thirring, 2 August 1917 [15, Doc. 369], Thirring to Einstein, 3 December 1917 [15, Doc. 401], and the letter cited in note 44.

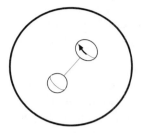 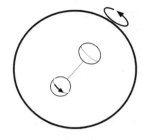

Fig. 5 Einstein's Machian solution to the problem of the two globes.

Not surprisingly, given the above, Einstein's account of the two globes rotating with respect to one another in Sect. 2 of the 1916 review article is modelled on his Machian account of Newton's bucket experiment. This is illustrated in Fig. 5, the analogue of Fig. 4. Recall the puzzle that Einstein drew attention to: why do the two globes take on different shapes, one becoming an ellipsoid, the other retaining its spherical shape?[48] Einstein [35, p. 772] identifies distant masses as the cause of this difference. He does not elaborate but after the discussion of Einstein's account of the bucket experiment, it is easy to fill in the details.

For Einstein, the distant masses (once again represented by a large shell in Fig. 5) function as the source of the metric field in the vicinity of the two globes. The globe that is rotating with respect to this metric field is the one that bulges out at the equator. As with the motion of the bucket with respect to the shell, the motion of the bulging globe with respect to the shell is not relative: the situation on the left in Fig. 5 with the bulging globe rotating and the shell (and the other globe) at rest is not equivalent to the situation on the right with the bulging globe at rest and the shell (and the other globe) rotating in the opposite direction. The relation between matter and metric field is different in these two cases. For one thing, the boundary conditions at infinity are different.

Still, Einstein's idiosyncratic relativity principle based on the equivalence principle – or, what amounts to the same thing, the relativity of the gravitational field – is satisfied in this case. Depending on which perspective we adopt in the situation depicted on the left in Fig. 5, that of an observer on the bulging globe or that of an observer on the other globe, we will interpret the forces on the bulging globe either as gravitational or as inertial forces. It is important that all perspectives are equally justified. Otherwise, as Einstein points out, we would still have a violation of the principle of sufficient reason. Einstein's analysis of the example of the two globes thus becomes an argument for general covariance. Note that general covariance in this context serves the purpose not of making rotation relative but of making the presence or absence of the gravitational field relative.[49]

As in the case of the bucket experiment, Einstein overlooked the role of boundary conditions. He proceeded as if the distant matter fully determines the metric field. References to motion with respect to the metric field could then be interpreted as shorthand for motion with respect to the sources of the field. But the metric field is determined by material sources plus boundary conditions. General relativity thus retains vestiges of absolute motion. This point was driven home by Willem de Sitter in discussions with Einstein in Leyden in the fall of 1916,[50] although Einstein's letters to Thirring a full year later give no indication that their author was aware of the problem. This is all the more puzzling since by that time Einstein had come up with an ingenious response to De Sitter.

In his paper on cosmology published in February 1917, Einstein [39] circumvented the need for boundary conditions by eliminating boundaries! He proposed a cosmological model that is spatially closed. The metric

---

[48] In terms of the somewhat more technical language that has meanwhile been introduced the simple answer of [19] to this puzzle is that the spherical globe moves on a geodesic, while the ellipsoidal one does not. Hence, the symmetry between the two globes in Einstein's example is illusory, like the symmetry between the two twins in the twin paradox, and there is nothing puzzling about them behaving differently.

[49] A similar way of interpreting the covariance properties of the "Entwurf" theory and general relativity in terms of the relativity of the gravitational potential or the gravitational field rather than in terms of the relativity of motion was proposed in [77, 78]. See Gustav Mie to Einstein, 30 May 1917 [15, Doc. 346].

[50] See the editorial note, "The Einstein–De Sitter–Weyl–Klein Debate", in [15, pp. 351–357].

field of such a model could thus be attributed in full to matter. He picked a model that was not only closed but static as well. To prevent this model from collapsing he had to modify the Einstein field equations and add a term with what came to be known as the cosmological constant. This term produces a gravitational repulsion, which exactly balances the gravitational attraction in the model.

De Sitter [18] promptly produced an alternative cosmological model that is also allowed by Einstein's modified field equations. This model is completely empty. Absolute motion thus returned with a vengeance. Einstein's modified field equations still allow space-times with no matter to explain why test particles prefer to move on geodesics. Before publishing his new solution, De Sitter reported it to Einstein.[51] In his response Einstein finally articulated the principle that he had tacitly been using in his Machian accounts of Newton's bucket experiment and of his own variant on Newton's thought experiment with the two globes. He wrote:

> It would be unsatisfactory, in my opinion, if a world without matter were possible. Rather, the $g^{\mu\nu}$-field should be *fully determined by matter and not be able to exist without it*.[52]

This passage is quoted in the postscript of [18]. Einstein rephrased and published it as "Mach's principle" in [41]: "The [metric] field is *completely* determined by the masses of bodies."[53] In a footnote he conceded that he had not been careful in the past to distinguish this principle from general covariance. A day after submitting [41], he submitted [42] in which he argued that there was matter tugged away on a singular surface in De Sitter's cosmological model. In that case the De Sitter solution would not be a counter-example to Mach's principle. The following June, Einstein had to admit that this singular surface is nothing but an artifact of the coordinates used.[54] The De Sitter solution is a perfectly regular vacuum solution and thus a genuine counter-example to Mach's principle after all. Einstein never retracted his earlier claim to the contrary, but he gradually lost his enthusiasm for Mach's principle over the next few years. The principle is still prominently discussed in the Princeton lectures, but its limitations are also emphasized ([46, pp. 64–70], [47, pp. 99–108]).

Much of the appeal of Mach's ideas disappears when one switches from a particle to a field ontology. Among the sources of the metric field in general relativity is the electromagnetic field. Mach's principle then amounts to the requirement that the metric field be reduced to the electromagnetic field. But why privilege one field over another? Einstein, to my knowledge, never explicitly raised this question, but by the early 1920s he was trying to unify the electromagnetic field and the metric field rather than trying to reduce one to the other.

Einstein thus accepted that the shape of the water surface in Newton's bucket and the bulging out of one of the globes in his own thought experiment is caused by the rotation of the water and the globe with respect to the metric field and that the metric field cannot be reduced to matter. Even in general relativity these effects are the result of acceleration with respect to space(-time) just as in Newtonian theory and in special relativity. That does not mean, however, that Einstein's objection in the 1916 review article, that Newtonian theory and special relativity violate the principle of sufficient reason, now also applies to general relativity. Space-time in general relativity, Einstein argued in his Princeton lectures, is a *bona fide* physical entity to which causal efficacy can be ascribed.[55] Unlike Newtonian absolute space or Minkowski space-time, he pointed out, space-time in general relativity both acts and is acted upon ([46, p. 36], [47, pp. 55–56]). As Misner et al. [80, p. 5] put it: "*Space acts on matter, telling it how to move. In turn, matter reacts back on space, telling it how to curve.*" Newtonian absolute space and Minkowski space-time only do the former. This is how Einstein was able to accept that general relativity did not eradicate absolute motion (in the sense of motion with respect to space-time rather than with respect to other matter).

---

[51] Willem de Sitter to Einstein, 20 March 1917 [15, Doc. 313].

[52] "Es wäre nach meiner Meinung unbefriedigend, wenn es eine denkbare Welt ohne Materie gäbe. Das $g^{\mu\nu}$-Feld soll vielmehr *durch die Materie bedingt sein, ohne dieselbe nicht bestehen können.*" Einstein to Willem de Sitter, 24 March 1917 [15, Doc. 317].

[53] "Das G-Feld ist restlos durch die Massen der Körper bestimmt" [41, p. 241, note].

[54] Einstein to Felix Klein, 20 June 1918 [15, Doc. 567].

[55] In 1920, Ehrenfest and Lorentz arranged for a special professorship for Einstein in Leyden. In his inaugural lecture, Einstein [45] talked about the metric field as a new kind of ether.

As the simple solution of [19] to Einstein's problem of the rotating globes shows, it is not necessary to turn space-time into a causally efficacious substance to avoid violations of the principle of sufficient reason. In the course of developing general relativity, Einstein in fact provided ammunition for a strong argument *against* a substantival and in support of a relational ontology of space-time. This argument is based on the resolution of the hole argument against generally-covariant field equations.

Einstein first explained what was wrong with the hole argument, which can be found in four of his papers of 1914,[56] in a letter to Ehrenfest written about a month after reaffirming general covariance in November 1915.[57] He told Ehrenfest that the hole argument should be *replaced* by a new argument that has come to be known as the "point-coincidence argument".[58] A week later he told Besso the same thing.[59] The field equations, Einstein argued, need not determine the metric field uniquely, only such things as the intersections of worldlines, i. e., the "point coincidences" from which the new argument derives its name. Generally-covariant field equations will certainly do that. Two years earlier Besso had suggested that the escape from what was to become the hole argument might be that only worldlines need to be determined uniquely, but that suggestion had immediately been rejected.[60] In August 1913, Einstein had no use for an escape from the hole argument. The argument was mainly a fig leaf at that point for Einstein's inability to find generally-covariant field equations. Now that he had found and published such equations, however, he did need a escape from the hole argument. It is probably no coincidence (no pun intended) that five days before the letter to Ehrenfest in which the point-coincidence argument makes its first appearance a paper by Kretschmann [69] was published in which the notion of point coincidences, if not the term, is introduced [59, p. 54]. Kretschmann thus provided Einstein with just the right tools at just the right time.

In his letters to Ehrenfest and Besso, Einstein did more than substitute the point-coincidence argument for the hole argument. He also explained in these letters, albeit rather cryptically, where the hole argument went wrong.[61] The notion of point coincidences almost certainly helped Einstein put his finger on the problem with the hole argument. Once again consider the transformation $x \to x'$ used in the hole construction. Suppose two geodesics of the metric field $g_{\mu\nu}(x)$ intersect one another at a point inside the hole with coordinates $x = a$. Let the primed coordinates of that point be $x' = b$. In the metric field $g'_{\mu\nu}(x)$, obtained from $g'_{\mu\nu}(x')$ by reading x for x', the two corresponding geodesics will intersect at the point $x = b$. If the two points labeled $x = a$ and $x = b$ can somehow be identified before we assign values of the metric field to them, $g_{\mu\nu}(x)$ and $g'_{\mu\nu}(x)$ describe different situations. This suggests that the escape from the hole argument is simply to deny that bare manifold points can be individuated independently of the metric field. Solutions such as $g_{\mu\nu}(x)$ and $g'_{\mu\nu}(x)$ related to one another through Einstein's hole construction dress up the bare manifold differently to become a space-time. The original solution $g_{\mu\nu}(x)$ may dress up the bare manifold point p to become the space-time point P where two geodesics intersect, whereas the alternative solution $g'_{\mu\nu}(x)$ dresses up the bare manifold point q to become that same space-time point P. If the bare manifold points p and q have their identities only by virtue of having the properties of the space-time point P, there is no difference between $g_{\mu\nu}(x)$ and $g'_{\mu\nu}(x)$. Point coincidences can be used to individuate such space-time points. Most modern commentators read Einstein's comments on the hole argument in his letters to Ehrenfest and Besso in this way.[62]

---

[56] [51, p. 260], [52, pp. 217–218], [29, p. 178], [30, p. 1067].

[57] Einstein to Paul Ehrenfest, 26 December 1915 [15, Doc. 173].

[58] For discussions of the point-coincidence argument, see [59, 83, 97–99], [65, Sect. 4], and, especially, [58].

[59] Einstein to Michele Besso, 3 January 1916 [15, Doc. 178].

[60] Immediately following the passage quoted in note 32, Besso's memo of 28 August 1913 says: "It is, however, not necessary that the g themselves are determined uniquely, only the observable phenomena in the gravitation space, e. g., the motion of a material point, must be" ("Es ist nun allerdings nicht nötig, dass die g selbst eindeutig bestimmt sind, sondern nur die im Gravitationsraum beobachtbaren Erscheinungen, z.B. die Bewegung des materiellen Punktes, müssen es sein"). Appended to this passage is the following comment: "Of no use, since with a solution a motion is also fully given" ("Nützt nichts, denn durch eine Lösung ist auch eine Bewegung voll gegeben"). For further discussion, see [65, Sect. 3].

[61] A clearer version can be found in a follow-up letter: Einstein to Paul Ehrenfest, 5 January 1916 [15, Doc. 180].

[62] See the papers cited in note 58.

This resolution of the hole argument amounts to an argument against space-time substantivalism. What it shows is that there are many indistinguishable ways of assigning spatio-temporal properties to bare manifold points. According to Leibniz's "Principle of the Identity of Indiscernibles" all such assignments must be physically identical. But then the points themselves cannot be physically real for that would make the indistinguishable ways of ascribing properties to them physically distinct. This argument can be seen as a stronger version of a famous argument due to Leibniz himself against Newton's substantival ontology of space [23]. In his correspondence with Clarke, Leibniz objected that on Newton's view of space as a container for matter God, in creating the universe, had to violate the principle of sufficient reason [2, p. 26].[63] Without any discernible difference, He could have switched East and West, to use Leibniz's own example, or shifted the whole world to a different place in Newton's container. The Principle of the Identity of Indiscernibles tells us that these indistinguishable creations should all be identical. That in turn leaves no room for the container. In the hole argument, a violation of determinism takes over the role of the violation of the principle of sufficient reason in Leibniz's argument. This is what makes the hole argument the stronger of the two. In this secular age it is hardly a source of great distress that God for no apparent reason had to actualize one member of a class of empirically equivalent worlds rather than another. Ruling out determinism, however, as the substantivalist seems forced to do is clearly less palatable. To give determinism as much as "a fighting chance" [20, p. 180] in general relativity, we had thus better adopt a relational rather than a substantival ontology of space-time.

After his return to general covariance, Einstein never mentioned the hole argument again in any of his publications, but he did use the point coincidence argument in the 1916 review paper. He did not use it as part of an argument against space-time substantivalism, however, but as another argument against preferred frames of reference [35, pp. 776–777]. Since all our measurements eventually consist of observations of point coincidences, he argued,[64] and since such point coincidences are preserved under arbitrary coordinate transformations, physical laws should be generally covariant. Reformulated in the spirit of Klein's Erlangen program:[65] the group of general point transformations preserves the set of point coincidences, which supposedly exhaust the essential content of the geometries allowed by general relativity. Of course, this group of transformations also preserves the difference between geodesics and non-geodesics. Einstein nonetheless continued to tie general covariance to the relativity of motion. Revisiting the foundations of general relativity two years later, Einstein no longer used the point-coincidence argument to *argue* for a relativity principle but to *define* it:

> *Relativity principle*: The laws of nature are merely statements about point coincidences; the only natural way to express them is therefore in terms of generally-covariant equations.[66]

As Einstein realized at this point, this principle will only give full relativity of motion in conjunction with the other two principles on which he based his theory in 1918, the equivalence principle and Mach's principle. Only a few months later, as we saw earlier, Einstein had to concede that the latter does not hold in general relativity.

---

[63] For detailed analysis of this argument, see [20, Chap. 6]. For an abridged annotated version of the Leibniz-Clarke correspondence, see [61, Chap. 8].

[64] Earman [20, p. 186] thus charges Einstein with "a crude verificationism and an impoverished conception of physical reality." For a detailed critique of this reading of the point-coincidence argument, see [58].

[65] In one of his physics textbooks, Sommerfeld [95, pp. 316–317] citing [46], explicitly endorses such a reading of Einstein's presentation of general relativity.

[66] "*Relativitätsprinzip*: Die Naturgesetze sind nur Aussagen über zeiträumliche Koinzidenzen; sie finden deshalb ihren einzig natürlichen Ausdruck in allgemein kovarianten Gleichungen" [41, p. 241].

The "fateful prejudice" and the "key to the solution"[67]

In his search for satisfactory field equations for $g_{\mu\nu}$ in 1912–1913, Einstein had consciously pursued the analogy with Maxwell's theory of electrodynamics.[68] He continued to pursue this analogy in developing a variational formalism for the "Entwurf" theory in 1914 [52,30].[69] In his lecture on gravity at the 85th *Naturforscherversammlung* in Vienna in the fall of 1913 [28], he had already shown that the "Entwurf" field equations, like Maxwell's equations, can be cast in the form "divergence of field = source." In Maxwell's equations, $\partial_\nu F^{\mu\nu} = j^\mu$, the tensor $F^{\mu\nu}$ represents the electromagnetic field and the charge-current density j^μ its source. In Einstein's gravitational theory, the source is represented by the sum of $T^{\mu\nu}$, the energy-momentum tensor of 'matter' (which can be anything from particles to an electromagnetic field) and $t^{\mu\nu}$, the energy-momentum pseudo-tensor of the gravitational field itself.

Einstein used the energy-momentum balance law,

$$T^{\mu\nu}{}_{;\nu} = 0 \tag{1}$$

(with the semi-colon indicating a covariant derivative), to identify both the pseudo-tensor $t^{\mu\nu}$ and the expression for the gravitational field. He had derived this equation early on in his work on the metric theory of gravity as the natural generalization of the special-relativistic law of energy-momentum conservation, $T^{\mu\nu}{}_{,\nu} = 0$ (with the comma indicating an ordinary coordinate derivative).[70] For a charge distribution described by the four-current density j^μ in an electromagnetic field $F^{\mu\nu}$, we have

$$T^{\mu\nu}{}_{,\nu} = j_\nu F^{\mu\nu}, \tag{2}$$

where $T^{\mu\nu}$ is the energy-momentum tensor of the electromagnetic field.[71] The right-hand side has the form "source × field." It gives the density of the four-force that the electromagnetic field exerts on the charges, or, equivalently, the energy-momentum transfer from field to charges. The equation $T^{\mu\nu}{}_{;\nu} = 0$ (where $T^{\mu\nu}$ is the energy-momentum tensor of arbitrary 'matter' again) can be interpreted in the same way. Eq. (1) can be rewritten as

$$\mathfrak{T}^\alpha_{\mu,\alpha} - \begin{Bmatrix} \beta \\ \mu\alpha \end{Bmatrix} \mathfrak{T}^\alpha_\beta = 0, \tag{3}$$

where $\mathfrak{T}^\mu_\nu \equiv \sqrt{-g}\, T^\mu_\nu$ is a mixed tensor density, and where

$$\begin{Bmatrix} \beta \\ \mu\alpha \end{Bmatrix} \equiv \frac{1}{2} g^{\beta\rho}(g_{\rho\mu,\alpha} + g_{\rho\alpha,\mu} - g_{\mu\alpha,\rho}) \tag{4}$$

are the Christoffel symbols of the second kind. Because $T^{\mu\nu}$ is symmetric, eq. (3) can be further reduced to:

$$\mathfrak{T}^\alpha_{\mu,\alpha} - \frac{1}{2} g^{\beta\rho} g_{\rho\alpha,\mu} \mathfrak{T}^\alpha_\beta = 0. \tag{5}$$

Compare the second term on the left-hand sides of eqs. (3) and (5) to the term on the right-hand side of eq. (2). All three terms are of the form "source × field." In eqs. (3) and (5), these terms represent the density of the four-force that the gravitational field exerts on matter (the source $\mathfrak{T}^\alpha_\beta$), or, equivalently, the energy-momentum transfer from field to matter. One can thus read off an expression for the gravitational field from

---

[67] This section is based on [66].
[68] See [90] for a detailed reconstruction of Einstein's reliance on this analogy.
[69] See Sect. 3 of [66] for a detailed analysis of this variational formalism the use of which runs like a red thread through Einstein's work on general relativity from 1914 through 1918.
[70] See the "Zurich Notebook", [p. 5R] [11, Doc. 10, p. 10]; a facsimile of this page graces the dust cover of this volume. For analysis of this page, see [87, Appendix C] and [67, Sect. 3].
[71] Cf., e. g., [35, Sect. 20, 814–815], eqs. (65), (65a), (66), and (66a) for $g_{\mu\nu} = \eta_{\mu\nu}$, the flat Minkowski metric.

these terms. Using the field equations to eliminate $\mathfrak{T}_\beta^\alpha$ from these terms and writing the resulting expression as a divergence, $\partial_\alpha \mathfrak{t}_\mu^\alpha$, one can identify the gravitational energy-momentum pseudo-tensor ($\mathfrak{t}_\nu^\mu \equiv \sqrt{-g} t_\nu^\mu$).

Einstein used the notation $\Gamma^\mu_{\alpha\beta}$ for the components of the gravitational field. Unfortunately, different forms of eq. (1) lead to different choices for $\Gamma^\mu_{\alpha\beta}$. Eq. (5) gives

$$\Gamma^\mu_{\alpha\beta} \equiv -\frac{1}{2} g^{\mu\rho} g_{\rho\alpha,\beta} , \qquad (6)$$

which is essentially the gradient of the metric tensor. This makes perfect sense since the metric is the gravitational potential in Einstein's theory. But the relation between field and potential could also be the one suggested by eq. (3):

$$\Gamma^\mu_{\alpha\beta} \equiv -\left\{ \begin{matrix} \mu \\ \alpha\ \beta \end{matrix} \right\}. \qquad (7)$$

With this definition of the field there are three terms with a gradient of the potential. In fact, $\Gamma^\mu_{\alpha\beta}$ in eq. (6) is nothing but a truncated version of $\Gamma^\mu_{\alpha\beta}$ in eq. (7).

For the "Entwurf" theory Einstein chose definition (6) (omitting the minus sign).[72] It was only in the fall of 1915 that he realized that he should have gone with definition (7) instead. In Einstein's own estimation this was a crucial mistake. In the first of his four papers of November 1915, he wrote:

> This conservation law [in the form of eq. (5)] has led me in the past to look upon the quantities [in eq. (6)] as the natural expressions of the components of the gravitational field, even though the formulas of the absolute differential calculus suggest the Christoffel symbols [...] instead. *This was a fateful prejudice.*[73]

Later that month, after he had completed the theory, he wrote in a letter:

> *The key to this solution* was my realization that not [the quantities in eq. (6)] but the related Christoffel symbols [...] are to be regarded as the natural expression for the "components" of the gravitational field.[74]

To understand why eq. (6) was a "fateful prejudice" and eq. (7) was "the key to [the] solution," we need to look at Einstein's 1914 derivation of the "Entwurf" field equations from the action principle $\delta J = 0$. The action functional,

$$J = \int H \sqrt{-g} d^4 x , \qquad (8)$$

is determined by the Lagrangian H (Hamilton's function in Einstein's terminology). Drawing on the electrodynamical analogy, Einstein modelled H on $-\frac{1}{4} F^{\mu\nu} F_{\mu\nu}$, the Lagrangian for the free Maxwell field:

$$H = -g^{\mu\nu} \Gamma^\alpha_{\beta\mu} \Gamma^\beta_{\alpha\nu} . \qquad (9)$$

Inserting (minus) eq. (6) for the gravitational field and evaluating the Euler-Lagrange equations, he recovered the vacuum "Entwurf" field equations

$$\partial_\alpha \left(\sqrt{-g} g^{\alpha\beta} \Gamma^\lambda_{\mu\beta} \right) = -\kappa \mathfrak{t}_\mu^\lambda , \qquad (10)$$

---

[72] See [30, p. 1058, eq. (46)]. In a footnote on p. 1060, Einstein explains why he used eq. (5) rather than eq. (3) to identify $\Gamma^\mu_{\alpha\beta}$

[73] "Diese Erhaltungsgleichung hat mich früher dazu verleitet, die Größen [...] als den natürlichen Ausdruck für die Komponenten des Gravitationsfeldes anzusehen, obwohl es im Hinblick auf die Formeln des absoluten Differentialkalküls näher liegt, die Christoffelschen Symbole statt jener Größen einzuführen. *Dies war ein verhängnisvolles Vorurteil*" [31, p. 782]; my emphasis.

[74] "*Den Schlüssel zu dieser Lösung* lieferte mir die Erkenntnis, dass nicht [...] sondern die damit verwandten Christoffel'schen Symbole [...] als natürlichen Ausdruck für die "Komponente" des Gravitationsfeldes anzusehen ist." This comment comes from the letter to Sommerfeld cited in note 3. The emphasis is mine.

where the left-hand side is essentially the divergence of the gravitational field and where

$$\kappa t^\lambda_\mu = \sqrt{-g}\left(g^{\lambda\rho}\Gamma^\alpha_{\tau\mu}\Gamma^\tau_{\alpha\rho} - \frac{1}{2}\delta^\lambda_\mu g^{\rho\tau}\Gamma^\alpha_{\beta\rho}\Gamma^\beta_{\alpha\tau}\right) \tag{11}$$

is the energy-momentum pseudo-tensor for the gravitational field ([30, p. 1077, eq. (81b)]). This quantity was chosen is such a way that the energy-momentum balance law (5) can be written as a proper conservation law:

$$\partial_\lambda(\mathfrak{T}^\lambda_\mu + t^\lambda_\mu) = 0 \,. \tag{12}$$

The field equations in the presence of matter are found by adding $-\kappa T^\lambda_\mu$ to the right-hand side of eq. (10) on the argument that all energy-momentum – of matter and of the gravitational field itself – should enter the field equations the same way. One thus arrives at:

$$\partial_\alpha\left(\sqrt{-g}g^{\alpha\beta}\Gamma^\lambda_{\mu\beta}\right) = -\kappa\left(\mathfrak{T}^\lambda_\mu + t^\lambda_\mu\right). \tag{13}$$

Having the four-divergence operator ∂_λ act on both sides of eq. (13), one sees that energy-momentum is conserved if and only if[75]

$$B_\mu \equiv \partial_\lambda\partial_\alpha\left(\sqrt{-g}g^{\alpha\beta}\Gamma^\lambda_{\mu\beta}\right) = 0 \tag{14}$$

Einstein showed that these same conditions also determine the covariance properties of the action (8).[76] He argued that the corresponding Euler-Lagrange equations – i. e., the vacuum "Entwurf" field equations (10) – will inherit these covariance properties from the action. Since $T^{\mu\nu}$ is a generally-covariant tensor, the four conditions $B_\mu = 0$ would then determine the covariance properties of the full "Entwurf" field equations (13) as well.

How do these conditions select transformations that leave the "Entwurf" field equations (13) invariant? Start with a metric field $g_{\mu\nu}$ given in coordinates x^α that satisfies both the field equations (13) and conditions (14). Now consider a transformation from x^α to x'^α under which $g_{\mu\nu}$ goes to $g'_{\mu\nu}$. Einstein believed that $g'_{\mu\nu}$ would also be solution of the field equations (13) *if and only if* $g'_{\mu\nu}$ satisfies conditions (14). The transformations picked out by the conditions $B_\mu = 0$ are thus of a somewhat peculiar nature. The condition selects transformations from x^α to x'^α leaving the "Entwurf" field equations invariant *given* a metric field that is a solution of the field equations in the original -coordinates. Because of their dependence on the metric, Einstein called such transformations "non-autonomous" ("unselbständig") at one point.[77] The Italian mathematician Tullio Levi-Civita wrote the Minkowski metric in two different coordinate systems and showed that both forms satisfy the condition $B_\mu = 0$, while only one form is a (vacuum) solution of the "Entwurf" field equations.[78] Despite this clear-cut counter-example, Einstein stubbornly continued to believe that the condition guaranteeing energy-momentum conservation was also the necessary and sufficient condition for a solution of the "Entwurf" field equations in one coordinate system to be a solution in some other coordinate system. In Einstein's defense, it must be said that he was on to an important result even if his math did not quite add up. The connection between the invariance of the action for the "Entwurf" field equations (as opposed to the field equations themselves) and energy-momentum conservation is a special case of one of Noether's celebrated theorems connecting symmetries and conservation laws.[79] Einstein had intuitively recognized this special case almost five years before Emmy Noether [81] published the general

---

[75] Einstein had learned the hard way that he had better make sure that the field equations be compatible with energy-momentum conservation. In 1912 he had been forced to modify the field equation of his theory for static gravitational fields because the original one violated energy-momentum conservation [27, Sect. 4].

[76] See [66, Sect. 3.3].

[77] Einstein to H. A. Lorentz, 14 August 1913 [12, Doc. 467].

[78] Tullio Levi-Civita to Einstein, 28 March 1915 [15, Doc. 67].

[79] See [4] for an insightful analysis of Noether's theorems. For historical discussion, see [91, 92], and [94].

theorem. "What can be more beautiful," he had written in 1913 when he believed that energy-momentum conservation restricted the covariance of acceptable field equations to linear transformations (see Sect. 2), "than that the necessary specialization [of admissible coordinate systems] follows from the conservation laws?"[80]

Given this clarification of the structure of the "Entwurf" theory, one can understand why Einstein felt that the time was ripe for an authoritative self-contained exposition of the theory. The result was [30], which appeared in November 1914.

In the fall of 1915, a number of worrisome cracks were beginning to show in the "Entwurf" edifice. Most importantly, Einstein was finally forced to accept that the "Entwurf" field equations are not invariant under the (non-autonomous) transformation to rotating coordinates in the special case of the standard diagonal Minkowski metric (see the discussion in Sect. 2 and the letter cited in note 40). His Machian solution to the problem of Newton's rotating bucket experiment required the Minkowski metric in rotating coordinates to be a vacuum solution of the field equations. The "Entwurf" field equations were no longer acceptable now that it had become clear that they do not meet this requirement. Einstein needed new field equations.

If we take him at his word when he identified definition (6) of the gravitational field as a "fateful prejudice" and definition (7) as "the key to [the] solution," a plausible scenario of how Einstein found the successor(s) to the "Entwurf" field equations suggests itself.[81] The scenario runs as follows. Einstein decided to keep the Maxwell-inspired Lagrangian (9), with the exception of the immaterial minus sign,

$$H = g^{\mu\nu} \Gamma^{\alpha}_{\beta\mu} \Gamma^{\beta}_{\alpha\nu} \,, \tag{15}$$

and change only the definition of the gravitational field entering into it. Inserting eq. (7) into eq. (15), setting $\sqrt{-g} = 1$ in eq. (8) for the action J, and evaluating the Euler-Lagrange equations for the resulting variational problem $\delta J = 0$, he arrived at:[82]

$$-\partial_\alpha \left\{ \begin{array}{c} \alpha \\ \mu\nu \end{array} \right\} + \left\{ \begin{array}{c} \alpha \\ \beta\mu \end{array} \right\} \left\{ \begin{array}{c} \beta \\ \nu\alpha \end{array} \right\} = 0 \,. \tag{16}$$

Einstein had encountered these two terms before. They are two of the four terms in the Ricci tensor, a direct descendant of the Riemann curvature tensor. The other two terms are[83]

$$\partial_\mu \left\{ \begin{array}{c} \alpha \\ \alpha\nu \end{array} \right\} - \left\{ \begin{array}{c} \alpha \\ \mu\nu \end{array} \right\} \left\{ \begin{array}{c} \beta \\ \alpha\beta \end{array} \right\} . \tag{17}$$

Introducing the quantity

$$T_\nu \equiv \left\{ \begin{array}{c} \alpha \\ \alpha\nu \end{array} \right\} = \partial_\nu \left(\lg\sqrt{-g} \right), \tag{18}$$

which transforms as a vector under unimodular transformations (i. e., transformations with a Jacobian equal to one), one recognizes that expression (17) is the covariant derivative of T_ν,

$$\partial_\mu T_\nu - \left\{ \begin{array}{c} \alpha \\ \mu\nu \end{array} \right\} T_\alpha = T_{\nu;\mu} \,, \tag{19}$$

---

[80] "Was kann es schöneres geben, als dies, dass jene nötige Spezialisierung aus den Erhaltungssätzen fliesst?" Einstein to Paul Ehrenfest, before 7 November 1913 [12, Doc. 481].
[81] The bulk of [66] is concerned with making the case for this scenario on the basis of all extant primary source material.
[82] This calculation can be found in Sect. 15 of [35].
[83] See [35, p. 801, eq. (44)] for this decomposition of the Ricci tensor.

which transforms as a tensor under unimodular transformations. Since the full Ricci tensor is a generally-covariant tensor and one half transforms as a tensor under unimodular transformations, the other half (i. e., the left-hand side of eq. (16)) must transform as a tensor under unimodular transformations as well.

In the "Zurich Notebook" Einstein had actually looked carefully into the possibility of using this unimodular tensor as the basis for gravitational field equations.[84] One of the problems that had defeated him back then was that he could not show that such equations would be compatible with energy-momentum conservation. In the course of developing his variational formalism for the "Entwurf" theory in 1914, Einstein had learned how to deal with that problem (cf. eqs. (12)–(13) above). This made field equations based on the unimodular tensor in eq. (16) extremely attractive.

By this time, October 1915, Einstein had been struggling with the intractable covariance properties of the "Entwurf" field equations for almost three years. Changing the definition of the gravitational field in the Lagrangian for the "Entwurf" theory had now led him back to field equations covariant under a broad class of transformations. The only fly in the ointment was the hole argument, according to which there could be no such field equations. Sooner or later he would have to deal with this objection of his own making. But that could wait. As we saw Sect. 2, he only addressed this issue in correspondence of late December and early January.

So Einstein went ahead and decided on the vacuum field equations

$$\partial_\alpha \Gamma^\alpha_{\mu\nu} + \Gamma^\alpha_{\beta\mu} \Gamma^\beta_{\alpha\nu} = 0 \,, \tag{20}$$

which are obtained by replacing the Christoffel symbols in eq. (16) by minus the components $\Gamma^\alpha_{\mu\nu}$ of the gravitational field (see eq. (7)). He generalized these equations to situations with matter present by putting (minus) the energy-momentum tensor for matter on the right-hand side

$$\partial_\alpha \Gamma^\alpha_{\mu\nu} + \Gamma^\alpha_{\beta\mu} \Gamma^\beta_{\alpha\nu} = -\kappa T_{\mu\nu} \,. \tag{21}$$

With Hilbert in hot pursuit, he rushed these equations into print [31]. He realized soon afterwards that they were still not quite right. Within a three-week span, he published two modifications of the equations [32, 34]. The second time he got it right. He had found the generally-covariant field equations still bearing his name.

In Sects. 14–18 on the field equations and energy-momentum conservation in [35], the reader is spared the detour through the erroneous field equations of November 1915. Using the derivation rehearsed in one of his letters to Ehrenfest (see note 12), Einstein introduced the correct equations right away, albeit not in their generally-covariant form, but, as in the November 1915 papers, in unimodular coordinates (picked out by the condition $\sqrt{-g} = 1$). It turns out that Einstein's generalization of eq. (20) to eq. (21) violates the requirement that all energy-momentum enter the field equations the same way. This becomes clear when the vacuum field equations are rewritten in terms of the energy-momentum pseudo-tensor for the gravitational field in the new theory. This pseudo-tensor is found in the same way as the one for the "Entwurf" theory in eq. (11) and has the exact same structure:

$$\kappa t^\lambda_\sigma = \frac{1}{2} \delta^\lambda_\sigma g^{\mu\nu} \Gamma^\alpha_{\beta\mu} \Gamma^\beta_{\alpha\nu} - g^{\mu\nu} \Gamma^\alpha_{\mu\sigma} \Gamma^\lambda_{\alpha\nu} \tag{22}$$

[35, p. 806, eq. (50)]. Eq. (22) is obtained from eq. (11) by setting $\sqrt{-g} = 1$ (reflecting the restriction to unimodular coordinates), introducing an overall minus sign (since the Lagrangians (9) and (15) have opposite signs), and – most importantly – replacing definition (6) of the components of the gravitational field $\Gamma^\alpha_{\mu\nu}$ by definition (7). With the help of eq. (22) and its trace, $\kappa t \equiv \kappa t^\lambda_\lambda = g^{\mu\nu} \Gamma^\alpha_{\beta\mu} \Gamma^\beta_{\alpha\nu}$, the vacuum field equations (20) can be rewritten as (Ibid., 806, eq. (51)):

$$\partial_\alpha (g^{\nu\sigma} \Gamma^\alpha_{\mu\nu}) = -\kappa \left(t^\sigma_\mu - \frac{1}{2} \delta^\sigma_\mu t \right) . \tag{23}$$

---

[84] See [67], Sect. 5.5, for a detailed analysis of the relevant pages of the notebook, [pp. 22L–24L] and [pp. 42L–43L] [11, Doc. 10, pp. 43–47 and pp. 7–9].

Notice how closely this equation (along with eq. (22) for t^σ_μ) resembles the vacuum "Entwurf" field equations (10) (along with eq. (11) for $t^\lambda_\mu = \sqrt{-g}t^\lambda_\mu$). The crucial difference – besides immaterial minus signs, factors of $\sqrt{-g}$, and a slightly different ordering of indices – is the presence of the trace term $(1/2)\delta^\sigma_\mu \kappa t$ on the right-hand side of eq. (23). On the by now familiar argument that all energy-momentum enter the field equations the same way, this means that the field equations in the presence of matter should likewise have a term with the trace of the energy-momentum tensor of matter (Ibid., 807, eq. (52)):

$$\partial_\alpha(g^{\nu\sigma}\Gamma^\alpha_{\mu\nu}) = -\kappa \left([t^\sigma_\mu + T^\sigma_\mu] - \frac{1}{2}\delta^\sigma_\mu[t+T] \right) \tag{24}$$

Now recall that eq. (23) is just an alternative way of writing eq. (20). Eq. (24) can thus also be written as:

$$\partial_\alpha \Gamma^\alpha_{\mu\nu} + \Gamma^\alpha_{\beta\mu}\Gamma^\beta_{\alpha\nu} = -\kappa \left(T_{\mu\nu} - \frac{1}{2}g_{\mu\nu}T \right) \tag{25}$$

(Ibid., 808, eq. (53)). In November 1915, Einstein had found his way from eq. (21) to eq. (25) following a more circuitous route.

In the unimodular coordinates used in this calculation $\sqrt{-g} = 1$ and the quantity T_μ in eq. (18) vanishes. This means that expression (17) vanishes as well and that the Ricci tensor reduces to the left-hand side of eq. (25) (cf. eq. (16)). The field equations (25) can thus be looked upon as generally covariant equations expressed in unimodular coordinates. The corresponding generally-covariant equations are:

$$R_{\mu\nu} = -\kappa \left(T_{\mu\nu} - \frac{1}{2}g_{\mu\nu}T \right), \tag{26}$$

where $R_{\mu\nu}$ is the Ricci tensor (denoted by $B_{\mu\nu}$ in ibid., 801, eq. (44)). The reader, I trust, will immediately recognize eq. (26) as the Einstein field equations.

Returning now to eq. (24), the original form of the field equations in unimodular coordinates, one sees that energy-momentum is conserved, i. e., $T^{\mu\nu}{}_{;\nu} = 0$, or, equivalently, $\partial_\sigma(t^\sigma_\mu + T^\sigma_\mu) = 0$, if and only if

$$\partial_\sigma \left[\partial_\alpha(g^{\nu\sigma}\Gamma^\alpha_{\mu\nu}) - \frac{1}{2}\kappa\delta^\sigma_\mu(t+T) \right] = 0. \tag{27}$$

These conditions are the analogues of the conditions $B_\mu = 0$ in the "Entwurf" theory (cf. eqs. (12)–(14)). Using the trace of eq. (24), $\partial_\alpha(g^{\nu\sigma}\Gamma^\alpha_{\sigma\nu}) = \kappa(t+T)$, to replace $\kappa(t+T)$ by an expression in terms of the metric and its derivatives, one can rewrite eq. (27) as

$$\partial_\sigma \left[\partial_\alpha \left(g^{\nu\sigma}\Gamma^\alpha_{\mu\nu} \right) - \frac{1}{2}\delta^\sigma_\mu \partial_\alpha \left(g^{\nu\beta}\Gamma^\alpha_{\beta\nu} \right) \right] = 0. \tag{28}$$

Einstein showed that these four relations, unlike the conditions $B_\mu = 0$, hold identically. In other words, the field equations (24) guarantee energy-momentum conservation without the need for restrictions on admissible coordinates over and above the condition $\sqrt{-g} = 1$ for unimodular coordinates (ibid., 808–810, Sect. 17–18).

This came as no surprise to Einstein. He expected there to be a close connection between covariance of the field equations and energy-momentum conservation. In [30] he had shown that the four conditions $B_\mu = 0$ (eq. (14)) that together with the "Entwurf" field equations guarantee energy-momentum conservation double as the conditions restricting the range of coordinate systems in which these field equations hold. The dual role of such conditions played a central role in the breakthrough of November 1915.[85] What made the field equations (21) replacing the "Entwurf" field equations in [31] so attractive was that the range of

---

[85] This is one of the central claims argued for in [66, see, in particular, Sects. 6–7].

coordinate systems in which they hold was restricted only by one condition, viz. that the determinant of the metric transform as a scalar, as opposed to the four conditions restricting the range of coordinate systems in which the "Entwurf" equations hold. Given the dual role of these conditions in the "Entwurf" theory, this suggested to Einstein that it would suffice to add this one condition on g to the new field equations to guarantee energy-momentum conservation in the new theory. In [31] he showed that the analogues of the four conditions $B_\mu = 0$ in the new theory can indeed be replaced by one condition on g. He presumably expected this condition to be that g transform as a scalar, since this expresses the restriction to unimodular transformations. Instead he found that g cannot be a constant, which though compatible with unimodular transformations is a stronger condition and rules out unimodular *coordinates* (with $g = -1$).

The transition from the field equations of the first November paper [31] to those of the second and the fourth [32, 34] was driven by the desire to change the field equations in such a way that the condition that g cannot be a constant can be replaced by the more congenial condition $\sqrt{-g} = 1$ for unimodular coordinates. These amended field equations could then be looked upon as generally-covariant field equations expressed in special coordinates. In [32] this goal is reached at the expense of the assumption, largely discredited at that point, that all matter (represented by $T_{\mu\nu}$) somehow consists of electromagnetic fields (in which case the trace T vanishes). In [34] generally-covariant field equations in unimodular coordinates are obtained without specifying $T_{\mu\nu}$ but by adding a term with the trace T to the right-hand side of eq. (20).

The "Zurich Notebook" shows that Einstein had already considered adding such a term three years earlier to render field equations based on the Ricci tensor compatible with energy-momentum conservation in the weak-field limit.[86] What had stopped him from doing so was that the resulting weak-field equations rule out a spatially flat metric of the form $g_{\mu\nu} = (-1, -1, -1, c^2(x, y, z))$. For such a metric the ten components of the gravitational potential reduce to one component, the variable speed of light $c(x, y, z)$ of the theory for static fields of [26, 27]. Einstein firmly believed that this was how weak static fields had to be represented in his theory.[87]

This changed only when he calculated the perihelion advance of Mercury on the basis of the field equations of [32]. He realized that if $\sqrt{-g} = 1$ and g_{44} is variable, the components g_{ij} cannot all be constants [82, p. 147]; [21, pp. 144–145]. This removed his old objection to adding a term with the trace T to the field equations. At that point Einstein realized that such a trace term was needed anyway to make sure that all energy-momentum enter the field equations in exactly the same way. This told him that he had finally got it right. He had found the Einstein field equations in unimodular coordinates (see eq. (25)).

In his 1916 review article Einstein still derived the field equations in unimodular coordinates only. The manuscript for an unpublished appendix to the article [13, Doc. 31] shows that he at least started an alternative discussion of the field equations and energy-momentum conservation in arbitrary coordinates. The numbering of the sections in this document suggests that at one point he considered substituting this discussion for the one in unimodular coordinates. He then considered adding it as an appendix. In the end he did neither. Instead he published the generally-covariant treatment separately a few months later [37].

In this paper he derived the generally-covariant field equations from an action principle with the Riemann curvature scalar as the Lagrangian. Terms with second-order derivatives of the metric in this quantity do not contribute to the action integral, so the effective Lagrangian becomes:

$$\sqrt{-g}g^{\mu\nu}\left[\left\{\begin{matrix}\beta\\ \mu\,\alpha\end{matrix}\right\}\left\{\begin{matrix}\alpha\\ \nu\beta\end{matrix}\right\} - \left\{\begin{matrix}\alpha\\ \mu\nu\end{matrix}\right\}\left\{\begin{matrix}\alpha\\ \alpha\,\beta\end{matrix}\right\}\right]. \qquad (29)$$

For $\sqrt{-g} = 1$ the second term vanishes (cf. eq. (18)) and the expression reduces to the Lagrangian (15) used in the November 1915 papers and in the 1916 review article. [37] fills two important gaps in [35]. First, Einstein derived the generally-covariant version of the identities (28), which in conjunction with the

---

[86] See the "Zurich Notebook", [p. 20L] [11, Doc. 10, p. 39]. For analysis of this page, see [67, Sect. 5.4.3].
[87] Einstein checked and confirmed this assumption on at least two occasions, in the "Zurich Notebook", [p. 21R] [11, Doc. 10, p. 42] (for analysis see [67, Sect. 5.4.4 and 5.4.6]), and in Einstein to Erwin Freundlich, 19 March 1915 [15, Doc. 63]

field equations imply energy-momentum conservation. These generally-covariant identities are the now famous contracted Bianchi identities. Second, Einstein showed – proceeding exactly the way he did in the premature review article [30] – that the identities guaranteeing energy-momentum conservation are a direct consequence of the covariance of the action functional. Einstein had thus, in a mathematically impeccable way, found a special case of one of Noether's theorems published two years later.

From a purely mathematical point of view, the discussion of the field equations and energy-momentum conservation in [37] is far more elegant than in [35]. This more elegant treatment, however, obscures the way in which Einstein found the Einstein field equations. It makes it look as if it was a matter of picking the most obvious candidate for the Lagrangian, the Riemann curvature scalar, at which point everything else fell into place. Ironically, this is exactly what Einstein in his later years came to believe himself, in part no doubt because it made his successful search for the field equations of general relativity look so similar to his fruitless search for a unified field theory. The clumsier discussion in unimodular coordinates in [35], however, may serve as a reminder that – whatever he believed, said, or wrote about it later on – Einstein only discovered the mathematical high road to the Einstein field equations after he had already found these equations at the end of a bumpy road through physics. Serving as road signs were Newton's gravitational theory, Maxwell's electrodynamics, and such key results of special relativity as the law of energy-momentum conservation. Considerations of mathematical elegance played only a subsidiary role.

References

[1] S. Abiko, Hist. Stud. Phys. Biol. Sci. **31**, 1–35 (2000).
[2] H. G. Alexander (ed.), The Leibniz-Clarke Correspondence (Manchester University Press, Manchester and New York, 1956).
[3] J. B. Barbour and H. Pfister (eds.), Mach's Principle: from Newton's Bucket to Quantum Gravity (Einstein Studies, Vol. 6) (Birkhäuser, Boston, 1995).
[4] K. Brading, Stud. Hist. Phil. Mod. Phys. **33**, 3–22 (2002).
[5] J. Butterfield, Int. Stud. Phil. Sci. **2**, 10–32 (1987).
[6] F. Cajori, Sir Isaac Newton's Mathematical Principles of Natural Philosophy and his System of the World (University of California Press, Berkeley, Los Angeles, London, 1934).
[7] I. B. Cohen and A. Whitman, Isaac Newton. The Principia. Mathematical Principles of Natural Philosophy. A New Translation (University of California Press, Berkeley, Los Angeles, London, 1999).
[8] L. Corry, J. Renn, and J. Stachel, Science **278**, 1270–1273 (1997).
[9] CPAE 2, edited by J. Stachel, D. C. Cassidy, R. Schulmann, and J. Renn, The Collected Papers of Albert Einstein. Vol. 2. The Swiss Years: Writings, 1900–1909 (Princeton University Press, Princeton, 1989).
[10] CPAE 3, edited by M. J. Klein, A. J. Kox, J. Renn, and R. Schulmann, The Collected Papers of Albert Einstein. Vol. 3. The Swiss Years: Writings, 1909–1911 (Princeton University Press, Princeton, 1993).
[11] CPAE 4, edited by M. J. Klein, A. J. Kox, J. Renn, and R. Schulmann, The Collected Papers of Albert Einstein. Vol. 4, The Swiss Years: Writings, 1912–1914 (Princeton University Press, Princeton, 1995).
[12] CPAE 5, edited by M. J. Klein, A. J. Kox, and R. Schulmann, The Collected Papers of Albert Einstein. Vol. 5, The Swiss Years: Correspondence, 1902–1914 (Princeton University Press, Princeton, 1993).
[13] CPAE 6, edited by A. J. Kox, M. J. Klein, and R. Schulmann, The Collected Papers of Albert Einstein. Vol. 6, The Berlin Years: Writings, 1914–1917. (Princeton University Press, Princeton, 1996).
[14] CPAE 7, edited by M. Janssen, R. Schulmann, J. Illy, C. Lehner, and D. Kormos Buchwald, The Collected Papers of Albert Einstein. Vol. 7, The Berlin Years: Writings, 1918–1921 (Princeton University Press, Princeton, 2002).
[15] CPAE 8, edited by R. Schulmann, A. J. Kox, M. Janssen, and J. Illy, The Collected Papers of Albert Einstein. Vol. 8, The Berlin Years: Correspondence, 1914–1918 (Princeton University Press, Princeton, 1998).
[16] CPAE 9, edited by D. Kormos Buchwald, R. Schulmann, J. Illy, D. J. Kennefick, and T. Sauer, The Collected Papers of Albert Einstein. Vol. 9, The Berlin Years: Correspondence, January 1919–April 1920 (Princeton University Press, Princeton, 2004).
[17] W. De Sitter, K. Ned. Akad. Wet., Wis- en Natuurkundige Afdeeling. Verslagen van de Gewone Vergaderingen **25**, 499–504 (1916–17); K. Ned. Akad. Wet. Sect. Sci. Proc. **19**, 527–532 (1916–17).

[18] W. De Sitter, K. Ned. Akad. Wet., Wis- en Natuurkundige Afdeeling. Verslagen van de Gewone Vergaderingen **25**, 1268–1276 (1916–17); K. Ned. Akad. Wet. Sect. Sci. Proc. **19**, 1217–1225 (1916–17).

[19] J. Dorling, Br. J. Phil. Sci. **29**, 311–323 (1978).

[20] J. Earman, World Enough and Space-Time, Absolute versus Relational Theories of Space and Time (The MIT Press, Cambridge, MA, 1989).

[21] J. Earman and M. Janssen Einstein's Explanation of the Motion of Mercury's Perihelion, in: [22], pp. 129–172.

[22] J. Earman, M. Janssen, and J. Norton (eds.), The Attraction of Gravitation: New Studies in the History of General Relativity (Einstein Studies, Vol. 5) (Birkhäuser, Boston, 1993).

[23] J. Earman and J. Norton, Br. J. Phil. Sci. **38**, 515–525 (1987).

[24] A. Einstein, Jahrb. Radioaktivität Elektronik **4**, 411–462 (1907). [9, Doc. 47].

[25] A. Einstein, Ann. Phys. (Leipzig) **35**, 898–908 (1911), reprinted in this volume (pp. 425–435); [10, Doc. 23]; reprinted in translation in [73], pp. 99–108.

[26] A. Einstein, Ann. Phys. (Leipzig) **38**, 355–369 (1912), reprinted in this volume (pp. 445–459); [11, Doc. 3].

[27] A. Einstein, Ann. Phys. (Leipzig) **38**, 443–458 (1912), reprinted in this volume (pp. 460–475); [11, Doc. 4].

[28] A. Einstein, Phys. Z. **14**, 1249–1262 (1913); [11, Doc. 17].

[29] A. Einstein, Phys. Z. **15**, 176–180 (1914); [11, Doc. 25].

[30] A. Einstein, Sitz.ber., Kgl. Preuss. Akad. Wiss. (Berlin), 1914, pp. 1030–1085; [13, Doc. 9].

[31] A. Einstein, Sitz.ber., Kgl. Preuss. Akad. Wiss. (Berlin), 1915, pp. 778–786; [13, Doc. 21].

[32] A. Einstein, Sitz.ber., Kgl. Preuss. Akad. Wiss. (Berlin), 1915, pp. 799–801; [13, Doc. 22].

[33] A. Einstein, Sitz.ber., Kgl. Preuss. Akad. Wiss. (Berlin), 1915, pp. 831–839; [13, Doc. 23].

[34] A. Einstein, Sitz.ber., Kgl. Preuss. Akad. Wiss. (Berlin), 1915, pp. 844–847; [13, Doc. 25].

[35] A. Einstein, Ann. Phys. (Leipzig) **49**, 769–822 (1916), reprinted in this volume (pp. 518–571); [13, Doc. 30]; reprinted in translation in [73], pp. 111–164.

[36] A. Einstein, Sitz.ber., Kgl. Preuss. Akad. Wiss. (Berlin), 1916, pp. 688–696; [13, Doc. 32].

[37] A. Einstein, Sitz.ber., Kgl. Preuss. Akad. Wiss. (Berlin), 1916, pp. 1111–1116; [13, Doc. 41]; translation in [73], pp. 167–173.

[38] A. Einstein, Über die spezielle und die allgemeine Relativitätstheorie. (Gemeinverständlich) (Vieweg, Braunschweig, 1917); reprinted in translation as [48].

[39] A. Einstein, Sitz.ber., Kgl. Preuss. Akad. Wiss. (Berlin), 1917, pp. 142–152; [13, Doc. 43], reprinted in translation in [73], pp. 177–188.

[40] A. Einstein, Sitz.ber., Kgl. Preuss. Akad. Wiss. (Berlin), 1918, pp. 154–167; [14, Doc. 1].

[41] A. Einstein, Ann. Phys. (Leipzig) **55**, 241–244 (1918), reprinted in this volume (pp. 578–581); [14, Doc. 4].

[42] A. Einstein, Sitz.ber., Kgl. Preuss. Akad. Wiss. (Berlin), 1918, pp. 270–272; [14, Doc. 5].

[43] A. Einstein, Sitz.ber., Kgl. Preuss. Akad. Wiss. (Berlin), 1918, pp. 448–459; [14, Doc. 9].

[44] A. Einstein, Naturwissenschaften **6**, 697–702 (1918); [14, Doc. 13].

[45] A. Einstein, Äther und Relativitätstheorie (Springer, Berlin, 1920); reprinted in translation in [49], pp. 1–24.

[46] A. Einstein, Vier Vorlesungen über Relativitätstheorie (Vieweg, Braunschweig, 1922) [14, Doc. 71]; reprinted in translation as [47].

[47] A. Einstein, The Meaning of Relativity. 5th ed. (Princeton University Press, Princeton, 1956).

[48] A. Einstein, Relativity. The Special and the General Theory. A Clear Explanation that Anyone Can Understand (Crown Publishers, New York, 1959).

[49] A. Einstein, Sidelights on Relativity (Dover, New York, 1983).

[50] A. Einstein and M. Grossmann, Entwurf einer verallgemeinerten Relativitätstheorie und einer Theorie der Gravitation (Teubner, Leipzig, 1913); [11, Doc. 13].

[51] A. Einstein and M. Grossmann, Z. Math. Phys. **62**, 225–259 (1914). Reprint of [50] with additional "Comments" ("Bemerkungen" [11, Doc. 26]).

[52] A. Einstein and M. Grossmann, Z. Math. Phys. **63**, 215–225 (1914); [13, Doc. 2].

[53] J. Eisenstaedt and A.J. Kox (eds.), Studies in the History of General Relativity (Einstein Studies, Vol. 3) (Birkhäuser, Boston, 1992).

[54] A. Fölsing, Albert Einstein. Eine Biographie (Suhrkamp, Frankfurt a.M., 1993); Translation: Albert Einstein, A Biography (Viking, New York, 1997).

[55] H. Goenner, J. Renn, J. Ritter, and T. Sauer (eds.), The Expanding Worlds of General Relativity (Einstein Studies, Vol. 7) (Boston, Birkhäuser, 1999).

[56] J. Gray (ed.), The Symbolic Universe: Geometry and Physics 1890–1930 (Oxford University Press, Oxford, 1999).
[57] D. Hilbert, Nachr. Kgl. Akad. Wiss. Goett., Math.-Phys. Kl. (Göttingen), 1915, pp. 395–407.
[58] D. Howard, Point Coincidences and Pointer Coincidences: Einstein on the Invariant Content of Space-Time Theories, in: [55], pp. 463–500.
[59] D. Howard, and J. Norton, Out of the Labyrinth? Einstein, Hertz, and the Göttingen Response to the Hole Argument, in: [22], pp. 30–62.
[60] D. Howard, and J. Stachel (eds.) Einstein and the History of General Relativity (Einstein Studies, Vol. 1) (Birkhäuser, Boston, 1989).
[61] N. Huggett (ed.), Space from Zeno to Einstein. Classic Readings with a Contemporary Commentary (The MIT Press, Cambridge, MA, 2000).
[62] M. Janssen, Rotation as the Nemesis of Einstein's *Entwurf* Theory, in: [55], pp. 127–157.
[63] M. Janssen, Perspect. Sci. **10**, 457–522 (2002).
[64] M. Janssen, Relativity, in: New Dictionary of the History of Ideas, edited by M. C. Horowitz et al. (Charles Scribner's Sons/Thomson Gale, New York, 2004).
[65] M. Janssen, What Did Einstein Know and When Did He Know It? A Besso Memo Dated August 1913, in: [89].
[66] M. Janssen and J. Renn, Untying the Knot: How Einstein Found His Way Back to Field Equations Discarded in the Zurich Notebook, in: [89].
[67] M. Janssen, J. Renn, T. Sauer, J. Norton, and J. Stachel, A Commentary on Einstein's Zurich Notebook, in: [89].
[68] A. J. Kox, Arch. Hist. Exact Sci. **38**, 67–78 (1988); reprinted in: [60], pp. 201–212.
[69] E. Kretschmann, Ann. Phys. (Leipzig) **48**, 907–942 (1915).
[70] E. Kretschmann, Ann. Phys. (Leipzig) **53**, 575–614 (1917).
[71] R. Laymon, J. Hist. Phil. **16**, 399–413 (1978).
[72] C. Lehner, Einstein and the Principle of General Relativity, 1916–1921, in: The Universe of General Relativity (Einstein Studies, vol. 11), edited by A. J. Kox and Jean Eisenstaedt (Birkhäuser, Boston, to appear).
[73] H. Lorentz, A. Einstein, H. Minkowski, and H. Weyl, The Principle of Relativity (Dover, New York, 1952).
[74] E. Mach, The Science of Mechanics: A Critical and Historical Account of Its Development (Open Court, Lasalle, Illinois, 1960).
[75] E. Mach, Die Mechanik in ihrer Entwicklung historisch-kritisch dargestellt (Wissenschaftliche Buchgesellschaft, Darmstadt, 1988).
[76] T. Maudlin, Stud. Hist. Phil. Mod. Phys. **21**, 531–561 (1990).
[77] G. Mie, Das Prinzip von der Relativität des Gravitationspotentials, in: Arbeiten aus den Gebieten der Physik, Mathematik, Chemie. Festschrift, Julius Elster und Hans Geitel zum sechzigsten Geburtstag gewidmet von Freunden und Schülern (Vieweg, Braunschweig, 1915), pp. 251–268.
[78] G. Mie, Phys. Z. **18**, 551–556, 574–580, 596–602 (1917).
[79] H. Minkowski, Phys. Z. **20**, 104–11 (1909); English translation in [73], pp. 75–91.
[80] C. W. Misner, K. S. Thorne, and J. A. Wheeler, Gravitation (San Francisco, Freeman, 1973).
[81] E. Noether, Nachr. Kgl. Akad. Wiss. Goett., Math.-Phys. Kl. (Göttingen), 1918, pp. 235–257.
[82] J. Norton, Hist. Stud. Phys. Biol. Sci. **14**, 253–316 (1984); reprinted in: [60], pp. 101–159.
[83] J. Norton, Einstein, the Hole Argument and the Reality of Space, in: Measurement, Realism, and Objectivity, edited by J. Forge (Reidel, Dordrecht, 1987), pp. 153–188.
[84] J. Norton, The Physical Content of General Covariance, in [53], pp. 281–315.
[85] J. Norton, Rep. Prog. Phys. **56**, 791–858 (1993).
[86] J. Norton, Geometries in Collision: Einstein, Klein, and Riemann, in: [56], pp. 128–144.
[87] J. Norton, Stud. Hist. Philos. Mod. Phys. **31**, 135–170 (2000).
[88] A. Pais, 'Subtle is the Lord ...' The Science and the Life of Albert Einstein (Oxford, Oxford University Press, 1982).
[89] J. Renn (ed.), The Genesis of General Relativity, 4 Vols. (Kluwer, Dordrecht, to appear).
[90] J. Renn and T. Sauer, Pathways out of Classical Physics: Einstein's Double Strategy in Searching for the Gravitational Field Equation, in: [89].
[91] J. Renn, and J. Stachel, Hilbert's Foundation of Physics: From a Theory of Everything to a Constituent of General Relativity, Preprint 118, Max Planck Institute for the History of Science, Berlin. To appear in [89].

[92] D. Rowe, The Göttingen Response to General Relativity and Emmy Noether's Theorems, in: [56], pp. 189–234.
[93] R. Rynasiewicz, Kretschmann's Analysis of Covariance and Relativity Principles. in: [55], pp. 431–462.
[94] T. Sauer, Arch. Hist. Exact Sci. **53**, 529–575 (1999).
[95] A. Sommerfeld, Vorlesungen über theoretische Physik. Vol. 3, Elektrodynamik (Akademische Verlagsgesellschaft Geest & Portig K.-G., Leipzig, 1949).
[96] J. Stachel, The Rigidly Rotating Disk as the "Missing Link in the History of General Relativity." In: [60], pp. 48–62.
[97] J. Stachel, Einstein's Search for General Covariance, 1912–1915, in: [60], pp. 62–100.
[98] J. Stachel, The Meaning of General Covariance: the Hole Story, in: Philosophical Problems of the Internal and External World: Essays on the Philosophy of Adolf Grünbaum, edited by J. Earman, A. I. Janis, G. J. Massey, and N. Rescher (Universitätsverlag, Konstanz, 1993)/(University of Pittsburgh Press, Pittsburgh, 1993), pp. 129–160.
[99] J. Stachel, The Relations between Things versus The Things between Relations: The Deeper Meaning of the Hole Argument, in: Reading Natural Philosophy. Essays in the History and Philosophy of Science and Mathematics, edited by D. B. Malament (Open Court, Chicago and La Salle, 2002), pp. 231–266.
[100] J. Stachel, The Story of Newstein. Or: Is Gravity Just Another Pretty Force?, in: [89].
[101] H. Thirring, Phys. Z. **19**, 33–39 (1918).
[102] J. Van Dongen, Einstein's Unification: General Relativity and the Quest for Mathematical Naturalness, Ph.D. Thesis, University of Amsterdam (2002).

ANNALEN
DER
PHYSIK.

BEGRÜNDET UND FORTGEFÜHRT DURCH

F. A. C. GREN, L. W. GILBERT, J. C. POGGENDORFF, G. UND E. WIEDEMANN.

VIERTE FOLGE.

BAND 4.

DER GANZEN REIHE 309. BAND.

KURATORIUM:

F. KOHLRAUSCH, M. PLANCK, G. QUINCKE,
W. C. RÖNTGEN, E. WARBURG.

UNTER MITWIRKUNG

DER DEUTSCHEN PHYSIKALISCHEN GESELLSCHAFT

UND INSBESONDERE VON

M. PLANCK

HERAUSGEGEBEN VON

PAUL DRUDE.

MIT NEUN FIGURENTAFELN.

LEIPZIG, 1901.

VERLAG VON JOHANN AMBROSIUS BARTH.

5. *Folgerungen aus den Capillaritätserscheinungen;* von *Albert Einstein.*

Bezeichnen wir mit γ diejenige Menge mechanischer Arbeit, welche wir der Flüssigkeit zuführen müssen, um die freie Oberfläche um die Einheit zu vergrössern, so ist γ nicht etwa die gesamte Energiezunahme des Systems, wie folgender Kreisprocess lehrt. Sei eine bestimmte Flüssigkeitsmenge vorliegend von der (absoluten) Temperatur T_1 und der Oberfläche O_1. Wir vermehren nun isothermisch die Oberfläche O_1 auf O_2, erhöhen die Temperatur auf T_2 (bei constanter Oberfläche), vermindern dann die Oberfläche auf O_1 und kühlen dann die Flüssigkeit wieder auf T_1 ab. Nimmt man nun an, dass dem Körper ausser der ihm vermöge seiner specifischen Wärme zukommenden keine andere Wärmemenge zugeführt wird, so ist bei dem Kreisprocess die Summe der dem Körper zugeführten Wärme gleich der Summe der ihm entnommenen. Es muss also nach dem Princip von der Erhaltung der Energie auch die Summe der zugeführten mechanischen Arbeiten gleich Null sein.

Es gilt also die Gleichung:

$$(O_2 - O_1)\gamma_1 - (O_2 - O_1)\gamma_1 = 0 \quad \text{oder} \quad \gamma_1 = \gamma_2.$$

Dies widerspricht aber der Erfahrung.

Es bleibt also nichts anderes übrig als anzunehmen, dass mit der Aenderung der Oberfläche auch ein Austausch der Wärme verbunden sei, und dass der Oberfläche eine eigene specifische Wärme zukomme. Bezeichnen wir also mit U die Energie, mit S die Entropie der Oberflächeneinheit der Flüssigkeit, mit s die specifische Wärme der Oberfläche, mit w_0 die zur Bildung der Oberflächeneinheit erforderliche Wärme in mechanischem Maass, so sind die Grössen:

$$dU = s.O.dT + \{\gamma + w_0\}dO$$

und

$$dS = \frac{s.O.dT}{T} + \frac{w_0}{T}dO$$

vollständige Differentiale. Es gelten also die Gleichungen:

$$\frac{\partial (s \cdot O)}{\partial O} = \frac{\partial (\gamma + w_0)}{\partial T},$$

$$\frac{\partial}{\partial O}\left(\frac{sO}{T}\right) = \frac{\partial}{\partial T}\left(\frac{w_0}{T}\right).$$

Aus diesen Gleichungen folgt:

$$\gamma + w_0 = \gamma - T\frac{\partial \gamma}{\partial T}.$$

Dies aber ist die gesamte Energie, welche zur Bildung der Einheit der Oberfläche nötig ist.

Bilden wir noch:

$$\frac{d}{dT}(\gamma + w_0) = -T\frac{d^2 \gamma}{dT^2}.$$

Die Experimentaluntersuchungen haben nun ergeben, dass sich stets sehr nahe γ als lineare Function der Temperatur darstellen lässt, d. h.:

Die zur Bildung der Oberflächeneinheit einer Flüssigkeit nötige Energie ist unabhängig von der Temperatur.

Ebenso folgt:

$$s = \frac{d\gamma}{dT} + \frac{dw_0}{dT} = \frac{d\gamma}{dT} - \frac{d\gamma}{dT} - T\frac{d^2\gamma}{dT^2} = 0,$$

also: Der Oberfläche als solcher ist kein Wärmeinhalt zuzuschreiben, sondern die Energie der Oberfläche ist potentieller Natur. Man sieht schon jetzt, dass

$$\gamma - T\frac{d\gamma}{dT}$$

eine zu stöchiometrischen Untersuchungen sich geeignetere Grösse ist, als das bisher benutzte γ bei Siedetemperatur. Die Thatsache, dass die zur Bildung der Oberflächeneinheit erforderliche Energie kaum mit der Temperatur variirt, lehrt uns aber auch, dass die Configuration der Molecüle in der Oberflächenschicht mit der Temperatur nicht variiren wird (abgesehen von Aenderungen von der Grössenordnung der thermischen Ausdehnung).

Um nun für die Grösse

$$\gamma - T\frac{d\gamma}{dT}$$

eine stöchiometrische Beziehung aufzufinden, ging ich von den einfachsten Annahmen über die Natur der molecularen Anziehungskräfte aus, und prüfte deren Consequenzen auf ihre Uebereinstimmung mit dem Experiment hin. Ich liess mich dabei von der Analogie der Gravitationskräfte leiten:

Sei also das relative Potential zweier Molecüle von der Form:
$$P = P_\infty - c_1 \cdot c_2 \cdot \varphi(r),$$

wobei c eine für das betreffende Molecül charakteristische Constante ist, $\varphi(r)$ aber eine vom Wesen der Molecüle unabhängige Function ihrer Entfernung. Wir nehmen ferner an, dass
$$\tfrac{1}{2} \sum_{\alpha=1}^{n} \sum_{\beta=1}^{n} c_\alpha c_\beta \varphi(r_{\alpha,\beta})$$

der entsprechende Ausdruck für n Molecüle sei. Sind speciell alle Molecüle gleich beschaffen, so geht dieser Ausdruck in
$$\tfrac{1}{2} c^2 \sum_{\alpha=1}^{n} \sum_{\beta=1}^{n} \varphi(r_{\alpha,\beta})$$

über. Wir machen ferner noch die Annahme, dass das Potential der Molecularkräfte ebenso gross sei, wie wenn die Materie homogen im Raume verteilt wäre; es ist dies allerdings eine Annahme, von der wir nur angenähert die Richtigkeit erwarten dürfen. Mit ihrer Hülfe verwandelt sich der obige Ausdruck in:
$$P = P_\infty - \tfrac{1}{2} c^2 N^2 \iint d\tau \cdot d\tau' \, \varphi(r_{d\tau, d\tau'}),$$

wobei N die Anzahl der Molecüle in der Volumeneinheit ist. Ist das Molecül unserer Flüssigkeit aus mehreren Atomen zusammengesetzt, so soll analog wie bei den Gravitationskräften $c = \sum c_\alpha$ gesetzt werden können, wobei die c_α den Atomen der Elemente charakteristische Zahlen bedeuten. Setzt man noch $1/N = v$, wobei v das Molecularvolum bedeutet, so erhält man die endgültige Formel:
$$P = P_\infty - \tfrac{1}{2} \frac{(\sum c_\alpha)^2}{v^2} \iint d\tau \cdot d\tau' \, \varphi(r_{d\tau, d\tau'}).$$

Setzen wir nun noch voraus, dass die Dichte der Flüssigkeit bis zu deren Oberfläche constant ist, was ja durch die Thatsache wahrscheinlich gemacht wird, dass die Energie der Oberfläche von der Temperatur unabhängig ist, so sind wir nun im stande die potentielle Energie der Volumeneinheit im

Inneren der Flüssigkeit und die der Oberflächeneinheit zu berechnen.

Setzen wir nämlich

$$\tfrac{1}{2}\int_{x=-\infty}^{+\infty}\int_{y=-\infty}^{+\infty}\int_{z=-\infty}^{+\infty} dx\,dy\,dz \cdot \varphi\,(\sqrt{x^2+y^2+z^2}) = K,$$

so ist die potentielle Energie der Volumeneinheit

$$P = P_\infty - K\frac{(\sum c_a)^2}{v^2}.$$

Denken wir uns eine Flüssigkeit vom Volumen V und von der Oberfläche S, so erhalten wir durch Integration

$$P = P_\infty - K\frac{(\sum c_a)^2}{v^2} \cdot V - K'\frac{(\sum c_a)^2}{v^2} \cdot O,$$

wobei die Constante K' bedeutet:

$$\int_{x'=0}^{x'=1}\int_{y'=0}^{y'=1}\int_{z'=\infty}^{z'=0}\int_{x=-\infty}^{x=\infty}\int_{y=-\infty}^{y=\infty}\int_{z=0}^{z=\infty} dx.dy.dz.dx'.dy'.dz$$
$$\varphi\sqrt{(x-x')^2+(y-y')^2+(z-z')^2}.$$

Da über φ nichts bekannt ist, bekommen wir natürlich keine Beziehung zwischen K und K'.

Dabei ist zunächst im Auge zu behalten, dass wir nicht wissen können, ob das Flüssigkeitsmolecül nicht die n-fache Masse des Gasmolecüles besitzt, doch folgt aus unserer Herleitung, dass dadurch unser Ausdruck der potentiellen Energie der Flüssigkeit nicht geändert wird. Für die potentielle Energie der Oberfläche bekommen wir, auf Grund der eben gemachten Annahme, den Ausdruck:

$$P = K'\frac{(\sum c_a)^2}{v^2} = \gamma - T\frac{d\gamma}{dT},$$

oder

$$\sum c_a = v \cdot \sqrt{\gamma - T\frac{d\gamma}{dT}} \cdot \frac{1}{\sqrt{K'}}.$$

Da die rechts stehende Grösse für Siedetemperatur für viele Stoffe aus den Beobachtungen von R. Schiff berechenbar ist, so bekommen wir reichlichen Stoff zur Bestimmung der Grössen c_a. Ich entnahm das gesamte Material dem Buch über Allgemeine Chemie von W. Ostwald. Ich gebe hier zunächst das Material an, mittels dessen ich das c_α für C, H, O nach der Methode der kleinsten Quadrate berechnete. In der mit $\sum c_{\alpha_\text{ber.}}$ überschriebenen Columne sind die $\sum c_\alpha$ angegeben, wie sie mit Hülfe der so gewonnenen c_α aus den chemischen Formeln sich ergeben. Isomere Verbindungen wurden zu einem Wert vereinigt, weil die ihnen zugehörigen Werte der linken Seite nur unbedeutend voneinander abwichen. Die Einheit wurde willkürlich gewählt, weil, da K' unbekannt ist, eine absolute Bestimmung der c_α nicht möglich ist.

Ich fand:

$$c_\text{H} = -1{,}6, \quad c_\text{C} = 55{,}0, \quad c_\text{O} = 46{,}8.$$

Formel	$\sum c_\alpha$	$\sum c_{\alpha_\text{ber.}}$	Name der Verbindung
$C_{10}H_{16}$	510	524	Citronenterpen
CO_2H_2	140	145	Ameisensäure
$C_2H_4O_2$	193	197	Essigsäure
$C_3H_6O_2$	250	249	Propionsäure
$C_4H_8O_2$	309	301	Buttersäure und Isobuttersäure
$C_5H_{10}O_2$	365	352	Valeriansäure
$C_4H_6O_3$	350	350	Acetanhydrid
$C_6H_{10}O_4$	505	501	Aethyloxalat
$C_8H_8O_2$	494	520	Methylbenzoat
$C_9H_{10}O_2$	553	562	Aethylbenzoat
$C_6H_{10}O_3$	471	454	Acetessigäther
C_7H_8O	422	419	Anisol
$C_8H_{10}O$	479	470	Phenetol und Methylcresolat
$C_8H_{10}O_2$	519	517	Dimethylresorcin
$C_5H_4O_2$	345	362	Furfurol
$C_5H_{10}O$	348	305	Valeraldehyd
$C_{10}H_{14}O$	587	574	Carvol

Man sieht, dass die Abweichungen in fast allen Fällen die Versuchsfehler wohl kaum übersteigen und keinerlei Gesetzmässigkeit zeigen.

Hierauf berechnete ich gesondert die Werte für Cl, Br und J, welchen Bestimmungen natürlich eine geringere Sicherheit zukommt, und fand:

$$c_\text{Cl} = 60, \quad c_\text{Br} = 152, \quad c_\text{J} = 198.$$

Ich lasse nun in gleicher Weise wie oben das Material folgen:

Formel	$\sum c_a$	$\sum c_{a_\text{ber.}}$	Name der Verbindung
C_6H_5Cl	385	379	Chlorbenzol
C_7H_7Cl	438	434	Chlortoluol
C_7H_7Cl	450	434	Benzychlorid
C_3H_5OCl	270	270	Epichlorhydrin
C_2OHCl_3	358	335	Chloral
C_7H_5OCl	462	484	Benzoylchlorid
$C_7H_6Cl_2$	492	495	Benzylidenchlorid
Br_2	217	304	Brom
C_2H_5Br	251	254	Aethylbromid
C_3H_7Br	311	306	Propylbromid
C_3H_7Br	311	306	Isopropylbromid
C_3H_5Br	302	309	Allylbromid
C_4H_5Br	353	354	Isobutylbromid
$C_5H_{11}Br$	425	410	Isoamylbromid
C_6H_5Br	411	474	Brombenzol
C_7H_7Br	421	526	o-Bromtoluol
$C_2H_4Br_2$	345	409	Aethylenbromid
$C_3H_6Br_2$	395	461	Propylenbromid
C_2H_5J	288	300	Aethyljodid
C_3H_7J	343	352	Propyljodid
C_3H_7J	357	352	Isopropyljodid
C_3H_5J	338	355	Allyljodid
C_4H_9J	428	403	Isobutyljodid
$C_5H_{11}J$	464	455	Isoamyljodid

Es scheint mir, dass grössere Abweichungen von unserer Theorie bei solchen Stoffen eintreten, welche verhältnismässig grosse Molecularmaasse und kleines Molecularvolum haben.

Wir haben aus unseren Annahmen gefunden, dass die potentielle Energie der Volumeneinheit den Ausdruck besitzt:

$$P = P_\infty - K \frac{(\sum c_a)^2}{v^2},$$

dabei bedeutet K eine bestimmte Grösse, welche wir aber nicht berechnen können, da es überhaupt erst durch die Wahl der c_a vollkommen definirt wird. Wir können daher $K = 1$ setzen und gewinnen so eine Definition für die absoluten Werte der c_a. Berücksichtigen wir dies von nun an, so erhalten wir für die Grösse des Potentiales, welche dem Aequivalent (Molecül) zukommt, den Ausdruck:

$$P = P_\infty - K \frac{(\sum c_a)^2}{v},$$

wobei natürlich P_∞ eine andere Constante bedeutet. Nun könnten wir aber das zweite Glied der rechten Seite dieser Gleichung der Differenz $D_m J - A v_d$ gleich setzen — wobei D_m die moleculare Verdampfungswärme (Dampfwärme × Molecularmasse), J das mechanische Aequivalent der Calorie, A den Atmosphärendruck in absolutem Maass und v_d das Molecularvolum des Dampfes ist —, wenn die potentielle Energie des Dampfes Null wäre und wenn für Siedetemperatur der Inhalt an kinetischer Energie beim Uebergang vom flüssigen in den Gaszustand ungeändert bliebe. Die erste dieser Annahmen scheint mir unbedenklich. Da wir aber zu der letzteren Annahme keinen Grund haben, aber auch keine Möglichkeit die fragliche Grösse abzuschätzen, so bleibt uns nichts anderes übrig, als die obige Grösse selbst zur Rechnung zu benutzen.

In die erste Spalte der folgenden Tabelle habe ich die Grössen $\sqrt{D_m' \cdot v}$ im Wärmemaass eingetragen, wobei D_m' die um die äussere Verdampfungsarbeit (in Wärmemaass) verminderte Verdampfungswärme bedeutet. In die zweite setzte ich die Grössen $\sum c_a$, wie sie aus den Capillaritätsversuchen ermittelt sind; in der dritten finden sich die Quotienten beider Werte. Isomere Verbindungen sind wieder zu einer Zeile vereinigt.

Name der Verbindung	Formel	$\sqrt{D_m' \cdot v}$	$\sum c_{a\text{ber.}}$	Quotient
Isobutylpropionat	$C_7H_{14}O_2$	1157	456	2,54
Isoamilacetat	,,			
Propylacetat	,,			
Isobutylisobutyrat	$C_8H_{16}O_2$	1257	510	2,47
Propylvalerat	,,			
Isobutylbutyrat	,,			
Isoamylpropionat	,,			
Isoamylisobutyrat	$C_9H_{18}O_2$	1367	559	2,45
Isobutylvalerat	,,			
Isoamylvalerat	$C_{10}H_{10}O_2$	1464	611	2,51
Benzol	C_6H_6	795	310	2,57
Toluol	C_7H_8	902	372	2,48
Aethylbenzol	C_8H_{10}	1005	424	2,37
m-Xylol	,,			
Propylbenzol	C_9H_{12}	1122	475	2,36
Mesitylen	,,			
Cymol	$C_{10}H_{14}$	1213	527	2,30
Aethylformiat	$C_3H_6O_2$	719	249	2,89
Methylacetat	,,			

Name der Verbindung	Formel	$\sqrt{D'_m \cdot v}$	$\sum c_{a_\text{ber.}}$	Quotient
Aethylacetat	$C_4H_8O_2$	837	301	2,78
Methylpropionat	,,			
Propylformiat	,,			
Methylisobutyrat	$C_5H_{10}O_2$	882	353	2,50
Isobutylformiat	,,			
Aethylpropionat	,,			
Propylacetat	,,			
Methylbutyrat	,,			
Aethylisobutyrat	$C_6H_{12}O_2$	971	405	2,40
Methylvalerat	,,			
Isobutylacetat	,,			
Aethylbutyrat	,,			
Propylpropionat	,,			
Isoamylformiat	,,			

Trotzdem der in der fünften Columne eingetragene Quotient keineswegs eine Constante ist, sondern vielmehr deutlich von der Constitution der Stoffe abhängt, so können wir das vorliegende Material doch dazu benutzen, diejenige Zahl, wenigstens der Grössenordnung nach, zu ermitteln, mit der unsere c_α multiplicirt werden müssen, damit wir sie in der von uns gewählten absoluten Einheit erhalten. Der gesuchte Multiplicator ergiebt sich im Mittel:

$$2{,}51 \cdot \sqrt{4{,}17 \cdot 10^7} = 1{,}62 \cdot 10^4.$$

Da die vorhergehende Betrachtung zeigt, dass sich bei der Verdampfung die kinetischen Verhältnisse der Molecüle verändern (wenigstens wenn unser Ausdruck für die potentielle Energie richtig ist), unternahm ich es die absolute Grösse c_α noch auf eine andere Weise aufzusuchen. Dabei ging ich von der folgenden Idee aus:

Comprimirt man eine Flüssigkeit isothermisch und ändert sich dabei ihr Wärmeinhalt nicht, was wir nun voraussetzen wollen, so ist die bei der Compression entweichende Wärme gleich der Summe der Compressionsarbeit und der von den Molecularkräften geleisteten Arbeit. Wir können also letztere Arbeit berechnen, wenn wir die bei der Compression entweichende Wärmemenge eruiren können. Dazu aber verhilft uns das Carnot'sche Princip.

Sei nämlich der Zustand der Flüssigkeit durch den Druck p in absoluten Einheiten und die absolute Temperatur T bestimmt; ist nun bei einer unendlich kleinen Zustandsänderung dQ die dem Körper zugeführte Wärme in absolutem Maass, dA die ihm zugeführte mechanische Arbeit, und setzen wir

$$dQ = X\,dp + S\,.\,dT,$$
$$dA = -p\,.\,dv = -p\left\{\frac{\partial v}{\partial p}dp + \frac{\partial v}{\partial T}dT\right\}$$
$$= p\,.\,v\,.\,\varkappa\,dp - p\,.\,v\,.\,\alpha\,dT,$$

so liefert uns die Bedingung, dass dQ/T und $dQ + dA$ vollständige Differentiale sein müssen, die Gleichungen

$$\frac{\partial}{\partial T}\left(\frac{X}{T}\right) = \frac{\partial}{\partial p}\left(\frac{S}{T}\right)$$

und

$$\frac{\partial}{\partial T}(X + p\varkappa) = \frac{\partial}{\partial p}(S - p\alpha)$$

hierbei bedeuten, wie man sieht, X die bei isothermischer Compression durch den Druck $p = 1$ dem Körper zugeführte Wärme in mechanischem Maass, S die specifische Wärme bei constantem Druck, $\varkappa$ den Compressibilitätscoefficienten, α den thermischen Ausdehnungscoefficienten. Aus diesen Gleichungen findet man:

$$X\,dp = -T\left\{\alpha + p\frac{\partial \alpha}{\partial p} + p\frac{\partial \varkappa}{\partial T}\right\}dp\,.$$

Nun ist daran zu erinnern, dass der Atmosphärendruck, unter dem sich unsere Körper gewöhnlich finden, für Compressionserscheinungen von Flüssigkeiten unbedenklich als unendlich klein zu betrachten ist; ebenso sind die Compressionen in unseren Experimenten sehr nahe proportional den angewandten Compressionskräften. Die Erscheinungen gehen also so vor sich, wie wenn die Compressionskräfte unendlich klein wären. Berücksichtigt man dies, so geht unsere Gleichung über in:

$$X\,.\,dp = -T\,.\,\alpha\,.\,dp\,.$$

Wenden wir nun die Voraussetzung an, dass bei isothermischer Compression die kinetische Energie des Systems nicht geändert wird, so erhalten wir die Gleichung

$$X\,.\,dp + \text{Compressionsarbeit} + \text{Arbeit der Molecularkräfte} = 0.$$

522   *A. Einstein.*

Ist P das Potential der Molecularkräfte, so ist die letzte Arbeit:
$$\frac{\partial P}{\partial v} \cdot \frac{\partial v}{\partial p} \cdot dp.$$

Setzt man unseren Ausdruck für die Grösse des Potentiales der Molecularkräfte hierin ein und berücksichtigt, dass die Compressionsarbeit von der Ordnung dp^2 ist, so erhält man bei Vernachlässigung dieser unendlich kleinen Grösse zweiter Ordnung
$$\frac{T\alpha}{\varkappa} = \frac{(\sum c_a)^2}{v^2},$$

wobei $\varkappa$ den Compressibilitätscoefficienten in absolutem Maasse bezeichnet. Wir erhalten so abermals ein Mittel, den gesuchten Proportionalitätscoefficienten für die Grössen c_a zu bestimmen. Die Grössen α und $\varkappa$ für die Temperatur des Eises entnahm ich den Tabellen von Landolt und Börnstein. Man erhält so für den gesuchten Factor die Werte:

Xylol	$1{,}71 \cdot 10^4$	Aethylalkohol	$1{,}70 \cdot 10^4$
Cymol	$1{,}71 \cdot 10^4$	Methylalkohol	$1{,}74 \cdot 10^4$
Terpentinöl	$1{,}73 \cdot 10^4$	Propylalkohol	$1{,}82 \cdot 10^4$
Aethyläther	$1{,}70 \cdot 10^4$	Amylalkohol	$2{,}00 \cdot 10^4$

Zunächst ist zu bemerken, dass die beiden durch verschiedene Methoden erlangten Coefficienten recht befriedigend übereinstimmen, trotzdem sie aus ganz verschiedenen Phenomenen hergeleitet sind. Die letzte Tabelle zeigt sehr befriedigende Uebereinstimmung der Werte, nur die kohlenstoffreicheren Alkohole weichen ab. Es ist dies auch zu erwarten, denn aus den Abweichungen, welche die Alkohole von dem thermischen Ausdehnungsgesetz von Mendelejew und von dem stöchiometrischen Capillaritätsgesetz von R. Schiff zeigen, hat man schon früher geschlossen, dass bei diesen Verbindungen mit Temperaturänderungen Aenderungen der Grösse der Flüssigkeitsmolecüle verbunden sind. Es ist also auch zu erwarten, dass bei isothermischer Compression solche moleculare Veränderungen auftreten, sodass für solche Stoffe bei gleicher Temperatur der Wärmeinhalt Function des Volums sein wird.

Zusammenfassend können wir also sagen, dass sich unsere fundamentale Annahme bewährt hat: Jedem Atom entspricht ein moleculares Anziehungsfeld, welches unabhängig von der Temperatur und unabhängig von der Art ist, wie das Atom mit anderen Atomen chemisch verbunden ist.

Schliesslich ist noch darauf hinzuweisen, dass mit steigendem Atomgewicht im allgemeinen auch die Constanten c_a steigen, doch nichts stets und nicht in proportionaler Art. Die Frage, ob und wie unsere Kräfte mit den Gravitationskräften verwandt sind, muss also noch vollkommen offen gelassen werden. Es ist auch hinzuzufügen, dass die Einführung der Function $\varphi(r)$, welche unabhängig von der Natur der Molecüle sein sollte, nur als Näherungsannahme aufzufassen ist, ebenso die Ersetzung der Summen durch Integrale; in der That scheint sich unsere Theorie für Stoffe von kleinem Atomvolum nicht zu bewähren, wie das Beispiel des Wassers darthut. Ueber diese Fragen sind erst von eingehenden Specialforschungen Aufschlüsse zu hoffen.

Zürich, den 13. December 1900.

(Eingegangen 16. December 1900.)

ANNALEN DER PHYSIK.

BEGRÜNDET UND FORTGEFÜHRT DURCH

F. A. C. GREN, L. W. GILBERT, J. C. POGGENDORFF, G. UND E. WIEDEMANN.

VIERTE FOLGE.

BAND 8.

DER GANZEN REIHE 313. BAND.

KURATORIUM:

F. KOHLRAUSCH, M. PLANCK, G. QUINCKE, W. C. RÖNTGEN, E. WARBURG.

UNTER MITWIRKUNG

DER DEUTSCHEN PHYSIKALISCHEN GESELLSCHAFT

UND INSBESONDERE VON

M. PLANCK

HERAUSGEGEBEN VON

PAUL DRUDE.

MIT SIEBEN FIGURENTAFELN.

LEIPZIG, 1902.

VERLAG VON JOHANN AMBROSIUS BARTH.

5. Ueber die thermodynamische Theorie der Potentialdifferenz zwischen Metallen und vollständig dissociirten Lösungen ihrer Salze und über eine elektrische Methode zur Erforschung der Molecularkräfte; von A. Einstein.

§ 1. Eine hypothetische Erweiterung des zweiten Hauptsatzes der mechanischen Wärmetheorie.

Der zweite Hauptsatz der mechanischen Wärmetheorie kann auf solche physikalische Systeme Anwendung finden, die im stande sind, mit beliebiger Annäherung umkehrbare Kreisprocesse zu durchlaufen. Gemäss der Herleitung dieses Satzes aus der Unmöglichkeit der Verwandlung latenter Wärme in mechanische Energie, ist hierbei notwendige Voraussetzung, dass jene Processe realisirbar seien. Bei einer wichtigen Anwendung der mechanischen Wärmetheorie ist es aber zweifelhaft, ob dieses Postulat erfüllt ist, nämlich bei der Vermischung zweier oder mehrerer Gase mit Hülfe von semipermeabeln Wänden. Auf der Voraussetzung der Realisirbarkeit dieses Vorganges basirt die thermodynamische Theorie der Dissociation der Gase und die Theorie der verdünnten Lösungen.

Die einzuführende Voraussetzung ist bekanntlich folgende: Zu je zwei Gasen A und B sind zwei Scheidewände herstellbar, sodass die eine durchlässig für A, nicht aber für B, die andere durchlässig für B, nicht aber für A ist. Besteht die Mischung aus mehreren Componenten, so gestaltet sich diese Voraussetzung noch complicirter und unwahrscheinlicher. Da nun die Erfahrung die Resultate der Theorie vollständig bestätigt hat, trotzdem wir mit Processen operirt haben, deren Realisirbarkeit wohl bezweifelt werden kann, so erhebt sich die Frage, ob nicht vielleicht der zweite Hauptsatz auf ideale Processe gewisser Art angewendet werden kann, ohne dass man mit der Erfahrung in Widerspruch gerät.

In diesem Sinne können wir auf Grund der gewonnenen Erfahrung jedenfalls den Satz aussprechen: Man bleibt im Einklang mit der Erfahrung, wenn man den zweiten Hauptsatz auf physikalische Gemische ausdehnt, deren einzelne Componenten durch in gewissen Flächen wirkende conservative Kräfte auf gewisse Teilräume beschränkt werden. Diesen Satz verallgemeinern wir hypothetisch zu folgendem:

Man bleibt im Einklange mit der Erfahrung, wenn man den zweiten Hauptsatz auf physikalische Gemische anwendet, auf deren einzelne Componenten beliebige conservative Kräfte wirken.

Auf diese Hypothese werden wir uns im Folgenden stets stützen, auch wo es nicht absolut notwendig erscheint.

§ 2. Ueber die Abhängigkeit der elektrischen Potentialdifferenz einer vollkommen dissociirten Salzlösung und einer aus dem Lösungsmetall bestehenden Elektrode, von der Concentration der Lösung und vom hydrostatischen Druck.

In einem cylindrischen Gefässe, dessen Axe zusammenfalle mit der z-Axe eines cartesischen Coordinatensystems befinde sich ein vollkommen dissociirtes Salz in Lösung. $v\,do$ sei die Anzahl der Grammmolecüle des Salzes, welche sich im Volumenelemente do gelöst finden, $v_m\,do$ die Anzahl der Metallionen, $v_s\,do$ die Anzahl der Säureionen daselbst, wobei v_m und v_s ganzzahlige Vielfache von v sind, sodass die Gleichungen bestehen:

$$v_m = n_m \cdot v,$$
$$v_s = n_s \cdot v.$$

Ferner sei $n \cdot v \cdot E \cdot do$ die Grösse der gesamten positiven elektrischen Ionenladung in do, also auch, bis auf unendlich Kleines, die Grösse der negativen. n ist dabei die Summe der Wertigkeiten der Metallionen des Molecüls, E die Elektricitätsmenge, welche zur elektrolytischen Ausscheidung eines Grammmolecüles eines einwertigen Ions erforderlich ist.

Diese Gleichungen gelten jedenfalls, da die Anzahl der überzähligen Ionen einer Gattung zu vernachlässigen sein wird.

Wir wollen ferner annehmen, dass auf die Metall- bez. Säureionen eine äussere conservative Kraft wirke, deren Potential pro Ion die Grösse P_m bez. P_s besitze. Wir vernachlässigen ferner die Veränderlichkeit der Dichte des Lösungsmittels mit dem Druck und der Dichte des gelösten Salzes, und nehmen an, dass auf die Teile des Lösungsmittels ebenfalls eine conservative Kraft wirke, deren Potential pro Grammäquivalent des Lösungsmittels die Grösse P_0 besitze, wobei $\nu_0 \, do$ Grammmolecüle des Lösungsmittels in do vorhanden seien.

Alle die Kräftefunctionen seien lediglich von der z-Coordinate abhängig, und das System befinde sich im elektrischen, thermischen und mechanischen Gleichgewicht. Es werden dann die Grössen: Concentration ν, das elektrische Potential π, osmotische Drucke der beiden Ionengattungen p_m und p_s, hydrostatischer Druck p_o nur Functionen von z sein.

Es müssen nun an jeder Stelle des Elektrolyten die beiden Elektronengattungen für sich im Gleichgewicht sein, was durch die Gleichungen ausgedrückt wird:

$$-\frac{dp_m}{dz} \cdot \frac{1}{\nu} - n_m \frac{dP_m}{dz} - nE\frac{d\pi}{dz} = 0,$$

$$-\frac{dp_s}{dz} \cdot \frac{1}{\nu} - n_s \frac{dP_s}{dz} + nE\frac{d\pi}{dz} = 0,$$

dabei ist:
$$p_m = \nu \cdot n_m \cdot RT,$$
$$p_s = \nu \cdot n_s \cdot RT,$$

wo R eine für alle Ionenarten gemeinsame Constante ist. Die Gleichungen nehmen also die Form an:

(1) $$\begin{cases} n_m RT \frac{d\lg \nu}{dz} + n_m \frac{dP_m}{dz} + nE\frac{d\pi}{dz} = 0, \\ n_s RT \frac{d\lg \nu}{dz} + n_s \frac{dP_s}{dz} - nE\frac{d\pi}{dz} = 0, \end{cases}$$

Sind P_m und P_s für alle z, sowie ν und π für ein bestimmtes z bekannt, so liefern die Gleichungen (1) ν und π als Functionen von z. Auch ergäbe die Bedingung, dass sich die Lösung als Ganzes im Gleichgewicht befindet, eine Gleichung zur Bestimmung des hydrostatischen Druckes p_o, die nicht angeschrieben zu werden braucht. Wir bemerken nur, dass dp_o von $d\nu$ und $d\pi$ deshalb unabhängig ist, weil es uns freisteht, beliebige conservative Kräfte anzunehmen, welche auf die Molecüle des Lösungsmittels wirken.

Wir denken uns nun in $z = z_1$ und $z = z_2$ Elektroden in die Lösung eingeführt, welche aus dem Lösungsmetalle bestehen, und nur einen verschwindend kleinen Teil des Querschnittes des cylindrischen Gefässes ausfüllen sollen. Lösung und Elektroden zusammen bilden ein physikalisches System, welches wir folgenden umkehrbaren isothermischen Kreisprocess ausführen lassen:

1. Teilprocess: Wir lassen die Elektricitätsmenge nE *unendlich langsam* durch die Lösung passiren, indem wir die in $z = z_1$ bez. $z = z_2$ befindliche Elektrode als Anode bez. Kathode verwenden.

2. Teilprocess: Wir bewegen die hierbei elektrolytisch von z_1 nach z_2 bewegte Metallmenge mechanisch in der Lösung unendlich langsam wieder von z_2 nach z_1.

Man ersieht zunächst, dass der Process strenge umkehrbar ist, da alle Vorgänge unendlich langsam vor sich gehend gedacht werden, derselbe also aus (idealen) Gleichgewichtszuständen zusammengesetzt ist. Der zweite Hauptsatz verlangt für einen solchen Process, dass die Summe der dem System während des Kreisprocesses zugeführten Wärmemengen verschwinde. Der erste Hauptsatz verlangt in Verbindung mit dem zweiten, dass die Summe der übrigen Energien, welche dem System während des Kreisprocesses zugeführt werden, verschwinde.

Während des ersten Teilprocesses wird die elektrische Arbeitsmenge zugeführt:
$$-nE(\Pi_2 - \Pi_1),$$
wobei Π_2 und Π_1 die elektrischen Potentiale der Elektroden bedeuten.

Während des zweiten Teilprocesses wird:
$$\int_{z_2}^{z_1} K\,dz$$
zugeführt, wobei K die in der positiven z-Richtung wirkende Kraft bedeutet, welche notwendig ist, um die zu bewegenden n_m Metallionen, welche sich jetzt im metallischen Zustande befinden, an der beliebigen Stelle z in Ruhe zu erhalten. Für K gilt, wie leicht ersichtlich die Gleichung:
$$K - n_m \frac{dP_m}{dz} - n_m v_m \frac{dp_o}{dz} = 0.$$

Dabei bedeutet v_m das Volumen eines Metallions im metallischen Zustande. Jene Arbeit erhält also den Wert:

$$\int_{z_2}^{z_1} K \cdot dz = -\int_{z_1}^{z_2}\left(n_m \frac{dP_m}{dz} + n_m v_m \frac{dp_o}{dz}\right)dz$$
$$= -n_m[(P_{m_2} - P_{m_1}) + v_m(p_{o_2} - p_{o_1})],$$

wobei der zweite Index die Coordinate der Elektrode bezeichnet.

Wir erhalten also die Gleichung:

(2) $\quad n \cdot E \cdot (\Pi_2 - \Pi_1) = -n_m(P_{m_2} - P_{m_1}) - n_m v_m(p_{o_2} - p_{o_1}).$

Bezeichnet man mit π_1 und π_2 die elektrischen Potentiale, welche in den Elektrodenquerschnitten im Innern der Lösung herrschen, so erhält man durch Integration aus der ersten Gleichung (1):

$$-n \cdot E(\pi_2 - \pi_1) = n_m[P_{m_2} - P_{m_1}] + n_m R T \log\left(\frac{\nu_2}{\nu_1}\right),$$

wobei sich ν_1 und ν_2 wieder auf die Elektrodenquerschnitte beziehen. Durch Addition dieser Gleichungen erhält man:

(3) $\quad \begin{cases} (\Pi_2 - \pi_2) - (\Pi_1 - \pi_1) = (\varDelta\Pi)_2 - (\varDelta\Pi)_1 \\ \qquad = \dfrac{n_m R T}{n E} \log\left(\dfrac{\nu_2}{\nu_1}\right) - \dfrac{n_m v_m}{n E}(p_{o_2} - p_{o_1}). \end{cases}$

Da die ν und p_o vollständig unabhängig voneinander sind, so enthält diese Gleichung die Abhängigkeit der Potentialdifferenz $\varDelta\Pi$ zwischen Metall und Lösung von Concentration und hydrostatischem Druck. Es ist zu bemerken, dass die angenommenen Kräfte im Resultat nicht mehr vorkommen. Kämen sie vor, so wäre die § 1 aufgestellte Hypothese ad absurdum geführt. Die gefundene Gleichung lässt sich in zwei zerlegen, nämlich:

(4) $\quad \begin{cases} (\varDelta\Pi)_2 - (\varDelta\Pi)_1 = \dfrac{n_m}{n} \dfrac{R T}{E} \log\left(\dfrac{\nu_2}{\nu_1}\right) \text{ bei const. Druck,} \\ (\varDelta\Pi)_2 - (\varDelta\Pi)_1 = -\dfrac{n_m}{n} \dfrac{v_m}{E}(p_{o_2} - p_{o_1}) \text{ bei const. Concentration.} \end{cases}$

Man hätte die Endformel (3) auch erhalten, ohne die in § 1 vorgeschlagene Hypothese, wenn man die äusseren Kräfte mit der Erdschwere identificirt hätte. Dann wären aber ν und p_o nicht unabhängig voneinander und eine Zerlegung in die Gleichungen (4) wäre nicht erlaubt.

Es soll noch kurz erwähnt werden, dass die Nernst'sche Theorie der elektrischen Kräfte im Innern dissociirter Elektro-

Thermodynamische Theorie der Potentialdifferenz etc. 803

lyte in Verbindung mit der ersten der Gleichungen (4) die elektromotorische Kraft des Concentrationselementes zu berechnen gestattet. Man gelangt so zu einem bereits mehrfach geprüften Resultat, welches bis jetzt aus speciellen Annahmen hergeleitet wurde.

§ 3. Ueber die Abhängigkeit der Grösse $\Delta \Pi$ von der Natur der Säure.

Wir betrachten folgenden idealen Gleichgewichtszustand: Sei wieder ein cylindrisches Gefäss vorhanden. In den Teilen I und II mögen sich vollständig dissociirte Salzlösungen befinden mit identischem Metallion (gleiches Metall und gleiche elektrische Ladung), aber verschiedenem Säureion. Zwischen den beiden befinde sich der Verbindungsraum V, in welchem

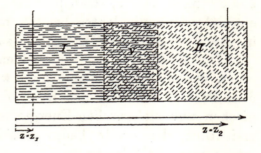

beide Salze gelöst vorkommen. In V mögen auf die Säureionen Kräfte wirken, deren Potentiale $P_s^{(1)}$ und $P_s^{(2)}$ nur von z abhängen, welche Kräfte bewirken sollen, dass nur unendlich wenig Säureionen erster Art in II, zweiter Art in I gelangen. Ausserdem seien $P_s^{(1)}$ und $P_s^{(2)}$ so gewählt, dass die Concentration der Metallionen in den beiden Teilen I und II die gleiche sei. Ebenso sei $p_{o_1} = p_{o_2}$.

Es seien $\nu_m^{(1)}$ Metallionen in der Volumeneinheit, welche der ersten, $\nu_m^{(2)}$, welche der zweiten Satzart entsprechen, dann ist:

(1) $\qquad \nu_{m_1}^{(1)} = \nu_{m_2}^{(2)}, \quad \nu_{s_1}^{(2)} = 0, \quad \nu_{s_2}^{(1)} = 0,$

wobei die unteren Indices die Zugehörigkeit zu Raum I bez. Raum II bezeichnet.

In V erhält man aber als Gleichgewichtsbedingung der Metallionen:

$$- RT \frac{d \log \left(\nu_m^{(1)} + \nu_m^{(2)} \right)}{dz} - \varepsilon E \frac{d\pi}{dz} = 0,$$

wobei ε die Wertigkeit des Metallions bedeutet.

Durch Integration über V und Berücksichtigung der Gleichungen (1) ergiebt sich:

(2) $$\pi_2 = \pi_1.$$

Wir bilden ferner, nachdem wir in I und II Elektroden aus Lösungsmetall eingesetzt denken, folgenden idealen Kreisprocess:

1. Teilprocess: Wir schicken durch das System unendlich langsam die Elektricitätsmenge εE, indem wir die im Raum I befindliche Elektrode als Anode, die andere als Kathode betrachten.

2. Teilprocess: Wir führen das so durch Elektrolyse von $z = z_1$ nach $z = z_2$ transportirte Metall, welches die Masse eines Grammäquivalentes besitzt, mechanisch wieder nach der in $z = z_1$ befindlichen Elektrode zurück.

Durch Anwendung der beiden Hauptsätze der mechanischen Wärmetheorie folgert man wieder, dass die Summe der dem System während des Kreisprocesses zugeführten mechanischen und elektrischen Energie verschwindet. Da, wie leicht ersichtlich, der zweite Teilprocess keine Energie erfordert, so erhält man die Gleichung

(3) $$\Pi_2 = \Pi_1,$$

wobei Π_2 und Π_1 wieder die Elektrodenpotentiale bedeuten. Durch Subtraction der Gleichungen (3) und (2) erhält man:

$$(\Pi_2 - \pi_2) - (\Pi_1 - \pi_1) = (\Delta \Pi)_2 - (\Delta \Pi)_1 = 0$$

und also folgenden Satz:

Die Potentialdifferenz zwischen einem Metall und einer vollständig dissociirten Lösung eines Salzes dieses Metalles in einem bestimmten Lösungsmittel ist unabhängig von der Natur des elektronegativen Bestandteiles, sie hängt lediglich von der Concentration der Metallionen ab. Voraussetzung ist dabei jedoch, dass bei den Salzen das Metallion mit derselben Elektricitätsmenge geladen ist.

§ 4.

Bevor wir dazu übergehen, die Abhängigkeit von $(\Delta \Pi)$ von der Natur des Lösungsmittels zu studiren, wollen wir kurz die Theorie der conservativen Molecularkräfte in Flüssigkeiten entwickeln. Ich entnehme dabei die Bezeichnungsweise einer früheren Abhandlung über diesen Gegenstand[1]), welche zugleich die einzuführenden Hypothesen einstweilen rechtfertigen soll.

Jedem Molecüle einer Flüssigkeit oder einer in einer Flüssigkeit gelösten Substanz komme eine gewisse Constante c zu, sodass der Ausdruck für das relative Potential der Molecularkräfte zweier Molecüle, welche durch die Indices $\ldots_1$ und $\ldots_2$ charakterisirt seien, lautet:

(a) $$P = P_\infty - c_1 c_2 \varphi(r),$$

wobei $\varphi(r)$ eine für alle Molecülarten gemeinsame Function der Entfernung sei. Jene Kräfte sollen sich einfach superponiren, sodass der Ausdruck des relativen Potentiales von n Molecülen die Form habe:

(b) $$\text{Const.} - \tfrac{1}{2} \sum_{\alpha=1}^{\alpha=n} \sum_{\beta=1}^{\beta=n} c_\alpha c_\beta \varphi(r_{\alpha\beta}).$$

Wären speciell alle Molecüle gleich beschaffen, so erhielten wir den Ausdruck:

(c) $$\text{Const.} - \tfrac{1}{2} c^2 \sum_{\alpha=1}^{\alpha=n} \sum_{\beta=1}^{\beta=n} \varphi(r_{\alpha\beta})$$

Ferner sei das Wirkungsgesetz und das Vertheilungsgesetz der Molecüle so beschaffen, dass die Summen in Integrale verwandelt werden dürfen, dann geht dieser Ausdruck über in:

$$\text{Const.} - \tfrac{1}{2} c^2 N^2 \int\int d\tau \cdot d\tau' \, \varphi(r_{d\tau, d\tau'}).$$

N bedeutet dabei die Zahl der Molecüle in der Volumeneinheit. Bezeichnet N_0 die Anzahl der Molecüle in einem Grammäquivalent, so ist $N_0/N = v$ das Molecularvolumen der Flüssigkeit, und nehmen wir an, dass ein Grammäquivalent zur Untersuchung vorliegt, so geht, wenn wir den Einfluss der Flüssigkeitsoberfläche vernachlässigen, unser Ausdruck über in:

$$\text{Const.} - \tfrac{1}{2} \frac{c^2}{v} N_0^2 \int_{-\infty}^{\infty} d\tau' \cdot \varphi(r_{0, d\tau'}).$$

---

1) A. Einstein, Ann. d. Phys. **4**. p. 513. 1901.

Wir wollen nun die Einheit der c so wählen, dass dieser Ausdruck übergeht in

(d) $\qquad \text{Const.} - \dfrac{c^2}{v}$, also $\quad \frac{1}{2} N_0^2 \int\limits_{-\infty}^{\infty} d\tau' \cdot \varphi(r_0, d\tau') = 1$.

Durch diese Festsetzung gewinnt man für die Grössen c ein absolutes Maass. In jener Abhandlung ist gezeigt, dass man mit der Erfahrung in Uebereinstimmung bleibt, wenn man setzt $c = \sum c_a$, wo sich die Grössen c_a auf die Atome beziehen, aus denen das Molecül zusammengesetzt ist.

Wir wollen nun das relative Anziehungspotential des Grammmolecüls eines Ions in Bezug auf sein Lösungsmittel berechnen, wobei wir ausdrücklich die Annahme machen, dass die Anziehungsfelder der Molecüle des Lösungsmittels nicht auf die elektrischen Ladungen der Ionen wirken. Später zu entwickelnde Methoden werden ein Mittel an die Hand geben, welches über die Zulässigkeit dieser Voraussetzung zu entscheiden gestattet.

Sei c_j die moleculare Constante des Ions, c_l die des Lösungsmittels, so hat das Potential eines Molecüles des Ions gegen das Lösungsmittel die Form:

$$\text{Const.} - \sum_l c_j c_l \cdot \varphi(r) = \text{const.} - c_j \cdot c_l N_l \int d\tau \cdot \varphi(r_0, d\tau),$$

wobei N_l die Zahl der Molecüle des Lösungsmittels pro Volumeneinheit bedeutet. Da $N_0/N_l = v_l$ ist, so geht dieser Ausdruck über in:

$$\text{Const.} - c_j \cdot c_l \cdot \dfrac{N_0}{v_l} \int d\tau \cdot \varphi(r_0, d\tau).$$

Das aber das Grammäquivalent N_0 Molecüle des Ions enthält, so erhalten wir für das relative Potential des Grammäquivalentes des Ions:

$$\text{Const.} - \dfrac{c_j \cdot c_l}{v_l} N_0^2 \int d\tau \cdot \varphi(r_0, d\tau) = \text{const.} - 2\dfrac{c_j \cdot c_l}{v_l}.$$

Führt man die Concentration des Lösungsmittels $1/v_l = \nu_l$ ein, so erhält man die Form:

(e) $\qquad\qquad P_{jl} = \text{const.} - 2\, c_j \cdot c_l \nu_l.$

Ist das Lösungsmittel eine Mischung mehrerer Flüssigkeiten, welche wir durch Indices unterscheiden wollen, erhalten wir

(e') $$P_{jl} = \text{const.} - 2\,c_j \sum \dot{}\, c_l \, \nu_l ,$$

wobei die ν_l die Anzahl der Grammolecüle der einzelnen Componenten des Lösungsmittels pro Volumeneinheit bedeuten. Die Formel (e') gilt angenähert auch in dem Falle, dass die Grössen ν_l mit dem Orte variiren.

§ 5. Ueber die Abhängigkeit der zwischen einem Metall und einer vollständig dissociirten Lösung eines Salzes dieses Metalles herrschenden elektrischen Potentialdifferenz von der Natur des Lösungsmittels.

Ein cylindrisches Gefäss zerfalle wieder, wie im § 3 angegeben wurde, in die Räume I, II und den Verbindungsraum V. In I befinde sich ein erstes, in II ein zweites Lösungsmittel, in V mögen beide gemischt vorkommen und es mögen in diesem Raume auf die Lösungsmittel Kräfte wirken, welche eine Diffusion verhindern. In dem Gefässe befinde sich ein gelöstes Salz im Zustande vollständiger Dissociation. Auf die Säureionen desselben sollen in V Kräfte wirken, deren Potential P_s heisse und so gewählt sei, dass das Salz in I und II gleiche Concentration besitze. Wir stellen nun die Bedingung für das Gleichgewicht der Metallionen auf. Die z-Axe führen wir wieder $\parallel$ der Cylinderaxe von I nach II.

Als Ausdruck der auf das Grammäquivalent wirkenden Kraft elektrischen Ursprunges ergibt sich:

$$-\frac{\varkappa}{n_m} E \frac{d\pi}{dz} .$$

Die auf das Aequivalent vom osmotischen Druck ausgeübte Kraft ist:

$$-RT \frac{d\log \nu}{dz} .$$

Die auf das Aequivalent ausgeübte Wirkung der Molecularkräfte ist:

$$-\frac{d}{dz}\left\{ -2\,c_m c_l^{(1)} \nu_l^{(1)} - 2\,c_m c_l^{(2)} \nu_l^{(2)} \right\} ,$$

wobei sich die oberen Indices auf die Lösungsmittel beziehen. Die gesuchte Gleichgewichtsbedingung ist also:

$$-\frac{n}{n_m} E \frac{d\pi}{dz} - RT \frac{d\log \nu}{dz} + \frac{d}{dz}\{2 c_m c_i^{(1)} v_i^{(1)} + 2 c_m c_i^{(2)} v_i^{(2)}\} = 0.$$

Integrirt man durch V hindurch und berücksichtigt, dass ν in I und II identisch ist, und dass $v_i^{(2)}$ in I und $v_i^{(1)}$ in II nach unserer Voraussetzung verschwindet, so erhält man:

$$\pi_2 - \pi_1 = \frac{n_m}{n} \frac{2 c_m}{E} \{c_i^{(2)} v_i^{(2)} - c_i^{(1)} v_i^{(1)}\},$$

wobei sich die oberen Indices auf Raum I bez. II beziehen.

Wir denken uns nun in I und II Elektroden angebracht, welche aus dem gelösten Metall bestehen, und bilden einen Kreisprocess, indem wir die Electricitätsmenge $n/n_m E$ durch das System schicken, und dann die transportirte Metallmenge mechanisch wieder zurückbewegen, was keine Arbeit erfordert, wenn wir annehmen, dass in I und II der hydrostatische Druck der nämliche sei. Durch Anwendung der beiden Hauptsätze der Wärmetheorie erhält man:

$$\Pi_2 - \Pi_1 = 0.$$

Durch Subtraction beider Resultate ergibt sich:

$$(\Pi_2 - \pi_2) - (\Pi_1 - \pi_1) = (\Delta \Pi)^{(2)} - (\Delta \Pi)^{(1)}$$
$$= -\frac{n_m}{n} \frac{2 c_m}{E} \{c_i^{(2)} v_i^{(2)} - c_i^{(1)} v_i^{(1)}\}.$$

Ist jedes der beiden Lösungsmittel eine Mischung mehrerer nichtleitender Flüssigkeiten, so erhält man etwas allgemeiner:

$$(\Delta \Pi)^{(2)} - (\Delta \Pi)^{(1)} = -\frac{n_m}{n} \frac{2 c_m}{E} \{\sum c_i^{(2)} v_i^{(2)} - \sum c_i^{(1)} v_i^{(1)}\},$$

in welcher Formel v_l die Zahl der Grammmoleküle einer Componente des Lösungsmittels in einem Volumelemente des gemischten Lösungsmittels bezeichnet.

Die Potentialdifferenz $\Delta \Pi$ ist also von der Natur des Lösungsmittels abhängig. Auf diese Abhängigkeit lässt sich eine Methode zur Erforschung der Molecularkräfte gründen.

§ 6. Methode zur Bestimmung der Constanten c für Metallionen und Lösungsmittel.

In einem cylindrischen Gefässe seien zwei vollständig dissociirte Salzlösungen in Diffusion begriffen; diese Salze seien durch untere Indices bezeichnet. Das Lösungsmittel sei im ganzen Gefäss dasselbe und werde durch den oberen Index bezeichnet. Das Gefäss zerfalle wieder in die Räume I, II und den Verbindungsraum V. Im Raume I sei nur das erste, im Raume II nur das zweite Salz vorhanden; im Raume V finde Diffusion beider Salze statt. In die Räume I und II seien Elektroden eingeführt, welche aus dem betreffenden Lösungsmetalle bestehen und die elektrischen Potentiale Π_1 bez. Π_2' besitzen; an die zweite Elektrode sei ein Stück des ersten Elektrodenmetalles angelötet, dessen Potential Π_2 sei. Wir bezeichnen ausserdem die elektrischen Potentiale im Innern der unvermischten, in I und II befindlichen Lösungen, mit π_1 und π_2, dann ist:

$$(\Pi_2 - \Pi_1)^{(1)} = (\Pi_2 - \Pi_2') + (\Pi_2' - \pi_2)^{(1)} + (\pi_2 - \pi_1)^{(1)} - (\Pi_1 - \pi_1)^{(1)}.$$

Stellt man ganz dieselbe Anordnung her, mit dem einzigen Unterschiede, dass man ein anderes Lösungsmittel benutzt, das durch den oberen Index$^{(2)}$ bezeichnet werde, so hat man:

$$(\Pi_2 - \Pi_1)^{(2)} = (\Pi_2 - \Pi_2') + (\Pi_2' - \pi_2)^{(2)} + (\pi_2 - \pi_1)^{(2)} - (\Pi_1 - \pi_1)^{(2)}.$$

Durch Subtraction dieser beiden Ausdrücke erhält man mit Berücksichtigung des in § 5 gefundenen Resultates:

$$(\Pi_2 - \Pi_1)^{(2)} - (\Pi_2 - \Pi_1)^{(1)} = \{(\pi_2 - \pi_1)^{(2)} - (\pi_2 - \pi_1)^{(1)}\}$$
$$- \frac{2}{E}\left\{\left(\frac{c_m n_m}{n}\right)_2 - \left(\frac{c_m n_m}{n}\right)_1\right\} \cdot \{c_l^{(2)} v_l^{(2)} - c_l^{(1)} v_l^{(1)}\}.$$

Die erforderliche Erweiterung für den Fall, dass die Lösungsmittel Mischungen sind, erhält man leicht wie in § 5.

Die Werte der linken Seite dieser Gleichung ergeben sich unmittelbar durch das Experiment. Mit der Bestimmung des ersten Gliedes der rechten Seite werden wir uns im folgenden Paragraph beschäftigen; es sei einstweilen gesagt, dass man dies Glied aus den angewandten Concentrationen und den molecularen Leitfähigkeiten der betreffenden Ionen für das betreffende Lösungsmittel berechnen kann, wenn man die Anordnung in geeigneter Weise wählt. Die Gleichung erlaubt daher die Berechnung des zweiten Gliedes der rechten Seite.

Dies benutzen wir zur Bestimmung der Constanten c für Metallionen und zur Prüfung unserer Hypothesen. Wir benutzen zu einer Reihe von Experimenten der geschilderten

Art immer dieselben beiden Lösungsmittel. Für die ganze Untersuchungsreihe ist dann die Grösse

$$\frac{2}{E}\{c_l^{(2)}\,v_l^{(2)} - c_l^{(1)}\,v_l^{(1)}\} = k = \text{const.}$$

Setzt man $n_1/n_{m_1} = E_1$ etc. gleich der Wertigkeit des ersten etc. Metallions, so ist also das berechnete letzte Glied der rechten Seite ein relatives Maass für die Grösse

$$\left(\frac{c_{m_2}}{\varepsilon_2} - \frac{c_{m_1}}{\varepsilon_1}\right).$$

Untersucht man so Combinationen aller Elektrodenmetalle zu Paaren, so erhält man in relativem Maass die Grössen

$$\left\{\frac{c_{m_j}}{\varepsilon_j} - \frac{c_{m_k}}{\varepsilon_k}\right\}.$$

Man erhält in demselben Maasse die Grössen c_m/ε selbst, wenn man bei einem Metall eine analoge Untersuchung in der Weise ausführt, dass man Salze und Elektroden in I und II von demselben Metall wählt, sodass jedoch ε, d. h. die Wertigkeit (elektrische Ladung) des Metallions auf beiden Seiten verschieden ist. Es sind dann in jenem Maasse die Werte für die Grössen c_m der einzelnen Metalle selbst ermittelbar. Eine Reihe von solchen Untersuchungen führt also auf die Verhältnisse der c_m, d. h. der Constanten für die Molecularattraction der Metallionen. Diese Reihe der c_m muss unabhängig sein von der Natur der benutzten Salze, und die Verhältnisse der so erhaltenen c_m unabhängig von der Natur der beiden Lösungsmittel, welche wir für die Untersuchung zu Grunde legten. Ferner muss verlangt werden, dass c_m unabhängig von der elektrischen Ladung (Wertigkeit), in welcher ein Ion auftritt, sich herausstelle. Ist dies der Fall, so ist die oben gemachte Voraussetzung richtig, dass die Molecularkräfte nicht auf die elektrischen Ladungen der Ionen wirken.

Will man den Wert der Grössen c_m wenigstens angenähert absolut bestimmen, so kann man dies, indem man die Grösse k angenähert für die beiden Lösungsmittel aus den Resultaten der oben angeführten Abhandlung entnimmt, indem man die Formel $c = \sum c_a$ anwendet. Freilich ist hier zu bemerken, dass sich gerade für die als Lösungsmittel am meisten naheliegenden Flüssigkeiten, Wasser und Alkohol, die Gültigkeit des Attractionsgesetzes aus den Erscheinungen der Capillarität, Verdampfung und Compressibilität nicht hat darthun lassen.

Es lässt sich auf Grund unseres Ergebnisses aber ebensogut eine Erforschung der Constanten c_l von Lösungsmitteln gründen, indem man der Untersuchung zwei Metallionen zu Grunde legt und das Lösungsmittel variiren lässt, sodass nun die Grösse

$$\frac{2}{E}\left\{\left(\frac{c_m n_m}{n}\right)_2 - \left(\frac{c_m n_m}{n}\right)_1\right\}$$

als constant zu betrachten ist. Indem man auch Mischungen als Lösungsmittel zulässt, kann so die Untersuchung auf alle elektrisch nicht leitenden Flüssigkeiten ausgedehnt werden. Es lassen sich aus solchen Versuchen relative Werte für die Grössen c_α herausrechnen, welche den die Flüssigkeitsmoleküle bildenden Atomen zukommen. Auch hier bietet sich eine Fülle von Prüfungen für die Theorie, indem die c_α beliebig überbestimmt werden können. Ebenso muss das Resultat unabhängig sein von der Wahl der Metallionen.

§ 7. Berechnung von $(\pi_2 - \pi_1)$.

Wir haben nun noch den Diffusionsvorgang im Raume V genauer zu studiren. Die variabeln Grössen seien nur von z abhängig, wobei die z-Axe des von uns gewählten cartesischen Coordinatensystems mit der Richtung der Axe unseres Gefässes zusammenfalle. $v_{m_1}, v_{s_1}, v_{m_2}, v_{s_2}$ seien die von z abhängigen Concentrationen (Grammäquivalente pro Volumeneinheit) der vier Ionengattungen, $\varepsilon_{m_1}E, -\varepsilon_{s_1}E, \varepsilon_{m_2}E, -\varepsilon_{s_2}E$ die elektrischen Ladungen, welche dieselben tragen; π sei das elektrische Potential. Da nirgends beträchtliche elektrische Ladungen auftreten, so ist für alle z nahezu:

(α) $\qquad v_{m_1}\varepsilon_{m_1} - v_{s_1}\varepsilon_{s_1} + v_{m_2}\varepsilon_{m_2} - v_{s_2}\varepsilon_{s_2} = 0\,.$

Ausserdem erhalten wir für jede Ionenart eine Gleichung, welche ausdrückt, dass die Vermehrung der Zahl der in einem Volumenelement befindlichen Ionen bestimmter Gattung pro Zeiteinheit gleich ist der Differenz der in dieser Zeit ins Volumenelement eintretenden und der in derselben Zeit aus ihm austretenden Moleküle:

$$(\beta) \begin{cases} v_{m_1} \cdot \dfrac{\partial}{\partial z}\left\{ RT \dfrac{\partial \nu_{m_1}}{\partial z} + \varepsilon_{m_1} \nu_{m_1} E \dfrac{\partial \pi}{\partial z} \right\} = \dfrac{\partial \nu_{m_1}}{\partial t}, \\ v_{s_1} \cdot \dfrac{\partial}{\partial z}\left\{ RT \dfrac{\partial \nu_{s_1}}{\partial z} - \varepsilon_{s_1} \nu_{s_1} E \dfrac{\partial \pi}{\partial z} \right\} = \dfrac{\partial \nu_{s_1}}{\partial t}, \\ \cdots \cdots \cdots \cdots \cdots \cdots \\ \cdots \cdots \cdots \cdots \cdots \cdots \end{cases}$$

wobei v mit dem betreffenden Index die constante Geschwindigkeit bedeutet, welche die mechanische Krafteinheit dem Grammäquivalent des betreffenden Ions in der Lösung erteilt.

Diese vier Gleichungen bestimmen im Verein mit den Grenzbedingungen den stattfindenden Vorgang vollständig, da sie für jeden Zeitmoment die fünf Grössen

$$\frac{\partial \pi}{\partial z}, \quad \frac{\partial \nu_{m_1}}{\partial t} \cdots \frac{\partial \nu_{s_2}}{\partial t}$$

in eindeutiger Weise zu berechnen gestatten. Die allgemeine Behandlung des Problemes wäre aber mit sehr grossen Schwierigkeiten verknüpft, zumal Gleichungen (β) nicht linear in den Unbekannten sind. Uns kommt es aber nur auf die Bestimmung von $\pi_2 - \pi_1$ an. Wir multipliciren daher die Gleichungen (β) der Reihe nach mit $\varepsilon_{m_1}, -\varepsilon_{s_1}, \varepsilon_{m_2}, -\varepsilon_{s_2}$ und erhalten mit Rücksicht auf (α)

$$\frac{\partial \varphi}{\partial z} = 0,$$

wobei

$$\varphi = RT \left\{ v_{m_1} \varepsilon_{m_1} \frac{\partial \nu_{m_1}}{\partial z} - v_{s_1} \varepsilon_{s_1} \frac{\partial \nu_{s_1}}{\partial z} + \cdot - \cdot \right\} \\ + \left\{ v_{m_1} \varepsilon_{m_1}^2 \nu_{m_1} + v_{s_1} \varepsilon_1^2 \nu_{s_1} + \cdot + \cdot \right\} \frac{\partial \pi}{\partial z}.$$

Durch Integration dieser Gleichung nach z ergiebt sich unter Berücksichtigung des Umstandes, dass überall, wo keine Diffusion stattfindet,

$$\frac{\partial \nu_{m_1}}{\partial z}, \quad \frac{\partial \nu_{s_1}}{\partial z} \cdots \frac{\partial \pi}{\partial z}$$

verschwinden:

$$\varphi = 0.$$

Thermodynamische Theorie der Potentialdifferenz etc. 813

Da die Zeit als constant zu betrachten ist, lässt sich schreiben:

$$d\pi = -\frac{RT\{v_{m_1}\varepsilon_{m_1} d\nu_{m_1} - v_{s_1}\cdot\varepsilon_{s_1}\cdot d\nu_{s_1} + v_{m_2}\varepsilon_{m_2} d\nu_{m_2} - v_{s_2}\varepsilon_{s_2} d\nu_{s_2}\}}{v_{m_1}\varepsilon_{m_1}^2 \nu_{m_1} + v_{s_1}\varepsilon_{s_1}^2 \nu_{s_1} + v_{m_2}\varepsilon_{m_2}^2 \nu_{m_2} + v_{s_2}\cdot\varepsilon_{s_2}^2 \nu_{s_2}}.$$

Der Ausdruck rechts ist im allgemeinen kein vollständiges Differential, was bedeutet, dass $\Delta\Pi$ nicht nur durch die an den diffusionslosen Bereichen herrschenden Concentrationen, sondern auch durch den Charakter des Diffusionsvorganges bestimmt wird. Es gelingt indessen durch einen Kunstgriff in der Anordnung, die Integration zu ermöglichen.

Wir denken uns den Raum V in drei Teile, Raum (1), Raum (2) und Raum (3) eingeteilt und dieselben vor Beginn des Experimentes durch zwei Scheidewände voneinander getrennt. (1) communicire mit I, (3) mit II, in (2) seien beide Salze gleichzeitig gelöst, mit genau denselben Concentrationen wie in I bez. II. Vor Beginn des Experimentes befindet sich also in I und (1) nur das erste, in II und (3) nur das zweite Salz in Lösung, in (2) eine Mischung beider. Die Concentration ist dabei allenthalben constant. Bei Beginn des Experimentes werden die Scheidewände weggenommen und gleich darauf die Potentialdifferenz zwischen den Elektroden gemessen. Für diese Zeit ist aber die Integration über die diffundirenden Schichten möglich, da in der ersten diffundirenden Schicht ν_{m_1} und ν_{s_1}, in der zweiten ν_{m_2} und ν_{s_2} constant sind. Die Integration liefert:

$$\pi_2 - \pi_1 = RT\left\{\frac{v_{m_1} - v_{s_1}}{v_{m_1}\varepsilon_{m_1} + v_{s_1}\varepsilon_{s_1}} \lg\left[1 + \frac{v_{m_1}\varepsilon_{m_1}^2 \nu_{m_1} + v_{s_1}\varepsilon_{s_1}^2 \nu_{s_1}}{v_{m_2}\varepsilon_{m_2}^2 \nu_{m_2} + v_{s_2}\varepsilon_{s_2}^2 \nu_{s_2}}\right]\right.$$

$$\left. - \frac{v_{m_2} - v_{s_2}}{v_{m_2}\varepsilon_{m_2} + v_{s_2}\varepsilon_{s_2}} \lg\left[1 + \frac{v_{m_2}\varepsilon_{m_2}^2 \nu_{m_2} + v_{s_2}\varepsilon_{s_2}^2 \nu_{s_2}}{v_{m_1}\varepsilon_{m_1}^2 \nu_{m_1} + v_{s_1}\varepsilon_{s_1}^2 \nu_{s_1}}\right]\right\}.$$

Eine Vereinfachung der Methode lässt sich erzielen, wenn es möglich ist, in I und II gleiches Säureion von gleicher Concentration zu wählen. Verbindet man nämlich in diesem Falle Raum I mit Raum II direct, so ist für den Anfang des Diffusionsvorganges zu setzen:

$$\frac{\partial(\nu_{s_1} + \nu_{s_2})}{\partial z} = 0;\quad \nu_{s_1} + \nu_{s_2} = \nu_s = \text{const.}$$

Ebenso ist nach Voraussetzung:

$$\varepsilon_{s_1} = \varepsilon_{s_2} = \varepsilon_s \quad \text{und} \quad v_{s_1} = v_{s_2} = v_s.$$

Gleichung (1) geht dann über in

(1') $$v_{m_1}\varepsilon_{m_1} + v_{m_2}\varepsilon_{m_2} - v_s\varepsilon_s = 0.$$

Von den Gleichungen (2) bleibt die erste und dritte unverändert bestehen, aus der zweiten und vierten ergiebt sich durch Addition:

$$v_s \frac{\partial}{\partial x}\left\{ RT \frac{\partial v_s}{\partial x} - \varepsilon_s v_s E \frac{\partial \pi}{\partial x} \right\} = \frac{\partial v_s}{\partial t}.$$

Eliminirt man aus den so veränderten Gleichungen (2) vermittelst der Gleichung (1') die Ableitungen nach der Zeit, so erhält man wie vorhin einen Ausdruck für $d\pi$, welcher ein vollständiges Differential ist. Durch Integration desselben erhält man:

$$\pi_2 - \pi_1 = -\frac{RT}{E} \frac{v_{m_2} - v_{m_1}}{v_{m_2}\varepsilon_2 - v_{m_1}\varepsilon_1} \lg \frac{\varepsilon_{m_2}^2 v_{m_2} \nu_{m_2} + \varepsilon_s^2 v_s \nu_s}{\varepsilon_{m_1}^2 v_{m_1} \nu_{m_1} + \varepsilon_s^2 v_s \nu_s},$$

wobei sich jetzt die Zahlenindices auf die Integrationsgrenzen beziehen. Infolge der Beziehung

$$\varepsilon_{m_1} v_{m_1} = \varepsilon_s v_s = \varepsilon_{m_2} v_{m_2}$$

erhalten wir noch einfacher

$$\pi_2 - \pi_1 = -\frac{RT}{E} \frac{v_{m_2} - v_{m_1}}{v_{m_2}\varepsilon_2 - v_{m_1}\varepsilon_1} \lg \frac{\varepsilon_{m_2} v_{m_2} + \varepsilon_s v_s}{\varepsilon_{m_1} v_{m_1} + \varepsilon_s v_s}.$$

Zum Schlusse empfinde ich noch das Bedürfnis, mich zu entschuldigen, dass ich hier nur einen dürftigen Plan für eine mühevolle Untersuchung entwerfe, ohne selbst zur experimentellen Lösung etwas beizutragen; ich bin jedoch dazu nicht in der Lage. Doch hat diese Arbeit ihr Ziel erreicht, wenn sie einen Forscher veranlasst, das Problem der Molecularkräfte von dieser Seite her in Angriff zu nehmen.

Bern, April 1902.

(Eingegangen 30. April 1902.)

ANNALEN
DER
PHYSIK.

BEGRÜNDET UND FORTGEFÜHRT DURCH
F. A. C. GREN, L. W. GILBERT, J. C. POGGENDORFF, G. UND E. WIEDEMANN.

VIERTE FOLGE.

BAND 9.
DER GANZEN REIHE 314. BAND.

KURATORIUM:
F. KOHLRAUSCH, M. PLANCK, G. QUINCKE,
W. C. RÖNTGEN, E. WARBURG.

UNTER MITWIRKUNG
DER DEUTSCHEN PHYSIKALISCHEN GESELLSCHAFT
UND INSBESONDERE VON
M. PLANCK

HERAUSGEGEBEN VON

PAUL DRUDE.

MIT FÜNF FIGURENTAFELN.

LEIPZIG, 1902.
VERLAG VON JOHANN AMBROSIUS BARTH.

6. *Kinetische Theorie des Wärmegleichgewichtes und des zweiten Hauptsatzes der Thermodynamik;* von *A. Einstein.*

So gross die Errungenschaften der kinetischen Theorie der Wärme auf dem Gebiete der Gastheorie gewesen sind, so ist doch bis jetzt die Mechanik nicht im stande gewesen, eine hinreichende Grundlage für die allgemeine Wärmetheorie zu liefern, weil es bis jetzt nicht gelungen ist, die Sätze über das Wärmegleichgewicht und den zweiten Hauptsatz unter alleiniger Benutzung der mechanischen Gleichungen und der Wahrscheinlichkeitsrechnung herzuleiten, obwohl Maxwell's und Boltzmann's Theorien diesem Ziele bereits nahe gekommen sind. Zweck der nachfolgenden Betrachtung ist es, diese Lücke auszufüllen. Dabei wird sich gleichzeitig eine Erweiterung des zweiten Hauptsatzes ergeben, welche für die Anwendung der Thermodynamik von Wichtigkeit ist. Ferner wird sich der mathematische Ausdruck für die Entropie vom mechanischen Standpunkt aus ergeben.

§ 1. Mechanisches Bild für ein physikalisches System.

Wir denken uns ein beliebiges physikalisches System darstellbar durch ein mechanisches System, dessen Zustand durch sehr viele Coordinaten $p_1, \ldots p_n$ und die dazu gehörigen Geschwindigkeiten

$$\frac{dp_1}{dt}, \ldots \frac{dp_n}{dt}$$

eindeutig bestimmt sei. Die Energie E derselben bestehe aus zwei Summanden, der potentiellen Energie V und der lebendigen Kraft L. Erstere sei eine Function der Coordinaten allein, letztere eine quadratische Function der

$$\frac{dp_\nu}{dt} = p'_\nu,$$

deren Coefficienten beliebige Function der p sind. Auf die Massen des Systems sollen zweierlei äussere Kräfte wirken.

Die einen seien von einem Potentiale V_a ableitbar und sollen die äusseren Bedingungen (Schwerkraft, Wirkung von festen Wänden ohne thermische Wirkung etc.) darstellen; ihr Potential kann die Zeit explicite enthalten, doch soll seine Ableitung nach derselben sehr klein sein. Die anderen Kräfte seien nicht von einem Potential ableitbar und seien schnell veränderlich. Sie sind als diejenigen Kräfte aufzufassen, welche die Wärmezufuhr bewirken. Wirken solche Kräfte nicht, ist aber V_a explicite von der Zeit abhängig, so haben wir einen adiabatischen Process vor uns.

Wir werden auch statt der Geschwindigkeiten, lineare Functionen derselben, die Momente $q_1, \ldots q_n$ als Zustandsvariable des System einführen, welche durch n Gleichungen von der Form

$$q_\nu = \frac{\partial L}{\partial p'_\nu}$$

definirt sind, wobei L als Function der $p_1, \ldots p_n$ und $p_1', \ldots p_n'$ zu denken ist.

§ 2. Ueber die Verteilung der möglichen Zustände unter N identischen adiabatischen stationären Systemen, bei nahezu gleichem Energieinhalt.

Seien unendlich viele (N) Systeme gleicher Art vorhanden, deren Energieinhalt zwischen den bestimmten sehr wenig verschiedenen Werten $\overline{E}$ und $\overline{E} + \delta E$ continuirlich verteilt sind. Aeussere Kräfte, welche nicht von einem Potential ableitbar sind, sollen nicht vorhanden sein und V_a möge die Zeit nicht explicite enthalten, sodass das System ein conservatives System ist. Wir untersuchen die Zustandsverteilung, von welcher wir voraussetzen, dass sie stationär sei.

Wir machen die Voraussetzung, dass ausser der Energie $E = L + V_a + V_i$ oder einer Function dieser Grösse, für das einzelne System keine Function der Zustandsvariabeln p und q allein vorhanden sei, welche mit der Zeit sich nicht ändert; auch fernerhin seien nur Systeme betrachtet, welche diese Bedingung erfüllen. Unsere Voraussetzung ist gleichbedeutend mit der Annahme, dass die Zustandsverteilung unserer Systeme durch den Wert von E bestimmt sei, und sich aus jeden beliebigen Anfangswerten der Zustandsvariabeln, welche nur unserer Bedingung für den Wert der Energie Genüge leisten, von selbst herstelle. Existirte nämlich für das System noch eine Bedingung von der Art $\varphi(p_1, \ldots q_n) = \text{const.}$, welche nicht auf die Form $\varphi(E) = \text{const.}$ gebracht werden kann, so wäre offenbar durch geeignete Wahl der Anfangsbedingungen zu erzielen, dass für jedes der N Systeme φ einen beliebigen vorgeschriebenen Wert hätte. Da sich diese Werte aber mit der Zeit nicht ändern, so folgt z. B., dass der Grösse $\sum \varphi$, erstreckt über alle Systeme, bei gegebenem Werte von E, durch geeignete Wahl der Anfangsbedingungen, jeder beliebige Wert erteilt werden könnte. $\sum \varphi$ ist nun andererseits aus der Zustandsverteilung eindeutig berechenbar, sodass anderen Werten von $\sum \varphi$ andere Zustandsverteilungen entsprechen. Man ersieht also, dass die Existenz eines zweiten solchen Integrals φ notwendig zur Folge hat, dass durch E allein die Zustandsverteilung noch nicht bestimmt wäre, sondern dass dieselbe notwendig vom Anfangszustande der Systeme abhängen müsste.

Bezeichnet man mit g ein unendlich kleines Gebiet aller Zustandsvariabeln $p_1, \ldots p_n, q_1, \ldots q_n$, welches so gewählt sein soll, dass $E(p_1, \ldots q_n)$ zwischen $\overline{E}$ und $\overline{E} + \delta E$ liegt, wenn die Zustandsvariabeln dem Gebiete g angehören, so ist die Verteilung der Zustände durch eine Gleichung von folgender Form zu charakterisiren

$$dN = \psi(p_1, \ldots q_n) \int_g dp_1 \ldots dq_n,$$

dN bedeutet die Anzahl der Systeme, deren Zustandsvariable zu einer bestimmten Zeit dem Gebiete g zugehören. Die Gleichung sagt die Bedingung aus, dass die Verteilung stationär ist.

Wir wählen nun ein solches unendlich kleines Gebiet G. Die Anzahl der Systeme, deren Zustandsvariable zu irgend einer bestimmten Zeit $t=0$ dem Gebiete G angehören, ist dann

$$dN = \psi(P_1, \ldots Q_n) \int_G dP_1 \ldots dQ_n,$$

wobei die grossen Buchstaben die Zugehörigkeit der abhängigen Variabeln zur Zeit $t=0$ andeuten sollen.

Wir lassen nun die beliebige Zeit t verstreichen. Besass ein System in $t=0$ die bestimmten Zustandsvariabeln $P_1, \ldots Q_n$, so besitzt es zur Zeit $t=t$ die bestimmten Zustandsvariabeln $p_1, \ldots q_n$. Die Systeme, deren Zustandsvariabeln in $t=0$ dem Gebiete G angehörten, und zwar nur diese, gehören zur Zeit $t=t$ einem bestimmten Gebiete g an, sodass also die Gleichung gilt:

$$dN = \psi(p_1, \ldots q_n) \int_g .$$

Für jedes derartige System gilt aber der Satz von Liouville, welcher die Form hat:

$$\int dP_1, \ldots dQ_n = \int dp_1, \ldots dq_n.$$

Aus den drei letzten Gleichungen folgt

$$\psi(P_1, \ldots Q_n) = \psi(p_1, \ldots q_n).\ ^1)$$

ψ ist also eine Invariante des Systems, welche nach dem obigen die Form haben muss $\psi(p_1, \ldots q_n) = \psi^*(E)$. Für alle betrachteten Systeme ist aber $\psi^*(E)$ nur unendlich wenig verschieden von $\psi^*(\bar{E}) = $ const., und unsere Zustandsgleichung lautet einfach

$$dN = A \int_g dp_1, \ldots dq_n,$$

wobei A eine von den p und q unabhängige Grösse bedeutet.

§ 3. Ueber die (stationäre) Wahrscheinlichkeit der Zustände eines Systems S, das mit einem System Σ von relativ unendlich grosser Energie mechanisch verbunden ist.

Wir betrachten wieder unendlich viele (N) mechanische Systeme, deren Energie zwischen zwei unendlich wenig verschiedenen Grenzen $\bar{E}$ und $\bar{E} + \delta\bar{E}$ liege. Jedes solche mechanische System sei wieder eine mechanische Verbindung eines Systems S mit den Zustandsvariabeln $p_1, \ldots q_n$ und eines Systems Σ mit den Zustandsvariabeln $\pi_1 \ldots \chi_n$. Der Ausdruck für die Gesamtenergie beider Systeme soll so beschaffen sein, dass jene Terme der Energie, welche durch Einwirkung der Massen eines Teilsystems auf die des anderen Teilsystems

---

1) Vgl. L. Boltzmann, Gastheorie, II. Teil. § 32 u. § 37.

hinzukommen, gegen die Energie E des Teilsystems S zu vernachlässigen seien. Ferner sei die Energie H des Teilsystems Σ unendlich gross gegen E. Bis auf unendlich Kleines höherer Ordnung lässt sich dann setzen:

$$\mathsf{E} = H + E.$$

Wir wählen nun ein in allen Zustandsvariabeln $p_1 \ldots q_n$, $\pi_1 \ldots \chi_n$ unendlich kleines Gebiet g, welches so beschaffen sei, dass E zwischen den constanten Werten $\overline{\mathsf{E}}$ und $\overline{\mathsf{E}} + \delta \overline{\mathsf{E}}$ liege. Die Anzahl dN der Systeme, deren Zustandsvariabeln dem Gebiet g angehören, ist dann nach dem Resultate des vorigen Paragraphen:

$$dN = A \int_g dp_1 \ldots d\chi_n.$$

Wir bemerken nun, dass es in unserem Belieben steht, statt A irgend eine stetige Function der Energie zu setzen, welche für $\mathsf{E} = \overline{\mathsf{E}}$ den Wert A annimmt. Dadurch ändert sich nämlich unser Resultat nur unendlich wenig. Als diese Function wählen wir $A' \cdot e^{-2h\mathsf{E}}$, wobei h eine vorläufig beliebige Constante bedeutet, über welche wir bald verfügen werden. Wir schreiben also:

$$dN = A' \int_g e^{-2h\mathsf{E}} dp_1 \ldots d\chi_n.$$

Wir fragen nun: Wie viele Systeme befinden sich in Zuständen, sodass p_1 zwischen p_1 und $p_1 + dp_1$, p_2 bez. p_2 und $p_2 + dp_2 \ldots q_n$ zwischen q_n und $q_n + dq_n$, $\pi_1 \ldots \chi_n$ aber beliebige, mit den Bedingungen unserer Systeme verträgliche Werte besitzen? Nennt man diese Anzahl dN', so erhält man:

$$dN' = A' e^{-2hE} dp_1 \ldots dq_n \int e^{-2hH} d\pi_1 \ldots d\chi_n.$$

Die Integration erstreckt sich dabei auf jene Werte der Zustandsvariabeln, für welche H zwischen $\overline{\mathsf{E}} - E$ und $\overline{\mathsf{E}} - E + \delta \overline{\mathsf{E}}$ liegt. Wir behaupten nun, der Wert von h sei auf eine und nur eine Weise so zu wählen, dass das in unserer Gleichung auftretende Integral von E unabhängig wird.

Das Integral $\int e^{-2hH} d\pi_1 \ldots d\chi_n$, wobei die Grenzen der Integration durch die Grenzen E und $\mathsf{E} + \delta \overline{\mathsf{E}}$ bestimmt sein mögen, ist nämlich bei bestimmtem $\delta \overline{\mathsf{E}}$ offenbar lediglich

Function von E allein; nennen wir dieselbe $\chi(\mathsf{E})$. Dass in dem Ausdruck für dN' auftretende Integral lässt sich dann in der Form schreiben:
$$\chi(\overline{\mathsf{E}} - E).$$
Da nun E gegen $\overline{\mathsf{E}}$ unendlich klein ist, so lässt sich dies bis auf unendlich Kleines höherer Ordnung in der Form schreiben:
$$\chi(\overline{\mathsf{E}} - E) = \chi(\overline{\mathsf{E}}) - E\chi'(\overline{\mathsf{E}}).$$
Die notwendige und hinreichende Bedingung dafür, dass jenes Integral von E unabhängig ist, lautet also
$$\chi'(\overline{\mathsf{E}}) = 0.$$
Nun lässt sich aber setzen
$$\chi(\mathsf{E}) = e^{-2h\mathsf{E}} \cdot \omega(\mathsf{E}),$$
wobei $\omega(\mathsf{E}) = \int d\pi_1 \ldots d\chi_n$, erstreckt über alle Werte der Variabeln, deren Energiefunction zwischen E und $\mathsf{E} + \delta\mathsf{E}$ liegt.

Die gefundene Bedingung für h nimmt also die Form an:
$$e^{-2h\overline{\mathsf{E}}} \cdot \omega(\overline{\mathsf{E}}) \cdot \left\{ -2h + \frac{\omega'(\overline{\mathsf{E}})}{\omega(\overline{\mathsf{E}})} \right\} = 0,$$
oder
$$h = \tfrac{1}{2} \frac{\omega'(\overline{\mathsf{E}})}{\omega(\overline{\mathsf{E}})}.$$

Es giebt also stets einen und nur einen Wert für h, welcher die gefundenen Bedingungen erfüllt. Da ferner, wie im nächsten Paragraphen gezeigt werden soll, $\omega(\mathsf{E})$ und $\omega'(\mathsf{E})$ stets positiv sind, ist auch h stets eine positive Grösse.

Wählen wir h in dieser Weise, so reducirt sich das Integral auf eine von E unabhängige Grösse, sodass wir für die Zahl der Systeme, deren Variabeln $p_1 \ldots q_n$ in den bezeichneten Grenzen liegen, den Ausdruck erhalten
$$dN' = A'' e^{-2hE} \cdot dp_1 \ldots dq_n.$$

Dies ist also auch bei anderer Bedeutung von A'' der Ausdruck für die Wahrscheinlichkeit, dass die Zustandsvariabeln eines mit einem System von relativ unendlich grosser Energie mechanisch verbundenen Systems zwischen unendlich nahen Grenzen liegen, wenn der Zustand stationär geworden ist.

§ 4. Beweis dafür, dass die Grösse h positiv ist.

Sei $\varphi(x)$ eine homogene, quadratische Function der Variabeln $x_1 \ldots x_n$. Wir betrachten die Grösse $z = \int dx_1 \ldots dx_n$, wobei die Integrationsgrenzen dadurch bestimmt sein mögen, dass $\varphi(x)$ zwischen einem gewissen Wert y und $y + \Delta$ liege, wobei Δ eine Constante sei. Wir behaupten, dass z, welches allein von y Function ist, stets mit wachsendem y zunimmt, wenn $n > 2$.

Führen wir die neuen Variabeln ein $x_1 = \alpha x_1' \ldots x_n = \alpha x_n'$, wobei $\alpha = $ const., dann ist:
$$z = \alpha^n \int dx_1' \ldots dx_n'.$$
Ferner erhalten wir $\varphi(x) = \alpha^2 \varphi(x')$.

Die Integrationsgrenzen des gewonnenen Integrals lauten also für $\varphi(x')$
$$\frac{y}{\alpha^2} \quad \text{und} \quad \frac{y}{\alpha^2} + \frac{\Delta}{\alpha^2}.$$

Ist ferner Δ unendlich klein, was wir annehmen, so erhalten wir
$$z = \alpha^{n-2} \int dx_1' \ldots dx_n'.$$
Hierbei ist y' zwischen den Grenzen
$$\frac{y}{\alpha^2} \quad \text{und} \quad \frac{y}{\alpha^2} + \Delta.$$
Obige Gleichung lässt sich auch schreiben
$$z(y) = \alpha^{n-2} z\left(\frac{y}{\alpha^2}\right).$$
Wählt man α positiv und $n > 2$, so ist also stets
$$\frac{z(y)}{z\left(\frac{y}{\alpha^2}\right)} > 1,$$
was zu beweisen war.

Dieses Resultat benutzen wir, um zu beweisen, dass h positiv ist.

Wir fanden
$$h = \tfrac{1}{2} \frac{\omega'(E)}{\omega(E)},$$
wobei
$$\omega(E) = \int dp_1 \ldots dq_n,$$

und E zwischen E und $E + \delta \overline{E}$. $\omega(E)$ ist der Definition nach notwendig positiv, wir haben nur zu zeigen, dass auch $\omega'(E)$ stets positiv ist.

Wir wählen E_1 und E_2, sodass $E_2 > E_1$, und beweisen, dass $\omega(E_2) > \omega(E_1)$ und zerlegen $\omega(E_1)$ in unendlich viele Summanden von der Form

$$d(\omega(E_1)) = dp_1 \ldots dp_n \int dq_1 \ldots dq_n.$$

Bei dem angedeuteten Integral besitzen die p bestimmte und zwar solche Werte, dass $V \leqq E_1$. Die Integrationsgrenzen des Integrals sind so charakterisirt, dass L zwischen $E_1 - V$ und $E_1 + \delta \overline{E} - V$ liegt.

Jedem unendlich kleinen derartigen Summanden entspricht aus $\omega(E_2)$ ein Term von der Grösse

$$d[\omega(E_2)] = dp_1 \ldots dp_n \int dq_1 \ldots dq_n,$$

wobei die p und dp die nämlichen Werte haben wie in $d[\omega(E_1)]$, L aber zwischen den Grenzen $E_2 - V$ und $E_2 - V + \delta \overline{E}$ liegt.

Es ist also nach dem eben bewiesenen Satze

$$d[\omega(E_2)] > d[\omega(E_1)].$$

Folglich

$$\sum d[\omega(E_2)] > \sum d[\omega(E_1)],$$

wobei $\sum$ über alle entsprechende Gebiete der p zu erstrecken ist. Es ist aber

$$\sum d[\omega(E_1)] = \omega(E_1),$$

wenn das Summenzeichen über alle p erstreckt wird, sodass

$$V \leqq E_1.$$

Ferner ist

$$\sum d[\omega(E_2)] < \omega(E_2),$$

weil das Gebiet der p, welches durch die Gleichung

$$V \leqq E_2$$

bestimmt wird, das durch die Gleichung

$$V \leqq E_1$$

definirte Gebiet vollständig in sich einschliesst.

§ 5. Ueber das Temperaturgleichgewicht.

Wir wählen nun ein System S von ganz bestimmter Beschaffenheit und nennen es Thermometer. Es stehe mit dem System Σ von relativ unendlich grosser Energie in mechanischer Wechselwirkung. Ist der Zustand des Ganzen stationär, so ist der Zustand des Thermometers durch die Gleichung definirt

$$dW = A e^{-2hE} dp_1 \ldots dq_n,$$

wobei dW die Wahrscheinlichkeit dafür bedeutet, dass die Werte der Zustandsvariabeln des Thermometers innerhalb der angedeuteten Grenzen liegen. Dabei besteht zwischen den Constanten A und h die Gleichung

$$1 = A \cdot \int e^{-2hE} dp_1 \ldots dq_n,$$

wobei die Integration über alle möglichen Werte der Zustandsvariabeln erstreckt ist. Die Grösse h bestimmt also den Zustand des Thermometers vollkommen. Wir nennen h die Temperaturfunction, indem wir bemerken, dass nach dem Gesagten jede an dem System S beobachtbare Grösse H Function von h allein sein muss, solange V_a unverändert bleibt, was wir angenommen haben. Die Grösse h aber hängt lediglich vom Zustande des Systems Σ ab (§ 3), ist also unabhängig davon, wie Σ mit S thermisch verbunden ist. Es folgt daraus unmittelbar der Satz: Ist ein System Σ mit zwei unendlich kleinen Thermometern S und S' verbunden, so kommt diesen beiden Thermometern dieselbe Grösse h zu. Sind S und S' identische Systeme, so kommt ihnen auch noch derselbe Wert der beobachtbaren Grösse H zu.

Wir führen nun nur identische Thermometer S ein und nennen H das beobachtbare Temperaturmaass. Wir erhalten also den Satz: Das an S beobachtbare Temperaturmaass H ist unabhängig von der Art, wie Σ mit S mechanisch verbunden ist; die Grösse H bestimmt h, dieses die Energie E des Systems Σ und diese dessen Zustand nach unserer Voraussetzung.

Aus dem Bewiesenen folgt sofort, dass zwei Systeme Σ_1 und Σ_2 im Falle mechanischer Verbindung kein im stationären Zustand befindliches System bilden können, wenn nicht zwei mit ihnen verbundene Thermometer S gleiches Temperaturmaass oder, was dasselbe bedeutet, sie selbst gleiche Temperaturfunction besitzen. Da der Zustand der Systeme Σ_1 und Σ_2 durch die Grössen h_1 und h_2 oder H_1 und H_2 vollständig definirt wird, so folgt, dass das Temperaturgleichgewicht lediglich durch die Bedingungen $h_1 = h_2$ oder $H_1 = H_2$ bestimmt sein kann.

Es bleibt jetzt noch übrig, zu zeigen, dass zwei Systeme von gleicher Temperaturfunction h (oder gleichem Temperaturmaass H) mechanisch verbunden werden können zu einem einzigen System von gleicher Temperaturfunction.

Seien zwei mechanische Systeme Σ_1 und Σ_2 mechanisch zu einem System verschmolzen, so jedoch, dass die Terme der Energie unendlich klein sind, welche Zustandsvariabeln beider Systeme enthalten. Sowohl Σ_1 als Σ_2 seien verknüpft mit einem unendlich kleinen Thermometer S. Die Angaben H_1 und H_2 desselben sind bis auf unendlich Kleines jedenfalls dieselben, weil sie sich nur auf verschiedene Stellen, eines einzigen, im stationären Zustande befindlichen Systems beziehen. Ebenso natürlich die Grössen h_1 und h_2. Wir denken uns nun unendlich langsam die beiden Systemen gemeinsame Terme der Energie gegen Null hin abnehmen. Hierbei ändern sich sowohl die Grössen H und h, als auch die Zustandsverteilungen beider Systeme unendlich wenig, da diese allein durch die Energie bestimmt sind. Ist dann die vollständige mechanische Trennung von Σ_1 und Σ_2 ausgeführt, so bleiben gleichwohl die Beziehungen

$$H_1 = H_2, \quad h_1 = h_2.$$

bestehen und die Zustandsverteilung ist unendlich wenig verändert. H_1 und h_1 beziehen sich aber nur mehr auf Σ_1, H_2 und h_2 nur mehr auf Σ_2. Unser Process ist streng umkehrbar, da er sich aus einer Aufeinanderfolge von stationären Zuständen zusammensetzt. Wir erhalten also den Satz:

Zwei Systeme von der gleichen Temperaturfunction h lassen sich zu einem einzigen System von der Temperaturfunction h verknüpfen, sodass sich deren Zustandsverteilung unendlich wenig ändert.

Gleichheit der Grössen h ist also die notwendige und hinreichende Bedingung für die stationäre Verknüpfung (Wärmegleichgewicht) zweier Systeme. Daraus folgt sofort: Sind die Systeme Σ_1 und Σ_2, und die Systeme Σ_1 und Σ_3 stationär mechanisch verknüpfbar (im Wärmegleichgewichte), so sind es auch Σ_2 und Σ_3.

Ich will hier bemerken, dass wir bis jetzt von der Voraussetzung, dass unsere Systeme mechanische seien, nur insofern Gebrauch gemacht haben, als wir den Liouville'schen Satz und das Energieprincip verwendet haben. Wahrscheinlich lassen sich die Fundamente der Wärmetheorie für noch weit allgemeiner definirte Systeme entwickeln. Solches wollen wir hier jedoch nicht versuchen, sondern uns auf die mechanischen Gleichungen stützen. Die wichtige Frage, inwiefern sich der Gedankengang von dem benutzten Bilde loslösen und verallgemeinern lässt, werden wir hier nicht behandeln.

§ 6. Ueber die mechanische Bedeutung der Grösse h.[1]

Die lebendige Kraft L eines Systems ist eine homogene quadratische Function der Grössen q. Durch eine lineare Substitution lassen sich stets Variable r einführen, sodass die lebendige Kraft in der Form erscheint

$$L = \tfrac{1}{2}(\alpha_1 r_1^2 + \alpha_2 r_2^2 + \ldots + \alpha_n r_n^2)$$

und dass

$$\int dq_1 \ldots dq_n = \int dr_1 \ldots dr_n,$$

wenn man die Integrale über entsprechende unendlich kleine Gebiete ausdehnt. Die Grössen r nennt Boltzmann Momentoiden. Die mittlere lebendige Kraft, welche einer Momentoide entspricht, wenn das System mit einem anderen, von viel grösserer Energie, ein System bildet, nimmt die Form an:

$$\frac{\int A'' e^{-2h[V + \alpha_1 r_1^2 + \alpha_2 r_2^2 + \ldots + \alpha_n r_n^2]} \cdot \frac{\alpha_\nu r_\nu^2}{2} \cdot dp_1 \ldots dp_n \cdot dr_1 \ldots dr_n}{\int A'' e^{-2h[V + \alpha_1 r_1^2 + \alpha_2 r_2^2 + \ldots + \alpha_n r_n^2]} \cdot dp_1 \ldots dp_n dr_1 \ldots dr_n} = \frac{1}{4h}.$$

---

[1] Vgl. L. Boltzmann, Gastheorie, II. Teil, §§ 33, 34, 42.

Die mittlere lebendige Kraft aller Momentoiden eines Systems ist also dieselbe und gleich:

$$\frac{1}{4h} = \frac{L}{n},$$

wobei L die lebendige Kraft des Systems bedeutet.

§ 7. Ideale Gase. Absolute Temperatur.

Die entwickelte Theorie enthält als speciellen Fall die Maxwell'sche Zustandsverteilung der idealen Gase. Verstehen wir nämlich in § 3 unter dem System S ein Gasmolecül, unter Σ die Gesamtheit aller anderen, so folgt für die Wahrscheinlichkeit, dass die Werte der Variabeln $p_1 \ldots q_n$ von S in einem in Bezug auf alle Variabeln unendlich kleinen Gebiet g liegen, der Ausdruck

$$dW = A e^{-2hE} \int_g dp_1 \ldots dq_n.$$

Auch erkennt man sogleich aus unserem, für die Grösse h in § 3 gefundenen Ausdruck, dass die Grösse h bis auf unendlich Kleines die nämliche wäre für ein Gasmolecül anderer Art, welches in dem Systeme vorkommt, in dem die Systeme Σ, welche h bestimmen, für beide Molecüle bis auf unendlich Kleines identisch sind. Damit ist die verallgemeinerte Maxwell'sche Zustandsverteilung für ideale Gase erwiesen. —

Ferner folgt sofort, dass die mittlere lebendige Kraft der Schwerpunktsbewegung eines Gasmolecüles, welches in einem System S vorkommt, den Wert $3/4h$ besitzt, weil dieselbe drei Momentoiden entspricht. Nun lehrt die kinetische Gastheorie, dass diese Grösse proportional dem vom Gase bei constanten Volumen ausgeübten Druck ist. Setzt man diesen definitionsgemäss der absoluten Temperatur T proportional, so hat man eine Beziehung von der Form

$$\frac{1}{4h} = \varkappa \cdot T = \tfrac{1}{2} \frac{\omega(\bar{E})}{\omega'(\bar{E})},$$

wobei $\varkappa$ eine universelle Constante, ω die in § 3 eingeführte Function bedeutet.

§ 8. Der zweite Hauptsatz der Wärmetheorie als Folgerung der mechanischen Theorie.

Wir betrachten ein gegebenes physikalisches System S als mechanisches System mit den Coordinaten $p_1 \ldots p_n$. Als Zustandsvariable in demselben führen wir ferner die Grössen

$$\frac{d p_1}{d t} = p_1' \ldots \frac{d p_n}{d t} = p_n'$$

ein. $P_1 \ldots P_n$ seien die äusseren Kräfte, welche die Coordinaten des Systems zu vergrössern streben. V_i sei die potentielle Energie des Systems, L dessen lebendige Kraft, welche eine homogene quadratische Function der p_ν' ist. Die Bewegungsgleichungen von Lagrange nehmen für ein solches System die Form an

$$\frac{\partial (V_i - L)}{\partial p_\nu} + \frac{d}{d t}\left[\frac{\partial L}{\partial p_\nu'}\right] - P_\nu = 0, \quad (\nu = 1, \ldots \nu = n).$$

Die äusseren Kräfte setzen sich aus zweierlei Kräften zusammen. Die einen, $P_\nu^{(1)}$, sind diejenigen Kräfte, welche die Bedingungen des Systems darstellen, und von einem Potential ableitbar sind, welches nur Function der $p_1 \ldots p_n$ ist (adiabatische Wände, Schwerkraft etc.):

$$P_\nu^{(1)} = \frac{\partial V_a}{\partial p_\nu}.$$

Da wir Processe zu betrachten haben, welche mit unendlicher Annäherung aus stationären Zuständen bestehen, haben wir anzunehmen, dass V_a die Zeit zwar explicite enthalte, dass aber die partiellen Ableitungen der Grössen $\partial V_a/\partial p_\nu$ nach der Zeit unendlich klein seien.

Die anderen Kräfte, $P_\nu^{(2)} = \Pi_\nu$, seien nicht von einem Potential ableitbar, welches nur von den p_ν abhängt. Die Kräfte Π_ν stellen die Kräfte dar, welche die Wärmezufuhr vermitteln.

Setzt man $V_a + V_i = V$, so gehen die Gleichungen (1) über in

$$\Pi_\nu = \frac{\partial (V - L)}{\partial p_\nu} + \frac{d}{d t}\left\{\frac{\partial L}{\partial p_\nu'}\right\}.$$

Die Arbeit, welche durch die Kräfte Π_ν in der Zeit dt dem System zugeführt wird, ist dann die Darstellung der vom

System S während dt aufgenommenen Wärmemenge dQ, welche wir im mechanischen Maass messen wollen.

$$dQ = \sum \Pi_\nu \, dp_\nu = \sum \frac{\partial V}{\partial p_\nu} dp_\nu - \sum \frac{\partial L}{\partial p_\nu} dp_\nu + \sum \frac{dp_\nu}{dt} \frac{d}{dt}\left\{\frac{\partial L}{\partial p_\nu}\right\} dt.$$

Da aber

$$\sum p'_\nu \frac{d}{dt}\left\{\frac{\partial L}{\partial p'_\nu}\right\} dt = d \sum p'_\nu \frac{\partial L}{\partial p'_\nu} - \sum \frac{\partial L}{\partial p'_\nu} dp'_\nu.$$

ferner

$$\sum \frac{\partial L}{\partial p'_\nu} p'_\nu = 2L, \quad \sum \frac{\partial L}{\partial p'_\nu} dp_\nu + \sum \frac{\partial L}{\partial p'_\nu} dp'_\nu = dL,$$

so ist

$$dQ = \sum \frac{\partial V}{\partial p'_\nu} dp_\nu + dL.$$

Da ferner

$$T = \frac{1}{4\varkappa h} = \frac{L}{n\varkappa},$$

so ist

(1) $$\frac{dQ}{T} = n\varkappa \frac{dL}{L} + 4\varkappa h \sum \frac{\partial V}{\partial p_\nu} dp_\nu.$$

Wir beschäftigen uns nun mit dem Ausdruck

$$\sum \frac{\partial V}{\partial p_\nu} dp_\nu.$$

Derselbe stellt die Zunahme des Systems an potentieller Energie dar, welche stattfinden würde während der Zeit dt, wenn V nicht explicite von der Zeit abhängig wäre. Das Zeitelement dt sei so gross gewählt, dass an die Stelle jener Summe deren Mittelwert für unendlich viele gleichtemperirte Systeme S gesetzt werden kann, aber doch so klein, dass die expliciten Aenderungen von h und V nach der Zeit unendlich klein seien.

Unendlich viele Systeme S im stationären Zustande, welche alle identische h und V_a besitzen, mögen übergehen in neue stationäre Zustände, welche durch die allen gemeinsamen Werte $h + \delta h$, $V + \delta V$ charakterisirt sein mögen. „δ" bezeichne allgemein die Aenderung einer Grösse beim Uebergang des Systems in den neuen Zustand; das Zeichen „d" bezeichne nicht mehr die Aenderung mit der Zeit, sondern Differentiale bestimmter Integrale. —

Die Anzahl der Systeme, deren Zustandsvariable vor der Aenderung innerhalb des unendlich kleinen Gebietes g sich befinden, ist durch die Formel gegeben

$$dN = A \cdot e^{-2h(V+L)} \int dp_1 \ldots dp_n,$$

dabei steht es in unserer Willkür, für jedes gegebene h und V_a die willkürliche Constante von V so zu wählen, dass die Constante A der Einheit gleich wird. Wir wollen dies thun, um die Rechnung einfacher zu gestalten, und die so genauer definirte Function V^* nennen.

Man sieht nun leicht, dass die von uns gesuchte Grösse den Wert erhält:

(2) $\quad \sum \dfrac{\partial V^*}{\partial p_n} dp_n = \dfrac{1}{N} \int \delta\{e^{-2h(V+L)}\} \cdot V^* dp_1 \ldots dq_n,$

wobei die Integration über alle Werte der Variabeln zu erstrecken ist. Dieser Ausdruck stellt nämlich die Vermehrung der mittleren potentiellen Energie des Systems dar, welche einträte, wenn zwar die Zustandsverteilung sich gemäss δV^* und δh änderte, V aber sich nicht explicite veränderte.

Ferner erhalten wir:

(3) $\begin{cases} 4 \varkappa h \sum \dfrac{\partial V}{\partial p_\nu} dp_\nu = 4\varkappa \dfrac{1}{N} \int \delta\{e^{-2h(V^*+L)}\} \cdot h \cdot V \cdot dp_1 \ldots dq_n \\ \qquad = 4\varkappa \delta[h\overline{V}] - \dfrac{4\varkappa}{N} \int e^{-2h(V^*+L)} \delta[hV] \\ \qquad \qquad\qquad\qquad\qquad\qquad\qquad dp_1 \ldots dq_n. \end{cases}$

Die Integrationen sind hier und im Folgenden über alle möglichen Werte der Variabeln zu erstrecken. Ferner hat man zu bedenken, dass die Anzahl der betrachteten Systeme sich nicht ändert. Dies liefert die Gleichung:

$$\int \delta(e^{-2h(V^*+L)}) dp_1 \ldots dq_n = 0,$$

oder

$$\int e^{-2h(V^*+L)} \delta(hV) dp_1 \ldots dq_n + \delta h \int e^{-2h(V^*+L)} \delta(L) \\ dp_1 \ldots dq_n = 0,$$

oder

(4) $\quad \dfrac{4\varkappa}{N} \int e^{-2h(V^*+L)} \delta(hV) dp_1 \ldots dq_n + 4\varkappa \overline{L} \delta h = 0.$

$\overline{V}$ und $\overline{L}$ bezeichnen die Mittelwerte der potentiellen Energie und der lebendigen Kraft der N-Systeme. Durch Addition von (3) und (4) erhält man:

$$4\varkappa h \sum \frac{\partial V^*}{\partial p_\nu} dp_\nu = 4\varkappa \delta[h\overline{V}] + 4\varkappa \overline{L}.\delta h,$$

oder, weil

$$h = \frac{n}{4\overline{L}}, \quad \delta h = -\frac{n}{4L^2}\cdot \delta L,$$

$$4\varkappa h \sum \frac{\partial V}{\partial p_\nu} dp_\nu = 4\varkappa \delta[h\overline{V}] - n\varkappa \frac{\delta L}{L}.$$

Setzt man diese Formel in (1) ein, so erhält man

$$\frac{dQ}{T} = \delta[4\varkappa h \overline{V}^*] = \delta\left[\frac{\overline{V}^*}{T}\right].$$

dQ/T ist also ein vollständiges Differential. Da

$$\frac{\overline{L}}{T} = n\varkappa, \quad \text{also} \quad \delta\left(\frac{L}{T}\right) = 0$$

ist, so lässt sich auch setzen

$$\frac{dQ}{T} = \delta\left(\frac{E^*}{T}\right).$$

E^*/T ist also bis auf eine willkürliche additive Constante der Ausdruck für die Entropie des Systems, wobei $E^* = V^* + L$ gesetzt ist. Der zweite Hauptsatz erscheint also als notwendige Folge des mechanischen Weltbildes.

§ 9. Berechnung der Entropie.

Der für die Entropie ε gefundene Ausdruck $\varepsilon = E^*/T$ ist nur scheinbar so einfach, da E^* aus den Bedingungen des mechanischen Systems erst berechnet werden muss. Es ist nämlich

$$E^* = E + E_0,$$

wobei E unmittelbar gegeben, E_0 aber durch die Bedingung

$$\int e^{-2h(E-E_0)} dp_1 \ldots dq_n = N$$

als Function von E und h zu bestimmen ist. Man erhält so:

$$\varepsilon = \frac{E^*}{T} = \frac{E}{T} + 2\varkappa \log\left\{\int e^{-2hE} dp_1 \ldots dq_n\right\} + \text{const.}$$

In dem so gefundenen Ausdruck ist die der Grösse E zuzufügende willkürliche Constante ohne Einfluss auf das Resultat, und das als „const" bezeichnete dritte Glied ist von V und T unabhängig.

Der Ausdruck für die Entropie ε ist darum merkwürdig, weil er lediglich von E und T abhängt, die specielle Form von E als Summe potentieller Energie und lebendiger Kraft aber nicht mehr hervortreten lässt. Diese Thatsache lässt vermuten, dass unsere Resultate allgemeiner sind als die benutzte mechanische Darstellung, zumal der in § 3 für h gefundene Ausdruck dieselbe Eigenschaft aufweist.

§ 10. Erweiterung des zweiten Hauptsatzes.

Ueber die Natur der Kräfte, welche dem Potential V_a entsprechen, brauchte nichts vorausgesetzt zu werden, auch nicht, dass solche Kräfte in der Natur vorkommen. Die mechanische Theorie der Wärme verlangt also, dass wir zu richtigen Resultaten gelangen, wenn wir das Carnot'sche Princip auf ideale Processe anwenden, welche aus den beobachteten durch Einführung beliebiger V_a erzeugt werden können. Natürlich haben die aus der theoretischen Betrachtung jener Processe gewonnenen Resultate nur dann reale Bedeutung, wenn in ihnen die idealen Hülfskräfte V_a nicht mehr vorkommen.

Bern, Juni 1902.

(Eingegangen 26. Juni 1902.)

ANNALEN
DER
PHYSIK.

BEGRÜNDET UND FORTGEFÜHRT DURCH

F. A. C. GREN, L. W. GILBERT, J. C. POGGENDORFF, G. UND E. WIEDEMANN.

VIERTE FOLGE.

BAND 11.

DER GANZEN REIHE 316. BAND.

KURATORIUM:

F. KOHLRAUSCH, M. PLANCK, G. QUINCKE,
W. C. RÖNTGEN, E. WARBURG.

UNTER MITWIRKUNG

DER DEUTSCHEN PHYSIKALISCHEN GESELLSCHAFT

UND INSBESONDERE VON

M. PLANCK

HERAUSGEGEBEN VON

PAUL DRUDE.

MIT SECHS FIGURENTAFELN.

LEIPZIG, 1903.

VERLAG VON JOHANN AMBROSIUS BARTH.

9. *Eine Theorie der Grundlagen der Thermodynamik;* von *A. Einstein.*

In einer neulich erschienenen Arbeit habe ich gezeigt, daß die Sätze vom Temperaturgleichgewicht und der Entropiebegriff mit Hülfe der kinetischen Theorie der Wärme hergeleitet werden können. Es drängt sich nun naturgemäß die Frage auf, ob die kinetische Theorie auch wirklich notwendig ist, um jene Fundamente der Wärmetheorie herleiten zu können, oder ob vielleicht bereits Voraussetzungen allgemeinerer Art dazu genügen können. Daß dieses letztere der Fall ist, und durch welche Art von Überlegungen man zum Ziele gelangen kann, soll in dieser Abhandlung gezeigt werden.

§ 1. Über eine allgemeine mathematische Darstellung der Vorgänge in isolierten physikalischen Systemen.

Der Zustand irgend eines von uns betrachteten physikalischen Systems sei eindeutig bestimmt durch sehr viele (n) skalare Größen $p_1, p_2 \ldots p_n$, welche wir Zustandsvariabeln nennen. Die Änderung des Systems in einem Zeitelement dt ist dann durch die Änderungen $dp_1, dp_2 \ldots dp_n$ bestimmt, welche die Zustandsvariabeln in jenem Zeitelement erleiden.

Das System sei isoliert, d. h. das betrachtete System stehe mit anderen Systemen nicht in Wechselwirkung. Es ist dann klar, daß der Zustand des Systems in einem bestimmten Zeitmoment in eindeutiger Weise die Veränderung des Systems im nächsten Zeitelement dt, d. h. die Größen $dp_1, dp_2 \ldots dp_n$ bestimmt. Diese Aussage ist gleichbedeutend mit einem System von Gleichungen von der Form:

$$(1) \quad \frac{dp_i}{dt} = \varphi_i(p_1 \ldots p_n) \quad (i = 1 \ldots i = n),$$

wobei die φ eindeutige Funktionen ihrer Argumente sind.

Für ein solches System von linearen Differentialgleichungen existiert im allgemeinen keine Integralgleichung von der Form

$$\psi(p_1 \ldots p_n) = \text{konst.},$$

welche die Zeit nicht explizite enthält. Für das Gleichungssystem aber, welches die Veränderungen eines nach außen abgeschlossenen, physikalischen Systems darstellt, müssen wir annehmen, daß mindestens eine solche Gleichung besteht, nämlich die Energiegleichung:

$$E(p_1 \ldots p_n) = \text{konst.}$$

Wir nehmen zugleich an, daß keine weitere, von dieser unabhängige Integralgleichung solcher Art vorhanden sei.

§ 2. Über die stationäre Zustandsverteilung unendlich vieler isolierter physikalischer Systeme, welche nahezu gleiche Energie besitzen.

Die Erfahrung zeigt, daß ein isoliertes physikalisches System nach einer gewissen Zeit einen Zustand annimmt, in welchem sich keine wahrnehmbare Größe des Systems mehr mit der Zeit ändert; wir nennen diesen Zustand den stationären. Es wird also offenbar nötig sein, daß die Funktionen φ_i eine gewisse Bedingung erfüllen, damit die Gleichungen (1) ein solches physikalisches System darstellen können.

Nehmen wir nun an, daß eine wahrnehmbare Größe stets durch einen zeitlichen Mittelwert einer gewissen Funktion der Zustandsvariabeln $p_1 \ldots p_n$ bestimmt sei, und daß diese Zustandsvariabeln $p_1 \ldots p_n$ immer wieder dieselben Wertsysteme mit stets gleichbleibender Häufigkeit annehmen, so folgt aus dieser Bedingung, welche wir zur Voraussetzung erheben wollen, mit Notwendigkeit die Konstanz der Mittelwerte aller Funktionen der Größen $p_1 \ldots p_n$; nach dem obigen also auch die Konstanz jeder wahrnehmbaren Größe.

Diese Voraussetzung wollen wir genau präzisieren. Wir betrachten ein physikalisches System, welches durch die Gleichungen (1) dargestellt und dessen Energie E sei, von einem beliebigen Zeitpunkte an die Zeit T hindurch. Denken wir uns ein beliebiges Gebiet $\varGamma$ der Zustandsvariabeln $p_1 \ldots p_n$ gewählt, so werden in einem bestimmten Zeitpunkt der Zeit T die Werte der Variabeln $p_1 \ldots p_n$ in diesem Gebiete $\varGamma$ gelegen sein, oder sie liegen außerhalb desselben; sie werden also während eines Bruchteiles der Zeit T, welchen wir τ nennen wollen, in dem gewählten Gebiete $\varGamma$ liegen. Unsere Bedingung lautet dann folgendermaßen: Wenn $p_1 \ldots p_n$ Zustandsvariable eines physikalischen Systems sind, also eines Systems, welches einen stationären Zustand annimmt, so besitzt die Größe τ/T für $T=\infty$ für jedes Gebiet Γ einen bestimmten Grenzwert. Dieser Grenzwert ist für jedes unendlich kleine Gebiet unendlich klein.

Auf diese Voraussetzung kann man folgende Betrachtung gründen. Seien sehr viele (N) unabhängige physikalische Systeme vorhanden, welche sämtlich durch das nämliche Gleichungssystem (1) dargestellt seien. Wir greifen einen beliebigen Zeitpunkt t heraus und fragen nach der Verteilung der möglichen Zustände unter diesen N Systemen, unter der Voraussetzung, daß die Energie E aller Systeme zwischen E^* und dem unendlich benachbarten Werte $E^* + \delta E^*$ liege. Aus der oben eingeführten Voraussetzung folgt sofort, daß die Wahrscheinlichkeit dafür, daß die Zustandsvariabeln eines zufällig herausgegriffenen der N Systeme in der Zeit t innerhalb des Gebietes Γ liegen, den Wert

$$\lim_{T=\infty} \frac{\tau}{T} = \text{konst.}$$

habe. Die Zahl der Systeme, deren Zustandsvariable in der Zeit t innerhalb des Gebietes Γ liegen, ist also:

$$N . \lim_{T=\infty} \frac{\tau}{T},$$

also eine von der Zeit unabhängige Größe. Bezeichnet g ein in allen Variabeln unendlich kleines Gebiet der Koordinaten $p_1 \ldots p_n$, so ist also die Anzahl der Systeme, deren Zustandsvariable zu einer beliebigen Zeit das beliebig gewählte unendlich kleine Gebiet g erfüllen:

(2) $$dN = \varepsilon(p_1 \ldots p_n) \int_g dp_1 \ldots dp_n.$$

Die Funktion ε gewinnt man, indem man die Bedingung in Zeichen faßt, daß die durch die Gleichung (2) ausgedrückte Zustandsverteilung eine stationäre ist. Es sei im speziellen das Gebiet g so gewählt, daß p_1 zwischen den bestimmten Werten p_1 und $p_1 + dp_1$, p_2 zwischen p_2 und $p_2 + dp_2 \ldots p_n$ zwischen p_n und $p_n + dp_n$ gelegen ist, dann ist für die Zeit t

$$dN_t = \varepsilon(p_1 \ldots p_n) . dp_1 . dp_2 \ldots dp_n,$$

Theorie der Grundlagen der Thermodynamik.

wobei der Index von dN die Zeit bezeichnet. Mit Berücksichtigung der Gleichung (1) erhält man ferner für die Zeit $t + dt$ und dasselbe Gebiet der Zustandsvariabeln

$$dN_{t+dt} = dN_t - \sum_{\nu=1}^{\nu=n} \frac{\partial(\varepsilon \varphi_\nu)}{\partial p_\nu} \cdot dp_1 \ldots dp_n \cdot dt.$$

Da aber $dN_t = dN_{t+dt}$ ist, da die Verteilung eine stationäre ist, so ist

$$\sum \frac{\partial(\varepsilon \varphi_\nu)}{\partial p_\nu} = 0.$$

Daraus ergibt sich

$$-\sum \frac{\partial \varphi_\nu}{\partial p_\nu} = \sum \frac{\partial(\log \varepsilon)}{\partial p_\nu} \cdot \varphi_\nu = \sum \frac{\partial(\log \varepsilon)}{\partial p_\nu} \cdot \frac{dp_\nu}{dt} = \frac{d(\log \varepsilon)}{dt},$$

wobei $d(\log \varepsilon)/dt$ die Veränderung der Funktion $\log \varepsilon$ für ein einzelnes System nach der Zeit unter Berücksichtigung der zeitlichen Veränderung der Größen p_ν bezeichnet.

Man erhält ferner:

$$\varepsilon = e^{-\int dt \sum_{\nu=1}^{\nu=n} \frac{\partial \varphi_\nu}{\partial p_\nu} + \psi(E)} = e^{-m + \psi(E)}.$$

Die unbekannte Funktion ψ ist die von der Zeit unabhängige Integrationskonstante, welche von den Variabeln $p_1 \ldots p_n$ zwar abhängen, sie jedoch, nach der im § 1 gemachten Voraussetzung, nur in der Kombination, wie sie in der Energie E auftreten, enthalten kann.

Da aber $\psi(E) = \psi(E^*) =$ konst. für alle N betrachteten Systeme ist, reduziert sich für unseren Fall der Ausdruck für ε auf:

$$\varepsilon = \text{konst.}\, e^{-\int dt \sum_{\nu=1}^{\nu=n} \frac{\partial \varphi_\nu}{\partial p_\nu}} = \text{konst.}\, e^{-m}.$$

Nach dem obigen ist nun:

$$dN = \text{konst.}\, e^{-m} \int_g dp_1 \ldots dp_n.$$

Der Einfachheit halber führen wir nun neue Zustandsvariabeln für die betrachteten Systeme ein; sie mögen mit π_ν bezeichnet werden. Es ist dann:
$$dN = \frac{e^{-m}}{\frac{D(\pi_1 \ldots \pi_n)}{D(p_1 \ldots p_n)}} \int_g d\pi_1 \ldots d\pi_n,$$

wobei das Symbol D die Funktionaldeterminante bedeutet. — Wir wollen nun die neuen Koordinaten so wählen, daß
$$e^{-m} = \frac{D(\pi_1 \ldots \pi_n)}{D(p_1 \ldots p_n)}$$

werde. Diese Gleichung läßt sich auf unendlich viele Arten befriedigen, z. B. wenn man setzt:
$$\pi_2 = p_2$$
$$\pi_3 = p_3 \qquad \pi_1 = \int e^{-m} \cdot dp_1.$$
$$\ldots$$
$$\pi_n = p_n$$

Wir erhalten also unter Benutzung der neuen Variabeln
$$dN = \text{konst.} \int d\pi_1 \ldots d\pi_n.$$

Im folgenden wollen wir uns stets solche Variabeln eingeführt denken.

§ 3. Über die Zustandsverteilung eines Systems, welches ein System von relativ unendlich großer Energie berührt.

Wir nehmen nun an, daß jedes der N isolierten Systeme, aus zwei Teilsystemen Σ und σ, welche in Wechselwirkung stehen, zusammengesetzt sei. Der Zustand des Teilsystems Σ möge durch die Werte der Variabeln $\Pi_1 \ldots \Pi_\lambda$, der Zustand des Systems σ durch die Werte der Variabeln $\pi_1 \ldots \pi_l$ bestimmt sein. Ferner setze sich die Energie E, welche für jedes System zwischen den Werten E^* und $E^* + \delta E^*$ liegen mag, also bis auf unendlich kleines gleich E^* sein soll, bis auf unendlich kleines, aus zwei Termen zusammen, von denen der erste H nur durch die Werte der Zustandsvariabeln von Σ, der zweite η nur durch die der Zustandsvariabeln von σ bestimmt sei, sodaß bis auf relativ unendlich kleines gilt:
$$E = H + \eta.$$

Zwei in Wechselwirkung stehende Systeme, welche diese Bedingung erfüllen, nennen wir zwei sich berührende Systeme. Wir setzen noch voraus, daß η gegen H unendlich klein sei.

Für die Anzahl dN_1 der N-Systeme, deren Zustandsvariabeln $\Pi_1 \ldots \Pi_\lambda$ und $\pi_1 \ldots \pi_l$ in den Grenzen zwischen Π_1 und $\Pi_1 + d\Pi_1$, Π_2 und $\Pi_2 + d\Pi_2 \ldots \Pi_\lambda$ und $\Pi_\lambda + d\Pi_\lambda$ und π_1 und $\pi_1 + d\pi_1$, π_2 und $\pi_2 + d\pi_2 \ldots \pi_l$ und $\pi_l + d\pi_l$ liegen, ergibt sich der Ausdruck:

$$dN_1 = C . d\Pi_1 \ldots d\Pi_\lambda . d\pi_1 \ldots d\pi_l,$$

wobei C eine Funktion von $E = H + \eta$ sein kann.

Da aber nach der obigen Annahme die Energie eines jeden betrachteten Systems bis auf unendlich kleines den Wert E^* besitzt, so können wir, ohne an dem Resultat etwas zu ändern, C durch konst. e^{-2hE^*} = konst. $e^{-2h(H+\eta)}$ ersetzen, wobei h eine noch näher zu definierende Konstante bedeutet. Der Ausdruck für dN_1 geht also über in:

$$dN_1 = \text{konst.}\, e^{-2h(H+\eta)} . d\Pi_1 \ldots d\Pi_\lambda . d\pi_1 \ldots d\pi_l.$$

Die Anzahl der Systeme, deren Zustandsvariabeln π zwischen den angedeuteten Grenzen liegen, während die Werte der Variabeln Π keiner beschränkenden Bedingung unterworfen sind, wird sich also in der Form

$$dN_2 = \text{konst.}\, e^{-2h\eta} . d\pi_1 \ldots d\pi_l \int e^{-2hH} d\Pi_1 \ldots d\Pi_\lambda$$

darstellen lassen, wobei das Integral über alle Werte der Π auszudehnen ist, denen Werte der Energie H zukommen, welche zwischen $E^* - \eta$ und $E^* + \delta E^* - \eta$ gelegen sind. Wäre die Integration ausgeführt, so hätten wir die Zustandsverteilung der Systeme σ gefunden. Dies ist nun tatsächlich möglich.

Wir setzen:

$$\int e^{-2hH} . d\Pi_1 \ldots d\Pi_\lambda = \chi(E),$$

wobei die Integration auf der linken Seite über alle Werte der Variabeln zu erstrecken ist, für welche H zwischen den bestimmten Werten E und $E + \delta E^*$ liegt. Das Integral, welches im Ausdruck dN_2 auftritt, nimmt dann die Form an

$$\chi(E^* - \eta),$$

oder, da η gegen E^* unendlich klein ist:

$$\chi(E^*) - \chi'(E^*) . \eta.$$

Läßt sich also h so wählen, daß $\chi'(E^*) = 0$, so reduziert sich das Integral auf eine vom Zustand von σ unabhängige Größe.

Es läßt sich bis auf unendlich kleines setzen:
$$\chi(E) = e^{-2hE} \int d\Pi_1 \ldots d\Pi_\lambda = e^{-2hE} \cdot \omega(E),$$
wo die Grenzen der Integration gleich sind wie oben, und ω eine neue Funktion von E bedeutet.

Die Bedingung für h nimmt nun die Form an:
$$\chi'(E^*) = e^{-2hE^*} \cdot \{\omega'(E^*) - 2h\,\omega(E^*)\} = 0,$$
folglich:
$$h = \tfrac{1}{2} \frac{\omega'(E^*)}{\omega(E^*)}.$$

Es sei h in dieser Weise gewählt, dann wird der Ausdruck für dN_2 die Form annehmen:

(3) $\qquad dN_2 = \text{konst.}\, e^{-2h\eta}\, d\pi_1 \ldots d\pi_l.$

Bei geeigneter Wahl der Konstanten stellt dieser Ausdruck die Wahrscheinlichkeit dafür dar, daß die Zustandsvariabeln eines Systems, welches ein anderes von relativ unendlich großer Energie berührt, innerhalb der angedeuteten Grenzen liegen. Die Größe h hängt dabei lediglich vom Zustande jenes Systems Σ von relativ unendlich großer Energie ab.

§ 4. Über absolute Temperatur und Wärmegleichgewicht.

Der Zustand des Systems σ hängt also lediglich von der Größe h ab, und diese lediglich vom Zustande des Systems Σ. Wir nennen die Größe $1/4h\varkappa = T$ die absolute Temperatur des Systems Σ, wobei $\varkappa$ eine universelle Konstante bedeutet.

Nennen wir das System σ „Thermometer", so können wir sofort die Sätze aussprechen:

1. Der Zustand des Thermometers hängt nur ab von der absoluten Temperatur des Systems Σ, nicht aber von der Art der Berührung der Systeme Σ und σ.

2. Erteilen zwei Systeme Σ_1 und Σ_2 einem Thermometer σ gleichen Zustand im Falle der Berührung, so besitzen sie gleiche absolute Temperatur, und erteilen folglich

einem anderen Thermometer σ' im Falle der Berührung ebenfalls gleichen Zustand.

Seien ferner zwei Systeme Σ_1 und Σ_2 in Berührung miteinander und Σ_1 außerdem in Berührung mit einem Thermometer σ. Es hängt dann die Zustandsverteilung von σ lediglich von der Energie des Systems $(\Sigma_1 + \Sigma_2)$, bez. von der Größe $h_{1,2}$ ab. Denkt man sich die Wechselwirkung von Σ_1 und Σ_2 unendlich langsam abnehmend, so ändert sich dadurch der Ausdruck für die Energie $H_{1,2}$ des Systems $(\Sigma_1 + \Sigma_2)$ nicht, wie leicht aus unserer Definition von der Berührung und dem im letzten Paragraphen aufgestellten Ausdruck für die Größe h zu ersehen ist. Hat endlich die Wechselwirkung ganz aufgehört, so hängt die Zustandsverteilung von σ, welche sich während der Trennung von Σ_1 und Σ_2 nicht ändert, nunmehr von Σ_1 ab, also von der Größe h_1; wobei der Index die Zugehörigkeit zum System Σ_1 allein andeuten soll. Es ist also:

$$h_1 = h_{1\,2}.$$

Durch eine analoge Schlußweise hätte man erhalten können:

$$h_2 = h_{1\,2},$$

also

$$h_1 = h_2,$$

oder in Worten: Trennt man zwei sich berührende Systeme Σ_1 und Σ_2, welche ein isoliertes System $(\Sigma_1 + \Sigma_2)$ von der absoluten Temperatur T bilden, so besitzen nach der Trennung die nunmehrigen isolierten Systeme Σ_1 und Σ_2 gleiche Temperatur. Wir denken uns ein gegebenes System mit einem idealen Gase in Berührung. Dieses Gas sei unter dem Bilde der kinetischen Gastheorie vollkommen darstellbar. Als System σ betrachten wir ein einziges einatomiges Gasmolekül von der Masse μ, dessen Zustand durch seine rechtwinkligen Koordinaten x, y, z und die Geschwindigkeiten ξ, η, ζ vollkommen bestimmt sei. Wir erhalten dann nach § 3 für die Wahrscheinlichkeit, daß die Zustandsvariabeln dieses Moleküles zwischen den Grenzen x und $x + dx \ldots \zeta$ und $\zeta + d\zeta$ liegen, den bekannten Maxwellschen Ausdruck:

$$dW = \text{konst.}\, e^{-h\mu(\xi^2 + \eta^2 + \zeta^2)} \cdot dx \ldots d\zeta.$$

Daraus erhält man durch Integration für den Mittelwert der lebendigen Kraft dieses Moleküles

$$\overline{\frac{\mu}{2}(\xi^2 + \eta^2 + \zeta^2)} = \frac{1}{4h}.$$

Die kinetische Gastheorie lehrt aber, daß diese Größe bei konstantem Volumen des Gases proportional dem vom Gase ausgeübten Drucke ist. Dieser ist definitionsgemäß der in der Physik als absolute Temperatur bezeichneten Größe proportional. Die von uns als absolute Temperatur bezeichnete Größe ist also nichts anderes als die mit dem Gasthermometer gemessene Temperatur eines Systems.

§ 5. Über unendlich langsame Prozesse.

Wir haben bisher nur Systeme ins Auge gefaßt, welche sich im stationären Zustande befanden. Wir wollen nun auch Veränderungen von stationären Zuständen untersuchen, jedoch nur solche, welche sich so langsam vollziehen, daß die in einem beliebigen Momente herrschende Zustandsverteilung von der stationären nur unendlich wenig abweicht; oder genauer gesprochen, daß in jedem Momente die Wahrscheinlichkeit, daß die Zustandsvariabeln in einem gewissen Gebiete G liegen, bis auf unendlich kleines durch die oben gefundene Formel dargestellt sei. Eine solche Veränderung nennen wir einen unendlich langsamen Prozeß.

Wenn die Funktionen φ_ν (Gleichung (1)) und die Energie E eines Systems bestimmt sind, so ist nach dem vorigen auch seine stationäre Zustandsverteilung bestimmt. Ein unendlich langsamer Prozeß wird also dadurch bestimmt sein, daß sich entweder E ändert oder die Funktionen φ_ν die Zeit explizite enthalten, oder beides zugleich, jedoch so, daß die entsprechenden Differentialquotienten nach der Zeit sehr klein sind.

Wir haben angenommen, daß die Zustandsvariabeln eines isolierten Systems sich nach Gleichungen (1) verändern. Umgekehrt wird aber nicht stets, wenn ein System von Gleichungen (1) existiert, nach denen sich die Zustandsvariabeln eines Systems ändern, dieses System ein isoliertes sein müssen. Es kann nämlich der Fall eintreten, daß ein betrachtetes System derart unter dem Einfluß anderer Systeme sich befindet, daß dieser Einfluß lediglich von Funktionen von veränderlichen Koordinaten beeinflussender Systeme abhängt, die sich bei konstanter Zustandsverteilung der beeinflussenden Systeme nicht ändern. In diesem Falle wird die Veränderung der Koordinaten p_ν des betrachteten Systems auch durch ein System von der Form der Gleichungen (1) darstellbar sein. Die Funktionen φ_ν werden aber dann nicht nur von der physikalischen Natur des betreffenden Systems, sondern auch von gewissen Konstanten abhängen, welche durch die beeinflussenden Systeme und deren Zustandsverteilungen definiert sind. Wir nennen diese Art von Beeinflussung des betrachteten Systems eine adiabatische. Es ist leicht einzusehen, daß für die Gleichungen (1) auch in diesem Falle eine Energiegleichung existiert, solange die Zustandsverteilungen der adiabatisch beeinflussenden Systeme sich nicht ändern. Ändern sich die Zustände adiabatisch beeinflussender Systeme, so ändern sich die Funktionen φ_ν des betrachteten Systems explizite mit der Zeit, wobei in jedem Moment die Gleichungen (1) ihre Gültigkeit behalten. Wir nennen eine solche Änderung der Zustandsverteilung des betrachteten Systems eine adiabatische.

Wir betrachten nun eine zweite Art von Zustandsveränderungen eines Systems Σ. Es liege ein System Σ zu Grunde, welches adiabatisch beeinflußt sein kann. Wir nehmen an, daß das System Σ in der Zeit $t = 0$ mit einem System P von verschiedener Temperatur in solche Wechselwirkung trete, wie wir sie oben als „Berührung" bezeichnet haben, und entfernen das System P nach der zum Ausgleich der Temperaturen von Σ und P nötigen Zeit. Es hat sich dann die Energie von Σ geändert. Während des Prozesses sind die Gleichungen (1) von Σ ungültig, vor und nach dem Prozesse aber gültig, wobei die Funktionen φ_ν vor und nach dem Prozesse dieselben sind. Einen solchen Prozeß nennen wir einen „isopyknischen" und die Σ zugeführte Energie „zugeführte Wärme".

Bis auf relativ unendlich kleines läßt sich nun offenbar jeder unendlich langsame Prozeß eines Systems Σ aus einer Aufeinanderfolge von unendlich kleinen adiabatischen und isopyknischen Prozessen konstruieren, sodaß wir, um einen Gesamtüberblick zu erhalten, nur die letzteren zu studieren haben.

§ 6. Über den Entropiebegriff.

Es liege ein physikalisches System vor, dessen momentaner Zustand durch die Werte der Zustandsvariabeln $p_1 \ldots p_n$ vollkommen bestimmt sei. Dieses System mache einen kleinen, unendlich langsamen Prozeß durch, indem die das System adiabatisch beeinflussenden Systeme eine unendlich kleine Zustandsveränderung erfahren, und außerdem dem betrachteten System durch berührende Systeme Energie zugeführt wird. Wir tragen den adiabatisch beeinflussenden Systemen dadurch Rechnung, daß wir festsetzen, die Energie E des betrachteten Systems sei außer von $p_1 \ldots p_n$ noch von gewissen Parametern $\lambda_1, \lambda_2 \ldots$ abhängig, deren Werte durch die Zustandsverteilungen der das System adiabatisch beeinflussenden Systeme bestimmt seien. Bei rein adiabatischen Prozessen gilt in jedem Moment ein Gleichungssystem (1), dessen Funktionen φ_ν außer von den Koordinaten p_ν auch von den langsam veränderlichen Größen λ abhängen; es gilt dann auch bei adiabatischen Prozessen in jedem Moment die Energiegleichung, welche die Form besitzt:

$$\sum \frac{\partial E}{\partial p_\nu} \varphi_\nu = 0.$$

Wir untersuchen nun die Energiezunahme des Systems während eines beliebigen unendlich kleinen, unendlich langsamen Prozesses.

Für jedes Zeitelement dt des Prozesses gilt:

$$(4) \qquad dE = \sum \frac{\partial E}{\partial \lambda} d\lambda + \sum \frac{\partial E}{\partial p_\nu} dp_\nu.$$

Für einen unendlich kleinen isopyknischen Prozeß verschwinden in jedem Zeitelement sämtliche $d\lambda$, mithin auch das erste Glied der rechten Seite dieser Gleichung. Da aber dE nach dem vorigen Paragraphen für einen isopyknischen Prozeß als zugeführte Wärme zu betrachten ist, so ist für einen solchen Prozeß die zugeführte Wärme dQ durch den Ausdruck:

$$dQ = \sum \frac{\partial E}{\partial p_\nu} dp_\nu$$

dargestellt.

Für einen adiabatischen Prozeß aber, während dessen stets die Gleichungen (1) gelten, ist nach der Energiegleichung

$$\sum \frac{\partial E}{\partial p_\nu} dp_\nu = \sum \frac{\partial E}{\partial p_\nu} \varphi_\nu dt = 0.$$

Andererseits ist nach dem vorigen Paragraphen für einen adiabatischen Prozeß $dQ = 0$, sodaß auch für einen adiabatischen Prozeß

$$dQ = \sum \frac{\partial E}{\partial p_\nu} dp_\nu$$

gesetzt werden kann. Diese Gleichung muß also für einen beliebigen Prozeß in jedem Zeitelement als gültig betrachtet werden. Die Gleichung (4) geht also über in

(4') $$dE = \sum \frac{\partial E}{\partial \lambda} d\lambda + dQ.$$

Dieser Ausdruck stellt auch bei veränderten Werten von $d\lambda$ und von dQ die während des ganzen unendlich kleinen Prozesses stattfindende Veränderung der Energie des Systems dar.

Am Anfang und am Ende des Prozesses ist die Zustandsverteilung des betrachteten Systems eine stationäre und wird, wenn das System vor und nach dem Prozesse mit einem Systeme von relativ unendlich großer Energie in Berührung steht, welche Annahme nur von formaler Bedeutung ist, durch die Gleichung definirt von der Form:

$$dW = \text{konst.} \, e^{-2hE} . dp_1 \ldots dp_n$$
$$= e^{c-2hE} . dp_1 \ldots dp_n,$$

wobei dW die Wahrscheinlichkeit dafür bedeutet, daß die Werte der Zustandsvariabeln des Systems in einem beliebig herausgegriffenen Zeitmoment zwischen den angedeuteten Grenzen liegen. Die Konstante c ist durch die Gleichung definirt:

(5) $$\int e^{c-2hE} . dp_1 \ldots dp_n = 1,$$

wobei die Integration über alle Werte der Variabeln zu erstrecken ist.

Gelte Gleichung (5) speziell vor dem betrachteten Prozesse, so gilt nach demselben:

(5') $$\int e^{(c+dc) - 2(h+dh)\left(E + \sum \frac{\partial E}{\partial \lambda} d\lambda\right)} . dp_1 \ldots dp_n = 1$$

und aus den beiden letzten Gleichungen ergibt sich:

$$\int \left(dc - 2E\,dh - 2h \sum \frac{\partial E}{\partial \lambda} . d\lambda\right) . e^{c-2hE} . dp_1 \ldots dp_n = 0,$$

oder, da bei der Integration der Klammerausdruck als eine Konstante gelten kann, da die Energie E des Systems vor und nach dem Prozesse sich nie merklich von einem bestimmten Mittelwerte unterscheidet, und unter Berücksichtigung von Gleichung (5):

(5'') $$dc - 2E\,dh - 2h \sum \frac{\partial E}{\partial \lambda} d\lambda = 0.$$

Nach Gleichung (4') ist aber:
$$-2h\,dE + 2h \sum \frac{\partial E}{\partial \lambda} d\lambda + 2h\,dQ = 0$$

und durch Addition dieser beiden Gleichungen erhält man:
$$2h \cdot dQ = d(2hE - c)$$

oder, da $1/4h = \varkappa \cdot T$
$$\frac{dQ}{T} = d\left(\frac{E}{T} - 2\varkappa c\right) = dS.$$

Diese Gleichung sagt aus, das dQ/T ein vollständiges Differential einer Größe ist, welche wir die Entropie S des Systems nennen wollen. Unter Berücksichtigung von Gleichung (5) erhält man:

$$S = 2\varkappa(2hE - c) = \frac{E}{T} + 2\varkappa \log \int e^{-2hE} dp_1 \ldots dp_n,$$

wobei die Integration über alle Werte der Variabeln zu erstrecken ist.

§ 7. Über die Wahrscheinlichkeit von Zustandsverteilungen.

Um den zweiten Hauptsatz in seiner allgemeinsten Form herzuleiten, müssen wir die Wahrscheinlichkeit von Zustandsverteilungen untersuchen.

Wir betrachten eine sehr große Zahl (N) isolierte Systeme, welche alle durch das nämliche Gleichungssystem (1) darstellbar seien, und deren Energie bis auf unendlich kleines übereinstimme. Die Zustandsverteilung dieser N Systeme läßt sich dann jedenfalls darstellen durch eine Gleichung von der Form:

(2') $$dN = \varepsilon(p_1 \ldots p_n, t)\,dp_1 \ldots dp_n,$$

wobei ε im allgemeinen von den Zustandsvariabeln $p_1 \ldots p_n$ und außerdem von der Zeit explizite abhängt. Die Funktion ε charakterisiert hierbei die Zustandsverteilung vollständig.

Aus § 2 geht hervor, daß, wenn die Zustandsverteilung konstant ist, was bei sehr großen Werten von t nach unseren

Theorie der Grundlagen der Thermodynamik. 183

Voraussetzungen stets der Fall ist, $\varepsilon =$ konst. sein muß, sodaß also für eine stationäre Zustandsverteilung
$$dN = \text{konst.}\, dp_1 \ldots dp_n$$
ist.

Daraus folgt sofort, daß die Wahrscheinlichkeit dW dafür, daß die Werte der Zustandsvariabeln eines zufällig herausgegriffenen der N Systeme, in dem unendlich kleinen, innerhalb der angenommenen Energiegrenzen gelegenen Gebiete g der Zustandsvariabeln gelegen sind, der Ausdruck:
$$dW = \text{konst.} \int_g dp_1 \ldots dp_n.$$

Dieser Satz läßt sich auch so aussprechen: Teilt man das ganze in Betracht kommende, durch die angenommenen Energiegrenzen bestimmte Gebiet der Zustandsvariabeln in l Teilgebiete $g_1, g_2 \ldots g_l$, derart, daß
$$\int_{g_1} = \int_{g_2} = \ldots = \int_{g_l},$$
und bezeichnet man mit W_1, W_2 etc. die Wahrscheinlichkeiten dafür, daß die Werte der Zustandsvariabeln des beliebig herausgegriffenen Systems in einem gewissen Zeitpunkt innerhalb $g_1, g_2 \ldots$ liegen, so ist
$$W_1 = W_2 = \ldots = W_l = \frac{1}{l}.$$

Das momentane Zugehören des betrachteten Systems zu einem bestimmten dieser Gebiete $g_1 \ldots g_l$ ist also genau ebenso wahrscheinlich, als das Zugehören zu irgend einem anderen dieser Gebiete.

Die Wahrscheinlichkeit dafür, daß von N betrachteten Systeme zu einer zufällig herausgegriffenen Zeit ε_1 zum Gebiete g_1, ε_2 zum Gebiete $g_2 \ldots \varepsilon_l$ zum Gebiete g_l gehören, ist also
$$W = \left(\frac{1}{l}\right)^N \frac{N!}{\varepsilon_1!\, \varepsilon_2! \ldots \varepsilon_n!},$$
oder auch, da $\varepsilon_1, \varepsilon_2 \ldots \varepsilon_n$ als sehr große Zahlen zu denken sind:
$$\log W = \text{konst.} - \sum_{\varepsilon=1}^{\varepsilon=l} \varepsilon \log \varepsilon.$$

Ist l groß genug, so kann man hierfür ohne merklichen Fehler setzen:

$$\log W = \text{konst.} - \int \varepsilon \log \varepsilon \, dp_1 \ldots dp_n.$$

In dieser Gleichung bedeutet W die Wahrscheinlichkeit dafür, daß die bestimmte, durch die Zahlen $\varepsilon_1, \varepsilon_2 \ldots \varepsilon_l$, bez. durch eine bestimmte Funktion ε von $p_1 \ldots p_n$ gemäß Gleichung (2') ausgedrückte Zustandsverteilung zu einer bestimmten Zeit herrscht.

Wäre in dieser Gleichung $\varepsilon = \text{konst.}$, d. h. von den p_ν unabhängig zwischen den betrachteten Energiegrenzen, so wäre die betrachtete Zustandsverteilung stationär, und, wie leicht zu beweisen, der Ausdruck für die Wahrscheinlichkeit W der Zustandsverteilung ein Maximum. Ist ε von den Werten der p_ν abhängig, so läßt sich zeigen, daß der Ausdruck für $\log W$ für die betrachtete Zustandsverteilung kein Extremum besitzt, d. h. es gibt dann von der betrachteten Zustandsverteilung unendlich wenig verschiedene, für welche W größer ist.

Verfolgen wir die betrachteten N Systeme eine beliebige Zeit hindurch, so wird sich die Zustandsverteilung, also auch W beständig mit der Zeit ändern, und wir werden anzunehmen haben, daß immer wahrscheinlichere Zustandsverteilungen auf unwahrscheinliche folgen werden, d. h. daß W stets zunimmt, bis die Zustandsverteilung konstant und W ein Maximum geworden ist.

In den folgenden Paragraphen wird gezeigt, daß aus diesem Satze der zweite Hauptsatz der Thermodynamik gefolgert werden kann.

Zunächst ist:

$$-\int \varepsilon' \log \varepsilon' \, dp_1 \ldots dp_n \geqq -\int \varepsilon \log \varepsilon \, dp_1 \ldots dp_n,$$

wobei durch die Funktion ε die Zustandsverteilung der N Systeme zu einer gewissen Zeit t, durch die Funktion ε' die Zustandsverteilung zu einer gewissen späteren Zeit t' bestimmt, und die Integration beiderseits über alle Werte der Variabeln zu erstrecken ist. Wenn ferner die Größen $\log \varepsilon$ und $\log \varepsilon'$ der

einzelnen unter den N Systemen sich nicht merklich von einander unterscheiden, so geht, da
$$\int \varepsilon \, dp_1 \ldots dp_n = \int \varepsilon' \, dp_1 \ldots dp_n = N,$$
die letzte Gleichung über in:
$$\text{(6)} \qquad -\log \varepsilon' \gtreqless -\log \varepsilon.$$

§ 8. Anwendung der gefundenen Resultate auf einen bestimmten Fall.

Wir betrachten eine endliche Zahl von physikalischen Systemen $\sigma_1, \sigma_2 \ldots$, welche zusammen ein isoliertes System bilden, welches wir Gesamtsystem nennen wollen. Die Systeme $\sigma_1, \sigma_2 \ldots$ sollen thermisch nicht merklich in Wechselwirkung stehen, wohl aber können sie sich adiabatisch beeinflussen. Die Zustandsverteilung eines jeden der Systeme $\sigma_1, \sigma_2 \ldots$, die wir Teilsysteme nennen wollen, sei bis auf unendlich kleines eine stationäre. Die absoluten Temperaturen der Teilsysteme können beliebig und voneinander verschieden sein.

Die Zustandsverteilung des Systems σ_1 wird sich nicht merklich von derjenigen Zustandsverteilung unterscheiden, welche gelten würde, wenn σ_1 mit einem physikalischen System von derselben Temperatur in Berührung stände. Wir können daher dessen Zustandsverteilung durch die Gleichung darstellen:
$$dw_1 = e^{c_{(1)} - 2h_{(1)} E_{(1)}} \int_g dp_1^{(1)} \ldots dp_{(n)}^{(1)},$$
wobei die Indizes (1) die Zugehörigkeit zum Teilsystem σ_1 andeuten sollen.

Analoge Gleichungen gelten für die übrigen Teilsysteme. Da die augenblicklichen Werte der Zustandsvariabeln der einzelnen Teilsysteme von denen der anderen unabhängig sind, so erhalten wir für die Zustandsverteilung des Gesamtsystems eine Gleichung von der Form:
$$\text{(7)} \qquad dw = dw_1 . dw_2 \ldots = e^{\sum c_\nu - 2h_\nu E_\nu} \int_g dp_1 \ldots dp_n,$$
wobei die Summation über alle Systeme, die Integration über das beliebige in allen Variabeln des Gesamtsystems unendlich kleine Gebiet g zu erstrecken ist.

Wir nehmen nun an, daß die Teilsysteme $\sigma_1, \sigma_2 \ldots$ nach einer gewissen Zeit in beliebige Wechselwirkung zueinander treten, bei welchem Prozesse aber das Gesamtsystem stets ein isoliertes bleiben möge. Nach Verlauf einer gewissen Zeit möge ein Zustand des Gesamtsystems eingetreten sein, bei welchem die Teilsysteme $\sigma_1, \sigma_2 \ldots$ einander thermisch nicht beeinflussen und bis auf unendlich kleines sich im stationären Zustand befinden.

Es gilt dann für die Zustandsverteilung des Gesamtsystems eine Gleichung, welche der vor dem Prozesse gültigen vollkommen analog ist:

$$(7') \quad dw' = dw_1' \cdot dw_2' \ldots = e^{\sum(c_\nu' - 2h_\nu' E_\nu')} \int_g dp_1 \ldots dp_n.$$

Wir betrachten nun N solcher Gesamtsysteme. Für jedes derselben gelte bis auf unendlich kleines zur Zeit t die Gleichung (7), zur Zeit t' die Gleichung (7'). Es wird dann die Zustandsverteilung der betrachteten N Gesamtsysteme zu den Zeiten t und t' gegeben sein durch die Gleichungen:

$$dN_t = N \cdot e^{\sum(c_\nu - 2h_\nu E_\nu)} \cdot dp_1 \ldots dp_n.$$

$$dN_{t'} = N \cdot e^{\sum(c_\nu' - 2h_\nu' E_\nu')} \cdot dp_1 \ldots dp_n.$$

Auf diese beiden Zustandsverteilungen wenden wir nun die Resultate des vorigen Paragraphen an. Es sind hier sowohl die

$$\varepsilon = N \cdot e^{\sum(c_\nu - 2h_\nu E_\nu)}$$

als auch die

$$\varepsilon' = N \cdot e^{\sum(c_\nu' - 2h_\nu' E_\nu')}$$

für die einzelnen der N Systeme nicht merklich verschieden, sodaß wir Gleichung (6) anwenden können, welche liefert

$$\sum(2h'E' - c') \geqq \sum(2hE - c),$$

oder indem man beachtet, daß die Größen $2h_1 E_1 - c_1$, $2h_2 E_2 - c_2, \ldots$ nach § 6 bis auf eine universelle Konstante mit den Entropien $S_1, S_2 \ldots$ der Teilsysteme übereinstimmen:

$$(8) \quad S_1' + S_2' + \ldots \geqq S_1 + S_2 + \ldots,$$

d. h. die Summe der Entropien der Teilsysteme eines isolierten Systems ist nach einem beliebigen Prozesse gleich oder größer als die Summe der Entropien der Teilsysteme vor dem Prozesse.

§ 9. Herleitung des zweiten Hauptsatzes.

Es liege nun ein isoliertes Gesamtsystem vor, dessen Teilsysteme W, M und Σ_1, Σ_2 ... heißen mögen. Das System W, welches wir Wärmereservoir nennen wollen, besitze gegen das System M (Maschine) eine unendlich große Energie. Ebenso sei die Energie der miteinander in adiabatischer Wechselwirkung stehenden Systeme Σ_1, Σ_2 ... gegen diejenige der Maschine M unendlich groß. Wir nehmen an, daß die sämtlichen Teilsysteme M, W, Σ_1, Σ_2 ... sich im stationären Zustand befinden.

Es durchlaufe nun die Maschine M einen beliebigen Kreisprozeß, wobei sie die Zustandsverteilungen der Systeme Σ_1, Σ_2 ... durch adiabatische Beeinflussung unendlich langsam ändere, d. h. Arbeit leiste, und von dem Systeme W die Wärmemenge Q aufnehme. Am Ende des Prozesses wird dann die gegenseitige adiabatische Beeinflussung der Systeme Σ_1, Σ_2 ... eine andere sein als vor dem Prozesse. Wir sagen, die Maschine M hat die Wärmemenge Q in Arbeit verwandelt.

Wir berechnen nun die Zunahme der Entropie der einzelnen Teilsysteme, welche bei dem betrachteten Prozeß eintritt. Die Zunahme der Entropie des Wärmereservoirs W beträgt nach den Resultaten des § 6 $-Q/T$, wenn T die absolute Temperatur bedeutet. Die Entropie von M ist vor und nach dem Prozeß dieselbe, da das System M einen Kreisprozeß durchlaufen hat. Die Systeme Σ_1, Σ_2 ... ändern ihre Entropie während des Prozesses überhaupt nicht, da diese Systeme nur unendlich langsame adiabatische Beeinflussung erfahren. Die Entropievermehrung $S' - S$ des Gesamtsystems erhält also den Wert

$$S' - S = -\frac{Q}{T}.$$

Da nach dem Resultate des vorigen Paragraphen diese Größe $S' - S$ stets $\geqq 0$ ist, so folgt

$$Q \leqq 0.$$

Diese Gleichung spricht die Unmöglichkeit der Existenz eines Perpetuum mobile zweiter Art aus.

Bern, Januar 1903.

(Eingegangen 26. Januar 1903.)

ANNALEN DER PHYSIK.

BEGRÜNDET UND FORTGEFÜHRT DURCH

F. A. C. GREN, L. W. GILBERT, J. C. POGGENDORFF, G. UND E. WIEDEMANN.

VIERTE FOLGE.

BAND 14.

DER GANZEN REIHE 319. BAND.

KURATORIUM:

F. KOHLRAUSCH, M. PLANCK, G. QUINCKE,
W. C. RÖNTGEN, E. WARBURG.

UNTER MITWIRKUNG

DER DEUTSCHEN PHYSIKALISCHEN GESELLSCHAFT

UND INSBESONDERE VON

M. PLANCK

HERAUSGEGEBEN VON

PAUL DRUDE.

MIT DREI FIGURENTAFELN.

LEIPZIG, 1904.

VERLAG VON JOHANN AMBROSIUS BARTH.

354

6. *Zur allgemeinen molekularen Theorie der Wärme;* von *A. Einstein.*

Im folgenden gebe ich einige Ergänzungen zu einer letztes Jahr von mir publizierten Abhandlung.[1])

Wenn ich von „allgemeiner molekularer Wärmetheorie" spreche, so meine ich damit eine Theorie, welche im wesentlichen auf den in § 1 der zitierten Abhandlung genannten Voraussetzungen beruht. Ich setze jene Abhandlung als bekannt voraus, um unnütze Wiederholungen zu vermeiden, und bediene mich der dort gebrauchten Bezeichnungen.

Zuerst wird ein Ausdruck für die Entropie eines Systems abgeleitet, welcher dem von Boltzmann für ideale Gase gefundenen und von Planck in seiner Theorie der Strahlung vorausgesetzten vollständig analog ist. Dann wird eine einfache Herleitung des zweiten Hauptsatzes gegeben. Hierauf wird die Bedeutung einer universellen Konstanten untersucht, welche in der allgemeinen molekularen Theorie der Wärme eine wichtige Rolle spielt. Schließlich folgt eine Anwendung der Theorie auf die Strahlung schwarzer Körper, wobei sich zwischen der erwähnten, durch die Größen der Elementarquanta der Materie und der Elektrizität bestimmten universellen Konstanten und der Größenordnung der Strahlungswellenlängen, ohne Zuhilfenahme spezieller Hypothesen, eine höchst interessante Beziehung ergibt.

§ 1. Über den Ausdruck der Entropie.

Für ein System, welches Energie nur in Form von Wärme aufnehmen kann, oder mit anderen Worten, für ein System, welches von anderen Systemen nicht adiabatisch beeinflußt wird, gilt zwischen der absoluten Temperatur T und der Energie E, nach § 3 und § 4, l. c., die Gleichung:

$$(1) \quad h = \tfrac{1}{2} \frac{\omega'(E)}{\omega(E)} = \frac{1}{4 \varkappa T},$$

---

[1] A. Einstein, Ann. d. Phys. **11**. p. 170. 1903.

wobei $\varkappa$ eine absolute Konstante bedeutet und ω (etwas abweichend von der zitierten Abhandlung) durch die Gleichung definiert sei:
$$\omega(E) \cdot \delta E = \int_E^{E+\delta E} dp_1 \ldots dp_n.$$

Das Integral rechts ist hierbei über alle Werte der den momentanen Zustand des Systems vollkommen und eindeutig definierenden Zustandsvariabeln zu erstrecken, denen Werte der Energie entsprechen, die zwischen E und $E + \delta E$ liegen.

Aus Gleichung (1) folgt:
$$S = \int \frac{dE}{T} = 2\varkappa \log[\omega(E)].$$

Der Ausdruck stellt also (unter Weglassung der willkürlichen Integrationskonstanten) die Entropie des Systems dar. Dieser Ausdruck für die Entropie eines Systems gilt übrigens keineswegs nur für Systeme, welche nur rein thermische Zustandsänderungen erfahren, sondern auch für solche, welche beliebige adiabatische und isopyknische Zustandsänderungen durchlaufen.

Der Beweis kann aus der letzten Gleichung von § 6, l. c., geführt werden; ich unterlasse dies, da ich hier keine Anwendung des Satzes in seiner allgemeinen Bedeutung zu machen beabsichtige.

§ 2. Herleitung des zweiten Hauptsatzes.

Befindet sich ein System in einer Umgebung von bestimmter konstanter Temperatur T_0 und steht es mit dieser Umgebung in thermischer Wechselwirkung („Berührung"), so nimmt es ebenfalls erfahrungsgemäß die Temperatur T_0 an und behält die Temperatur T_0 für alle Zeiten bei.

Nach der molekularen Theorie der Wärme gilt jedoch dieser Satz nicht streng, sondern nur mit gewisser — wenn auch für alle der direkten Untersuchung zugänglichen Systeme mit sehr großer — Annäherung. Hat sich vielmehr das betrachtete System unendlich lange in der genannten Umgebung befunden, so ist die Wahrscheinlichkeit W dafür, daß in einem beliebig herausgegriffenen Zeitpunkt der Wert der Energie des Systems sich zwischen den Grenzen E und $E+1$ befindet (§ 3, l c.):

$$W = C\, e^{-\frac{E}{2 \varkappa T_0}}\, \omega(E),$$

wobei C eine Konstante bedeutet. Dieser Wert ist für jedes E ein von Null verschiedener, hat jedoch für ein bestimmtes E ein Maximum und nimmt — wenigstens für alle der direkten Untersuchung zugänglichen Systeme — für jedes merklich größere oder kleinere E einen sehr kleinen Wert an. Wir nennen das System „Wärmereservoir" und sagen kurz: obiger Ausdruck stellt die Wahrscheinlichkeit dafür dar, daß die Energie des betrachteten Wärmereservoirs in der genannten Umgebung den Wert E hat. Nach dem Ergebnis des vorigen Paragraphen kann man auch schreiben:

$$W = C\, e^{\frac{1}{2\varkappa}\left(S - \frac{E}{T_0}\right)},$$

wobei S die Entropie des Wärmereservoirs bedeutet.

Es mögen nun eine Anzahl Wärmereservoirs vorliegen, welche sich sämtlich in der Umgebung von der Temperatur T_0 befinden. Die Wahrscheinlichkeit dafür, daß die Energie des ersten Reservoirs den Wert E_1, des zweiten den Wert $E_2 \ldots$ des letzten den Wert E_l besitzt, ist dann in leicht verständlicher Bezeichnung:

(a) $$\mathfrak{W} = W_1 \cdot W_2 \ldots W_l = C_1 \cdot C_2 \ldots C_l\, e^{\frac{1}{2\varkappa}\left\{\sum_1^l S - \frac{\sum_1^l E}{T_0}\right\}}.$$

Diese Reservoirs mögen nun in Wechselwirkung treten mit einer Maschine, wobei letztere einen Kreisprozeß durchläuft. Bei diesem Vorgange finde weder zwischen Wärmereservoirs und Umgebung noch zwischen Maschine und Umgebung ein Wärmeaustausch statt. Nach dem betrachteten Vorgange seien die Energien und Entropien der Systeme:

$$E_1',\ E_2' \ldots E_l',$$

bez.

$$S_1',\ S_2' \ldots S_l'.$$

Dem Gesamtzustande der Wärmereservoirs, welcher durch diese Werte definiert ist, kommt die Wahrscheinlichkeit zu:

(b) $$\mathfrak{W}' = C_1 \cdot C_2 \ldots C_l \, e^{\frac{1}{2\varkappa} \left(\sum_1^l S' - \frac{\sum_1^l E'}{T_0} \right)}$$

Bei dem Vorgange hat sich weder der Zustand der Umgebung noch der Zustand der Maschine geändert, da letztere einen Kreisprozeß durchlief.

Nehmen wir nun an, daß nie unwahrscheinlichere Zustände auf wahrscheinlichere folgen, so ist:
$$\mathfrak{W}' \geqq \mathfrak{W}.$$

Es ist aber auch nach dem Energieprinzip:
$$\sum_1^l E = \sum_1^l E'.$$

Berücksichtigt man dies, so folgt aus Gleichungen (a) und (b):
$$\sum S' \geqq \sum S.$$

§ 3. Über die Bedeutung der Konstanten $\varkappa$ in der kinetischen Atomtheorie.

Es werde ein physikalisches System betrachtet, dessen momentaner Zustand durch die Werte der Zustandsvariabeln
$$p_1, p_2 \ldots p_n$$
vollständig bestimmt sei.

Wenn das betrachtete System mit einem System von relativ unendlich großer Energie und der absoluten Temperatur T_0 in „Berührung" steht, so ist dessen Zustandsverteilung durch die Gleichung bestimmt:
$$dW = C\, e^{-\frac{E}{2\varkappa T_0}} dp_1 \ldots dp_n.$$

In dieser Gleichung ist $\varkappa$ eine universelle Konstante, deren Bedeutung nun untersucht werden soll.

Unter Zugrundelegung der kinetischen Atomtheorie gelangt man auf folgendem, aus Boltzmanns Arbeiten über Gastheorie geläufigen Wege zu einer Deutung dieser Konstanten.

Es seien die p_ν die rechtwinkligen Koordinaten $x_1\, y_1\, z_1$, $x_2\, y_2 \ldots, x_n\, y_n\, z_n$ und $\xi_1\, \eta_1\, \zeta_1, \xi_2\, \eta_2\, \zeta_2 \ldots, \xi_n\, \eta_n\, \zeta_n$ die Geschwindigkeiten der einzelnen (punktförmig gedachten) Atome des Systems. Diese Zustandsvariabeln können gewählt werden, weil sie der Bedingung $\sum \partial \varphi_\nu / \partial p_\nu = 0$ Genüge leisten (l. c., § 2). Man hat dann:

$$E = \Phi(x_1 \ldots z_n) + \sum_1^n \frac{m_\nu}{2}(\xi_\nu^2 + \eta_\nu^2 + \zeta_\nu^2),$$

wobei der erste Summand die potentielle Energie, der zweite die lebendige Kraft des Systems bezeichnet. Sei nun ein unendlich kleines Gebiet $dx_1 \ldots dz_n$ gegeben. Wir finden den Mittelwert der Größe

$$\frac{m_\nu}{2}(\xi_\nu^2 + \eta_\nu^2 + \zeta_\nu^2),$$

welcher diesem Gebiete entspricht:

$$\overline{L}_\nu = \overline{\frac{m}{2}(\xi_\nu^2 + \eta_\nu^2 + \zeta_\nu^2)}$$

$$= \frac{e^{-\frac{\Phi(x_1\ldots z_n)}{4\varkappa T_0}} dx_1 \ldots dx_n \int \frac{m_\nu}{2}(\xi_\nu^2 + \eta_\nu^2 + \zeta_\nu^2) e^{\frac{\sum_1^n \frac{m_\nu}{2}(\xi_\nu^2 + \eta_\nu^2 + \zeta_\nu^2)}{2\varkappa T_0}} d\xi_1 \ldots d\zeta_n}{e^{-\frac{\Phi(x_1\ldots z_n)}{4\varkappa T_0}} dx_1 \ldots dx_n \int e^{\frac{\sum \frac{m_\nu}{2}(\xi_\nu^2 + \eta_\nu^2 + \zeta_\nu^2)}{2\varkappa T_0}} d\xi_1 \ldots d\zeta_n}$$

$$= 3 \frac{\int_{-\infty}^{+\infty} m_\nu \xi_\nu^2 e^{\frac{m_\nu \xi_\nu^2}{4\varkappa T_0}} d\xi_\nu}{\int_{-\infty}^{+\infty} e^{\frac{m_\nu \xi_\nu^2}{4\varkappa T_0}} d\xi_\nu} = 3\varkappa T_0.$$

Diese Größe ist also unabhängig von der Wahl des Gebietes und von der Wahl des Atoms, ist also überhaupt der Mittelwert des Atoms bei der absoluten Temperatur T_0. Die Größe $3\varkappa$ ist gleich dem Quotienten aus der mittleren lebendigen Kraft eines Atoms in die absolute Temperatur.[1]

Die Konstante $\varkappa$ ist ferner aufs engste verknüpft mit der Anzahl N der wirklichen Moleküle, welche in einem Molekül

---

[1] Vgl. L. Boltzmann, Vorl. über Gastheorie 2. § 42. 1898.

im Sinne des Chemikers (Äquivalentgewicht bezogen auf 1 g Wasserstoff als Einheit) enthalten sind.

Liege nämlich eine solche Quantität eines idealen Gases vor, so ist bekanntlich, wenn Gramm und Zentimeter als Einheiten benutzt werden

$$pv = RT, \quad \text{wobei} \quad R = 8{,}31 \cdot 10^7.$$

Nach der kinetischen Gastheorie ist aber:

$$pv = \tfrac{2}{3} N \overline{L},$$

wobei $\overline{L}$ den Mittelwert der lebendigen Kraft der Schwerpunktsbewegung eines Moleküles bedeutet. Berücksichtigt man noch, daß

$$\overline{L} = \overline{L}_\nu,$$

so erhält man:

$$N \cdot 2\varkappa = R.$$

Die Konstante $2\varkappa$ ist also gleich dem Quotienten der Konstanten R in Anzahl der in einem Äquivalent enthaltenen Moleküle.

Setzt man mit O. E. Meyer $N = 6{,}4 \cdot 10^{23}$, so erhält man $\varkappa = 6{,}5 \cdot 10^{-17}$.

§ 4. Allgemeine Bedeutung der Konstanten $\varkappa$.

Ein gegebenes System berühre ein System von relativ unendlich großer Energie und der Temperatur T. Die Wahrscheinlichkeit dW dafür, daß der Wert seiner Energie in einem beliebig herausgegriffenen Zeitpunkte zwischen E und $E + dE$ liegt, ist:

$$dW = C e^{-\frac{E}{2\varkappa T}} \omega E\, dE.$$

Für den Mittelwert $\overline{E}$ von E erhält man:

$$\overline{E} = \int_0^\infty C E e^{-\frac{E}{2\varkappa T}} \omega E\, dE.$$

Da ferner

$$1 = \int_0^\infty C e^{-\frac{E}{2\varkappa T}} \omega E\, dE,$$

so ist

$$\int_0^\infty (\bar{E} - E) e^{-\frac{E}{2\varkappa T}} \omega(E) dE = 0.$$

Differenziert man diese Gleichung nach T, so erhält man:

$$\int_0^\infty \left(2\varkappa T^2 \frac{d\bar{E}}{dT} + \bar{E}E - E^2\right) e^{-\frac{E}{2\varkappa T}} \omega E dE = 0.$$

Diese Gleichung besagt, daß der Mittelwert der Klammer verschwindet, also:

$$2\varkappa T^2 \frac{d\bar{E}}{dT} = \bar{E^2} - \bar{E}\bar{E}.$$

Im allgemeinen unterscheidet sich der Momentanwert E der Energie von $\bar{E}$ um eine gewisse Größe, welche wir „Energieschwankung" nennen; wir setzen:

$$E = \bar{E} + \varepsilon.$$

Man erhält dann

$$\bar{E^2} - \bar{E}\bar{E} = \overline{\varepsilon^2} = 2\varkappa T^2 \frac{d\bar{E}}{dT}.$$

Die Größe $\overline{\varepsilon^2}$ ist ein Maß für die thermische Stabilität des Systems; je größer $\overline{\varepsilon^2}$, desto kleiner diese Stabilität.

Die absolute Konstante $\varkappa$ bestimmt also die thermische Stabilität der Systeme. Die zuletzt gefundene Beziehung ist darum interessant, weil in derselben keine Größe mehr vorkommt, welche an die der Theorie zugrunde liegenden Annahmen erinnert.

Durch wiederholtes Differenzieren kann man ohne Schwierigkeit die Größen $\overline{\varepsilon^3}$, $\overline{\varepsilon^4}$ etc. berechnen.

§ 5. Anwendung auf die Strahlung.

Die zuletzt gefundene Gleichung würde eine exakte Bestimmung der universellen Konstanten $\varkappa$ zulassen, wenn es möglich wäre, den Mittelwert des Quadrates der Energieschwankung eines Systems zu bestimmen; dies ist jedoch bei dem gegenwärtigen Stande unseres Wissens nicht der Fall.

Wir können überhaupt nur bei einer einzigen Art physikalischer Systeme aus der Erfahrung vermuten, daß ihnen eine Energieschwankung zukomme; es ist dies der mit Temperaturstrahlung erfüllte leere Raum.

Ist nämlich ein mit Temperaturstrahlung erfüllter Raum von Lineardimensionen, welche sehr groß gegen die Wellenlänge ist, der das Energiemaximum der Strahlung bei der betreffenden Temperatur zukommt, so wird offenbar der Betrag der Energieschwankung im Mittel im Vergleich zur mittleren Strahlungsenergie dieses Raumes sehr klein sein. Wenn dagegen der Strahlungsraum von der Größenordnung jener Wellenlänge ist, so wird die Energieschwankung von derselben Größenordnung sein, wie die Energie der Strahlung des Strahlungsraumes.

Es ist allerdings einzuwenden, daß wir nicht behaupten können, daß ein Strahlungs*raum* als ein *System* von der von uns vorausgesetzten Art zu betrachten sei, auch dann nicht, wenn die Anwendbarkeit der allgemeinen molekularen Theorie zugestanden wird. Vielleicht müßte man zum Beispiel die Grenzen des Raumes als mit den elektromagnetischen Zuständen desselben veränderlich annehmen. Diese Umstände kommen indessen hier, wo es sich nur um Größenordnungen handelt, nicht in Betracht.

Setzen wir also in der im vorigen Paragraphen gefundenen Gleichung

$$\overline{\varepsilon^2} = \overline{E^2},$$

und nach dem Stefan-Boltzmannschen Gesetze

$$\overline{E} = c v T^4,$$

wobei v das Volumen in cm³ und c die Konstante dieses Gesetzes bedeutet, so müssen wir für $\sqrt[3]{v}$ einen Wert von der Größenordnung der Wellenlänge maximaler Strahlungsenergie erhalten, welche der betreffenden Temperatur entspricht.

Man erhält:

$$\sqrt[3]{v} = \frac{2\sqrt[3]{\frac{\varkappa}{c}}}{T} = \frac{0{,}42}{T},$$

wobei für $\varkappa$ der aus der kinetischen Gastheorie gefundene Wert und für c der Wert $7{,}06 \cdot 10^{-15}$ gesetzt ist.

362 A. Einstein. *Allgemeine molekulare Theorie der Wärme.*

Ist λ_m die Wellenlänge des Energiemaximums der Strahlung, so liefert die Erfahrung:

$$\lambda_m = \frac{0{,}293}{T}.$$

Man sieht, daß sowohl die Art der Abhängigkeit von der Temperatur als auch die Größenordnung von λ_m mittels der allgemeinen molekularen Theorie der Wärme richtig bestimmt werden kann, und ich glaube, daß diese Übereinstimmung bei der großen Allgemeinheit unserer Voraussetzungen nicht dem Zufall zugeschrieben werden darf.

Bern, den 27. März 1904.

(Eingegangen 29. März 1904.)

ANNALEN
DER
PHYSIK.

BEGRÜNDET UND FORTGEFÜHRT DURCH

F. A. C. GREN, L. W. GILBERT, J. C. POGGENDORFF, G. UND E. WIEDEMANN.

VIERTE FOLGE.

BAND 17.

DER GANZEN REIHE 322. BAND.

KURATORIUM:

F. KOHLRAUSCH, M. PLANCK, G. QUINCKE,
W. C. RÖNTGEN, E. WARBURG.

UNTER MITWIRKUNG

DER DEUTSCHEN PHYSIKALISCHEN GESELLSCHAFT

UND INSBESONDERE VON

M. PLANCK

HERAUSGEGEBEN VON

PAUL DRUDE.

MIT FÜNF FIGURENTAFELN.

LEIPZIG, 1905.

VERLAG VON JOHANN AMBROSIUS BARTH.

6. Über einen die Erzeugung und Verwandlung des Lichtes betreffenden heuristischen Gesichtspunkt; von A. Einstein.

Zwischen den theoretischen Vorstellungen, welche sich die Physiker über die Gase und andere ponderable Körper gebildet haben, und der Maxwellschen Theorie der elektromagnetischen Prozesse im sogenannten leeren Raume besteht ein tiefgreifender formaler Unterschied. Während wir uns nämlich den Zustand eines Körpers durch die Lagen und Geschwindigkeiten einer zwar sehr großen, jedoch endlichen Anzahl von Atomen und Elektronen für vollkommen bestimmt ansehen, bedienen wir uns zur Bestimmung des elektromagnetischen Zustandes eines Raumes kontinuierlicher räumlicher Funktionen, so daß also eine endliche Anzahl von Größen nicht als genügend anzusehen ist zur vollständigen Festlegung des elektromagnetischen Zustandes eines Raumes. Nach der Maxwellschen Theorie ist bei allen rein elektromagnetischen Erscheinungen, also auch beim Licht, die Energie als kontinuierliche Raumfunktion aufzufassen, während die Energie eines ponderabeln Körpers nach der gegenwärtigen Auffassung der Physiker als eine über die Atome und Elektronen erstreckte Summe darzustellen ist. Die Energie eines ponderabeln Körpers kann nicht in beliebig viele, beliebig kleine Teile zerfallen, während sich die Energie eines von einer punktförmigen Lichtquelle ausgesandten Lichtstrahles nach der Maxwellschen Theorie (oder allgemeiner nach jeder Undulationstheorie) des Lichtes auf ein stets wachsendes Volumen sich kontinuierlich verteilt.

Die mit kontinuierlichen Raumfunktionen operierende Undulationstheorie des Lichtes hat sich zur Darstellung der rein optischen Phänomene vortrefflich bewährt und wird wohl nie durch eine andere Theorie ersetzt werden. Es ist jedoch im Auge zu behalten, daß sich die optischen Beobachtungen auf zeitliche Mittelwerte, nicht aber auf Momentanwerte beziehen, und es ist trotz der vollständigen Bestätigung der Theorie der Beugung, Reflexion, Brechung, Dispersion etc. durch das

Experiment wohl denkbar, daß die mit kontinuierlichen Raumfunktionen operierende Theorie des Lichtes zu Widersprüchen mit der Erfahrung führt, wenn man sie auf die Erscheinungen der Lichterzeugung und Lichtverwandlung anwendet.

Es scheint mir nun in der Tat, daß die Beobachtungen über die „schwarze Strahlung", Photolumineszenz, die Erzeugung von Kathodenstrahlen durch ultraviolettes Licht und andere die Erzeugung bez. Verwandlung des Lichtes betreffende Erscheinungsgruppen besser verständlich erscheinen unter der Annahme, daß die Energie des Lichtes diskontinuierlich im Raume verteilt sei. Nach der hier ins Auge zu fassenden Annahme ist bei Ausbreitung eines von einem Punkte ausgehenden Lichtstrahles die Energie nicht kontinuierlich auf größer und größer werdende Räume verteilt, sondern es besteht dieselbe aus einer endlichen Zahl von in Raumpunkten lokalisierten Energiequanten, welche sich bewegen, ohne sich zu teilen und nur als Ganze absorbiert und erzeugt werden können.

Im folgenden will ich den Gedankengang mitteilen und die Tatsachen anführen, welche mich auf diesen Weg geführt haben, in der Hoffnung, daß der darzulegende Gesichtspunkt sich einigen Forschern bei ihren Untersuchungen als brauchbar erweisen möge.

§ 1. Über eine die Theorie der „schwarzen Strahlung" betreffende Schwierigkeit.

Wir stellen uns zunächst auf den Standpunkt der Maxwellschen Theorie und Elektronentheorie und betrachten folgenden Fall. In einem von vollkommen reflektierenden Wänden eingeschlossenen Raumes befinde sich eine Anzahl Gasmoleküle und Elektronen, welche freibeweglich sind und aufeinander konservative Kräfte ausüben, wenn sie einander sehr nahe kommen, d. h. miteinander wie Gasmoleküle nach der kinetischen Gastheorie zusammenstoßen können.[1] Eine Anzahl

---

[1] Diese Annahme ist gleichbedeutend mit der Voraussetzung, daß die mittleren kinetischen Energien von Gasmolekülen und Elektronen bei Temperaturgleichgewicht einander gleich seien. Mit Hilfe letzterer Voraussetzung hat Hr. Drude bekanntlich das Verhältnis von thermischem und elektrischem Leitungsvermögen der Metalle auf theoretischem Wege abgeleitet.

Elektronen sei ferner an voneinander weit entfernte Punkte des Raumes gekettet durch nach diesen Punkten gerichtete, den Elongationen proportionale Kräfte. Auch diese Elektronen sollen mit den freien Molekülen und Elektronen in konservative Wechselwirkung treten, wenn ihnen letztere sehr nahe kommen. Wir nennen die an Raumpunkte geketteten Elektronen „Resonatoren"; sie senden elektromagnetische Wellen bestimmter Periode aus und absorbieren solche.

Nach der gegenwärtigen Ansicht über die Entstehung des Lichtes müßte die Strahlung im betrachteten Raume, welche unter Zugrundelegung der Maxwellschen Theorie für den Fall des dynamischen Gleichgewichtes gefunden wird, mit der „schwarzen Strahlung" identisch sein — wenigstens wenn Resonatoren aller in Betracht zu ziehenden Frequenzen als vorhanden angesehen werden.

Wir sehen vorläufig von der von den Resonatoren emittierten und absorbierten Strahlung ab und fragen nach der der Wechselwirkung (den Zusammenstößen) von Molekülen und Elektronen entsprechenden Bedingung für das dynamische Gleichgewicht. Die kinetische Gastheorie liefert für letzteres die Bedingung, daß die mittlere lebendige Kraft eines Resonatorelektrons gleich der mittleren kinetischen Energie der fortschreitenden Bewegung eines Gasmoleküles sein muß. Zerlegen wir die Bewegung des Resonatorelektrons in drei aufeinander senkrechte Schwingungsbewegungen, so finden wir für den Mittelwert $\bar{E}$ der Energie einer solchen geradlinigen Schwingungsbewegung

$$\bar{E} = \frac{R}{N} T,$$

wobei R die absolute Gaskonstante, N die Anzahl der „wirklichen Moleküle" in einem Grammäquivalent und T die absolute Temperatur bedeutet. Die Energie $\bar{E}$ ist nämlich wegen der Gleichheit der zeitlichen Mittelwerte von kinetischer und potentieller Energie des Resonators $2/3$ mal so groß wie die lebendige Kraft eines freien, einatomigen Gasmoleküles. Würde nun durch irgend eine Ursache — in unserem Falle durch Strahlungsvorgänge — bewirkt, daß die Energie eines Resonators einen größeren oder kleineren zeitlichen Mittelwert als $\bar{E}$ besitzt, so würden die Zusammenstöße mit den freien Elektronen und Molekülen zu einer im Mittel von Null verschiedenen Energieabgabe an das Gas bez. Energieaufnahme von dem Gas führen. Es ist also in dem von uns betrachteten Falle dynamisches Gleichgewicht nur dann möglich, wenn jeder Resonator die mittlere Energie $\bar{E}$ besitzt.

Eine ähnliche Überlegung machen wir jetzt bezüglich der Wechselwirkung der Resonatoren und der im Raume vorhandenen Strahlung. Hr. Planck hat für diesen Fall die Bedingung des dynamischen Gleichgewichtes abgeleitet[1]) unter der Voraussetzung, daß die Strahlung als ein denkbar ungeordnetster Prozeß[2]) betrachtet werden kann. Er fand:

$$\bar{E}_\nu = \frac{L^3}{8\pi\nu^2}\varrho_\nu.$$

$\bar{E}_\nu$ ist hierbei die mittlere Energie eines Resonators von der Eigenfrequenz ν (pro Schwingungskomponente), L die Lichtgeschwindigkeit, ν die Frequenz und $\varrho_\nu\,d\nu$ die Energie pro Volumeinheit desjenigen Teiles der Strahlung, dessen Schwingungszahl zwischen ν und $\nu + d\nu$ liegt.

---

1) M. Planck, Ann. d. Phys. **1**. p. 99. 1900.

2) Diese Voraussetzung läßt sich folgendermaßen formulieren. Wir entwickeln die Z-Komponente der elektrischen Kraft (Z) in einem beliebigen Punkte des betreffenden Raumes zwischen den Zeitgrenzen $t=0$ und $t=T$ (wobei T eine relativ zu allen in Betracht zu ziehenden Schwingungsdauern sehr große Zeit bedeute) in eine Fouriersche Reihe

$$Z = \sum_{\nu=1}^{\nu=\infty} A_\nu \sin\left(2\pi\nu\frac{t}{T} + \alpha_\nu\right),$$

wobei $A_\nu \geqq 0$ und $0 \leqq \alpha_\nu \leqq 2\pi$. Denkt man sich in demselben Raumpunkte eine solche Entwickelung beliebig oft bei zufällig gewählten Anfangspunkten der Zeit ausgeführt, so wird man für die Größen A_ν und α_ν verschiedene Wertsysteme erhalten. Es existieren dann für die Häufigkeit der verschiedenen Wertekombinationen der Größen A_ν und α_ν (statistische) Wahrscheinlichkeiten dW von der Form:

$$dW = f(A_1 A_2 \ldots \alpha_1 \alpha_2 \ldots)\,dA_1\,dA_2 \ldots d\alpha_1\,d\alpha_2 \ldots$$

Die Strahlung ist dann eine denkbar ungeordnetste, wenn

$$f(A_1, A_2 \ldots \alpha_1, \alpha_2 \ldots) = F_1(A_1) F_2(A_2) \ldots f_1(\alpha_1)\cdot f_2(\alpha_2)\ldots,$$

d. h. wenn die Wahrscheinlichkeit eines bestimmten Wertes einer der Größen A bez. α von den Werten, welche die anderen Größen A bez. x besitzen, unabhängig ist. Mit je größerer Annäherung die Bedingung erfüllt ist, daß die einzelnen Paare von Größen A_ν und α_ν von Emissions- und Absorptionsprozessen *besonderer* Resonatorengruppen abhängen, mit desto größerer Annäherung wird also in dem von uns betrachteten Falle die Strahlung als eine „denkbar ungeordnetste" anzusehen sein.

Soll die Strahlungsenergie von der Frequenz ν nicht beständig im Ganzen weder vermindert noch vermehrt werden, so muß gelten:

$$\frac{R}{N}T = \bar{E} = \bar{E}_\nu = \frac{L^3}{8\pi\nu^2}\varrho_\nu,$$

$$\varrho_\nu = \frac{R}{N}\frac{8\pi\nu^2}{L^3}T.$$

Diese als Bedingung des dynamischen Gleichgewichtes gefundene Beziehung entbehrt nicht nur der Übereinstimmung mit der Erfahrung, sondern sie besagt auch, daß in unserem Bilde von einer bestimmten Energieverteilung zwischen Äther und Materie nicht die Rede sein kann. Je weiter nämlich der Schwingungszahlenbereich der Resonatoren gewählt wird, desto größer wird die Strahlungsenergie des Raumes, und wir erhalten in der Grenze

$$\int_0^\infty \varrho_\nu\, d\nu = \frac{R}{N}\frac{8\pi}{L^3}T\int_0^\infty \nu^2\, d\nu = \infty.$$

§ 2. Über die Plancksche Bestimmung der Elementarquanta.

Wir wollen im folgenden zeigen, daß die von Hrn. Planck gegebene Bestimmung der Elementarquanta von der von ihm aufgestellten Theorie der „schwarzen Strahlung" bis zu einem gewissen Grade unabhängig ist.

Die allen bisherigen Erfahrungen genügende Plancksche Formel[1]) für ϱ_ν lautet

$$\varrho_\nu = \frac{\alpha\nu^3}{e^{\frac{\beta\nu}{T}} - 1},$$

wobei

$$\alpha = 6{,}10 \cdot 10^{-56},$$

$$\beta = 4{,}866 \cdot 10^{-11}.$$

Für große Werte von T/ν, d. h. für große Wellenlängen und Strahlungsdichten geht diese Formel in der Grenze in folgende über:

$$\varrho_\nu = \frac{\alpha}{\beta}\nu^2 T.$$

---

1) M. Planck, Ann. d. Phys. **4**. p. 561. 1901.

Man erkennt, daß diese Formel mit der in § 1 aus der Maxwellschen und der Elektronentheorie entwickelten übereinstimmt. Durch Gleichsetzung der Koeffizienten beider Formeln erhält man:

oder
$$\frac{R}{N}\frac{8\pi}{L^3} = \frac{\alpha}{\beta}$$

$$N = \frac{\beta}{\alpha}\frac{8\pi R}{L^3} = 6{,}17 \cdot 10^{23},$$

d. h. ein Atom Wasserstoff wiegt $1/N$ Gramm $= 1{,}62 \cdot 10^{-24}$ g. Dies ist genau der von Hrn. Planck gefundene Wert, welcher mit den auf anderen Wegen gefundenen Werten für diese Größe befriedigend übereinstimmt.

Wir gelangen daher zu dem Schlusse: Je größer die Energiedichte und die Wellenlänge einer Strahlung ist, als um so brauchbarer erweisen sich die von uns benutzten theoretischen Grundlagen; für kleine Wellenlängen und kleine Strahlungsdichten aber versagen dieselben vollständig.

Im folgenden soll die „schwarze Strahlung" im Anschluß an die Erfahrung ohne Zugrundelegung eines Bildes über die Erzeugung und Ausbreitung der Strahlung betrachtet werden.

§ 3. Über die Entropie der Strahlung.

Die folgende Betrachtung ist in einer berühmten Arbeit des Hrn. W. Wien enthalten und soll hier nur der Vollständigkeit halber Platz finden.

Es liege eine Strahlung vor, welche das Volumen v einnehme. Wir nehmen an, daß die wahrnehmbaren Eigenschaften der vorliegenden Strahlung vollkommen bestimmt seien, wenn die Strahlungsdichte $\varrho(\nu)$ für alle Frequenzen gegeben ist.[1]) Da Strahlungen von verschiedenen Frequenzen als ohne Arbeitsleistung und ohne Wärmezufuhr voneinander trennbar anzusehen sind, so ist die Entropie der Strahlung in der Form

$$S = v \int_0^\infty \varphi(\varrho, \nu)\, d\nu$$

darstellbar, wobei φ eine Funktion der Variabeln ϱ und ν

---

1) Diese Annahme ist eine willkürliche. Man wird naturgemäß an dieser einfachsten Annahme so lange festhalten, als nicht das Experiment dazu zwingt, sie zu verlassen.

bedeutet. Es kann φ auf eine Funktion von nur einer Variabeln reduziert werden durch Formulierung der Aussage, daß durch adiabatische Kompression einer Strahlung zwischen spiegelnden Wänden, deren Entropie nicht geändert wird. Wir wollen jedoch hierauf nicht eintreten, sondern sogleich untersuchen, wie die Funktion φ aus dem Strahlungsgesetz des schwarzen Körpers ermittelt werden kann.

Bei der „schwarzen Strahlung" ist ϱ eine solche Funktion von ν, daß die Entropie bei gegebener Energie ein Maximum ist, d. h. daß

$$\delta \int_0^\infty \varphi(\varrho, \nu)\, d\nu = 0,$$

wenn

$$\delta \int_0^\infty \varrho\, d\nu = 0.$$

Hieraus folgt, daß für jede Wahl des $\delta\varrho$ als Funktion von ν

$$\int_0^\infty \left(\frac{\partial \varphi}{\partial \varrho} - \lambda\right) \delta \varrho\, d\nu = 0,$$

wobei λ von ν unabhängig ist. Bei der schwarzen Strahlung ist also $\partial\varphi/\partial\varrho$ von ν unabhängig.

Für die Temperaturzunahme einer schwarzen Strahlung vom Volumen $v = 1$ um dT gilt die Gleichung:

$$dS = \int_{\nu=0}^{\nu=\infty} \frac{\partial \varphi}{\partial \varrho}\, d\varrho\, d\nu,$$

oder, da $\partial\varphi/\partial\varrho$ von ν unabhängig ist:

$$dS = \frac{\partial \varphi}{\partial \varrho}\, dE.$$

Da dE gleich der zugeführten Wärme und der Vorgang umkehrbar ist, so gilt auch:

$$dS = \frac{1}{T}\, dE.$$

Durch Vergleich erhält man:

$$\frac{\partial \varphi}{\partial \varrho} = \frac{1}{T}.$$

Dies ist das Gesetz der schwarzen Strahlung. Man kann also

aus der Funktion φ das Gesetz der schwarzen Strahlung und umgekehrt aus letzterem die Funktion φ durch Integration bestimmen mit Rücksicht darauf, daß φ für $\varrho = 0$ verschwindet.

§ 4. Grenzgesetz für die Entropie der monochromatischen Strahlung bei geringer Strahlungsdichte.

Aus den bisherigen Beobachtungen über die „schwarze Strahlung" geht zwar hervor, daß das ursprünglich von Hrn. W. Wien für die „schwarze Strahlung" aufgestellte Gesetz

$$\varrho = \alpha \nu^3 e^{-\beta \frac{\nu}{T}}$$

nicht genau gültig ist. Dasselbe wurde aber für große Werte von ν/T sehr vollkommen durch das Experiment bestätigt. Wir legen diese Formel unseren Rechnungen zugrunde, behalten aber im Sinne, daß unsere Resultate nur innerhalb gewisser Grenzen gelten.

Aus dieser Formel ergibt sich zunächst:

$$\frac{1}{T} = -\frac{1}{\beta \nu} \lg \frac{\varrho}{\alpha \nu^3}$$

und weiter unter Benutzung der in dem vorigen Paragraphen gefundenen Beziehung:

$$\varphi(\varrho, \nu) = -\frac{\varrho}{\beta \nu} \left\{ \lg \frac{\varrho}{\alpha \nu^3} - 1 \right\}.$$

Es sei nun eine Strahlung von der Energie E gegeben, deren Frequenz zwischen ν und $\nu + d\nu$ liegt. Die Strahlung nehme das Volumen v ein. Die Entropie dieser Strahlung ist:

$$S = v\, \varphi(\varrho, \nu)\, d\nu = -\frac{E}{\beta \nu} \left\{ \lg \frac{E}{v\, \alpha \nu^3\, d\nu} - 1 \right\}.$$

Beschränken wir uns darauf, die Abhängigkeit der Entropie von dem von der Strahlung eingenommenen Volumen zu untersuchen, und bezeichnen wir die Entropie der Strahlung mit S_0, falls dieselbe das Volumen v_0 besitzt, so erhalten wir:

$$S - S_0 = \frac{E}{\beta \nu} \lg \left(\frac{v}{v_0} \right).$$

Diese Gleichung zeigt, daß die Entropie einer monochromatischen Strahlung von genügend kleiner Dichte nach dem gleichen Gesetze mit dem Volumen variiert wie die Entropie eines idealen Gases oder die einer verdünnten Lösung. Die soeben gefundene Gleichung soll im folgenden interpretiert werden unter Zugrundelegung des von Hrn. Boltzmann in die Physik eingeführten Prinzips, nach welchem die Entropie eines Systems eine Funktion der Wahrscheinlichkeit seines Zustandes ist.

§ 5. Molekulartheoretische Untersuchung der Abhängigkeit der Entropie von Gasen und verdünnten Lösungen vom Volumen.

Bei Berechnung der Entropie auf molekulartheoretischem Wege wird häufig das Wort „Wahrscheinlichkeit" in einer Bedeutung angewendet, die sich nicht mit der Definition der Wahrscheinlichkeit deckt, wie sie in der Wahrscheinlichkeitsrechnung gegeben wird. Insbesondere werden die „Fälle gleicher Wahrscheinlichkeit" häufig hypothetisch festgesetzt in Fällen, wo die angewendeten theoretischen Bilder bestimmt genug sind, um statt jener hypothetischen Festsetzung eine Deduktion zu geben. Ich will in einer besonderen Arbeit zeigen, daß man bei Betrachtungen über thermische Vorgänge mit der sogenannten „statistischen Wahrscheinlichkeit" vollkommen auskommt und hoffe dadurch eine logische Schwierigkeit zu beseitigen, welche der Durchführung des Boltzmannschen Prinzips noch im Wege steht. Hier aber soll nur dessen allgemeine Formulierung und dessen Anwendung auf ganz spezielle Fälle gegeben werden.

Wenn es einen Sinn hat, von der Wahrscheinlichkeit eines Zustandes eines Systems zu reden, wenn ferner jede Entropiezunahme als ein Übergang zu einem wahrscheinlicheren Zustande aufgefaßt werden kann, so ist die Entropie S_1 eines Systems eine Funktion der Wahrscheinlichkeit W_1 seines momentanen Zustandes. Liegen also zwei nicht miteinander in Wechselwirkung stehende Systeme S_1 und S_2 vor, so kann man setzen:

$$S_1 = \varphi_1(W_1),$$
$$S_2 = \varphi_2(W_2).$$

Betrachtet man diese beiden Systeme als ein einziges System von der Entropie S und der Wahrscheinlichkeit W, so ist:

$$S = S_1 + S_2 = \varphi(W)$$

und

$$W = W_1 \cdot W_2.$$

Die letztere Beziehung sagt aus, daß die Zustände der beiden Systeme voneinander unabhängige Ereignisse sind.

Aus diesen Gleichungen folgt:
$$\varphi(W_1 \cdot W_2) = \varphi_1(W_1) + \varphi_2(W_2)$$
und hieraus endlich
$$\varphi_1(W_1) = C \lg(W_1) + \text{konst.},$$
$$\varphi_2(W_2) = C \lg(W_2) + \text{konst.},$$
$$\varphi(W) = C \lg(W) + \text{konst.}$$

Die Größe C ist also eine universelle Konstante; sie hat, wie aus der kinetischen Gastheorie folgt, den Wert R/N, wobei den Konstanten R und N dieselbe Bedeutung wie oben beizulegen ist. Bedeutet S_0 die Entropie bei einem gewissen Anfangszustande eines betrachteten Systems und W die relative Wahrscheinlichkeit eines Zustandes von der Entropie S, so erhalten wir also allgemein:
$$S - S_0 = \frac{R}{N} \lg W.$$

Wir behandeln zunächst folgenden Spezialfall. In einem Volumen v_0 sei eine Anzahl (n) beweglicher Punkte (z. B. Moleküle) vorhanden, auf welche sich unsere Überlegung beziehen soll. Außer diesen können in dem Raume noch beliebig viele andere bewegliche Punkte irgendwelcher Art vorhanden sein. Über das Gesetz, nach dem sich die betrachteten Punkte in dem Raume bewegen, sei nichts vorausgesetzt, als daß in bezug auf diese Bewegung kein Raumteil (und keine Richtung) von den anderen ausgezeichnet sei. Die Anzahl der betrachteten (ersterwähnten) beweglichen Punkte sei ferner so klein, daß von einer Wirkung der Punkte aufeinander abgesehen werden kann.

Dem betrachteten System, welches z. B. ein ideales Gas oder eine verdünnte Lösung sein kann, kommt eine gewisse Entropie S_0 zu. Wir denken uns einen Teil des Volumens v_0 von der Größe v und alle n beweglichen Punkte in das Volumen v versetzt, ohne daß an dem System sonst etwas geändert wird. Diesem Zustand kommt offenbar ein anderer Wert der Entropie (S) zu, und wir wollen nun die Entropiedifferenz mit Hilfe des Boltzmannschen Prinzips bestimmen.

Wir fragen: Wie groß ist die Wahrscheinlichkeit des letzterwähnten Zustandes relativ zum ursprünglichen? Oder: Wie groß ist die Wahrscheinlichkeit dafür, daß sich in einem zufällig herausgegriffenen Zeitmoment alle n in einem gegebenen Volumen v_0 unabhängig voneinander beweglichen Punkte (zufällig) in dem Volumen v befinden?

Für diese Wahrscheinlichkeit, welche eine „statistische Wahrscheinlichkeit" ist, erhält man offenbar den Wert:

$$W = \left(\frac{v}{v_0}\right)^n;$$

man erhält hieraus durch Anwendung des Boltzmannschen Prinzipes:

$$S - S_0 = R\left(\frac{n}{N}\right) \lg\left(\frac{v}{v_0}\right).$$

Es ist bemerkenswert, daß man zur Herleitung dieser Gleichung, aus welcher das Boyle-Gay-Lussacsche Gesetz und das gleichlautende Gesetz des osmotischen Druckes leicht thermodynamisch ableiten kann[1]); keine Voraussetzung über das Gesetz zu machen braucht, nachdem sich die Moleküle bewegen.

§ 6. **Interpretation des Ausdruckes für die Abhängigkeit der Entropie der monochromatischen Strahlung vom Volumen nach dem Boltzmannschen Prinzip.**

Wir haben in § 4 für die Abhängigkeit der Entropie der monochromatischen Strahlung vom Volumen den Ausdruck gefunden:

$$S - S_0 = \frac{E}{\beta \nu} \lg\left(\frac{v}{v_0}\right).$$

Schreibt man diese Formel in der Gestalt:

$$S - S_0 = \frac{R}{N} \lg\left[\left(\frac{v}{v_0}\right)^{\frac{N}{R}\frac{E}{\beta \nu}}\right]$$

---

1) Ist E die Energie des Systems, so erhält man:

$$-d(E - TS) = p\,dv = T\,dS = R\frac{n}{N}\frac{dv}{v};$$

also

$$p\,v = R\frac{n}{N}T.$$

und vergleicht man sie mit der allgemeinen, das Boltzmannsche Prinzip ausdrückenden Formel

$$S - S_0 = \frac{R}{N} \lg W$$

so gelangt man zu folgendem Schluß:

Ist monochromatische Strahlung von der Frequenz v und der Energie E in das Volumen v_0 (durch spiegelnde Wände) eingeschlossen, so ist die Wahrscheinlichkeit dafür, daß sich in einem beliebig herausgegriffenen Zeitmoment die ganze Strahlungsenergie in dem Teilvolumen v des Volumens v_0 befindet:

$$W = \left(\frac{v}{v_0}\right)^{\frac{N}{R}\frac{E}{\beta v}}.$$

Hieraus schließen wir weiter:

Monochromatische Strahlung von geringer Dichte (innerhalb des Gültigkeitsbereiches der Wienschen Strahlungsformel) verhält sich in wärmetheoretischer Beziehung so, wie wenn sie aus voneinander unabhängigen Energiequanten von der Größe $R\beta v/N$ bestünde.

Wir wollen noch die mittlere Größe der Energiequanten der „schwarzen Strahlung" mit der mittleren lebendigen Kraft der Schwerpunktsbewegung eines Moleküls bei der nämlichen Temperatur vergleichen. Letztere ist $\frac{3}{2}(R/N)T$, während man für die mittlere Größe des Energiequantums unter Zugrundelegung der Wienschen Formel erhält:

$$\frac{\int\limits_0^\infty \alpha v^3 e^{-\frac{\beta v}{T}} dv}{\int\limits_0^\infty \frac{N}{R\beta v} \alpha v^3 e^{-\frac{\beta v}{T}} dv} = 3\frac{R}{N}T$$

Wenn sich nun monochromatische Strahlung (von hinreichend kleiner Dichte) bezüglich der Abhängigkeit der Entropie vom Volumen wie ein diskontinuierliches Medium verhält, welches aus Energiequanten von der Größe $R\beta v/N$ besteht, so liegt es nahe, zu untersuchen, ob auch die Gesetze der

Erzeugung und Verwandlung des Lichtes so beschaffen sind, wie wenn das Licht aus derartigen Energiequanten bestünde. Mit dieser Frage wollen wir uns im folgenden beschäftigen.

§ 7. Über die Stokessche Regel.

Es werde monochromatisches Licht durch Photolumineszenz in Licht anderer Frequenz verwandelt und gemäß dem eben erlangten Resultat angenommen, daß sowohl das erzeugende wie das erzeugte Licht aus Energiequanten von der Größe $(R/N)\beta v$ bestehe, wobei v die betreffende Frequenz bedeutet. Der Verwandlungsprozeß wird dann folgendermaßen zu deuten sein. Jedes erzeugende Energiequant von der Frequenz v_1 wird absorbiert und gibt — wenigstens bei genügend kleiner Verteilungsdichte der erzeugenden Energiequanten — für sich allein Anlaß zur Entstehung eines Lichtquants von der Frequenz v_2; eventuell können bei der Absorption des erzeugenden Lichtquants auch gleichzeitig Lichtquanten von den Frequenzen v_3, v_4 etc. sowie Energie anderer Art (z. B. Wärme) entstehen. Unter Vermittelung von was für Zwischenprozessen dies Endresultat zustande kommt, ist gleichgültig. Wenn die photolumineszierende Substanz nicht als eine beständige Quelle von Energie anzusehen ist, so kann nach dem Energieprinzip die Energie eines erzeugten Energiequants nicht größer sein als die eines erzeugenden Lichtquants; es muß also die Bezeichnung gelten:

$$\frac{R}{N}\beta v_2 \leq \frac{R}{N}\beta v_1$$

oder

$$v_2 \leq v_1.$$

Dies ist die bekannte Stokessche Regel.

Besonders hervorzuheben ist, daß bei schwacher Belichtung die erzeugte Lichtmenge der erregenden unter sonst gleichen Umständen nach unserer Auffassung der erregenden Lichtstärke proportional sein muß, da jedes erregende Energiequant einen Elementarprozeß von der oben angedeuteten Art verursachen wird, unabhängig von der Wirkung der anderen erregenden Energiequanten. Insbesondere wird es keine untere Grenze für die Intensität des erregenden Lichtes geben, unterhalb welcher das Licht unfähig wäre, lichterregend zu wirken.

Abweichungen von der Stokesschen Regel sind nach der dargelegten Auffassung der Phänomene in folgenden Fällen denkbar:

1. wenn die Anzahl der gleichzeitig in Umwandlung begriffenen Energiequanten pro Volumeneinheit so groß ist, daß ein Energiequant des erzeugten Lichtes seine Energie von mehreren erzeugenden Energiequanten erhalten kann;

2. wenn das erzeugende (oder erzeugte) Licht nicht von derjenigen energetischen Beschaffenheit ist, die einer „schwarzen Strahlung" aus dem Gültigkeitsbereich des Wienschen Gesetzes zukommt, wenn also z. B. das erregende Licht von einem Körper so hoher Temperatur erzeugt ist, daß für die in Betracht kommende Wellenlänge das Wiensche Gesetz nicht mehr gilt.

Die letztgenannte Möglichkeit verdient besonderes Interesse. Nach der entwickelten Auffassung ist es nämlich nicht ausgeschlossen, daß eine „nicht Wiensche Strahlung" auch in großer Verdünnung sich in energetischer Beziehung anders verhält als eine „schwarze Strahlung" aus dem Gültigkeitsbereich des Wienschen Gesetzes.

§ 8. Über die Erzeugung von Kathodenstrahlen durch Belichtung fester Körper.

Die übliche Auffassung, daß die Energie des Lichtes kontinuierlich über den durchstrahlten Raum verteilt sei, findet bei dem Versuch, die lichtelektrischen Erscheinungen zu erklären, besonders große Schwierigkeiten, welche in einer bahnbrechenden Arbeit von Hrn. Lenard dargelegt sind.[1]

Nach der Auffassung, daß das erregende Licht aus Energiequanten von der Energie $(R/N)\beta\nu$ bestehe, läßt sich die Erzeugung von Kathodenstrahlen durch Licht folgendermaßen auffassen. In die oberflächliche Schicht des Körpers dringen Energiequanten ein, und deren Energie verwandelt sich wenigstens zum Teil in kinetische Energie von Elektronen. Die einfachste Vorstellung ist die, daß ein Lichtquant seine ganze Energie an ein einziges Elektron abgibt; wir wollen annehmen, daß dies vorkomme. Es soll jedoch nicht ausgeschlossen sein, daß Elektronen die Energie von Lichtquanten nur teilweise aufnehmen. Ein im Innern des Körpers mit kinetischer Energie

---

[1] P. Lenard, Ann. d. Phys. 8. p. 169 u. 170. 1902.

versehenes Elektron wird, wenn es die Oberfläche erreicht hat, einen Teil seiner kinetischen Energie eingebüßt haben. Außerdem wird anzunehmen sein, daß jedes Elektron beim Verlassen des Körpers eine (für den Körper charakteristische) Arbeit P zu leisten hat, wenn es den Körper verläßt. Mit der größten Normalgeschwindigkeit werden die unmittelbar an der Oberfläche normal zu dieser erregten Elektronen den Körper verlassen. Die kinetische Energie solcher Elektronen ist

$$\frac{R}{N}\beta v - P.$$

Ist der Körper zum positiven Potential Π geladen und von Leitern vom Potential Null umgeben und ist Π eben imstande, einen Elektrizitätsverlust des Körpers zu verhindern, so muß sein:

$$\Pi \varepsilon = \frac{R}{N}\beta v - P,$$

wobei ε die elektrische Masse des Elektrons bedeutet, oder

$$\Pi E = R\beta v - P',$$

wobei E die Ladung eines Grammäquivalentes eines einwertigen Ions und P' das Potential dieser Menge negativer Elektrizität in bezug auf den Körper bedeutet.[1]

Setzt man $E = 9{,}6 \cdot 10^3$, so ist $\Pi \cdot 10^{-8}$ das Potential in Volts, welches der Körper bei Bestrahlung im Vakuum annimmt.

Um zunächst zu sehen, ob die abgeleitete Beziehung der Größenordnung nach mit der Erfahrung übereinstimmt, setzen wir $P' = 0$, $v = 1{,}03 \cdot 10^{15}$ (entsprechend der Grenze des Sonnenspektrums nach dem Ultraviolett hin) und $\beta = 4{,}866 \cdot 10^{-11}$. Wir erhalten $\Pi \cdot 10^7 = 4{,}3$ Volt, welches Resultat der Größenordnung nach mit den Resultaten von Hrn. Lenard übereinstimmt.[2]

Ist die abgeleitete Formel richtig, so muß Π, als Funktion der Frequenz des erregenden Lichtes in kartesischen Koordinaten dargestellt, eine Gerade sein, deren Neigung von der Natur der untersuchten Substanz unabhängig ist.

---

[1] Nimmt man an, daß das einzelne Elektron durch das Licht aus einem neutralen Molekül unter Aufwand einer gewissen Arbeit losgelöst werden muß, so hat man an der abgeleiteten Beziehung nichts zu ändern; nur ist dann P' als Summe von zwei Summanden aufzufassen.

[2] P. Lenard, Ann. d. Phys. 8. p. 165 u. 184. Taf. I, Fig. 2. 1902.

Mit den von Hrn. Lenard beobachteten Eigenschaften der lichtelektrischen Wirkung steht unsere Auffassung, soweit ich sehe, nicht im Widerspruch. Wenn jedes Energiequant des erregenden Lichtes unabhängig von allen übrigen seine Energie an Elektronen abgibt, so wird die Geschwindigkeitsverteilung der Elektronen, d. h. die Qualität der erzeugten Kathodenstrahlung von der Intensität des erregenden Lichtes unabhängig sein; andererseits wird die Anzahl der den Körper verlassenden Elektronen der Intensität des erregenden Lichtes unter sonst gleichen Umständen proportional sein.[1]

Über die mutmaßlichen Gültigkeitsgrenzen der erwähnten Gesetzmäßigkeiten wären ähnliche Bemerkungen zu machen wie bezüglich der mutmaßlichen Abweichungen von der Stokesschen Regel.

Im vorstehenden ist angenommen, daß die Energie wenigstens eines Teiles der Energiequanten des erregenden Lichtes je an ein einziges Elektron vollständig abgegeben werde. Macht man diese naheliegende Voraussetzung nicht, so erhält man statt obiger Gleichung die folgende:

$$\Pi E + P' \leqq R\beta v.$$

Für die Kathodenlumineszenz, welche den inversen Vorgang zu dem eben betrachteten bildet, erhält man durch eine der durchgeführten analoge Betrachtung:

$$\Pi E + P' \geqq R\beta v.$$

Bei den von Hrn. Lenard untersuchten Substanzen ist PE stets bedeutend größer als $R\beta v$, da die Spannung, welche die Kathodenstrahlen durchlaufen haben müssen, um eben sichtbares Licht erzeugen zu können, in einigen Fällen einige Hundert, in anderen Tausende von Volts beträgt.[2] Es ist also anzunehmen, daß die kinetische Energie eines Elektrons zur Erzeugung vieler Lichtenergiequanten verwendet wird.

§ 9. Über die Ionisierung der Gase durch ultraviolettes Licht.

Wir werden anzunehmen haben, daß bei der Ionisierung eines Gases durch ultraviolettes Licht je ein absorbiertes Licht-

---

1) P. Lenard, l. c. p. 150 und p. 166–168.
2) P. Lenard, Ann. d. Phys. **12**. p. 469. 1903.

energiequant zur Ionisierung je eines Gasmoleküles verwendet wird. Hieraus folgt zunächst, daß die Ionisierungsarbeit (d. h. die zur Ionisierung theoretisch nötige Arbeit) eines Moleküles nicht größer sein kann als die Energie eines absorbierten wirksamen Lichtenergiequantes. Bezeichnet man mit J die (theoretische) Ionisierungsarbeit pro Grammäquivalent, so muß also sein:

$$R \beta \nu \geqq J.$$

Nach Messungen Lenards ist aber die größte wirksame Wellenlänge für Luft ca. $1,9 \cdot 10^{-5}$ cm, also

$$R \beta \nu = 6,4 \cdot 10^{12} \text{ Erg} \geqq J.$$

Eine obere Grenze für die Ionisierungsarbeit gewinnt man auch aus den Ionisierungsspannungen in verdünnten Gasen. Nach J. Stark[1]) ist die kleinste gemessene Ionisierungsspannung (an Platinanoden) für Luft ca. 10 Volt.[2]) Es ergibt sich also für J die obere Grenze $9,6 \cdot 10^{12}$, welche nahezu gleich der eben gefundenen ist. Es ergibt sich noch eine andere Konsequenz, deren Prüfung durch das Experiment mir von großer Wichtigkeit zu sein scheint. Wenn jedes absorbierte Lichtenergiequant ein Molekül ionisiert, so muß zwischen der absorbierten Lichtmenge L und der Anzahl j der durch dieselbe ionisierten Grammoleküle die Beziehung bestehen:

$$j = \frac{L}{R \beta \nu}.$$

Diese Beziehung muß, wenn unsere Auffassung der Wirklichkeit entspricht, für jedes Gas gelten, welches (bei der betreffenden Frequenz) keine merkliche nicht von Ionisation begleitete Absorption aufweist.

Bern, den 17. März 1905.

---

1) J. Stark, Die Elektrizität in Gasen p. 57. Leipzig 1902.
2) Im Gasinnern ist die Ionisierungsspannung für negative Ionen allerdings fünfmal größer.

(Eingegangen 18. März 1905.)

5. Über die von der molekularkinetischen Theorie der Wärme geforderte Bewegung von in ruhenden Flüssigkeiten suspendierten Teilchen; von A. Einstein.

In dieser Arbeit soll gezeigt werden, daß nach der molekularkinetischen Theorie der Wärme in Flüssigkeiten suspendierte Körper von mikroskopisch sichtbarer Größe infolge der Molekularbewegung der Wärme Bewegungen von solcher Größe ausführen müssen, daß diese Bewegungen leicht mit dem Mikroskop nachgewiesen werden können. Es ist möglich, daß die hier zu behandelnden Bewegungen mit der sogenannten „Brownschen Molekularbewegung" identisch sind; die mir erreichbaren Angaben über letztere sind jedoch so ungenau, daß ich mir hierüber kein Urteil bilden konnte.

Wenn sich die hier zu behandelnde Bewegung samt den für sie zu erwartenden Gesetzmäßigkeiten wirklich beobachten läßt, so ist die klassische Thermodynamik schon für mikroskopisch unterscheidbare Räume nicht mehr als genau gültig anzusehen und es ist dann eine exakte Bestimmung der wahren Atomgröße möglich. Erwiese sich umgekehrt die Voraussage dieser Bewegung als unzutreffend, so wäre damit ein schwerwiegendes Argument gegen die molekularkinetische Auffassung der Wärme gegeben.

§ 1. Über den suspendierten Teilchen zuzuschreibenden osmotischen Druck.

Im Teilvolumen V^* einer Flüssigkeit vom Gesamtvolumen V seien z-Gramm-Moleküle eines Nichtelektrolyten gelöst. Ist das Volumen V^* durch eine für das Lösungsmittel, nicht aber für die gelöste Substanz durchlässige Wand vom reinen Lösungsmittel getrennt, so wirkt auf diese Wand der sogenannte osmotische Druck, welcher bei genügend großen Werten von V^*/z der Gleichung genügt:

$$p V^* = R T z\,.$$

Sind hingegen statt der gelösten Substanz in dem Teilvolumen V^* der Flüssigkeit kleine suspendierte Körper vorhanden, welche ebenfalls nicht durch die für das Lösungsmittel durchlässige Wand hindurchtreten können, so hat man nach der klassischen Theorie der Thermodynamik — wenigstens bei Vernachlässigung der uns hier nicht interessierenden Schwerkraft — nicht zu erwarten, daß auf die Wand eine Kraft wirke; denn die „freie Energie" des Systems scheint nach der üblichen Auffassung nicht von der Lage der Wand und der suspendierten Körper abzuhängen, sondern nur von den Gesamtmassen und Qualitäten der suspendierten Substanz, der Flüssigkeit und der Wand, sowie von Druck und Temperatur. Es kämen allerdings für die Berechnung der freien Energie noch Energie und Entropie der Grenzflächen in Betracht (Kapillarkräfte); hiervon können wir jedoch absehen, indem bei den ins Auge zu fassenden Lagenänderungen der Wand und der suspendierten Körper Änderungen der Größe und Beschaffenheit der Berührungsflächen nicht eintreten mögen.

Vom Standpunkte der molekularkinetischen Wärmetheorie aus kommt man aber zu einer anderen Auffassung. Nach dieser Theorie unterscheidet sich eingelöstes Molekül von einem suspendierten Körper *lediglich* durch die Größe, und man sieht nicht ein, warum einer Anzahl suspendierter Körper nicht derselbe osmotische Druck entsprechen sollte, wie der nämlichen Anzahl gelöster Moleküle. Man wird anzunehmen haben, daß die suspendierten Körper infolge der Molekularbewegung der Flüssigkeit eine wenn auch sehr langsame ungeordnete Bewegung in der Flüssigkeit ausführen; werden sie durch die Wand verhindert, das Volumen V^* zu verlassen, so werden sie auf die Wand Kräfte ausüben, ebenso wie gelöste Moleküle. Sind also n suspendierte Körper im Volumen V^*, also $n/V^* = v$ in der Volumeneinheit vorhanden, und sind benachbarte unter ihnen genügend weit voneinander entfernt, so wird ihnen ein osmotischer Druck p entsprechen von der Größe:

$$p = \frac{RT}{V^*}\frac{n}{N} = \frac{RT}{N} \cdot v,$$

wobei N die Anzahl der in einem Gramm-Molekül enthaltenen wirklichen Moleküle bedeutet. Im nächsten Paragraph soll gezeigt werden, daß die molekularkinetische Theorie der Wärme wirklich zu dieser erweiterten Auffassung des osmotischen Druckes führt.

§ 2. Der osmotische Druck vom Standpunkte der molekularkinetischen Theorie der Wärme.[1]

Sind $p_1 p_2 \ldots p_l$ Zustandsvariable eines physikalischen Systems, welche den momentanen Zustand desselben vollkommen bestimmen (z. B. die Koordinaten und Geschwindigkeitskomponenten aller Atome des Systems) und ist das vollständige System der Veränderungsgleichungen dieser Zustandsvariabeln von der Form

$$\frac{\partial p_\nu}{\partial t} = \varphi_\nu(p_1 \ldots p_l) \quad (\nu = 1, 2 \ldots l)$$

gegeben, wobei $\sum \frac{\partial \varphi_\nu}{\partial p_\nu} = 0$, so ist die Entropie des Systems durch den Ausdruck gegeben:

$$S = \frac{\bar{E}}{T} + 2\varkappa \lg \int e^{-\frac{E}{2\varkappa T}} dp_1 \ldots dp_l.$$

Hierbei bedeutet T die absolute Temperatur, $\bar{E}$ die Energie des Systems, E die Energie als Funktion der p_ν. Das Integral ist über alle mit den Bedingungen des Problems vereinbaren Wertekombinationen der p_ν zu erstrecken. $\varkappa$ ist mit der oben erwähnten Konstanten N durch die Relation $2\varkappa N = R$ verbunden. Für die freie Energie F erhalten wir daher:

$$F = -\frac{R}{N}T \lg \int e^{-\frac{EN}{RT}} dp_1 \ldots dp_l = -\frac{RT}{N} \lg B.$$

---

[1] In diesem Paragraph sind die Arbeiten des Verfassers über die Grundlagen der Thermodynamik als bekannt vorausgesetzt (vgl. Ann. d. Phys. 9. p. 417. 1902; 11. p. 170. 1903). Für das Verständnis der Resultate der vorliegenden Arbeit ist die Kenntnis jener Arbeiten sowie dieses Paragraphen der vorliegenden Arbeit entbehrlich.

Wir denken uns nun eine in dem Volumen V eingeschlossene Flüssigkeit; in dem Teilvolumen V^* von V mögen sich n gelöste Moleküle bez. suspendierte Körper befinden, welche im Volumen V^* durch eine semipermeabele Wand festgehalten seien; es werden hierdurch die Integrationsgrenzen des in den Ausdrücken für S und F auftretenden Integrales B beeinflußt. Das Gesamtvolumen der gelösten Moleküle bez. suspendierten Körper sei klein gegen V^*. Dies System werde im Sinne der erwähnten Theorie durch die Zustandsvariabeln $p_1 \ldots p_l$ vollständig dargestellt.

Wäre nun auch das molekulare Bild bis in alle Einzelheiten festgelegt, so böte doch die Ausrechnung des Integrales B solche Schwierigkeiten, daß an eine exakte Berechnung von F kaum gedacht werden könnte. Wir brauchen jedoch hier nur zu wissen, wie F von der Größe des Volumens V^* abhängt, in welchem alle gelösten Moleküle bez. suspendierten Körper (im folgenden kurz „Teilchen" genannt) enthalten sind.

Wir nennen x_1, y_1, z_1 die rechtwinkligen Koordinaten des Schwerpunktes des ersten Teilchens, x_2, y_2, z_2 die des zweiten etc., x_n, y_n, z_n die des letzten Teilchens und geben für die Schwerpunkte der Teilchen die unendlich kleinen parallelepipedförmigen Gebiete $dx_1 \, dy_1 \, dz_1, \, dx_2 \, dy_2 \, dz_2 \ldots dx_n \, dy_n \, dz_n$, welche alle in V^* gelegen seien. Gesucht sei der Wert des im Ausdruck für F auftretenden Integrales mit der Beschränkung, daß die Teilchenschwerpunkte in den ihnen soeben zugewiesenen Gebieten liegen. Dies Integral läßt sich jedenfalls auf die Form

$$dB = dx_1 \, dy_1 \ldots dz_n . J$$

bringen, wobei J von $dx_1 \, dy_1$ etc., sowie von V^*, d. h. von der Lage der semipermeabeln Wand, unabhängig ist. J ist aber auch unabhängig von der speziellen Wahl *der Lagen* der Schwerpunktsgebiete und von dem Werte von V^*, wie sogleich gezeigt werden soll. Sei nämlich ein zweites System von unendlich kleinen Gebieten für die Teilchenschwerpunkte gegeben und bezeichnet durch $dx_1' \, dy_1' \, dz_2', \, dx_2' \, dy_2' \, dz_2' \ldots dx_n' \, dy_n' \, dz_n'$, welche Gebiete sich von den ursprünglich gegebenen nur durch ihre Lage, nicht aber durch ihre Größe unterscheiden mögen und ebenfalls alle in V^* enthalten seien, so gilt analog:

$$dB' = dx_1' \, dy_1' \ldots dz_n' . J',$$

wobei
$$dx_1 dy_1 \ldots dz_n = dx'_1 dy'_1 \ldots dz'_n.$$
Es ist also:
$$\frac{dB}{dB'} = \frac{J}{J'}.$$

Aus der in den zitierten Arbeiten gegebenen molekularen Theorie der Wärme läßt sich aber leicht folgern[1]), daß dB/B bez. dB'/B gleich ist der Wahrscheinlichkeit dafür, daß sich in einem beliebig herausgegriffenen Zeitpunkte die Teilchenschwerpunkte in den Gebieten $(dx_1 \ldots dz_n)$ bez. in den Gebieten $(dx'_1 \ldots dz'_n)$ befinden. Sind nun die Bewegungen der einzelnen Teilchen (mit genügender Annäherung) voneinander unabhängig, ist die Flüssigkeit homogen und wirken auf die Teilchen keine Kräfte, so müssen bei gleicher Größe der Gebiete die den beiden Gebietssystemen zukommenden Wahrscheinlichkeiten einander gleich sein, so daß gilt:
$$\frac{dB}{B} = \frac{dB'}{B}.$$

Aus dieser und aus der zuletzt gefundenen Gleichung folgt aber
$$J = J'.$$

Es ist somit erwiesen, daß J weder von V^* noch von $x_1, y_1 \ldots z_n$ abhängig ist. Durch Integration erhält man
$$B = \int J dx_1 \ldots dz_n = J V^{*n}$$
und daraus
$$F = -\frac{RT}{N}\{\lg J + n \lg V^*\}$$
und
$$p = -\frac{\partial F}{\partial V^*} = \frac{RT}{V^*}\frac{n}{N} = \frac{RT}{N} \nu.$$

Durch diese Betrachtung ist gezeigt, daß die Existenz des osmotischen Druckes eine Konsequenz der molekularkinetischen Theorie der Wärme ist, und daß nach dieser Theorie gelöste Moleküle und suspendierte Körper von gleicher Anzahl sich in bezug auf osmotischen Druck bei großer Verdünnung vollkommen gleich verhalten.

---

1) A. Einstein, Ann. d. Phys. 11. p. 170. 1903.

§ 3. Theorie der Diffusion kleiner suspendierter Kugeln.

In einer Flüssigkeit seien suspendierte Teilchen regellos verteilt. Wir wollen den dynamischen Gleichgewichtszustand derselben untersuchen unter der Voraussetzung, daß auf die einzelnen Teilchen eine Kraft K wirkt, welche vom Orte, nicht aber von der Zeit abhängt. Der Einfachheit halber werde angenommen, daß die Kraft überall die Richtung der X-Achse habe.

Es sei ν die Anzahl der suspendierten Teilchen pro Volumeneinheit, so ist im Falle des thermodynamischen Gleichgewichtes ν eine solche Funktion von x, daß für eine beliebige virtuelle Verrückung δx der suspendierten Substanz die Variation der freien Energie verschwindet. Man hat also:

$$\delta F = \delta E - T \delta S = 0.$$

Es werde angenommen, daß die Flüssigkeit senkrecht zur X-Achse den Querschnitt 1 habe und durch die Ebenen $x = 0$ und $x = l$ begrenzt sei. Man hat dann:

$$\delta E = - \int_0^l K \nu \, \delta x \, dx$$

und

$$\delta S = \int_0^l R \frac{\nu}{N} \frac{\partial \delta x}{\partial x} dx = - \frac{R}{N} \int_0^l \frac{\partial \nu}{\partial x} \delta x \, dx.$$

Die gesuchte Gleichgewichtsbedingung ist also:

(1) $$- K \nu + \frac{RT}{N} \frac{\partial \nu}{\partial x} = 0$$

oder

$$K \nu - \frac{\partial p}{\partial x} = 0.$$

Die letzte Gleichung sagt aus, daß der Kraft K durch osmotische Druckkräfte das Gleichgewicht geleistet wird.

Die Gleichung (1) benutzen wir, um den Diffusionskoeffizienten der suspendierten Substanz zu ermitteln. Wir können den eben betrachteten dynamischen Gleichgewichtszustand als die Superposition zweier in umgekehrtem Sinne verlaufender Prozesse auffassen, nämlich

1. einer Bewegung der suspendierten Substanz unter der Wirkung der auf jedes einzelne suspendierte Teilchen wirkenden Kraft K,

2. eines Diffusionsvorganges, welcher als Folge der ungeordneten Bewegungen der Teilchen infolge der Molekularbewegung der Wärme aufzufassen ist.

Haben die suspendierten Teilchen Kugelform (Kugelradius P) und besitzt die Flüssigkeit den Reibungskoeffizienten k, so erteilt die Kraft K dem einzelnen Teilchen die Geschwindigkeit[1]

$$\frac{K}{6\pi k P},$$

und es treten durch die Querschnittseinheit pro Zeiteinheit

$$\frac{\nu K}{6\pi k P}$$

Teilchen hindurch.

Bezeichnet ferner D den Diffusionskoeffizienten der suspendierten Substanz und μ die Masse eines Teilchens, so treten pro Zeiteinheit infolge der Diffusion

$$-D\frac{\partial(\mu\nu)}{\partial x} \text{ Gramm}$$

oder

$$-D\frac{\partial\nu}{\partial x}$$

Teilchen durch die Querschnittseinheit. Da dynamisches Gleichgewicht herrschen soll, so muß sein:

(2) $$\frac{\nu K}{6\pi k P} - D\frac{\partial\nu}{\partial x} = 0.$$

Aus den beiden für das dynamische Gleichgewicht gefundenen Bedingungen (1) und (2) kann man den Diffusionskoeffizienten berechnen. Man erhält:

$$D = \frac{T}{N}\frac{1}{6\pi k P}.$$

Der Diffusionskoeffizient der suspendierten Substanz hängt also

---

[1] Vgl. z. B. G. Kirchhoff, Vorlesungen über Mechanik, 26. Vorlesung § 4.

außer von universellen Konstanten und der absoluten Temperatur nur vom Reibungskoeffizienten der Flüssigkeit und von der Größe der suspendierten Teilchen ab.

§ 4. Über die ungeordnete Bewegung von in einer Flüssigkeit suspendierten Teilchen und deren Beziehung zur Diffusion.

Wir gehen nun dazu über, die ungeordneten Bewegungen genauer zu untersuchen, welche, von der Molekularbewegung der Wärme hervorgerufen, Anlaß zu der im letzten Paragraphen untersuchten Diffusion geben.

Es muß offenbar angenommen werden, daß jedes einzelne Teilchen eine Bewegung ausführe, welche unabhängig ist von der Bewegung aller anderen Teilchen; es werden auch die Bewegungen eines und desselben Teilchens in verschiedenen Zeitintervallen als voneinander unabhängige Vorgänge aufzufassen sein, solange wir diese Zeitintervalle nicht zu klein gewählt denken.

Wir führen ein Zeitintervall τ in die Betrachtung ein, welches sehr klein sei gegen die beobachtbaren Zeitintervalle, aber doch so groß, daß die in zwei aufeinanderfolgenden Zeitintervallen τ von einem Teilchen ausgeführten Bewegungen als voneinander unabhängige Ereignisse aufzufassen sind.

Seien nun in einer Flüssigkeit im ganzen n suspendierte Teilchen vorhanden. In einem Zeitintervall τ werden sich die X-Koordinaten der einzelnen Teilchen um Δ vergrößern, wobei Δ für jedes Teilchen einen anderen (positiven oder negativen) Wert hat. Es wird für Δ ein gewisses Häufigkeitsgesetz gelten; die Anzahl dn der Teilchen, welche in dem Zeitintervall τ eine Verschiebung erfahren, welche zwischen Δ und $\Delta + d\Delta$ liegt, wird durch eine Gleichung von der Form

$$dn = n\,\varphi(\Delta)\,d\Delta$$

ausdrückbar sein, wobei

$$\int_{-\infty}^{+\infty} \varphi(\Delta)\,d\Delta = 1$$

und φ nur für sehr kleine Werte von Δ von Null verschieden ist und die Bedingung

$$\varphi(\Delta) = \varphi(-\Delta)$$

erfüllt.

Wir untersuchen nun, wie der Diffusionskoeffizient von φ abhängt, wobei wir uns wieder auf den Fall beschränken, daß die Anzahl ν der Teilchen pro Volumeneinheit nur von x und t abhängt.

Es sei $\nu = f(x, t)$ die Anzahl der Teilchen pro Volumeneinheit, wir berechnen die Verteilung der Teilchen zur Zeit $t + \tau$ aus deren Verteilung zur Zeit t. Aus der Definition der Funktion $\varphi(\Delta)$ ergibt sich leicht die Anzahl der Teilchen, welche sich zur Zeit $t + \tau$ zwischen zwei zur X-Achse senkrechten Ebenen mit den Abszissen x und $x + dx$ befinden. Man erhält:

$$f(x, t+\tau)\,dx = dx \cdot \int_{\Delta=-\infty}^{\Delta=+\infty} f(x+\Delta)\,\varphi(\Delta)\,d\Delta.$$

Nun können wir aber, da τ sehr klein ist, setzen:

$$f(x, t+\tau) = f(x, t) + \tau\,\frac{\partial f}{\partial t}.$$

Ferner entwickeln wir $f(x + \Delta, t)$ nach Potenzen von Δ:

$$f(x+\Delta, t) = f(x, t) + \Delta\,\frac{\partial f(x,t)}{\partial x} + \frac{\Delta^2}{2!}\,\frac{\partial^2 f(x,t)}{\partial x^2} \ldots \text{ in inf.}$$

Diese Entwicklung können wir unter dem Integral vornehmen, da zu letzterem nur sehr kleine Werte von Δ etwas beitragen. Wir erhalten:

$$f + \frac{\partial f}{\partial t}\cdot\tau = f\cdot\int_{-\infty}^{+\infty}\varphi(\Delta)\,d\Delta + \frac{\partial f}{\partial x}\int_{-\infty}^{+\infty}\Delta\,\varphi(\Delta)\,d\Delta$$
$$+ \frac{\partial^2 f}{\partial x^2}\int_{-\infty}^{+\infty}\frac{\Delta^2}{2}\varphi(\Delta)\,d\Delta \ldots$$

Auf der rechten Seite verschwindet wegen $\varphi(x) = \varphi(-x)$ das zweite, vierte etc. Glied, während von dem ersten, dritten, fünften etc. Gliede jedes folgende gegen das vorhergehende sehr klein ist. Wir erhalten aus dieser Gleichung, indem wir berücksichtigen, daß

$$\int_{-\infty}^{+\infty}\varphi(\Delta)\,d\Delta = 1,$$

und indem wir

$$\frac{1}{\tau} \int_{-\infty}^{+\infty} \frac{\Delta^2}{2} \varphi(\Delta) \, d\Delta = D$$

setzen und nur das erste und dritte Glied der rechten Seite berücksichtigen:

(1) $$\frac{\partial f}{\partial t} = D \frac{\partial^2 f}{\partial x^2}.$$

Dies ist die bekannte Differentialgleichung der Diffusion, und man erkennt, daß D der Diffusionskoeffizient ist.

An diese Entwicklung läßt sich noch eine wichtige Überlegung anknüpfen. Wir haben angenommen, daß die einzelnen Teilchen alle auf dasselbe Koordinatensystem bezogen seien. Dies ist jedoch nicht nötig, da die Bewegungen der einzelnen Teilchen voneinander unabhängig sind. Wir wollen nun die Bewegung jedes Teilchens auf ein Koordinatensystem beziehen, dessen Ursprung mit der Lage des Schwerpunktes des betreffenden Teilchens zur Zeit $t = 0$ zusammenfällt, mit dem Unterschiede, daß jetzt $f(x,t)\,dx$ die Anzahl der Teilchen bedeutet, deren X-Koordinaten von der Zeit $t = 0$ bis zur Zeit $t = t$ um eine Größe *gewachsen* ist, welche zwischen x und $x + dx$ liegt. Auch in diesem Falle ändert sich also die Funktion f gemäß Gleichung (1). Ferner muß offenbar für $x \gtreqless 0$ und $t = 0$

$$f(x,t) = 0 \quad \text{und} \quad \int_{-\infty}^{+\infty} f(x,t)\,dx = n$$

sein. Das Problem, welches mit dem Problem der Diffusion von einem Punkte aus (unter Vernachlässigung der Wechselwirkung der diffundierenden Teilchen) übereinstimmt, ist nun mathematisch vollkommen bestimmt; seine Lösung ist:

$$f(x,t) = \frac{n}{\sqrt{4\pi D}} \frac{e^{-\frac{x^2}{4Dt}}}{\sqrt{t}}.$$

Die Häufigkeitsverteilung der in einer beliebigen Zeit t erfolgten Lagenänderungen ist also dieselbe wie die der zufälligen Fehler, was zu vermuten war. Von Bedeutung aber ist, wie die Konstante im Exponenten mit dem Diffusionskoeffizienten zusammenhängt. Wir berechnen nun mit Hilfe dieser Gleichung die Verrückung λ_x in Richtung der X-Achse, welche ein Teilchen im Mittel erfährt, oder — genauer ausgedrückt — die Wurzel aus dem arithmetischen Mittel der Quadrate der Verrückungen in Richtung der X-Achse; es ist:

$$\lambda_x = \sqrt{\overline{x^2}} = \sqrt{2 D t}.$$

Die mittlere Verschiebung ist also proportional der Quadratwurzel aus der Zeit. Man kann leicht zeigen, daß die Wurzel aus dem Mittelwert der Quadrate der *Gesamtverschiebungen* der Teilchen den Wert $\lambda_x \sqrt{3}$ besitzt.

§ 5. Formel für die mittlere Verschiebung suspendierter Teilchen. Eine neue Methode zur Bestimmung der wahren Größe der Atome.

In § 3 haben wir für den Diffusionskoeffizienten D eines in einer Flüssigkeit in Form von kleinen Kugeln vom Radius P suspendierten Stoffes den Wert gefunden:

$$D = \frac{RT}{N} \frac{1}{6 \pi k P}.$$

Ferner fanden wir in § 4 für den Mittelwert der Verschiebungen der Teilchen in Richtung der X-Achse in der Zeit t:

$$\lambda_x = \sqrt{2 D t}.$$

Durch Eliminieren von D erhalten wir:

$$\lambda_x = \sqrt{t} \cdot \sqrt{\frac{RT}{N} \frac{1}{3 \pi k P}}.$$

Diese Gleichung läßt erkennen, wie λ_x von T, k und P abhängen muß.

Wir wollen berechnen, wie groß λ_x für eine Sekunde ist, wenn N gemäß den Resultaten der kinetischen Gastheorie $6 \cdot 10^{23}$ gesetzt wird; es sei als Flüssigkeit Wasser von 17° C. gewählt ($k = 1{,}35 \cdot 10^{-2}$) und der Teilchendurchmesser sei $0{,}001$ mm. Man erhält:

$$\lambda_x = 8 \cdot 10^{-5} \text{ cm} = 0{,}8 \text{ Mikron.}$$

Die mittlere Verschiebung in 1 Min. wäre also ca. 6 Mikron.

Umgekehrt läßt sich die gefundene Beziehung zur Bestimmung von N benutzen. Man erhält:

$$N = \frac{t}{\lambda_x^2} \cdot \frac{RT}{3\pi k P}.$$

Möge es bald einem Forscher gelingen, die hier aufgeworfene, für die Theorie der Wärme wichtige Frage zu entscheiden!

Bern, Mai 1905.

(Eingegangen 11. Mai 1905.)

3. *Zur Elektrodynamik bewegter Körper;* von *A. Einstein.*

Daß die Elektrodynamik Maxwells — wie dieselbe gegenwärtig aufgefaßt zu werden pflegt — in ihrer Anwendung auf bewegte Körper zu Asymmetrien führt, welche den Phänomenen nicht anzuhaften scheinen, ist bekannt. Man denke z. B. an die elektrodynamische Wechselwirkung zwischen einem Magneten und einem Leiter. Das beobachtbare Phänomen hängt hier nur ab von der Relativbewegung von Leiter und Magnet, während nach der üblichen Auffassung die beiden Fälle, daß der eine oder der andere dieser Körper der bewegte sei, streng voneinander zu trennen sind. Bewegt sich nämlich der Magnet und ruht der Leiter, so entsteht in der Umgebung des Magneten ein elektrisches Feld von gewissem Energiewerte, welches an den Orten, wo sich Teile des Leiters befinden, einen Strom erzeugt. Ruht aber der Magnet und bewegt sich der Leiter, so entsteht in der Umgebung des Magneten kein elektrisches Feld, dagegen im Leiter eine elektromotorische Kraft, welcher an sich keine Energie entspricht, die aber — Gleichheit der Relativbewegung bei den beiden ins Auge gefaßten Fällen vorausgesetzt — zu elektrischen Strömen von derselben Größe und demselben Verlaufe Veranlassung gibt, wie im ersten Falle die elektrischen Kräfte.

Beispiele ähnlicher Art, sowie die mißlungenen Versuche, eine Bewegung der Erde relativ zum „Lichtmedium" zu konstatieren, führen zu der Vermutung, daß dem Begriffe der absoluten Ruhe nicht nur in der Mechanik, sondern auch in der Elektrodynamik keine Eigenschaften der Erscheinungen entsprechen, sondern daß vielmehr für alle Koordinatensysteme, für welche die mechanischen Gleichungen gelten, auch die gleichen elektrodynamischen und optischen Gesetze gelten, wie dies für die Größen erster Ordnung bereits erwiesen ist. Wir wollen diese Vermutung (deren Inhalt im folgenden „Prinzip der Relativität" genannt werden wird) zur Voraussetzung erheben und außerdem die mit ihm nur scheinbar unverträgliche

Voraussetzung einführen, daß sich das Licht im leeren Raume stets mit einer bestimmten, vom Bewegungszustande des emittierenden Körpers unabhängigen Geschwindigkeit V fortpflanze. Diese beiden Voraussetzungen genügen, um zu einer einfachen und widerspruchsfreien Elektrodynamik bewegter Körper zu gelangen unter Zugrundelegung der Maxwellschen Theorie für ruhende Körper. Die Einführung eines „Lichtäthers" wird sich insofern als überflüssig erweisen, als nach der zu entwickelnden Auffassung weder ein mit besonderen Eigenschaften ausgestatteter „absolut ruhender Raum" eingeführt, noch einem Punkte des leeren Raumes, in welchem elektromagnetische Prozesse stattfinden, ein Geschwindigkeitsvektor zugeordnet wird.

Die zu entwickelnde Theorie stützt sich — wie jede andere Elektrodynamik — auf die Kinematik des starren Körpers, da die Aussagen einer jeden Theorie Beziehungen zwischen starren Körpern (Koordinatensystemen), Uhren und elektromagnetischen Prozessen betreffen. Die nicht genügende Berücksichtigung dieses Umstandes ist die Wurzel der Schwierigkeiten, mit denen die Elektrodynamik bewegter Körper gegenwärtig zu kämpfen hat.

I. Kinematischer Teil.

§ 1. Definition der Gleichzeitigkeit.

Es liege ein Koordinatensystem vor, in welchem die Newtonschen mechanischen Gleichungen gelten. Wir nennen dies Koordinatensystem zur sprachlichen Unterscheidung von später einzuführenden Koordinatensystemen und zur Präzisierung der Vorstellung das „ruhende System".

Ruht ein materieller Punkt relativ zu diesem Koordinatensystem, so kann seine Lage relativ zu letzterem durch starre Maßstäbe unter Benutzung der Methoden der euklidischen Geometrie bestimmt und in kartesischen Koordinaten ausgedrückt werden.

Wollen wir die *Bewegung* eines materiellen Punktes beschreiben, so geben wir die Werte seiner Koordinaten in Funktion der Zeit. Es ist nun wohl im Auge zu behalten, daß eine derartige mathematische Beschreibung erst dann einen physikalischen Sinn hat, wenn man sich vorher darüber klar geworden ist, was hier unter „Zeit" verstanden wird.

Wir haben zu berücksichtigen, daß alle unsere Urteile, in welchen die Zeit eine Rolle spielt, immer Urteile über *gleichzeitige Ereignisse* sind. Wenn ich z. B. sage: „Jener Zug kommt hier um 7 Uhr an," so heißt dies etwa: „Das Zeigen des kleinen Zeigers meiner Uhr auf 7 und das Ankommen des Zuges sind gleichzeitige Ereignisse."[1])

Es könnte scheinen, daß alle die Definition der „Zeit" betreffenden Schwierigkeiten dadurch überwunden werden könnten, daß ich an Stelle der „Zeit" die „Stellung des kleinen Zeigers meiner Uhr" setze. Eine solche Definition genügt in der Tat, wenn es sich darum handelt, eine Zeit zu definieren ausschließlich für den Ort, an welchem sich die Uhr eben befindet; die Definition genügt aber nicht mehr, sobald es sich darum handelt, an verschiedenen Orten stattfindende Ereignisreihen miteinander zeitlich zu verknüpfen, oder — was auf dasselbe hinausläuft — Ereignisse zeitlich zu werten, welche in von der Uhr entfernten Orten stattfinden.

Wir könnten uns allerdings damit begnügen, die Ereignisse dadurch zeitlich zu werten, daß ein samt der Uhr im Koordinatenursprung befindlicher Beobachter jedem von einem zu wertenden Ereignis Zeugnis gebenden, durch den leeren Raum zu ihm gelangenden Lichtzeichen die entsprechende Uhrzeigerstellung zuordnet. Eine solche Zuordnung bringt aber den Übelstand mit sich, daß sie vom Standpunkte des mit der Uhr versehenen Beobachters nicht unabhängig ist, wie wir durch die Erfahrung wissen. Zu einer weit praktischeren Festsetzung gelangen wir durch folgende Betrachtung.

Befindet sich im Punkte A des Raumes eine Uhr, so kann ein in A befindlicher Beobachter die Ereignisse in der unmittelbaren Umgebung von A zeitlich werten durch Aufsuchen der mit diesen Ereignissen gleichzeitigen Uhrzeigerstellungen. Befindet sich auch im Punkte B des Raumes eine Uhr — wir wollen hinzufügen, „eine Uhr von genau derselben Beschaffenheit wie die in A befindliche" — so ist auch eine zeitliche Wertung der Ereignisse in der unmittelbaren Umgebung von

---

1) Die Ungenauigkeit, welche in dem Begriffe der Gleichzeitigkeit zweier Ereignisse an (annähernd) demselben Orte steckt und gleichfalls durch eine Abstraktion überbrückt werden muß, soll hier nicht erörtert werden.

B durch einen in B befindlichen Beobachter möglich. Es ist aber ohne weitere Festsetzung nicht möglich, ein Ereignis in A mit einem Ereignis in B zeitlich zu vergleichen; wir haben bisher nur eine „A-Zeit" und eine „B-Zeit", aber keine für A und B gemeinsame „Zeit" definiert. Die letztere Zeit kann nun definiert werden, indem man *durch Definition* festsetzt, daß die „Zeit", welche das Licht braucht, um von A nach B zu gelangen, gleich ist der „Zeit", welche es braucht, um von B nach A zu gelangen. Es gehe nämlich ein Lichtstrahl zur „A-Zeit" t_A von A nach B ab, werde zur „B-Zeit" t_B in B gegen A zu reflektiert und gelange zur „A-Zeit" t'_A nach A zurück. Die beiden Uhren laufen definitionsgemäß synchron, wenn

$$t_B - t_A = t'_A - t_B.$$

Wir nehmen an, daß diese Definition des Synchronismus in widerspruchsfreier Weise möglich sei, und zwar für beliebig viele Punkte, daß also allgemein die Beziehungen gelten:

1. Wenn die Uhr in B synchron mit der Uhr in A läuft, so läuft die Uhr in A synchron mit der Uhr in B.

2. Wenn die Uhr in A sowohl mit der Uhr in B als auch mit der Uhr in C synchron läuft, so laufen auch die Uhren in B und C synchron relativ zueinander.

Wir haben so unter Zuhilfenahme gewisser (gedachter) physikalischer Erfahrungen festgelegt, was unter synchron laufenden, an verschiedenen Orten befindlichen, ruhenden Uhren zu verstehen ist und damit offenbar eine Definition von „gleichzeitig" und „Zeit" gewonnen. Die „Zeit" eines Ereignisses ist die mit dem Ereignis gleichzeitige Angabe einer am Orte des Ereignisses befindlichen, ruhenden Uhr, welche mit einer bestimmten, ruhenden Uhr, und zwar für alle Zeitbestimmungen mit der nämlichen Uhr, synchron läuft.

Wir setzen noch der Erfahrung gemäß fest, daß die Größe

$$\frac{2\,\overline{AB}}{t'_A - t_A} = V$$

eine universelle Konstante (die Lichtgeschwindigkeit im leeren Raume) sei.

Wesentlich ist, daß wir die Zeit mittels im ruhenden System

ruhender Uhren definiert haben; wir nennen die eben definierte Zeit wegen dieser Zugehörigkeit zum ruhenden System „die Zeit des ruhenden Systems".

§ 2. Über die Relativität von Längen und Zeiten.

Die folgenden Überlegungen stützen sich auf das Relativitätsprinzip und auf das Prinzip der Konstanz der Lichtgeschwindigkeit, welche beiden Prinzipien wir folgendermaßen definieren.

1. Die Gesetze, nach denen sich die Zustände der physikalischen Systeme ändern, sind unabhängig davon, auf welches von zwei relativ zueinander in gleichförmiger Translationsbewegung befindlichen Koordinatensystemen diese Zustandsänderungen bezogen werden.

2. Jeder Lichtstrahl bewegt sich im „ruhenden" Koordinatensystem mit der bestimmten Geschwindigkeit V, unabhängig davon, ob dieser Lichtstrahl von einem ruhenden oder bewegten Körper emittiert ist. Hierbei ist

$$\text{Geschwindigkeit} = \frac{\text{Lichtweg}}{\text{Zeitdauer}},$$

wobei „Zeitdauer" im Sinne der Definition des § 1 aufzufassen ist.

Es sei ein ruhender starrer Stab gegeben; derselbe besitze, mit einem ebenfalls ruhenden Maßstabe gemessen, die Länge l. Wir denken uns nun die Stabachse in die X-Achse des ruhenden Koordinatensystems gelegt und dem Stabe hierauf eine gleichförmige Paralleltranslationsbewegung (Geschwindigkeit v) längs der X-Achse im Sinne der wachsenden x erteilt. Wir fragen nun nach der Länge des *bewegten* Stabes, welche wir uns durch folgende zwei Operationen ermittelt denken:

a) Der Beobachter bewegt sich samt dem vorher genannten Maßstabe mit dem auszumessenden Stabe und mißt direkt durch Anlegen des Maßstabes die Länge des Stabes, ebenso, wie wenn sich auszumessender Stab, Beobachter und Maßstab in Ruhe befänden.

b) Der Beobachter ermittelt mittels im ruhenden Systeme aufgestellter, gemäß § 1 synchroner, ruhender Uhren, in welchen Punkten des ruhenden Systems sich Anfang und Ende des auszumessenden Stabes zu einer bestimmten Zeit t befinden.

Die Entfernung dieser beiden Punkte, gemessen mit dem schon benutzten, in diesem Falle ruhenden Maßstabe ist ebenfalls eine Länge, welche man als „Länge des Stabes" bezeichnen kann.

Nach dem Relativitätsprinzip muß die bei der Operation a) zu findende Länge, welche wir „die Länge des Stabes im bewegten System" nennen wollen, gleich der Länge l des ruhenden Stabes sein.

Die bei der Operation b) zu findende Länge, welche wir „die Länge des (bewegten) Stabes im ruhenden System" nennen wollen, werden wir unter Zugrundelegung unserer beiden Prinzipien bestimmen und finden, daß sie von l verschieden ist.

Die allgemein gebrauchte Kinematik nimmt stillschweigend an, daß die durch die beiden erwähnten Operationen bestimmten Längen einander genau gleich seien, oder mit anderen Worten, daß ein bewegter starrer Körper in der Zeitepoche t in geometrischer Beziehung vollständig durch *denselben* Körper, wenn er in bestimmter Lage *ruht*, ersetzbar sei.

Wir denken uns ferner an den beiden Stabenden (A und B) Uhren angebracht, welche mit den Uhren des ruhenden Systems synchron sind, d. h. deren Angaben jeweilen der „Zeit des ruhenden Systems" an den Orten, an welchen sie sich gerade befinden, entsprechen; diese Uhren sind also „synchron im ruhenden System".

Wir denken uns ferner, daß sich bei jeder Uhr ein mit ihr bewegter Beobachter befinde, und daß diese Beobachter auf die beiden Uhren das im § 1 aufgestellte Kriterium für den synchronen Gang zweier Uhren anwenden. Zur Zeit[1] t_A gehe ein Lichtstrahl von A aus, werde zur Zeit t_B in B reflektiert und gelange zur Zeit t'_A nach A zurück. Unter Berücksichtigung des Prinzipes von der Konstanz der Lichtgeschwindigkeit finden wir:

$$t_B - t_A = \frac{r_{AB}}{V - v}$$

---

[1] „Zeit" bedeutet hier „Zeit des ruhenden Systems" und zugleich „Zeigerstellung der bewegten Uhr, welche sich an dem Orte, von dem die Rede ist, befindet".

und
$$t'_A - t_B = \frac{r_{AB}}{V+v},$$

wobei r_{AB} die Länge des bewegten Stabes — im ruhenden System gemessen — bedeutet. Mit dem bewegten Stabe bewegte Beobachter würden also die beiden Uhren nicht synchron gehend finden, während im ruhenden System befindliche Beobachter die Uhren als synchron laufend erklären würden.

Wir sehen also, daß wir dem Begriffe der Gleichzeitigkeit keine *absolute* Bedeutung beimessen dürfen, sondern daß zwei Ereignisse, welche, von einem Koordinatensystem aus betrachtet, gleichzeitig sind, von einem relativ zu diesem System bewegten System aus betrachtet, nicht mehr als gleichzeitige Ereignisse aufzufassen sind.

§ 3. Theorie der Koordinaten- und Zeittransformation von dem ruhenden auf ein relativ zu diesem in gleichförmiger Translationsbewegung befindliches System.

Seien im „ruhenden" Raume zwei Koordinatensysteme, d. h. zwei Systeme von je drei von einem Punkte ausgehenden, aufeinander senkrechten starren materiellen Linien, gegeben. Die X-Achsen beider Systeme mögen zusammenfallen, ihre Y- und Z-Achsen bezüglich parallel sein. Jedem Systeme sei ein starrer Maßstab und eine Anzahl Uhren beigegeben, und es seien beide Maßstäbe sowie alle Uhren beider Systeme einander genau gleich.

Es werde nun dem Anfangspunkte des einen der beiden Systeme (k) eine (konstante) Geschwindigkeit v in Richtung der wachsenden x des anderen, ruhenden Systems (K) erteilt, welche sich auch den Koordinatenachsen, dem betreffenden Maßstabe sowie den Uhren mitteilen möge. Jeder Zeit t des ruhenden Systems K entspricht dann eine bestimmte Lage der Achsen des bewegten Systems und wir sind aus Symmetriegründen befugt anzunehmen, daß die Bewegung von k so beschaffen sein kann, daß die Achsen des bewegten Systems zur Zeit t (es ist mit „t" immer eine Zeit des ruhenden Systems bezeichnet) den Achsen des ruhenden Systems parallel seien.

Wir denken uns nun den Raum sowohl vom ruhenden System K aus mittels des ruhenden Maßstabes als auch vom bewegten System k mittels des mit ihm bewegten Maßstabes ausgemessen und so die Koordinaten x, y, z bez. ξ, η, ζ ermittelt. Es werde ferner mittels der im ruhenden System befindlichen ruhenden Uhren durch Lichtsignale in der in § 1 angegebenen Weise die Zeit t des ruhenden Systems für alle Punkte des letzteren bestimmt, in denen sich Uhren befinden; ebenso werde die Zeit τ des bewegten Systems für alle Punkte des bewegten Systems, in welchen sich relativ zu letzterem ruhende Uhren befinden, bestimmt durch Anwendung der in § 1 genannten Methode der Lichtsignale zwischen den Punkten, in denen sich die letzteren Uhren befinden.

Zu jedem Wertsystem x, y, z, t, welches Ort und Zeit eines Ereignisses im ruhenden System vollkommen bestimmt, gehört ein jenes Ereignis relativ zum System k festlegendes Wertsystem ξ, η, ζ, τ, und es ist nun die Aufgabe zu lösen, das diese Größen verknüpfende Gleichungssystem zu finden.

Zunächst ist klar, daß die Gleichungen *linear* sein müssen wegen der Homogenitätseigenschaften, welche wir Raum und Zeit beilegen.

Setzen wir $x' = x - vt$, so ist klar, daß einem im System k ruhenden Punkte ein bestimmtes, von der Zeit unabhängiges Wertsystem x', y, z zukommt. Wir bestimmen zuerst τ als Funktion von x', y, z und t. Zu diesem Zwecke haben wir in Gleichungen auszudrücken, daß τ nichts anderes ist als der Inbegriff der Angaben von im System k ruhenden Uhren, welche nach der im § 1 gegebenen Regel synchron gemacht worden sind.

Vom Anfangspunkt des Systems k aus werde ein Lichtstrahl zur Zeit τ_0 längs der X-Achse nach x' gesandt und von dort zur Zeit τ_1 nach dem Koordinatenursprung reflektiert, wo er zur Zeit τ_2 anlange; so muß dann sein:
$$\tfrac{1}{2}(\tau_0 + \tau_2) = \tau_1$$
oder, indem man die Argumente der Funktion τ beifügt und das Prinzip der Konstanz der Lichtgeschwindigkeit im ruhenden Systeme anwendet:
$$\tfrac{1}{2}\left[\tau(0,0,0,t) + \tau\left(0,0,0,\left\{t + \frac{x'}{V-v} + \frac{x'}{V+v}\right\}\right)\right]$$
$$= \tau\left(x', 0, 0, t + \frac{x'}{V-v}\right).$$

Hieraus folgt, wenn man x' unendlich klein wählt:

$$\tfrac{1}{2}\left(\frac{1}{V-v}+\frac{1}{V+v}\right)\frac{\partial \tau}{\partial t}=\frac{\partial \tau}{\partial x'}+\frac{1}{V-v}\frac{\partial \tau}{\partial t},$$

oder

$$\frac{\partial \tau}{\partial x'}+\frac{v}{V^2-v^2}\frac{\partial \tau}{\partial t}=0.$$

Es ist zu bemerken, daß wir statt des Koordinatenursprunges jeden anderen Punkt als Ausgangspunkt des Lichtstrahles hätten wählen können und es gilt deshalb die eben erhaltene Gleichung für alle Werte von x', y, z.

Eine analoge Überlegung — auf die H- und Z-Achse angewandt — liefert, wenn man beachtet, daß sich das Licht längs dieser Achsen vom ruhenden System aus betrachtet stets mit der Geschwindigkeit $\sqrt{V^2-v^2}$ fortpflanzt:

$$\frac{\partial \tau}{\partial y}=0$$

$$\frac{\partial \tau}{\partial z}=0.$$

Aus diesen Gleichungen folgt, da τ eine *lineare* Funktion ist:

$$\tau=a\left(t-\frac{v}{V^2-v^2}x'\right),$$

wobei a eine vorläufig unbekannte Funktion $\varphi(v)$ ist und der Kürze halber angenommen ist, daß im Anfangspunkte von k für $\tau=0$ $t=0$ sei.

Mit Hilfe dieses Resultates ist es leicht, die Größen ξ, η, ζ zu ermitteln, indem man durch Gleichungen ausdrückt, daß sich das Licht (wie das Prinzip der Konstanz der Lichtgeschwindigkeit in Verbindung mit dem Relativitätsprinzip verlangt) auch im bewegten System gemessen mit der Geschwindigkeit V fortpflanzt. Für einen zur Zeit $\tau=0$ in Richtung der wachsenden ξ ausgesandten Lichtstrahl gilt:

$$\xi=V\tau,$$

oder

$$\xi=aV\left(t-\frac{v}{V^2-v^2}x'\right).$$

Nun bewegt sich aber der Lichtstrahl relativ zum Anfangs-

punkt von k im ruhenden System gemessen mit der Geschwindigkeit $V-v$, so daß gilt:
$$\frac{x'}{V-v} = t.$$

Setzen wir diesen Wert von t in die Gleichung für ξ ein, so erhalten wir:
$$\xi = a \frac{V^2}{V^2 - v^2} x'.$$

Auf analoge Weise finden wir durch Betrachtung von längs den beiden anderen Achsen bewegte Lichtstrahlen:
$$\eta = V\tau = aV\left(t - \frac{v}{V^2 - v^2} x'\right),$$
wobei
$$\frac{y}{\sqrt{V^2 - v^2}} = t; \quad x' = 0;$$
also
$$\eta = a \frac{V}{\sqrt{V^2 - v^2}} y$$
und
$$\zeta = a \frac{V}{\sqrt{V^2 - v^2}} z.$$

Setzen wir für x' seinen Wert ein, so erhalten wir:
$$\tau = \varphi(v)\beta\left(t - \frac{v}{V^2} x\right),$$
$$\xi = \varphi(v)\beta(x - vt),$$
$$\eta = \varphi(v) y,$$
$$\zeta = \varphi(v) z,$$
wobei
$$\beta = \frac{1}{\sqrt{1 - \left(\frac{v}{V}\right)^2}}$$

und φ eine vorläufig unbekannte Funktion von v ist. Macht man über die Anfangslage des bewegten Systems und über den Nullpunkt von τ keinerlei Voraussetzung, so ist auf den rechten Seiten dieser Gleichungen je eine additive Konstante zuzufügen.

Wir haben nun zu beweisen, daß jeder Lichtstrahl sich, im bewegten System gemessen, mit der Geschwindigkeit V fortpflanzt, falls dies, wie wir angenommen haben, im ruhenden

System der Fall ist; denn wir haben den Beweis dafür noch nicht geliefert, daß das Prinzip der Konstanz der Lichtgeschwindigkeit mit dem Relativitätsprinzip vereinbar sei.

Zur Zeit $t = \tau = 0$ werde von dem zu dieser Zeit gemeinsamen Koordinatenursprung beider Systeme aus eine Kugelwelle ausgesandt, welche sich im System K mit der Geschwindigkeit V ausbreitet. Ist (x, y, z) ein eben von dieser Welle ergriffener Punkt, so ist also

$$x^2 + y^2 + z^2 = V^2 t^2.$$

Diese Gleichung transformieren wir mit Hilfe unserer Transformationsgleichungen und erhalten nach einfacher Rechnung:

$$\xi^2 + \eta^2 + \zeta^2 = V^2 \tau^2.$$

Die betrachtete Welle ist also auch im bewegten System betrachtet eine Kugelwelle von der Ausbreitungsgeschwindigkeit V. Hiermit ist gezeigt, daß unsere beiden Grundprinzipien miteinander vereinbar sind.

In den entwickelten Transformationsgleichungen tritt noch eine unbekannte Funktion φ von v auf, welche wir nun bestimmen wollen.

Wir führen zu diesem Zwecke noch ein drittes Koordinatensystem K' ein, welches relativ zum System k derart in Paralleltranslationsbewegung parallel zur Ξ-Achse begriffen sei, daß sich dessen Koordinatenursprung mit der Geschwindigkeit $-v$ auf der Ξ-Achse bewege. Zur Zeit $t = 0$ mögen alle drei Koordinatenanfangspunkte zusammenfallen und es sei für $t = x = y = z = 0$ die Zeit t' des Systems K' gleich Null. Wir nennen x', y', z' die Koordinaten, im System K' gemessen, und erhalten durch zweimalige Anwendung unserer Transformationsgleichungen:

$$t' = \varphi(-v)\beta(-v)\left\{\tau + \frac{v}{V^2}\xi\right\} = \varphi(v)\varphi(-v)\,t,$$

$$x' = \varphi(-v)\beta(-v)\{\xi + v\tau\} \quad\ = \varphi(v)\varphi(-v)\,x,$$

$$y' = \varphi(-v)\eta \quad\quad\quad\quad\quad\quad\ \ = \varphi(v)\varphi(-v)\,y,$$

$$z' = \varphi(-v)\zeta \quad\quad\quad\quad\quad\quad\ \ = \varphi(v)\varphi(-v)\,z.$$

Da die Beziehungen zwischen x', y', z' und x, y, z die Zeit t nicht enthalten, so ruhen die Systeme K und K' gegeneinander, und es ist klar, daß die Transformation von K auf K' die identische Transformation sein muß. Es ist also:

$$\varphi(v)\varphi(-v) = 1.$$

Wir fragen nun nach der Bedeutung von $\varphi(v)$. Wir fassen das Stück der H-Achse des Systems k ins Auge, das zwischen $\xi=0$, $\eta=0$, $\zeta=0$ und $\xi=0$, $\eta=l$, $\zeta=0$ gelegen ist. Dieses Stück der H-Achse ist ein relativ zum System K mit der Geschwindigkeit v senkrecht zu seiner Achse bewegter Stab, dessen Enden in K die Koordinaten besitzen:

$$x_1 = vt, \quad y_1 = \frac{l}{\varphi(v)}, \quad z_1 = 0$$

und

$$x_2 = vt, \quad y_2 = 0, \quad z_2 = 0.$$

Die Länge des Stabes, in K gemessen, ist also $l/\varphi(v)$; damit ist die Bedeutung der Funktion φ gegeben. Aus Symmetriegründen ist nun einleuchtend, daß die im ruhenden System gemessene Länge eines bestimmten Stabes, welcher senkrecht zu seiner Achse bewegt ist, nur von der Geschwindigkeit, nicht aber von der Richtung und dem Sinne der Bewegung abhängig sein kann. Es ändert sich also die im ruhenden System gemessene Länge des bewegten Stabes nicht, wenn v mit $-v$ vertauscht wird. Hieraus folgt:

$$\frac{l}{\varphi(v)} = \frac{l}{\varphi(-v)},$$

oder

$$\varphi(v) = \varphi(-v).$$

Aus dieser und der vorhin gefundenen Relation folgt, daß $\varphi(v) = 1$ sein muß, so daß die gefundenen Transformationsgleichungen übergehen in:

$$\tau = \beta\left(t - \frac{v}{V^2}x\right),$$
$$\xi = \beta(x - vt),$$
$$\eta = y,$$
$$\zeta = z,$$

wobei

$$\beta = \frac{1}{\sqrt{1 - \left(\frac{v}{V}\right)^2}},$$

§ 4. Physikalische Bedeutung der erhaltenen Gleichungen, bewegte starre Körper und bewegte Uhren betreffend.

Wir betrachten eine starre Kugel[1]) vom Radius R, welche relativ zum bewegten System k ruht, und deren Mittelpunkt im Koordinatenursprung von k liegt. Die Gleichung der Oberfläche dieser relativ zum System K mit der Geschwindigkeit v bewegten Kugel ist:

$$\xi^2 + \eta^2 + \zeta^2 = R^2.$$

Die Gleichung dieser Oberfläche ist in x, y, z ausgedrückt zur Zeit $t = 0$:

$$\frac{x^2}{\left(\sqrt{1-\left(\frac{v}{V}\right)^2}\right)^2} + y^2 + z^2 = R^2.$$

Ein starrer Körper, welcher in ruhendem Zustande ausgemessen die Gestalt einer Kugel hat, hat also in bewegtem Zustande — vom ruhenden System aus betrachtet — die Gestalt eines Rotationsellipsoides mit den Achsen

$$R\sqrt{1-\left(\frac{v}{V}\right)^2},\ R,\ R.$$

Während also die Y- und Z-Dimension der Kugel (also auch jedes starren Körpers von beliebiger Gestalt) durch die Bewegung nicht modifiziert erscheinen, erscheint die X-Dimension im Verhältnis $1 : \sqrt{1-(v/V)^2}$ verkürzt, also um so stärker, je größer v ist. Für $v = V$ schrumpfen alle bewegten Objekte — vom „ruhenden" System aus betrachtet — in flächenhafte Gebilde zusammen. Für Überlichtgeschwindigkeiten werden unsere Überlegungen sinnlos; wir werden übrigens in den folgenden Betrachtungen finden, daß die Lichtgeschwindigkeit in unserer Theorie physikalisch die Rolle der unendlich großen Geschwindigkeiten spielt.

Es ist klar, daß die gleichen Resultate von im „ruhenden" System ruhenden Körpern gelten, welche von einem gleichförmig bewegten System aus betrachtet werden. —

Wir denken uns ferner eine der Uhren, welche relativ zum ruhenden System ruhend die Zeit t, relativ zum bewegten

---

[1]) Das heißt einen Körper, welcher ruhend untersucht Kugelgestalt besitzt.

System ruhend die Zeit τ anzugeben befähigt sind, im Koordinatenursprung von k gelegen und so gerichtet, daß sie die Zeit τ angibt. Wie schnell geht diese Uhr, vom ruhenden System aus betrachtet?

Zwischen die Größen x, t und τ, welche sich auf den Ort dieser Uhr beziehen, gelten offenbar die Gleichungen:
$$\tau = \frac{1}{\sqrt{1-\left(\frac{v}{V}\right)^2}}\left(t - \frac{v}{V^2}x\right)$$
und
$$x = vt.$$
Es ist also
$$\tau = t\sqrt{1-\left(\frac{v}{V}\right)^2} = t - \left(1 - \sqrt{1-\left(\frac{v}{V}\right)^2}\right)t,$$
woraus folgt, daß die Angabe der Uhr (im ruhenden System betrachtet) pro Sekunde um $(1-\sqrt{1-(v/V)^2})$ Sek. oder — bis auf Größen vierter und höherer Ordnung um $\frac{1}{2}(v/V)^2$ Sek. zurückbleibt.

Hieraus ergibt sich folgende eigentümliche Konsequenz. Sind in den Punkten A und B von K ruhende, im ruhenden System betrachtet, synchron gehende Uhren vorhanden, und bewegt man die Uhr in A mit der Geschwindigkeit v auf der Verbindungslinie nach B, so gehen nach Ankunft dieser Uhr in B die beiden Uhren nicht mehr synchron, sondern die von A nach B bewegte Uhr geht gegenüber der von Anfang an in B befindlichen um $\frac{1}{2}tv^2/V^2$ Sek. (bis auf Größen vierter und höherer Ordnung) nach, wenn t die Zeit ist, welche die Uhr von A nach B braucht.

Man sieht sofort, daß dies Resultat auch dann noch gilt, wenn die Uhr in einer beliebigen polygonalen Linie sich von A nach B bewegt, und zwar auch dann, wenn die Punkte A und B zusammenfallen.

Nimmt man an, daß das für eine polygonale Linie bewiesene Resultat auch für eine stetig gekrümmte Kurve gelte, so erhält man den Satz: Befinden sich in A zwei synchron gehende Uhren und bewegt man die eine derselben auf einer geschlossenen Kurve mit konstanter Geschwindigkeit, bis sie wieder nach A zurückkommt, was t Sek. dauern möge, so geht die letztere Uhr bei ihrer Ankunft in A gegenüber der unbewegt gebliebenen um $\frac{1}{2} t (v/V)^2$ Sek. nach. Man schließt daraus, daß eine am Erdäquator befindliche Unruhuhr um einen sehr kleinen Betrag langsamer laufen muß als eine genau gleich beschaffene, sonst gleichen Bedingungen unterworfene, an einem Erdpole befindliche Uhr.

§ 5. Additionstheorem der Geschwindigkeiten.

In dem längs der X-Achse des Systems K mit der Geschwindigkeit v bewegten System k bewege sich ein Punkt gemäß den Gleichungen:

$$\xi = w_\xi \tau,$$
$$\eta = w_\eta \tau,$$
$$\zeta = 0,$$

wobei w_ξ und w_η Konstanten bedeuten.

Gesucht ist die Bewegung des Punktes relativ zum System K. Führt man in die Bewegungsgleichungen des Punktes mit Hilfe der in § 3 entwickelten Transformationsgleichungen die Größen x, y, z, t ein, so erhält man:

$$x = \frac{w_\xi + v}{1 + \frac{v w_\xi}{V^2}} t,$$

$$y = \frac{\sqrt{1 - \left(\frac{v}{V}\right)^2}}{1 + \frac{v w_\xi}{V^2}} w_\eta t,$$

$$z = 0.$$

Das Gesetz vom Parallelogramm der Geschwindigkeiten gilt also nach unserer Theorie nur in erster Annäherung. Wir setzen:

$$U^2 = \left(\frac{dx}{dt}\right)^2 + \left(\frac{dy}{dt}\right)^2,$$
$$w^2 = w_\xi^2 + w_\eta^2$$

und

$$\alpha = \operatorname{arctg} \frac{w_y}{w_x};$$

α ist dann als der Winkel zwischen den Geschwindigkeiten v und w anzusehen. Nach einfacher Rechnung ergibt sich:

$$U = \frac{\sqrt{(v^2 + w^2 + 2vw\cos\alpha) - \left(\dfrac{vw\sin\alpha}{V}\right)^2}}{1 + \dfrac{vw\cos\alpha}{V^2}}.$$

Es ist bemerkenswert, daß v und w in symmetrischer Weise in den Ausdruck für die resultierende Geschwindigkeit eingehen. Hat auch w die Richtung der X-Achse (Ξ-Achse), so erhalten wir:

$$U = \frac{v + w}{1 + \dfrac{vw}{V^2}}.$$

Aus dieser Gleichung folgt, daß aus der Zusammensetzung zweier Geschwindigkeiten, welche kleiner sind als V, stets eine Geschwindigkeit kleiner als V resultiert. Setzt man nämlich $v = V - \varkappa$, $w = V - \lambda$, wobei $\varkappa$ und λ positiv und kleiner als V seien, so ist:

$$U = V \frac{2V - \varkappa - \lambda}{2V - \varkappa - \lambda + \dfrac{\varkappa\lambda}{V}} < V.$$

Es folgt ferner, daß die Lichtgeschwindigkeit V durch Zusammensetzung mit einer „Unterlichtgeschwindigkeit" nicht geändert werden kann. Man erhält für diesen Fall:

$$U = \frac{V + w}{1 + \dfrac{w}{V}} = V.$$

Wir hätten die Formel für U für den Fall, daß v und w gleiche Richtung besitzen, auch durch Zusammensetzen zweier Transformationen gemäß § 3 erhalten können. Führen wir neben den in § 3 figurierenden Systemen K und k noch ein drittes, zu k in Parallelbewegung begriffenes Koordinatensystem k' ein, dessen Anfangspunkt sich auf der Ξ-Achse mit der Geschwindigkeit w bewegt, so erhalten wir zwischen den Größen x, y, z, t und den entsprechenden Größen von k' Gleichungen, welche sich von den in § 3 gefundenen nur dadurch unterscheiden, daß an Stelle von „v" die Größe

$$\frac{v + w}{1 + \dfrac{vw}{V^2}}$$

tritt; man sieht daraus, daß solche Paralleltransformationen — wie dies sein muß — eine Gruppe bilden.

Wir haben nun die für uns notwendigen Sätze der unseren zwei Prinzipien entsprechenden Kinematik hergeleitet und gehen dazu über, deren Anwendung in der Elektrodynamik zu zeigen.

II. Eektrodynamischer Teil.

§ 6. Transformation der Maxwell-Hertzschen Gleichungen für den leeren Raum. Über die Natur der bei Bewegung in einem Magnetfeld auftretenden elektromotorischen Kräfte.

Die Maxwell-Hertzschen Gleichungen für den leeren Raum mögen gültig sein für das ruhende System K, so daß gelten möge:

$$\frac{1}{V}\frac{\partial X}{\partial t} = \frac{\partial N}{\partial y} - \frac{\partial M}{\partial z}, \quad \frac{1}{V}\frac{\partial L}{\partial t} = \frac{\partial Y}{\partial z} - \frac{\partial Z}{\partial y},$$

$$\frac{1}{V}\frac{\partial Y}{\partial t} = \frac{\partial L}{\partial z} - \frac{\partial N}{\partial x}, \quad \frac{1}{V}\frac{\partial M}{\partial t} = \frac{\partial Z}{\partial x} - \frac{\partial X}{\partial z},$$

$$\frac{1}{V}\frac{\partial Z}{\partial t} = \frac{\partial M}{\partial x} - \frac{\partial L}{\partial y}, \quad \frac{1}{V}\frac{\partial N}{\partial t} = \frac{\partial X}{\partial y} - \frac{\partial Y}{\partial x},$$

wobei (X, Y, Z) den Vektor der elektrischen, (L, M, N) den der magnetischen Kraft bedeutet.

Wenden wir auf diese Gleichungen die in § 3 entwickelte Transformation an, indem wir die elektromagnetischen Vorgänge auf das dort eingeführte, mit der Geschwindigkeit v bewegte Koordinatensystem beziehen, so erhalten wir die Gleichungen:

$$\frac{1}{V}\frac{\partial X}{\partial \tau} = \frac{\partial \beta\left(N - \frac{v}{V}Y\right)}{\partial \eta} - \frac{\partial \beta\left(M + \frac{v}{V}Z\right)}{\partial \zeta},$$

$$\frac{1}{V}\frac{\partial \beta\left(Y - \frac{v}{V}N\right)}{\partial \tau} = \frac{\partial L}{\partial \zeta} - \frac{\partial \beta\left(N - \frac{v}{V}Y\right)}{\partial \xi},$$

$$\frac{1}{V}\frac{\partial \beta\left(Z + \frac{v}{V}M\right)}{\partial \tau} = \frac{\partial \beta\left(M + \frac{v}{V}Z\right)}{\partial \xi} - \frac{\partial L}{\partial \eta},$$

$$\frac{1}{V}\frac{\partial L}{\partial \tau} = \frac{\partial \beta\left(Y - \frac{v}{V}N\right)}{\partial \zeta} - \frac{\partial \beta\left(Z + \frac{v}{V}M\right)}{\partial \eta},$$

$$\frac{1}{V}\frac{\partial \beta\left(M + \frac{v}{V}Z\right)}{\partial \tau} = \frac{\partial \beta\left(Z + \frac{v}{V}M\right)}{\partial \xi} - \frac{\partial X}{\partial \zeta},$$

$$\frac{1}{V}\frac{\partial \beta\left(N - \frac{v}{V}Y\right)}{\partial \tau} = \frac{\partial X}{\partial \eta} - \frac{\partial \beta\left(Y - \frac{v}{V}N\right)}{\partial \xi},$$

wobei

$$\beta = \frac{1}{\sqrt{1 - \left(\frac{v}{V}\right)^2}}.$$

Das Relativitätsprinzip fordert nun, daß die **Maxwell-Hertz**schen Gleichungen für den leeren Raum auch im System k gelten, wenn sie im System K gelten, d. h. daß für die im bewegten System k durch ihre ponderomotorischen Wirkungen auf elektrische bez. magnetische Massen definierten Vektoren der elektrischen und magnetischen Kraft ((X', Y' Z') und (L', M', N')) des bewegten Systems k die Gleichungen gelten:

$$\frac{1}{V}\frac{\partial X'}{\partial \tau} = \frac{\partial N'}{\partial \eta} - \frac{\partial M'}{\partial \zeta}, \quad \frac{1}{V}\frac{\partial L'}{\partial \tau} = \frac{\partial Y'}{\partial \zeta} - \frac{\partial Z'}{\partial \eta},$$

$$\frac{1}{V}\frac{\partial Y'}{\partial \tau} = \frac{\partial L'}{\partial \zeta} - \frac{\partial N'}{\partial \xi}, \quad \frac{1}{V}\frac{\partial M'}{\partial \tau} = \frac{\partial Z'}{\partial \xi} - \frac{\partial X'}{\partial \zeta},$$

$$\frac{1}{V}\frac{\partial Z'}{\partial \tau} = \frac{\partial M'}{\partial \xi} - \frac{\partial L'}{\partial \eta}, \quad \frac{1}{V}\frac{\partial N'}{\partial \tau} = \frac{\partial X'}{\partial \eta} - \frac{\partial Y'}{\partial \xi}.$$

Offenbar müssen nun die beiden für das System k gefundenen Gleichungssysteme genau dasselbe ausdrücken, da beide Gleichungssysteme den **Maxwell-Hertz**schen Gleichungen für das System K äquivalent sind. Da die Gleichungen beider Systeme ferner bis auf die die Vektoren darstellenden Symbole übereinstimmen, so folgt, daß die in den Gleichungssystemen an entsprechenden Stellen auftretenden Funktionen bis auf einen für alle Funktionen des einen Gleichungssystems gemeinsamen, von ξ, η, ζ und τ unabhängigen, eventuell von v abhängigen Faktor $\psi(v)$ übereinstimmen müssen. Es gelten also die Beziehungen:

$$X' = \psi(v)X, \qquad L' = \psi(v)L,$$
$$Y' = \psi(v)\beta\left(Y - \frac{v}{V}N\right), \quad M' = \psi(v)\beta\left(M + \frac{v}{V}Z\right),$$
$$Z' = \psi(v)\beta\left(Z + \frac{v}{V}M\right), \quad N' = \psi(v)\beta\left(N - \frac{v}{V}Y\right).$$

Bildet man nun die Umkehrung dieses Gleichungssystems, erstens durch Auflösen der soeben erhaltenen Gleichungen, zweitens durch Anwendung der Gleichungen auf die inverse Transformation (von k auf K), welche durch die Geschwindigkeit $-v$ charakterisiert ist, so folgt, indem man berücksichtigt, daß die beiden so erhaltenen Gleichungssysteme identisch sein müssen:
$$\varphi(v)\cdot\varphi(-v) = 1.$$
Ferner folgt aus Symmetriegründen[1])
$$\varphi(v) = \varphi(-v):$$
es ist also
$$\varphi(v) = 1,$$
und unsere Gleichungen nehmen die Form an:
$$X' = X, \qquad L' = L,$$
$$Y' = \beta\left(Y - \frac{v}{V}N\right), \qquad M' = \beta\left(M + \frac{v}{V}Z\right),$$
$$Z' = \beta\left(Z + \frac{v}{V}M\right), \qquad N' = \beta\left(N - \frac{v}{V}Y\right).$$

Zur Interpretation dieser Gleichungen bemerken wir folgendes. Es liegt eine punktförmige Elektrizitätsmenge vor, welche im ruhenden System K gemessen von der Größe „eins" sei, d. h. im ruhenden System ruhend auf eine gleiche Elektrizitätsmenge im Abstand 1 cm die Kraft 1 Dyn ausübe. Nach dem Relativitätsprinzip ist diese elektrische Masse auch im bewegten System gemessen von der Größe „eins". Ruht diese Elektrizitätsmenge relativ zum ruhenden System, so ist definitionsgemäß der Vektor (X, Y, Z) gleich der auf sie wirkenden Kraft. Ruht die Elektrizitätsmenge gegenüber dem bewegten System (wenigstens in dem betreffenden Augenblick), so ist die auf sie wirkende, in dem bewegten System gemessene Kraft gleich dem Vektor (X', Y', Z'). Die ersten drei der obigen Gleichungen lassen sich mithin auf folgende zwei Weisen in Worte kleiden:

1. Ist ein punktförmiger elektrischer Einheitspol in einem elektromagnetischen Felde bewegt, so wirkt auf ihn außer der

---

[1]) Ist z. B. $X = Y = Z = L = M = 0$ und $N \neq 0$, so ist aus Symmetriegründen klar, daß bei Zeichenwechsel von v ohne Änderung des numerischen Wertes auch Y' sein Vorzeichen ändern muß, ohne seinen numerischen Wert zu ändern.

elektrischen Kraft eine „elektromotorische Kraft", welche unter Vernachlässigung von mit der zweiten und höheren Potenzen von v/V multiplizierten Gliedern gleich ist dem mit der Lichtgeschwindigkeit dividierten Vektorprodukt der Bewegungsgeschwindigkeit des Einheitspoles und der magnetischen Kraft. (Alte Ausdrucksweise.)

2. Ist ein punktförmiger elektrischer Einheitspol in einem elektromagnetischen Felde bewegt, so ist die auf ihn wirkende Kraft gleich der an dem Orte des Einheitspoles vorhandenen elektrischen Kraft, welche man durch Transformation des Feldes auf ein relativ zum elektrischen Einheitspol ruhendes Koordinatensystem erhält. (Neue Ausdrucksweise.)

Analoges gilt über die „magnetomotorischen Kräfte". Man sieht, daß in der entwickelten Theorie die elektromotorische Kraft nur die Rolle eines Hilfsbegriffes spielt, welcher seine Einführung dem Umstande verdankt, daß die elektrischen und magnetischen Kräfte keine von dem Bewegungszustande des Koordinatensystems unabhängige Existenz besitzen.

Es ist ferner klar, daß die in der Einleitung angeführte Asymmetrie bei der Betrachtung der durch Relativbewegung eines Magneten und eines Leiters erzeugten Ströme verschwindet. Auch werden die Fragen nach dem „Sitz" der elektrodynamischen elektromotorischen Kräfte (Unipolarmaschinen) gegenstandslos.

§ 7. Theorie des Doppelerschen Prinzips und der Aberration.

Im Systeme K befinde sich sehr ferne vom Koordinatenursprung eine Quelle elektrodynamischer Wellen, welche in einem den Koordinatenursprung enthaltenden Raumteil mit genügender Annäherung durch die Gleichungen dargestellt sei:

$$X = X_0 \sin \Phi, \quad L = L_0 \sin \Phi,$$
$$Y = Y_0 \sin \Phi, \quad M = M_0 \sin \Phi, \quad \Phi = \omega\left(t - \frac{ax + by + cz}{V}\right).$$
$$Z = Z_0 \sin \Phi, \quad N = N_0 \sin \Phi,$$

Hierbei sind (X_0, Y_0, Z_0) und (L_0, M_0, N_0) die Vektoren, welche die Amplitude des Wellenzuges bestimmen, a, b, c die Richtungskosinus der Wellennormalen.

Wir fragen nach der Beschaffenheit dieser Wellen, wenn dieselben von einem in dem bewegten System k ruhenden

Beobachter untersucht werden. — Durch Anwendung der in § 6 gefundenen Transformationsgleichungen für die elektrischen und magnetischen Kräfte und der in § 3 gefundenen Transformationsgleichungen für die Koordinaten und die Zeit erhalten wir unmittelbar:

$$X' = X_0 \sin \Phi', \qquad L' = L_0 \sin \Phi',$$

$$Y' = \beta\left(Y_0 - \frac{v}{V} N_0\right) \sin \Phi', \qquad M' = \beta\left(M_0 + \frac{v}{V} Z_0\right) \sin \Phi',$$

$$Z' = \beta\left(Z_0 + \frac{v}{V} M_0\right) \sin \Phi', \qquad N' = \beta\left(N_0 - \frac{v}{V} Y_0\right) \sin \Phi',$$

$$\Phi' = \omega'\left(\tau - \frac{a'\xi + b'\eta + c'\zeta}{V}\right),$$

wobei

$$\omega' = \omega \beta \left(1 - a \frac{v}{V}\right),$$

$$a' = \frac{a - \frac{v}{V}}{1 - a \frac{v}{V}},$$

$$b' = \frac{b}{\beta\left(1 - a \frac{v}{V}\right)},$$

$$c' = \frac{c}{\beta\left(1 - a \frac{v}{V}\right)}$$

gesetzt ist.

Aus der Gleichung für ω' folgt: Ist ein Beobachter relativ zu einer unendlich fernen Lichtquelle von der Frequenz v mit der Geschwindigkeit v derart bewegt, daß die Verbindungslinie „Lichtquelle–Beobachter" mit der auf ein relativ zur Lichtquelle ruhendes Koordinatensystem bezogenen Geschwindigkeit des Beobachters den Winkel φ bildet, so ist die von dem Beobachter wahrgenommene Frequenz v' des Lichtes durch die Gleichung gegeben:

$$v' = v \frac{1 - \cos\varphi \frac{v}{V}}{\sqrt{1 - \left(\frac{v}{V}\right)^2}}.$$

Dies ist das Doppelersche Prinzip für beliebige Geschwindigkeiten. Für $\varphi = 0$ nimmt die Gleichung die übersichtliche Form an:

$$\nu' = \nu \sqrt{\frac{1 - \frac{v}{V}}{1 + \frac{v}{V}}}.$$

Man sieht, daß — im Gegensatz zu der üblichen Auffassung — für $v = -\infty$, $\nu = \infty$ ist.

Nennt man φ' den Winkel zwischen Wellennormale (Strahlrichtung) im bewegten System und der Verbindungslinie „Lichtquelle–Beobachter", so nimmt die Gleichung für a' die Form an:

$$\cos \varphi' = \frac{\cos \varphi - \frac{v}{V}}{1 - \frac{v}{V} \cos \varphi}.$$

Diese Gleichung drückt das Aberrationsgesetz in seiner allgemeinsten Form aus. Ist $\varphi = \pi/2$, so nimmt die Gleichung die einfache Gestalt an:

$$\cos \varphi' = -\frac{v}{V}.$$

Wir haben nun noch die Amplitude der Wellen, wie dieselbe im bewegten System erscheint, zu suchen. Nennt man A bez. A' die Amplitude der elektrischen oder magnetischen Kraft im ruhenden bez. im bewegten System gemessen, so erhält man:

$$A'^2 = A^2 \frac{\left(1 - \frac{v}{V} \cos \varphi\right)^2}{1 - \left(\frac{v}{V}\right)^2},$$

welche Gleichung für $\varphi = 0$ in die einfachere übergeht:

$$A'^2 = A^2 \frac{1 - \frac{v}{V}}{1 + \frac{v}{V}}.$$

Es folgt aus den entwickelten Gleichungen, daß für einen Beobachter, der sich mit der Geschwindigkeit V einer Lichtquelle näherte, diese Lichtquelle unendlich intensiv erscheinen müßte.

§ 8. Transformation der Energie der Lichtstrahlen. Theorie des auf vollkommene Spiegel ausgeübten Strahlungsdruckes.

Da $A^2/8\pi$ gleich der Lichtenergie pro Volumeneinheit ist, so haben wir nach dem Relativitätsprinzip $A'^2/8\pi$ als die Lichtenergie im bewegten System zu betrachten. Es wäre daher A'^2/A^2 das Verhältnis der „bewegt gemessenen" und „ruhend gemessenen" Energie eines bestimmten Lichtkomplexes, wenn das Volumen eines Lichtkomplexes in K gemessen und in k gemessen das gleiche wäre. Dies ist jedoch nicht der Fall. Sind a, b, c die Richtungskosinus der Wellennormalen des Lichtes im ruhenden System, so wandert durch die Oberflächenelemente der mit Lichtgeschwindigkeit bewegten Kugelfläche

$$(x - Vat)^2 + (y - Vbt)^2 + (z - Vct)^2 = R^2$$

keine Energie hindurch; wir können daher sagen, daß diese Fläche dauernd denselben Lichtkomplex umschließt. Wir fragen nach der Energiemenge, welche diese Fläche im System k betrachtet umschließt, d. h. nach der Energie des Lichtkomplexes relativ zum System k.

Die Kugelfläche ist — im bewegten System betrachtet — eine Ellipsoidfläche, welche zur Zeit $\tau = 0$ die Gleichung besitzt:

$$\left(\beta\xi - a\beta\frac{v}{V}\xi\right)^2 + \left(\eta - b\beta\frac{v}{V}\xi\right)^2 + \left(\zeta - c\beta\frac{v}{V}\xi\right)^2 = R^2.$$

Nennt man S das Volumen der Kugel, S' dasjenige dieses Ellipsoides, so ist, wie eine einfache Rechnung zeigt:

$$\frac{S'}{S} = \frac{\sqrt{1 - \left(\frac{v}{V}\right)^2}}{1 - \frac{v}{V}\cos\varphi}.$$

Nennt man also E die im ruhenden System gemessene, E' die im bewegten System gemessene Lichtenergie, welche von der betrachteten Fläche umschlossen wird, so erhält man:

$$\frac{E'}{E} = \frac{\dfrac{A'^2}{8\pi}S'}{\dfrac{A^2}{8\pi}S} = \frac{1 - \dfrac{v}{V}\cos\varphi}{\sqrt{1 - \left(\dfrac{v}{V}\right)^2}},$$

welche Formel für $\varphi = 0$ in die einfachere übergeht:

$$\frac{E'}{E} = \sqrt{\frac{1-\frac{v}{V}}{1+\frac{v}{V}}}.$$

Es ist bemerkenswert, daß die Energie und die Frequenz eines Lichtkomplexes sich nach demselben Gesetze mit dem Bewegungszustande des Beobachters ändern.

Es sei nun die Koordinatenebene $\xi = 0$ eine vollkommen spiegelnde Fläche, an welcher die im letzten Paragraph betrachteten ebenen Wellen reflektiert werden. Wir fragen nach dem auf die spiegelnde Fläche ausgeübten Lichtdruck und nach der Richtung, Frequenz und Intensität des Lichtes nach der Reflexion.

Das einfallende Licht sei durch die Größen A, $\cos\varphi$, ν (auf das System K bezogen) definiert. Von k aus betrachtet sind die entsprechenden Größen:

$$A' = A \frac{1 - \frac{v}{V}\cos\varphi}{\sqrt{1 - \left(\frac{v}{V}\right)^2}},$$

$$\cos\varphi' = \frac{\cos\varphi - \frac{v}{V}}{1 - \frac{v}{V}\cos\varphi},$$

$$\nu' = \nu \frac{1 - \frac{v}{V}\cos\varphi}{\sqrt{1 - \left(\frac{v}{V}\right)^2}}.$$

Für das reflektierte Licht erhalten wir, wenn wir den Vorgang auf das System k beziehen:

$$A'' = A',$$
$$\cos\varphi'' = -\cos\varphi',$$
$$\nu'' = \nu'.$$

Endlich erhält man durch Rücktransformieren aufs ruhende System K für das reflektierte Licht:

$$A''' = A'' \frac{1 + \frac{v}{V}\cos\varphi''}{\sqrt{1 - \left(\frac{v}{V}\right)^2}} = A \frac{1 - 2\frac{v}{V}\cos\varphi + \left(\frac{v}{V}\right)^2}{1 - \left(\frac{v}{V}\right)^2},$$

$$\cos\varphi''' = \frac{\cos\varphi'' + \frac{v}{V}}{1 + \frac{v}{V}\cos\varphi''} = -\frac{\left(1 + \left(\frac{v}{V}\right)^2\right)\cos\varphi - 2\frac{v}{V}}{1 - 2\frac{v}{V}\cos\varphi + \left(\frac{v}{V}\right)^2},$$

$$v''' = v'' \frac{1 + \frac{v}{V}\cos\varphi''}{\sqrt{1 - \left(\frac{v}{V}\right)^2}} = v \frac{1 - 2\frac{v}{V}\cos\varphi + \left(\frac{v}{V}\right)^2}{\left(1 - \frac{v}{V}\right)^2}.$$

Die auf die Flächeneinheit des Spiegels pro Zeiteinheit auftreffende (im ruhenden System gemessene) Energie ist offenbar $A^2/8\pi \, (V\cos\varphi - v)$. Die von der Flächeneinheit des Spiegels in der Zeiteinheit sich entfernende Energie ist $A'''^2/8\pi \, (-V\cos\varphi''' + v)$. Die Differenz dieser beiden Ausdrücke ist nach dem Energieprinzip die vom Lichtdrucke in der Zeiteinheit geleistete Arbeit. Setzt man die letztere gleich dem Produkt $P \cdot v$, wobei P der Lichtdruck ist, so erhält man:

$$P = 2\frac{A^2}{8\pi} \frac{\left(\cos\varphi - \frac{v}{V}\right)^2}{1 - \left(\frac{v}{V}\right)^2}.$$

In erster Annäherung erhält man in Übereinstimmung mit der Erfahrung und mit anderen Theorien

$$P = 2\frac{A^2}{8\pi}\cos^2\varphi.$$

Nach der hier benutzten Methode können alle Probleme der Optik bewegter Körper gelöst werden. Das Wesentliche ist, daß die elektrische und magnetische Kraft des Lichtes, welches durch einen bewegten Körper beeinflußt wird, auf ein relativ zu dem Körper ruhendes Koordinatensystem transformiert werden. Dadurch wird jedes Problem der Optik bewegter Körper auf eine Reihe von Problemen der Optik ruhender Körper zurückgeführt.

§ 9. Transformation der Maxwell-Hertzschen Gleichungen mit Berücksichtigung der Konvektionsströme.

Wir gehen aus von den Gleichungen:

$$\frac{1}{V}\left\{u_x \varrho + \frac{\partial X}{\partial t}\right\} = \frac{\partial N}{\partial y} - \frac{\partial M}{\partial z}, \quad \frac{1}{V}\frac{\partial L}{\partial t} = \frac{\partial Y}{\partial z} - \frac{\partial Z}{\partial y},$$

$$\frac{1}{V}\left\{u_y \varrho + \frac{\partial Y}{\partial t}\right\} = \frac{\partial L}{\partial z} - \frac{\partial N}{\partial x}, \quad \frac{1}{V}\frac{\partial M}{\partial t} = \frac{\partial Z}{\partial x} - \frac{\partial X}{\partial z},$$

$$\frac{1}{V}\left\{u_z \varrho + \frac{\partial Z}{\partial t}\right\} = \frac{\partial M}{\partial x} - \frac{\partial L}{\partial y}, \quad \frac{1}{V}\frac{\partial N}{\partial t} = \frac{\partial X}{\partial y} - \frac{\partial Y}{\partial x},$$

wobei

$$\varrho = \frac{\partial X}{\partial x} + \frac{\partial Y}{\partial y} + \frac{\partial Z}{\partial z}$$

die 4π-fache Dichte der Elektrizität und (u_x, u_y, u_z) den Geschwindigkeitsvektor der Elektrizität bedeutet. Denkt man sich die elektrischen Massen unveränderlich an kleine, starre Körper (Ionen, Elektronen) gebunden, so sind diese Gleichungen die elektromagnetische Grundlage der Lorentzschen Elektrodynamik und Optik bewegter Körper.

Transformiert man diese Gleichungen, welche im System K gelten mögen, mit Hilfe der Transformationsgleichungen von § 3 und § 6 auf das System k, so erhält man die Gleichungen:

$$\frac{1}{V}\left\{u_\xi \varrho' + \frac{\partial X'}{\partial \tau}\right\} = \frac{\partial N'}{\partial \eta} - \frac{\partial M'}{\partial \zeta}, \quad \frac{\partial L'}{\partial \tau} = \frac{\partial Y'}{\partial \zeta} - \frac{\partial Z'}{\partial \eta},$$

$$\frac{1}{V}\left\{u_\eta \varrho' + \frac{\partial Y'}{\partial \tau}\right\} = \frac{\partial L'}{\partial \zeta} - \frac{\partial N'}{\partial \xi}, \quad \frac{\partial M'}{\partial \tau} = \frac{\partial Z'}{\partial \xi} - \frac{\partial X'}{\partial \zeta},$$

$$\frac{1}{V}\left\{u_\zeta \varrho' + \frac{\partial Z'}{\partial \tau}\right\} = \frac{\partial M'}{\partial \xi} - \frac{\partial L'}{\partial \eta}, \quad \frac{\partial N'}{\partial \tau} = \frac{\partial X'}{\partial \eta} - \frac{\partial Y'}{\partial \xi},$$

wobei

$$\frac{u_x - v}{1 - \frac{u_x v}{V^2}} = u_\xi,$$

$$\frac{u_y}{\beta\left(1 - \frac{u_x v}{V^2}\right)} = u_\eta, \quad \varrho' = \frac{\partial X'}{\partial \xi} + \frac{\partial Y'}{\partial \eta} + \frac{\partial Z'}{\partial \zeta} = \beta\left(1 - \frac{v u_x}{V^2}\right)\varrho.$$

$$\frac{u_z}{\beta\left(1 - \frac{u_x v}{V^2}\right)} = u_\zeta.$$

Da — wie aus dem Additionstheorem der Geschwindigkeiten (§ 5) folgt — der Vektor (u_ξ, u_η, u_ζ) nichts anderes ist als die Geschwindigkeit der elektrischen Massen im System k gemessen, so ist damit gezeigt, daß unter Zugrundelegung unserer kinematischen Prinzipien die elektrodynamische Grundlage der Lorentzschen Theorie der Elektrodynamik bewegter Körper dem Relativitätsprinzip entspricht.

Es möge noch kurz bemerkt werden, daß aus den entwickelten Gleichungen leicht der folgende wichtige Satz gefolgert werden kann: Bewegt sich ein elektrisch geladener Körper beliebig im Raume und ändert sich hierbei seine Ladung nicht, von einem mit dem Körper bewegten Koordinatensystem aus betrachtet, so bleibt seine Ladung auch — von dem „ruhenden" System K aus betrachtet — konstant.

§ 10. Dynamik des (langsam beschleunigten) Elektrons.

In einem elektromagnetischen Felde bewege sich ein punktförmiges, mit einer elektrischen Ladung ε versehenes Teilchen (im folgenden „Elektron" genannt), über dessen Bewegungsgesetz wir nur folgendes annehmen:

Ruht das Elektron in einer bestimmten Epoche, so erfolgt in dem nächsten Zeitteilchen die Bewegung des Elektrons nach den Gleichungen

$$\mu \frac{d^2 x}{dt^2} = \varepsilon X$$

$$\mu \frac{d^2 y}{dt^2} = \varepsilon Y$$

$$\mu \frac{d^2 z}{dt^2} = \varepsilon Z,$$

wobei x, y, z die Koordinaten des Elektrons, μ die Masse des Elektrons bedeutet, sofern dasselbe langsam bewegt ist.

Es besitze nun zweitens das Elektron in einer gewissen Zeitepoche die Geschwindigkeit v. Wir suchen das Gesetz, nach welchem sich das Elektron im unmittelbar darauf folgenden Zeitteilchen bewegt.

Ohne die Allgemeinheit der Betrachtung zu beeinflussen, können und wollen wir annehmen, daß das Elektron in dem Momente, wo wir es ins Auge fassen, sich im Koordinaten-sprung befinde und sich längs der X-Achse des Systems K mit der Geschwindigkeit v bewege. Es ist dann einleuchtend, daß das Elektron im genannten Momente ($t = 0$) relativ zu einem längs der X-Achse mit der konstanten Geschwindigkeit v parallelbewegten Koordinatensystem k ruht.

Aus der oben gemachten Voraussetzung in Verbindung mit dem Relativitätsprinzip ist klar, daß sich das Elektron in der unmittelbar folgenden Zeit (für kleine Werte von t) vom System k aus betrachtet nach den Gleichungen bewegt:

$$\mu \frac{d^2 \xi}{d \tau^2} = \varepsilon X',$$

$$\mu \frac{d^2 \eta}{d \tau^2} = \varepsilon Y',$$

$$\mu \frac{d^2 \zeta}{d \tau^2} = \varepsilon Z',$$

wobei die Zeichen ξ, η, ζ, τ, X', Y', Z' sich auf das System k beziehen. Setzen wir noch fest, daß für $t = x = y = z = 0$ $\tau = \xi = \eta = \zeta = 0$ sein soll, so gelten die Transformationsgleichungen der §§ 3 und 6, so daß gilt:

$$\tau = \beta \left(t - \frac{v}{V^2} x \right),$$

$$\xi = \beta (x - v t), \qquad X' = X,$$

$$\eta = y, \qquad Y' = \beta \left(Y - \frac{v}{V} N \right),$$

$$\zeta = z, \qquad Z' = \beta \left(Z + \frac{v}{V} M \right).$$

Mit Hilfe dieser Gleichungen transformieren wir die obigen Bewegungsgleichungen vom System k auf das System K und erhalten:

(A)
$$\begin{cases} \dfrac{d^2 x}{d t^2} = \dfrac{\varepsilon}{\mu} \dfrac{1}{\beta^3} X, \\ \dfrac{d^2 y}{d t^2} = \dfrac{\varepsilon}{\mu} \dfrac{1}{\beta} \left(Y - \dfrac{v}{V} N \right), \\ \dfrac{d^2 z}{d t^2} = \dfrac{\varepsilon}{\mu} \dfrac{1}{\beta} \left(Z + \dfrac{v}{V} M \right). \end{cases}$$

Wir fragen nun in Anlehnung an die übliche Betrachtungsweise nach der „longitudinalen" und „transversalen" Masse des bewegten Elektrons. Wir schreiben die Gleichungen (A) in der Form

$$\mu \beta^3 \frac{d^2 x}{dt^2} = \varepsilon X = \varepsilon X',$$

$$\mu \beta^2 \frac{d^2 y}{dt^2} = \varepsilon \beta \left(Y - \frac{v}{V} N \right) = \varepsilon Y',$$

$$\mu \beta^2 \frac{d^2 z}{dt^2} = \varepsilon \beta \left(Z + \frac{v}{V} M \right) = \varepsilon Z'$$

und bemerken zunächst, daß $\varepsilon X'$, $\varepsilon Y'$, $\varepsilon Z'$ die Komponenten der auf das Elektron wirkenden ponderomotorischen Kraft sind, und zwar in einem in diesem Moment mit dem Elektron mit gleicher Geschwindigkeit wie dieses bewegten System betrachtet. (Diese Kraft könnte beispielsweise mit einer im letzten System ruhenden Federwage gemessen werden.) Wenn wir nun diese Kraft schlechtweg „die auf das Elektron wirkende Kraft" nennen und die Gleichung

Massenzahl × Beschleunigungszahl = Kraftzahl

aufrechterhalten, und wenn wir ferner festsetzen, daß die Beschleunigungen im ruhenden System K gemessen werden sollen, so erhalten wir aus obigen Gleichungen:

$$\text{Longitudinale Masse} = \frac{\mu}{\left(\sqrt{1 - \left(\frac{v}{V} \right)^2} \right)^3},$$

$$\text{Transversale Masse} = \frac{\mu}{1 - \left(\frac{v}{V} \right)^2}.$$

Natürlich würde man bei anderer Definition der Kraft und der Beschleunigung andere Zahlen für die Massen erhalten; man ersieht daraus, daß man bei der Vergleichung verschiedener Theorien der Bewegung des Elektrons sehr vorsichtig verfahren muß.

Wir bemerken, daß diese Resultate über die Masse auch für die ponderabeln materiellen Punkte gilt; denn ein ponderabler materieller Punkt kann durch Zufügen einer *beliebig kleinen* elektrischen Ladung zu einem Elektron (in unserem Sinne) gemacht werden.

Wir bestimmen die kinetische Energie des Elektrons. Bewegt sich ein Elektron vom Koordinatenursprung des Systems K aus mit der Anfangsgeschwindigkeit 0 beständig auf der

X-Achse unter der Wirkung einer elektrostatischen Kraft X, so ist klar, daß die dem elektrostatischen Felde entzogene Energie den Wert $\int \varepsilon X dx$ hat. Da das Elektron langsam beschleunigt sein soll und infolgedessen keine Energie in Form von Strahlung abgeben möge, so muß die dem elektrostatischen Felde entzogene Energie gleich der Bewegungsenergie W des Elektrons gesetzt werden. Man erhält daher, indem man beachtet, daß während des ganzen betrachteten Bewegungsvorganges die erste der Gleichungen (A) gilt:

$$W = \int \varepsilon X dx = \int_0^v \beta^3 v\, dv = \mu V^2 \left\{ \frac{1}{\sqrt{1 - \left(\frac{v}{V}\right)^2}} - 1 \right\}.$$

W wird also für $v = V$ unendlich groß. Überlichtgeschwindigkeiten haben — wie bei unseren früheren Resultaten — keine Existenzmöglichkeit.

Auch dieser Ausdruck für die kinetische Energie muß dem oben angeführten Argument zufolge ebenso für ponderable Massen gelten.

Wir wollen nun die aus dem Gleichungssystem (A) resultierenden, dem Experimente zugänglichen Eigenschaften der Bewegung des Elektrons aufzählen.

1. Aus der zweiten Gleichung des Systems (A) folgt, daß eine elektrische Kraft Y und eine magnetische Kraft N dann gleich stark ablenkend wirken auf ein mit der Geschwindigkeit v bewegtes Elektron, wenn $Y = N \cdot v/V$. Man ersieht also, daß die Ermittelung der Geschwindigkeit des Elektrons aus dem Verhältnis der magnetischen Ablenkbarkeit A_m und der elektrischen Ablenkbarkeit A_e nach unserer Theorie für beliebige Geschwindigkeiten möglich ist durch Anwendung des Gesetzes:

$$\frac{A_m}{A_e} = \frac{v}{V}.$$

Diese Beziehung ist der Prüfung durch das Experiment zugänglich, da die Geschwindigkeit des Elektrons auch direkt, z. B. mittels rasch oszillierender elektrischer und magnetischer Felder, gemessen werden kann.

2. Aus der Ableitung für die kinetische Energie des Elektrons folgt, daß zwischen der durchlaufenen Potential-

differenz und der erlangten Geschwindigkeit v des Elektrons die Beziehung gelten muß:

$$P = \int X\,dx = \frac{\mu}{\varepsilon} V^2 \left\{ \frac{1}{\sqrt{1-\left(\frac{v}{V}\right)^2}} - 1 \right\}.$$

3. Wir berechnen den Krümmungsradius R der Bahn, wenn eine senkrecht zur Geschwindigkeit des Elektrons wirkende magnetische Kraft N (als einzige ablenkende Kraft) vorhanden ist. Aus der zweiten der Gleichungen (A) erhalten wir:

$$-\frac{d^2 y}{dt^2} = \frac{v^2}{R} = \frac{\varepsilon}{\mu} \frac{v}{V} N \cdot \sqrt{1-\left(\frac{v}{V}\right)^2}$$

oder

$$R = V^2 \frac{\mu}{\varepsilon} \cdot \frac{\frac{v}{V}}{\sqrt{1-\left(\frac{v}{V}\right)^2}} \cdot \frac{1}{N}.$$

Diese drei Beziehungen sind ein vollständiger Ausdruck für die Gesetze, nach denen sich gemäß vorliegender Theorie das Elektron bewegen muß.

Zum Schlusse bemerke ich, daß mir beim Arbeiten an dem hier behandelten Probleme mein Freund und Kollege M. Besso treu zur Seite stand und daß ich demselben manche wertvolle Anregung verdanke.

Bern, Juni 1905.

(Eingegangen 30. Juni 1905.)

ANNALEN DER PHYSIK.

BEGRÜNDET UND FORTGEFÜHRT DURCH

F. A. C. GREN, L. W. GILBERT, J. C. POGGENDORFF, G. UND E. WIEDEMANN.

VIERTE FOLGE.

BAND 18.

DER GANZEN REIHE 323. BAND.

KURATORIUM:

F. KOHLRAUSCH, M. PLANCK, G. QUINCKE, W. C. RÖNTGEN, E. WARBURG.

UNTER MITWIRKUNG

DER DEUTSCHEN PHYSIKALISCHEN GESELLSCHAFT

UND INSBESONDERE VON

M. PLANCK

HERAUSGEGEBEN VON

PAUL DRUDE.

MIT ACHT FIGURENTAFELN.

LEIPZIG, 1905.

VERLAG VON JOHANN AMBROSIUS BARTH.

13. *Ist die Trägheit eines Körpers von seinem Energieinhalt abhängig?*
von A. Einstein.

Die Resultate einer jüngst in diesen Annalen von mir publizierten elektrodynamischen Untersuchung[1]) führen zu einer sehr interessanten Folgerung, die hier abgeleitet werden soll.

Ich legte dort die Maxwell-Hertzschen Gleichungen für den leeren Raum nebst dem Maxwellschen Ausdruck für die elektromagnetische Energie des Raumes zugrunde und außerdem das Prinzip:

Die Gesetze, nach denen sich die Zustände der physikalischen Systeme ändern, sind unabhängig davon, auf welches von zwei relativ zueinander in gleichförmiger Parallel-Translationsbewegung befindlichen Koordinatensystemen diese Zustandsänderungen bezogen werden (Relativitätsprinzip).

Gestützt auf diese Grundlagen[2]) leitete ich unter anderem das nachfolgende Resultat ab (l. c. § 8):

Ein System von ebenen Lichtwellen besitze, auf das Koordinatensystem (x, y, z) bezogen, die Energie l; die Strahlrichtung (Wellennormale) bilde den Winkel φ mit der x-Achse des Systems. Führt man ein neues, gegen das System (x, y, z) in gleichförmiger Paralleltranslation begriffenes Koordinatensystem (ξ, η, ζ) ein, dessen Ursprung sich mit der Geschwindigkeit v längs der x-Achse bewegt, so besitzt die genannte Lichtmenge — im System (ξ, η, ζ) gemessen — die Energie:

$$l^* = l \frac{1 - \frac{v}{V} \cos \varphi}{\sqrt{1 - \left(\frac{v}{V}\right)^2}},$$

wobei V die Lichtgeschwindigkeit bedeutet. Von diesem Resultat machen wir im folgenden Gebrauch.

---

[1]) A. Einstein, Ann. d. Phys. **17**. p. 891. 1905.
[2]) Das dort benutzte Prinzip der Konstanz der Lichtgeschwindigkeit ist natürlich in den Maxwellschen Gleichungen enthalten.

Es befinde sich nun im System (x, y, z) ein ruhender Körper, dessen Energie — auf das System (x, y, z) bezogen — E_0 sei. Relativ zu dem wie oben mit der Geschwindigkeit v bewegten System (ξ, η, ζ) sei die Energie des Körpers H_0.

Dieser Körper sende in einer mit der x-Achse den Winkel φ bildenden Richtung ebene Lichtwellen von der Energie $L/2$ (relativ zu (x, y, z) gemessen) und gleichzeitig eine gleich große Lichtmenge nach der entgegengesetzten Richtung. Hierbei bleibt der Körper in Ruhe in bezug auf das System (x, y, z). Für diesen Vorgang muß das Energieprinzip gelten und zwar (nach dem Prinzip der Relativität) in bezug auf beide Koordinatensysteme. Nennen wir E_1 bez. H_1 die Energie des Körpers nach der Lichtaussendung relativ zum System (x, y, z) bez. (ξ, η, ζ) gemessen, so erhalten wir mit Benutzung der oben angegebenen Relation:

$$E_0 = E_1 + \left[\frac{L}{2} + \frac{L}{2}\right],$$

$$H_0 = H_1 + \left[\frac{L}{2}\frac{1 - \frac{v}{V}\cos\varphi}{\sqrt{1 - \left(\frac{v}{V}\right)^2}} + \frac{L}{2}\frac{1 + \frac{v}{V}\cos\varphi}{\sqrt{1 - \left(\frac{v}{V}\right)^2}}\right]$$

$$= H_1 + \frac{L}{\sqrt{1 - \left(\frac{v}{V}\right)^2}}.$$

Durch Subtraktion erhält man aus diesen Gleichungen:

$$(H_0 - E_0) - (H_1 - E_1) = L\left\{\frac{1}{\sqrt{1 - \left(\frac{v}{V}\right)^2}} - 1\right\}.$$

Die beiden in diesem Ausdruck auftretenden Differenzen von der Form $H - E$ haben einfache physikalische Bedeutungen. H und E sind Energiewerte desselben Körpers, bezogen auf zwei relativ zueinander bewegte Koordinatensysteme, wobei der Körper in dem einen System (System (x, y, z)) ruht. Es ist also klar, daß die Differenz $H - E$ sich von der kinetischen Energie K des Körpers in bezug auf das andere System (System (ξ, η, ζ)) nur durch eine additive Konstante C unterscheiden kann, welche von der Wahl der willkürlichen addi-

tiven Konstanten der Energien H und E abhängt. Wir können also setzen:
$$H_0 - E_0 = K_0 + C,$$
$$H_1 - E_1 = K_1 + C,$$
da C sich während der Lichtaussendung nicht ändert. Wir erhalten also:
$$K_0 - K_1 = L \left\{ \frac{1}{\sqrt{1-\left(\frac{v}{V}\right)^2}} - 1 \right\}.$$

Die kinetische Energie des Körpers in bezug auf (ξ, η, ζ) nimmt infolge der Lichtaussendung ab, und zwar um einen von den Qualitäten des Körpers unabhängigen Betrag. Die Differenz $K_0 - K_1$ hängt ferner von der Geschwindigkeit ebenso ab wie die kinetische Energie des Elektrons (l. c. § 10).

Unter Vernachlässigung von Größen vierter und höherer Ordnung können wir setzen:
$$K_0 - K_1 = \frac{L}{V^2} \frac{v^2}{2}.$$

Aus dieser Gleichung folgt unmittelbar:

Gibt ein Körper die Energie L in Form von Strahlung ab, so verkleinert sich seine Masse um L/V^2. Hierbei ist es offenbar unwesentlich, daß die dem Körper entzogene Energie gerade in Energie der Strahlung übergeht, so daß wir zu der allgemeineren Folgerung geführt werden:

Die Masse eines Körpers ist ein Maß für dessen Energieinhalt; ändert sich die Energie um L, so ändert sich die Masse in demselben Sinne um $L/9 \cdot 10^{20}$, wenn die Energie in Erg und die Masse in Grammen gemessen wird.

Es ist nicht ausgeschlossen, daß bei Körpern, deren Energieinhalt in hohem Maße veränderlich ist (z. B. bei den Radiumsalzen), eine Prüfung der Theorie gelingen wird.

Wenn die Theorie den Tatsachen entspricht, so überträgt die Strahlung Trägheit zwischen den emittierenden und absorbierenden Körpern.

Bern, September 1905.

(Eingegangen 27. September 1905.)

ANNALEN
DER
PHYSIK.

BEGRÜNDET UND FORTGEFÜHRT DURCH

F. A. C. GREN, L. W. GILBERT, J. C. POGGENDORFF, G. UND E. WIEDEMANN.

VIERTE FOLGE.

BAND 19.

DER GANZEN REIHE 324. BAND.

KURATORIUM:

F. KOHLRAUSCH, M. PLANCK, G. QUINCKE,
W. C. RÖNTGEN, E. WARBURG.

UNTER MITWIRKUNG

DER DEUTSCHEN PHYSIKALISCHEN GESELLSCHAFT

UND INSBESONDERE VON

M. PLANCK

HERAUSGEGEBEN VON

PAUL DRUDE.

MIT FÜNF FIGURENTAFELN.

LEIPZIG, 1906.

VERLAG VON JOHANN AMBROSIUS BARTH.

3. *Eine neue Bestimmung der Moleküldimensionen; von A. Einstein.*

Die ältesten Bestimmungen der wahren Größe der Moleküle hat die kinetische Theorie der Gase ermöglicht, während die an Flüssigkeiten beobachteten physikalischen Phänomene bis jetzt zur Bestimmung der Molekülgrößen nicht gedient haben. Es liegt dies ohne Zweifel an den bisher unüberwindlichen Schwierigkeiten, welche der Entwickelung einer ins einzelne gehenden molekularkinetischen Theorie der Flüssigkeiten entgegenstehen. In dieser Arbeit soll nun gezeigt werden, daß man die Größe der Moleküle des gelösten Stoffs in einer nicht dissoziierten verdünnten Lösung aus der inneren Reibung der Lösung und des reinen Lösungsmittels und aus der Diffusion des gelösten Stoffes im Lösungsmittel ermitteln kann, wenn das Volumen eines Moleküls des gelösten Stoffs groß ist gegen das Volumen eines Moleküls des Lösungsmittels. Ein derartiges gelöstes Molekül wird sich nämlich bezüglich seiner Beweglichkeit im Lösungsmittel und bezüglich seiner Beeinflussung der inneren Reibung des letzteren annähernd wie ein im Lösungsmittel suspendierter fester Körper verhalten, und es wird erlaubt sein, auf die Bewegung des Lösungsmittels in unmittelbarer Nähe eines Moleküls die hydrodynamischen Gleichungen anzuwenden, in welchen die Flüssigkeit als homogen betrachtet, eine molekulare Struktur derselben also nicht berücksichtigt wird. Als Form der festen Körper, welche die gelösten Moleküle darstellen sollen, wählen wir die Kugelform.

§ 1. Über die Beeinflussung der Bewegung einer Flüssigkeit durch eine sehr kleine in derselben suspendierte Kugel.

Es liege eine inkompressible homogene Flüssigkeit mit dem Reibungskoeffizienten k der Betrachtung zugrunde, deren Geschwindigkeitskomponenten u, v, w als Funktionen der Koordinaten x, y, z und der Zeit gegeben seien. Von einem beliebigen Punkt x_0, y_0, z_0 aus denken wir uns die Funktionen u, v, w als Funktionen von $x - x_0, y - y_0, z - z_0$ nach dem Taylorschen Satze entwickelt und um diesen Punkt ein so kleines Gebiet G abgegrenzt, daß innerhalb desselben nur die linearen Glieder dieser Entwickelung berücksichtigt werden müssen. Die Bewegung der in G enthaltenen Flüssigkeit kann dann bekanntlich als die Superposition dreier Bewegungen aufgefaßt werden, nämlich

1. einer Parallelverschiebung aller Flüssigkeitsteilchen ohne Änderung von deren relativer Lage,

2. einer Drehung der Flüssigkeit ohne Änderung der relativen Lage der Flüssigkeitsteilchen,

3. einer Dilatationsbewegung in drei aufeinander senkrechten Richtungen (den Hauptdilatationsrichtungen).

Wir denken uns nun im Gebiete G einen kugelförmigen starren Körper, dessen Mittelpunkt im Punkte x_0, y_0, z_0 liege und dessen Dimensionen gegen diejenigen des Gebietes G sehr klein seien. Wir nehmen ferner an, daß die betrachtete Bewegung eine so langsame sei, daß die kinetische Energie der Kugel sowie diejenige der Flüssigkeit vernachlässigt werden können. Es werde ferner angenommen, daß die Geschwindigkeitskomponenten eines Oberflächenelementes der Kugel mit den entsprechenden Geschwindigkeitskomponenten der unmittelbar benachbarten Flüssigkeitsteilchen übereinstimme, d. h., daß auch die (kontinuierlich gedachte) Trennungsschicht überall einen nicht unendlich kleinen Koeffizienten der inneren Reibung aufweise.

Es ist ohne weiteres klar, daß die Kugel die Teilbewegungen 1. und 2. einfach mitmacht, ohne die Bewegung der benachbarten Flüssigkeit zu modifizieren, da sich bei diesen Teilbewegungen die Flüssigkeit wie ein starrer Körper bewegt, und da wir die Wirkungen der Trägheit vernachlässigt haben.

Die Bewegung 3. aber wird durch das Vorhandensein der Kugel modifiziert, und es wird unsere nächste Aufgabe sein, den Einfluß der Kugel auf diese Flüssigkeitsbewegung zu untersuchen. Beziehen wir die Bewegung 3. auf ein Koordinatensystem, dessen Achsen den Hauptdilatationsrichtungen parallel sind, und setzen wir

$$x - x_0 = \xi,$$
$$y - y_0 = \eta,$$
$$z - z_0 = \zeta,$$

so läßt sich jene Bewegung, falls die Kugel nicht vorhanden ist, durch die Gleichungen darstellen:

(1) $$\begin{cases} u_0 = A\xi, \\ v_0 = B\eta, \\ w_0 = C\zeta; \end{cases}$$

A, B, C sind Konstanten, welche wegen der Inkompressibilität der Flüssigkeit die Bedingung erfüllen:

(2) $$A + B + C = 0.$$

Befindet sich nun im Punkte x_0, y_0, z_0 die starre Kugel mit dem Radius P, so ändert sich in der Umgebung derselben die Flüssigkeitsbewegung. Im folgenden wollen wir der Bequemlichkeit wegen P als „endlich" bezeichnen, dagegen die Werte von ξ, η, ζ, für welche die Flüssigkeitsbewegung durch die Kugel nicht mehr merklich modifiziert wird, als „unendlich groß".

Zunächst ist wegen der Symmetrie der betrachteten Flüssigkeitsbewegung klar, daß die Kugel bei der betrachteten Bewegung weder eine Translation noch eine Drehung ausführen kann, und wir erhalten die Grenzbedingungen:

$$u = v = w = 0 \quad \text{für} \quad \varrho = P,$$

wobei

$$\varrho = \sqrt{\xi^2 + \eta^2 + \zeta^2} > 0$$

gesetzt ist. Hierbei bedeuten u, v, w die Geschwindigkeitskomponenten der nun betrachteten (durch die Kugel modifizierten) Bewegung. Setzt man

(3) $$\begin{cases} u = A\xi + u_1, \\ v = B\eta + v_1, \\ w = C\zeta + w_1, \end{cases}$$

so müßte, da die in Gleichungen (3) dargestellte Bewegung im Unendlichen in die in Gleichungen (1) dargestellte übergehen soll, die Geschwindigkeiten u_1, v_1, w_1 im Unendlichen verschwinden.

Die Funktionen u, v, w haben den Gleichungen der Hydrodynamik zu genügen unter Berücksichtigung der inneren Reibung und unter Vernachlässigung der Trägheit. Es gelten also die Gleichungen[1])

$$\begin{cases} \dfrac{\partial p}{\partial \xi} = k\,\Delta u \quad \dfrac{\partial p}{\partial \eta} = k\,\Delta v \quad \dfrac{\partial p}{\partial \zeta} = k\,\Delta w, \\ \dfrac{\partial u}{\partial \xi} + \dfrac{\partial v}{\partial \eta} + \dfrac{\partial w}{\partial \zeta} = 0, \end{cases} \quad (4)$$

wobei Δ den Operator

$$\frac{\partial^2}{\partial \xi^2} + \frac{\partial^2}{\partial \eta^2} + \frac{\partial^2}{\partial \zeta^2}$$

und p den hydrostatischen Druck bedeutet.

Da die Gleichungen (1) Lösungen der Gleichungen (4) und letztere linear sind, müssen nach (3) auch die Größen u_1, v_1, w_1 den Gleichungen (4) genügen. Ich bestimmte u_1, v_1, w_1 und p nach einer im § 4 der erwähnten Kirchhoffschen Vorlesung angegebenen Methode[2]) und fand:

---

[1]) G. Kirchhoff, Vorlesungen über Mechanik. 26. Vorl.

[2]) „Aus den Gleichungen (4) folgt $\Delta p = 0$. Ist p dieser Bedingung gemäß angenommen und eine Funktion V bestimmt, die der Gleichung

$$\Delta V = \frac{1}{k} p$$

genügt, so erfüllt man die Gleichungen (4), wenn man

$$u = \frac{\partial V}{\partial \xi} + u', \quad v = \frac{\partial V}{\partial \eta} + v', \quad w = \frac{\partial V}{\partial \zeta} + w'$$

setzt und u', v', w' so wählt, daß $\Delta u' = 0$, $\Delta v' = 0$ und $\Delta w' = 0$ und

$$\frac{\partial u'}{\partial \xi} + \frac{\partial v'}{\partial \eta} + \frac{\partial w'}{\partial \zeta} = -\frac{1}{k} p$$

ist."

Setzt man nun

$$\frac{p}{k} = 2c\,\frac{\partial^2 \tfrac{1}{\varrho}}{\partial \xi^2}$$

und im Einklang hiermit

$$V = c\,\frac{\partial^2 \varrho}{\partial \xi^2} + b\,\frac{\partial^2 \tfrac{1}{\varrho}}{\partial \xi^2} + \frac{a}{2}\left(\xi^2 - \frac{\eta^2}{2} - \frac{\zeta^2}{2}\right)$$

und

$$u' = -2c\,\frac{\partial \tfrac{1}{\varrho}}{\partial \xi}, \quad v' = 0, \quad w' = 0,$$

so lassen sich die Konstanten a, b, c so bestimmen, daß für $\varrho = P$ $u = v = w = 0$ ist. **Durch Superposition dreier derartiger Lösungen erhält man die in den Gleichungen (5) und (5a) angegebene Lösung.**

$$(5)\begin{cases} p = -\tfrac{5}{3} k P^3 \left\{ A \dfrac{\partial^2 \left(\tfrac{1}{\varrho}\right)}{\partial \xi^2} + B \dfrac{\partial^2 \left(\tfrac{1}{\varrho}\right)}{\partial \eta^2} + C \dfrac{\partial^2 \left(\tfrac{1}{\varrho}\right)}{\partial \zeta^2} \right\} + \text{konst.}, \\ u = A \xi - \tfrac{5}{3} P^3 A \dfrac{\xi}{\varrho^3} - \dfrac{\partial D}{\partial \xi}, \\ v = B \eta - \tfrac{5}{3} P^3 B \dfrac{\eta}{\varrho^3} - \dfrac{\partial D}{\partial \eta}, \\ w = C \zeta - \tfrac{5}{3} P^3 C \dfrac{\zeta}{\varrho^3} - \dfrac{\partial D}{\partial \zeta}, \end{cases}$$

wobei

$$(5\,\text{a})\begin{cases} D = A \left\{ \tfrac{5}{6} P^3 \dfrac{\partial^2 \varrho}{\partial \xi^2} + \tfrac{1}{6} P^5 \dfrac{\partial^2 \left(\tfrac{1}{\varrho}\right)}{\partial \xi^2} \right\} \\ \quad + B \left\{ \tfrac{5}{6} P^3 \dfrac{\partial^2 \varrho}{\partial \eta^2} + \tfrac{1}{6} P^5 \dfrac{\partial^2 \left(\tfrac{1}{\varrho}\right)}{\partial \eta^2} \right\} \\ \quad + C \left\{ \tfrac{5}{6} P^3 \dfrac{\partial^2 \varrho}{\partial \zeta^2} + \tfrac{1}{6} P^5 \dfrac{\partial^2 \left(\tfrac{1}{\varrho}\right)}{\partial \zeta^2} \right\}. \end{cases}$$

Es ist leicht zu beweisen, daß die Gleichungen (5) Lösungen der Gleichungen (4) sind. Denn da

$$\varDelta \xi = 0, \quad \varDelta \tfrac{1}{\varrho} = 0, \quad \varDelta \varrho = \tfrac{2}{\varrho}$$

und

$$\varDelta \left(\tfrac{\xi}{\varrho^3} \right) = - \tfrac{\partial}{\partial \xi} \left\{ \varDelta \left(\tfrac{1}{\varrho} \right) \right\} = 0,$$

erhält man

$$k \varDelta u = - k \tfrac{\partial}{\partial \xi} \{ \varDelta D \} = - k \tfrac{\partial}{\partial \xi} \left\{ \tfrac{5}{3} P^3 A \dfrac{\partial^2 \tfrac{1}{\varrho}}{\partial \xi^2} + \tfrac{5}{3} P^3 B \dfrac{\partial^2 \tfrac{1}{\varrho}}{\partial \eta^2} + \cdots \right\}.$$

Der zuletzt erhaltene Ausdruck ist aber nach der ersten der Gleichungen (5) mit $\partial n / \partial \xi$ identisch. Auf gleiche Weise zeigt man, daß die zweite und dritte der Gleichungen (4) erfüllt ist. Ferner erhält man

$$\dfrac{\partial u}{\partial \xi} + \dfrac{\partial v}{\partial \eta} + \dfrac{\partial w}{\partial \xi} = (A + B + C)$$
$$+ \tfrac{5}{3} P^3 \left\{ A \dfrac{\partial^2 \left(\tfrac{1}{\varrho}\right)}{\partial \xi^2} + B \dfrac{\partial^2 \left(\tfrac{1}{\varrho}\right)}{\partial \eta^2} + C \dfrac{\partial^2 \left(\tfrac{1}{\varrho}\right)}{\partial \zeta^2} \right\} - \varDelta D.$$

Da aber nach Gleichung (5a)

$$\Delta D = \tfrac{5}{3} A P^3 \left\{ A \frac{\partial^2 \left(\frac{1}{\varrho}\right)}{\partial \xi^2} + B \frac{\partial^2 \left(\frac{1}{\varrho}\right)}{\partial \eta^2} + C \frac{\partial^2 \left(\frac{1}{\varrho}\right)}{\partial \zeta^2} \right\},$$

so folgt, daß auch die letzte der Gleichungen (4) erfüllt ist. Was die Grenzbedingungen betrifft, so gehen zunächst für unendlich große ϱ unsere Gleichungen für u, v, w in die Gleichungen (1) über. Durch Einsetzen des Wertes von D aus Gleichung (5a) in die zweite der Gleichungen (5) erhält man:

$$(6) \quad \begin{cases} u = A\,\xi - \tfrac{5}{2}\dfrac{P^3}{\varrho^5}\xi\,(A\,\xi^2 + B\,\eta^2 + C\,\zeta^2) \\ \qquad + \tfrac{5}{2}\dfrac{P^5}{\varrho^7}\xi\,(A\,\xi^2 + B\,\eta^2 + C)\,\zeta^2 - \dfrac{P^5}{\varrho^5}A\,\xi. \end{cases}$$

Man erkennt, daß u für $\varrho = P$ verschwindet. Gleiches gilt aus Symmetriegründen für v und w. Es ist nun bewiesen, daß durch die Gleichungen (5) sowohl den Gleichungen (4) als auch den Grenzbedingungen der Aufgabe Genüge geleistet ist.

Es läßt sich auch beweisen, daß die Gleichungen (5) die einzige mit den Grenzbedingungen der Aufgabe verträgliche Lösung der Gleichungen (4) sind. Der Beweis soll hier nur angedeutet werden. Es mögen in einem endlichen Raume die Geschwindigkeitskomponenten u, v, w einer Flüssigkeit den Gleichungen (4) genügen. Existierte noch eine andere Lösung U, V, W der Gleichungen (4), bei welcher an den Grenzen des betrachteten Raumes $U = u$, $V = v$, $W = w$ ist, so ist ($U - u$, $V - v$, $W - w$) eine Lösung der Gleichungen (4), bei welcher die Geschwindigkeitskomponenten an der Grenze des Raumes verschwinden. Der in dem betrachteten Raume befindlichen Flüssigkeit wird also keine mechanische Arbeit zugeführt. Da wir die lebendige Kraft der Flüssigkeit vernachlässigt haben, so folgt daraus, daß auch die im betrachteten Raume in Wärme verwandelte Arbeit gleich Null ist. Hieraus folgert man, daß im ganzen Raume $u = u_1$, $v = v_1$, $w = w_1$ sein muß, falls der Raum wenigstens zum Teil durch ruhende Wände begrenzt ist. Durch Grenzübergang kann dies Resultat auch auf den Fall ausgedehnt werden, daß, wie in dem oben betrachteten Falle, der betrachtete Raum unendlich ist. Man kann so dartun, daß die oben gefundene Lösung die einzige Lösung der Aufgabe ist.

Wir legen nun um den Punkt x_0, y_0, z_0 eine Kugel vom Radius R, wobei R gegen P unendlich groß sei, und berechnen die Energie, welche in der innerhalb der Kugel befindlichen Flüssigkeit (in der Zeiteinheit) in Wärme verwandelt wird. Diese Energie W ist gleich der der Flüssigkeit mechanisch zugeführten Arbeit. Bezeichnet man die Komponenten des auf die Oberfläche der Kugel vom Radius R ausgeübten Druckes mit X_n, Y_n, Z_n, so ist:

$$W = \int (X_n u + Y_n v + Z_n w)\, ds,$$

wobei das Integral über die Oberfläche der Kugel vom Radius R zu erstrecken ist. Hierbei ist:

$$X_n = -\left(X_\xi \frac{\xi}{\varrho} + X_\eta \frac{\eta}{\varrho} + X_\zeta \frac{\zeta}{\varrho}\right),$$

$$Y_n = -\left(Y_\xi \frac{\xi}{\varrho} + Y_\eta \frac{\eta}{\varrho} + Y_\zeta \frac{\zeta}{\varrho}\right),$$

$$Z_n = -\left(Z_\xi \frac{\xi}{\varrho} + Z_\eta \frac{\eta}{\varrho} + Z_\zeta \frac{\zeta}{\varrho}\right),$$

wobei

$$X_\xi = p - 2k\frac{\partial u}{\partial \xi}, \qquad Y_\zeta = Z_\eta = -k\left(\frac{\partial v}{\partial \zeta} + \frac{\partial w}{\partial \eta}\right),$$

$$Y_\eta = p - 2k\frac{\partial v}{\partial \eta}, \qquad Z_\xi = X_\zeta = -k\left(\frac{\partial w}{\partial \xi} + \frac{\partial u}{\partial \zeta}\right),$$

$$Z_\zeta = p - 2k\frac{\partial w}{\partial \zeta}, \qquad X_\eta = Y_\xi = -k\left(\frac{\partial u}{\partial \eta} + \frac{\partial v}{\partial \xi}\right).$$

Die Ausdrücke für u, v, w vereinfachen sich, wenn wir beachten, daß für $\varrho = R$ die Glieder mit dem Faktor P^5/ϱ^5 gegenüber denen mit dem Faktor P^3/ϱ^3 verschwinden. Wir haben zu setzen:

(6a)
$$\begin{cases} u = A\xi - \frac{5}{2}P^3 \dfrac{\xi(A\xi^2 + B\eta^2 + C\zeta^2)}{\varrho^5}, \\ v = B\eta - \frac{5}{2}P^3 \dfrac{\eta(A\xi^2 + B\eta^2 + C\zeta^2)}{\varrho^5}, \\ w = C\zeta - \frac{5}{2}P^3 \dfrac{\zeta(A\xi^2 + B\eta^2 + C\zeta^2)}{\varrho^5}. \end{cases}$$

Für p erhalten wir aus der ersten der Gleichungen (5) durch die entsprechenden Vernachlässigungen

$$p = -5kP^3 \frac{A\xi^2 + B\eta^2 + C\zeta^2}{\varrho^5} + \text{konst.}$$

Wir erhalten zunächst:

$$X_\xi = -2kA + 10kP^3\frac{A\xi^2}{\varrho^5} - 25kP^3\frac{\xi^2(A\xi^2 + B\eta^2 + C\zeta^2)}{\varrho^7},$$

$$X_\eta = \qquad\quad + 10kP^3\frac{A\xi\eta}{\varrho^5} - 25kP^3\frac{\eta^2(A\xi^2 + B\eta^2 + C\zeta^2)}{\varrho^7},$$

$$X_\zeta = \qquad\quad + 10kP^3\frac{A\xi\zeta}{\varrho^5} - 25kP^3\frac{\zeta^2(A\xi^2 + B\eta^2 + C\zeta^2)}{\varrho^7},$$

und hieraus

$$X_n = 2Ak\frac{\xi}{\varrho} - 10AkP^3\frac{\xi}{\varrho^4} + 25kP^3\frac{\xi(A\xi^2 + B\eta^2 + C\zeta^2)}{\varrho^6}.$$

Mit Hilfe der durch zyklische Vertauschung abzuleitenden Ausdrücke für Y_n und Z_n erhält man unter Vernachlässigung aller Glieder, die das Verhältnis P/ϱ in einer höheren als der dritten Potenz enthalten:

$$X_n u + Y_n v + Z_n w + \frac{2k}{\varrho}(A^2\xi^2 + B^2\eta^2 + C^2\zeta^2)$$
$$- 10k\frac{P^3}{\varrho^4}(A^2\xi^2 + \cdot + \cdot) + 20k\frac{P^3}{\varrho^6}(A\xi^2 + \cdot + \cdot)^2.$$

Integriert man über die Kugel und berücksichtigt, daß

$$\int ds = 4R^2\pi,$$

$$\int \xi^2 ds = \int \eta^2 ds = \int \zeta^2 ds = \tfrac{4}{3}\pi R^4,$$

$$\int \xi^4 ds = \int \eta^4 ds = \int \zeta^4 ds = \tfrac{4}{5}\pi R^6,$$

$$\int \eta^2\zeta^2 ds = \int \zeta^2\xi^2 ds = \int \xi^2\eta^2 ds = \tfrac{4}{15}\pi R^6,$$

$$\int (A\xi^2 + B\eta^2 + C\zeta^2)^2 ds = \tfrac{4}{15}\pi R^6(A^2 + B^2 + C^2),$$

so erhält man:

(7) $\qquad W = \tfrac{8}{3}\pi R^3 k\delta^2 - \tfrac{8}{3}\pi P^3 k\delta^2 = 2\delta^2 k(V - \Phi),$

wobei

$$\delta = A^2 + B^2 + C^2,$$
$$\tfrac{4}{3}\pi R^3 = V$$

und
$$\tfrac{4}{3}\pi P^3 = \Phi$$

gesetzt ist. Wäre die suspendierte Kugel nicht vorhanden ($\Phi = 0$), so erhielte man für die im Volumen V verzehrte Energie

(7 a) $\qquad\qquad W_0 = 2\delta^2 kV.$

Durch das Vorhandensein der Kugel wird also die verzehrte Energie um $2\,\delta^2 k\,\Phi$ verkleinert. Es ist bemerkenswert, daß der Einfluß der suspendierten Kugel auf die Größe der verzehrten Energie gerade so groß ist, wie er wäre, wenn durch die Anwesenheit der Kugel die Bewegung der sie umgebenden Flüssigkeit gar nicht modifiziert würde.

§ 2. Berechnung des Reibungskoeffizienten einer Flüssigkeit, in welcher sehr viele kleine Kugeln in regelloser Verteilung suspendiert sind.

Wir haben im vorstehenden den Fall betrachtet, daß in einem Gebiete G von der oben definierten Größenordnung eine relativ zu diesem Gebiete sehr kleine Kugel suspendiert ist und untersucht, wie dieselbe die Flüssigkeitsbewegung beeinflußt. Wir wollen nun annehmen, daß in dem Gebiete G unendlich viele Kugeln von gleichem, und zwar so kleinem Radius regellos verteilt sind, daß das Volumen aller Kugeln zusammen sehr klein sei gegen das Gebiet G. Die Zahl der auf die Volumeneinheit entfallenden Kugeln sei n, wobei n allenthalben in der Flüssigkeit bis auf Vernachlässigbares konstant sei.

Wir gehen nun wieder aus von einer Bewegung einer homogenen Flüssigkeit ohne suspendierte Kugeln und betrachten wieder die allgemeinste Dilatationsbewegung. Sind keine Kugeln vorhanden, so können wir bei passender Wahl des Koordinatensystems die Geschwindigkeitskomponenten u_0, v_0, w_0 in dem beliebigen Punkte x, y, z des Gebietes G darstellen durch die Gleichungen:

$$u_0 = A\,x,$$
$$v_0 = B\,y,$$
$$w_0 = C\,z,$$

wobei
$$A + B + C = 0.$$

Eine im Punkte x_ν, y_ν, z_ν suspendierte Kugel beeinflußt nun diese Bewegung in der aus Gleichung (6) ersichtlichen Weise. Da wir den mittleren Abstand benachbarter Kugeln als sehr groß gegen deren Radius wählen, und folglich die von allen suspendierten Kugeln zusammen herrührenden zusätzlichen Geschwindigkeitskomponenten gegen u_0, v_0, w_0 sehr klein sind, so erhalten wir für die Geschwindigkeitskomponenten u, v, w in der Flüssigkeit unter Berücksichtigung der suspendierten Kugeln und unter Vernachlässigung von Gliedern höherer Ordnungen:

$$(8) \begin{cases} u = Ax - \sum \left\{ \tfrac{5}{2} \dfrac{P^3}{\varrho_\nu^2} \dfrac{\xi_\nu(A\xi_\nu^2 + B\eta_\nu^2 + C\zeta_\nu^2)}{\varrho_\nu^3} \right. \\ \qquad\qquad\qquad \left. - \tfrac{5}{2} \dfrac{P^5}{\varrho_\nu^4} \dfrac{\xi_\nu(A\xi_\nu^2 + B\eta_\nu^2 + C\zeta_\nu^2)}{\varrho_\nu^3} + \dfrac{P^5}{\varrho_\nu^4} \dfrac{A\xi_\nu}{\varrho_\nu} \right\}, \\ v = By - \sum \left\{ \tfrac{5}{2} \dfrac{P^3}{\varrho_\nu^2} \dfrac{\eta_\nu(A\xi_\nu^2 + B\eta_\nu^2 + C\zeta_\nu^2)}{\varrho_\nu^3} \right. \\ \qquad\qquad\qquad \left. - \tfrac{5}{2} \dfrac{P^5}{\varrho_\nu^4} \dfrac{\eta_\nu(A\xi_\nu^2 + B\eta_\nu^2 + C\zeta_\nu^2)}{\varrho_\nu^3} + \dfrac{P^5}{\varrho_\nu^4} \dfrac{B\eta_\nu}{\varrho_\nu} \right\}, \\ w = Cz - \sum \left\{ \tfrac{5}{2} \dfrac{P^3}{\varrho_\nu^2} \dfrac{\zeta_\nu(A\xi_\nu^2 + B\eta_\nu^2 + C\zeta_\nu^2)}{\varrho_\nu^3} \right. \\ \qquad\qquad\qquad \left. - \tfrac{5}{2} \dfrac{P^5}{\varrho_\nu^4} \dfrac{\zeta_\nu(A\xi_\nu^2 + B\eta_\nu^2 + C\zeta_\nu^2)}{\varrho_\nu^3} + \dfrac{P^5}{\varrho_\nu^4} \dfrac{C\zeta_\nu}{\varrho_\nu} \right\}, \end{cases}$$

wobei die Summation über alle Kugeln des Gebietes G zu erstrecken ist und

$$\xi_\nu = x - x_\nu,$$
$$\eta_\nu = y - y_\nu, \quad \varrho_\nu = \sqrt{\xi_\nu^2 + \eta_\nu^2 + \zeta_\nu^2},$$
$$\zeta_\nu = z - z_\nu$$

gesetzt ist. x_ν, y_ν, z_ν sind die Koordinaten der Kugelmittelpunkte. Aus den Gleichungen (7) und (7a) schließen wir ferner, daß die Anwesenheit jeder der Kugeln bis auf unendlich Kleines höherer Ordnung eine Verringerung der Wärmeproduktion pro Zeiteinheit um $2\delta^2 k \Phi$ zum Gefolge hat und daß im Gebiete G die pro Volumeneinheit in Wärme verwandelte Energie den Wert hat:

$$W = 2\delta^2 k - 2n\delta^2 k \Phi,$$

oder

(7b) $$W = 2\delta^2 k(1 - \varphi),$$

wobei φ den von den Kugeln eingenommenen Bruchteil des Volumens bedeutet.

Gleichung (7b) erweckt den Anschein, als ob der Reibungskoeffizient der von uns betrachteten inhomogenen Mischung von Flüssigkeit und suspendierten Kugeln (im folgenden kurz „Mischung" genannt) kleiner sei als der Reibungskoeffizient k der Flüssigkeit. Dies ist jedoch nicht der Fall, da A, B, C nicht die Werte der Hauptdilatationen der in Gleichungen (8) dargestellten Flüssigkeitsbewegung sind; wir wollen die Hauptdilatationen der Mischung A^x, B^x, C^x nennen. Aus Symmetriegründen folgt, daß die Hauptdilatationsrichtungen der Mischung den Richtungen der Hauptdilatationen A, B, C, also den Koordinatenrichtungen parallel sind. Schreiben wir die Gleichungen (8) in der Form:

$$u = A x + \sum u_\nu,$$
$$v = B y + \sum v_\nu,$$
$$w = C z + \sum w_\nu,$$

so erhalten wir:

$$A^x = \left(\frac{\partial u}{\partial x}\right)_{x=0} = A + \sum \left(\frac{\partial u_\nu}{\partial x}\right)_{x=0} = A - \sum \left(\frac{\partial u_\nu}{\partial x_0}\right)_{x=0}$$

Schließen wir die unmittelbaren Umgebungen der einzelnen Kugeln von der Betrachtung aus, so können wir die zweiten und dritten Glieder der Ausdrücke von u, v, w weglassen und erhalten für $x = y = z = 0$:

$$(9) \quad \begin{cases} u_\nu = -\frac{5}{2} \frac{P^3}{r_\nu^2} \frac{x_\nu (A x_\nu^2 + B y_\nu^2 + C z_\nu^2)}{r_\nu^3}, \\ v_\nu = -\frac{5}{2} \frac{P^3}{r_\nu^2} \frac{y_\nu (A x_\nu^2 + B y_\nu^2 + C z_\nu^2)}{r_\nu^3}, \\ w_\nu = -\frac{5}{2} \frac{P^3}{r_\nu^2} \frac{z_\nu (A x_\nu^2 + B y_\nu^2 + C z_\nu^2)}{r_\nu^3}, \end{cases}$$

wobei

$$r_\nu = \sqrt{x_\nu^2 + y_\nu^2 + z_\nu^2} > 0$$

gesetzt ist. Die Summierung erstrecken wir über das Volumen einer Kugel K von sehr großem Radius R, deren Mittelpunkt im Koordinatenursprung liegt. Betrachten wir ferner die *regellos* verteilten Kugeln als *gleichmäßig* verteilt und setzen an Stelle der Summe ein Integral, so erhalten wir:

$$A^* = A - n \int_K \frac{\partial u_\nu}{\partial x_\nu} dx_\nu\, dy_\nu\, dz_\nu,$$

$$= A - n \int \frac{u_\nu x_\nu}{r_\nu} ds,$$

wobei das letzte Integral über die Oberfläche der Kugel K zu erstrecken ist. Wir finden unter Berücksichtigung von (9):

$$A^* = A - \tfrac{5}{2} \frac{P^3}{R^5} n \int x_0^2 (A x_0^2 + B y_0^2 + C z_0^2) ds,$$

$$= A - n (\tfrac{4}{3} P^3 \pi) A = A(1 - \varphi).$$

Analog ist

$$B^* = B(1 - \varphi),$$
$$C^* = C(1 - \varphi).$$

Setzen wir

$$\delta^* = A^{*2} + B^{*2} + C^{*2},$$

so ist bis auf unendlich Kleines höherer Ordnung:

$$\delta^{*2} = \delta^2 (1 - 2\varphi).$$

Wir haben für die Wärmeentwickelung pro Zeit- und Volumeneinheit gefunden:

$$W^* = 2\,\delta^2 k\,(1 - \varphi).$$

Bezeichnen wir mit k^* den Reibungskoeffizienten des Gemisches, so ist:

$$W^* = 2\,\delta^{*2} k^*.$$

Aus den drei letzten Gleichungen erhält man unter Vernachlässigung von unendlich Kleinem höherer Ordnung:

$$k^* = k(1 + \varphi).$$

Wir erhalten also das Resultat:

Werden in einer Flüssigkeit sehr kleine starre Kugeln suspendiert, so wächst dadurch der Koeffizient der inneren Reibung um einen Bruchteil, der gleich ist dem Gesamtvolumen der in der Volumeneinheit suspendierten Kugeln, vorausgesetzt, daß dieses Gesamtvolumen sehr klein ist.

§ 3. Über das Volumen einer gelösten Substanz von im Vergleich zum Lösungsmittel großem Molekularvolumen.

Es liege eine verdünnte Lösung vor eines Stoffes, welcher in der Lösung nicht dissoziiert. Ein Molekül des gelösten Stoffes sei groß gegenüber einem Molekül des Lösungsmittels und werde als starre Kugel vom Radius P aufgefaßt. Wir können dann das in § 2 gewonnene Resultat anwenden. Bedeutet k^* den Reibungskoeffizienten der Lösung, k denjenigen des reinen Lösungsmittels, so ist:

$$\frac{k^*}{k} = 1 + \varphi,$$

wobei φ das Gesamtvolumen der in Lösung befindlichen Moleküle pro Volumeinheit ist.

Wir wollen φ für eine 1 proz. wässerige Zuckerlösung berechnen. Nach Beobachtungen von Burkhard (Tabellen von Landolt und Börnstein) ist bei einer 1proz. wässerigen Zuckerlösung $k^*/k = 1{,}0245$ (bei 20^0 C.), also $\varphi = 0{,}0245$ für (beinahe genau) 0,01 g Zucker. Ein Gramm in Wasser gelöster Zucker hat also auf den Reibungskoeffizienten denselben Einfluß wie kleine suspendierte starre Kugeln vom Gesamtvolumen 2,45 cm³.

Es ist nun daran zu erinnern, daß 1 g festen Zuckers das Volumen 0,61 cm³ besitzt. Dasselbe Volumen findet man auch für das spezifische Volumen s des in Lösung befindlichen Zuckers, wenn man die Zuckerlösung als eine *Mischung* von Wasser und Zucker in gelöster Form auffaßt. Die Dichte einer 1 proz. wässerigen Zuckerlösung (bezogen auf Wasser von derselben Temperatur) bei $17{,}5^0$ ist nämlich 1,00388. Man hat also (unter Vernachlässigung des Dichteunterschiedes von Wasser von 4^0 und Wasser von $17{,}5^0$):

$$\frac{1}{1{,}00388} = 0{,}99 + 0{,}01\, s;$$

also $\qquad s = 0{,}61$.

Während also die Zuckerlösung, was ihre Dichte anbelangt, sich wie eine Mischung von Wasser und festem Zucker verhält, ist der Einfluß auf die innere Reibung viermal größer, als er aus der Suspendierung der gleichen Zuckermenge resultieren würde. Es scheint mir dies Resultat im Sinne der Molekulartheorie kaum anders gedeutet werden zu können, als indem man annimmt, daß das in Lösung befindliche Zuckermolekül die Beweglichkeit des unmittelbar angrenzenden Wassers hemme, so daß ein Quantum Wasser, dessen Volumen ungefähr das Dreifache des Volums des Zuckermoleküls ist, an das Zuckermolekül gekettet ist.

Wir können also sagen, daß ein gelöstes Zuckermolekül (bez. das Molekül samt dem durch dasselbe festgehaltene Wasser) in hydrodynamischer Beziehung sich verhält wie eine Kugel vom Volumen $2{,}45 \cdot 342/N \, \text{cm}^3$, wobei 342 das Molekulargewicht des Zuckers und N die Anzahl der wirklichen Moleküle in einem Grammolekül ist.

§ 4. Über die Diffusion eines nicht dissoziierten Stoffes in flüssiger Lösung.

Es liege eine Lösung vor, wie sie in § 3 betrachtet wurde. Wirkt auf das Molekül, welches wir als eine Kugel vom Radius P betrachten, eine Kraft K, so bewegt sich das Molekül mit einer Geschwindigkeit ω, welche durch P und den Reibungskoeffizienten k des Lösungsmittels bestimmt ist. Es besteht nämlich die Gleichung[1]:

$$(1) \qquad \omega = \frac{\overline{}}{6 \pi k P}.$$

Diese Beziehung benutzen wir zur Berechnung des Diffusionskoeffizienten einer nicht dissoziierten Lösung. Bedeutet p den osmotischen Druck der gelösten Substanz, welcher bei der betrachteten verdünnten Lösung als die einzige bewegende Kraft anzusehen sei, so ist die auf die gelöste Substanz pro Volumeneinheit der Lösung in Richtung der X-Achse ausgeübte Kraft $= -\partial p/\partial x$. Befinden sich ϱ Gramm in der Volumeneinheit und ist m das Molekulargewicht des gelösten Stoffes, N die Anzahl wirklicher Moleküle in einem Grammolekül, so ist $(\varrho/m) N$ die Anzahl der (wirklichen) Moleküle in der Vo-

---

[1] G. Kirchhoff, Vorlesungen über Mechanik. 26. Vorl., Gl. (22).

lumeneinheit und die auf ein Molekül infolge des Konzentrationsgefälles wirkende Kraft:

$$(2) \quad K = - \frac{m}{\varrho N} \frac{\partial p}{\partial x}.$$

Ist die Lösung genügend verdünnt, so ist der osmotische Druck durch die Gleichung gegeben:

$$(3) \quad p = \frac{R}{m} \varrho T,$$

wobei T die absolute Temperatur und $R = 8{,}31 \cdot 10^7$ ist. Aus den Gleichungen (1), (2) und (3) erhalten wir für die Geschwindigkeit der Wanderung der gelösten Substanz:

$$\omega = - \frac{RT}{6\pi k} \frac{1}{NP} \frac{1}{\varrho} \frac{\partial \varrho}{\partial x}.$$

Die pro Zeiteinheit durch die Einheit des Querschnittes in Richtung der X-Achse hindurchtretende Stoffmenge ist endlich:

$$(4) \quad \omega \varrho = - \frac{RT}{6\pi k} \cdot \frac{1}{NP} \frac{\partial \varrho}{\partial x}.$$

Wir erhalten also für den Diffusionskoeffizienten D:

$$D = \frac{RT}{6\pi k} \cdot \frac{1}{NP}.$$

Man kann also aus dem Diffusionskoeffizienten und dem Koeffizienten der inneren Reibung des Lösungsmittels das Produkt aus der Anzahl N der wirklichen Moleküle in einem Grammolekül und dem hydrodynamisch wirksamen Molekularradius P berechnen.

In dieser Ableitung ist der osmotische Druck wie eine auf die einzelnen Moleküle wirkende Kraft behandelt worden, was offenbar der Auffassung der kinetischen Molekulartheorie nicht entspricht, da gemäß letzterer in dem vorliegenden Falle der osmotische Druck nur als eine scheinbare Kraft aufzufassen ist. Diese Schwierigkeit verschwindet jedoch, wenn man bedenkt, daß den (scheinbaren) osmotischen Kräften, welche den Konzentrationsverschiedenheiten der Lösung entsprechen, durch ihnen numerisch gleiche, entgegengesetzt gerichtete, auf die einzelnen Moleküle wirkende Kräfte das (dynamische) Gleichgewicht geleistet werden kann, wie auf thermodynamischem Wege leicht eingesehen werden kann.

Der auf die Masseneinheit wirkenden osmotischen Kraft $-\frac{1}{\varrho}\frac{\partial p}{\partial x}$ kann durch die (an den einzelnen gelösten Molekülen angreifende) Kraft $-P_x$ das Gleichgewicht geleistet werden, wenn

$$-\frac{1}{\varrho}\frac{\partial p}{\partial x} - P_x = 0.$$

Denkt man sich also an der gelösten Substanz (pro Masseneinheit) die zwei sich gegenseitig aufhebenden Kräftesysteme P_x und $-P_x$ angreifend, so leistet $-P_x$ dem osmotischen Drucke das Gleichgewicht und es bleibt nur die dem osmotischen Drucke numerisch gleiche Kraft P_x als Bewegungsursache übrig. Damit ist die erwähnte Schwierigkeit beseitigt.[1])

§ 5. Bestimmung der Moleküldimensionen mit Hilfe der erlangten Relationen.

Wir haben in § 3 gefunden:

$$\frac{k^*}{k} = 1 + \varphi = 1 + n \cdot \tfrac{4}{3}\pi P^3,$$

wobei n die Anzahl der gelösten Moleküle pro Volumeneinheit und P den hydrodynamisch wirksamen Molekülradius bedeutet. Berücksichtigt man, daß

$$\frac{n}{N} = \frac{\varrho}{m},$$

wobei ϱ die in der Volumeneinheit befindliche Masse des gelösten Stoffes und m dessen Molekulargewicht bedeutet, so erhält man:

$$NP^3 = \frac{3}{4\pi}\frac{m}{\varrho}\left(\frac{k^*}{k} - 1\right).$$

Andererseits wurde in § 4 gefunden:

$$NP = \frac{RT}{6\pi k}\frac{1}{D}.$$

Diese beiden Gleichungen setzen uns in den Stand, die Größen P und N einzeln zu berechnen, von welchen sich N als un-

---

1) Eine ausführliche Darlegung dieses Gedankenganges findet sich in Ann. d. Phys. **17.** p. 549. 1905.

abhängig von der Natur des Lösungsmittels, der gelösten Substanz und der Temperatur herausstellen muß, wenn unsere Theorie den Tatsachen entspricht.

Wir wollen die Rechnung für wässerige Zuckerlösung durchführen. Nach den oben mitgeteilten Angaben über die innere Reibung der Zuckerlösung folgt zunächst für 20° C.:

$$NP^3 = 200.$$

Nach Versuchen von Graham (berechnet von Stefan) ist der Diffusionskoeffizient von Zucker in Wasser bei 9,5° C. 0,384, wenn der Tag als Zeiteinheit gewählt wird. Die Zähigkeit des Wassers bei 9,5° ist 0,0135. Wir wollen diese Daten in unsere Formel für den Diffusionskoeffizienten einsetzen, trotzdem sie an 10 proz. Lösungen gewonnen sind und eine genaue Gültigkeit unserer Formel bei so hohen Konzentrationen nicht zu erwarten ist. Wir erhalten

$$NP = 2{,}08 \cdot 10^{16}.$$

Aus den für NP^3 und NP gefundenen Werten folgt, wenn wir die Verschiedenheit von P bei 9,5° und 20° vernachlässigen,

$$P = 9{,}9 \cdot 10^{-8} \text{ cm},$$
$$N = 2{,}1 \cdot 10^{23}.$$

Der für N gefundene Wert stimmt der Größenordnung nach mit den durch andere Methoden gefundenen Werten für diese Größe befriedigend überein.

Bern, den 30. April 1905.

(Eingegangen 19. August 1905.)

Nachtrag.

In der neuen Auflage der physikalisch-chemischen Tabellen von Landolt und Börnstein finden sich weit brauchbarere Angaben zur Berechnung der Größe des Zuckermoleküls und der Anzahl N der wirklichen Moleküle in einem Grammmolekül.

Thovert fand (Tab. p. 372) für den Diffusionskoeffizienten von Zucker in Wasser bei 18,5° C. und der Konzentration

0,005 Mol./Liter den Wert 0,33 cm²/Tage. Aus einer Tabelle mit Beobachtungsresultaten von Hosking (Tab. p. 81) findet man ferner durch Interpolation, daß bei verdünnter Zuckerlösung einer Zunahme des Zuckergehaltes um 1 Proz. bei 18,5° C. eine Zunahme des Viskositätskoeffizienten um 0,00025 entspricht.

Unter Zugrundelegung dieser Angaben findet man
$$P = 0{,}78 \cdot 10^{-6} \text{ mm}$$
und
$$N = 4{,}15 \cdot 10^{23}.$$

Bern, Januar 1906.

7. *Zur Theorie der Brownschen Bewegung;* von *A. Einstein.*

Kurz nach dem Erscheinen meiner Arbeit über die durch die Molekulartheorie der Wärme geforderte Bewegung von in Flüssigkeiten suspendierten Teilchen[1]) teilte mir Hr. Siedentopf (Jena) mit, daß er und andere Physiker — zuerst wohl Hr. Prof. Gouy (Lyon) — durch direkte Beobachtung zu der Überzeugung gelangt seien, daß die sogenannte Brownsche Bewegung durch die ungeordnete Wärmebewegung der Flüssigkeitsmoleküle verursacht sei.[2]) Nicht nur die qualitativen Eigenschaften der Brownschen Bewegung, sondern auch die Größenordnung der von den Teilchen zurückgelegten Wege entspricht durchaus den Resultaten der Theorie. Ich will hier nicht eine Vergleichung des mir zur Verfügung stehenden dürftigen Erfahrungsmaterials mit den Resultaten der Theorie anstellen, sondern diese Vergleichung denjenigen überlassen, welche das Thema experimentell behandeln.

Die nachfolgende Arbeit soll meine oben genannte Arbeit in einigen Punkten ergänzen. Wir leiten hier nicht nur die fortschreitende, sondern auch die Rotationsbewegung suspendierter Teilchen ab für den einfachsten Spezialfall, daß die Teilchen Kugelgestalt besitzen. Wir zeigen ferner bis zu wie kurzen Beobachtungszeiten das in jener Abhandlung gegebene Resultat gilt.

Für die Herleitung wollen wir uns hier einer allgemeineren Methode bedienen, teils um zu zeigen, wie die Brownsche Bewegung mit den Grundlagen der molekularen Theorie der Wärme zusammenhängt, teils um die Formeln für die fortschreitende und für die rotierende Bewegung durch eine einheitliche Untersuchung entwickeln zu können. Es sei nämlich α ein beobachtbarer Parameter eines im Temperatur-

---

1) A. Einstein, Ann. d. Phys. **17**. p. 549. 1905.
2) M. Gouy, Journ. de Phys. (2) **7**. p. 561. 1888.

gleichgewicht befindlichen physikalischen Systems und es sei angenommen, daß das System bei jedem (möglichen) Wert von α im sogenannten indifferenten Gleichgewicht sich befinde. Nach der klassischen Thermodynamik, die zwischen Wärme und anderen Energiearten *prinzipiell* unterscheidet, finden spontane Änderungen von α nicht statt, wohl aber nach der molekularen Theorie der Wärme. Wir wollen im nachfolgenden untersuchen, nach welchen Gesetzen jene Änderungen gemäß der letzteren Theorie stattfinden müssen. Wir haben dann jene Gesetze auf folgende Spezialfälle anzuwenden:

1. α ist die x-Koordinate des Schwerpunktes eines in einer (der Schwerkraft nicht unterworfenen) homogenen Flüssigkeit suspendierten Teilchens von Kugelgestalt.

2. α ist der Drehwinkel, welcher die Lage eines in einer Flüssigkeit suspendierten, um einen Durchmesser drehbaren Teilchens von Kugelgestalt bestimmt.

§ 1. Über einen Fall thermodynamischen Gleichgewichtes.

In einer Umgebung von der absoluten Temperatur T befinde sich ein physikalisches System, das mit dieser Umgebung in thermischer Wechselwirkung stehe und im Zustand des Temperaturgleichgewichtes sei. Dies System, das also ebenfalls die absolute Temperatur T besitzt, sei im Sinne der molekularen Theorie der Wärme vollständig bestimmt[1]) durch die Zustandsvariabeln $p_1 \ldots p_n$. Als Zustandsvariable $p_1 \ldots p_n$ können in den zu behandelnden Spezialfällen die Koordinaten und Geschwindigkeitskomponenten aller das betrachtete System bildender Atome gewählt werden.

Es gilt für die Wahrscheinlichkeit dafür, daß in einem zufällig herausgegriffenen Zeitpunkt die Zustandsvariabeln $p_1 \ldots p_n$ in dem n-fach unendlich kleinen Gebiete $(dp_1 \ldots dp_n)$ liegen, die Gleichung[2]):

$$(1) \qquad dw = C e^{-\frac{N}{RT}E} dp_1 \ldots dp_n,$$

wobei C eine Konstante, R die universelle Konstante der Gasgleichung, N die Anzahl der wirklichen Moleküle in einem Grammmolekül und E die Energie bedeutet.

---

1) Vgl. Ann. d. Phys. **17**. p. 549. 1905.
2) l. c. § 3 und 4.

Es sei α ein beobachtbarer Parameter des Systems und es entspreche jedem Wertsystem $p_1 \ldots p_n$ ein bestimmter Wert α. Wir bezeichnen mit $A\,d\alpha$ die Wahrscheinlichkeit dafür, daß in einem zufällig herausgegriffenen Zeitpunkt der Wert des Parameters α zwischen α und $\alpha + d\alpha$ liege. Es ist dann

$$(2) \quad A\,d\alpha = \int_{d\alpha} C e^{-\frac{N}{RT}E} dp_1 \ldots dp_n,$$

wenn das Integral der rechten Seite über alle Wertkombinationen der Zustandsvariabeln erstreckt wird, deren α-Wert zwischen α und $\alpha + d\alpha$ liegt.

Wir beschränken uns auf den Fall, daß aus der Natur des Problems ohne weiteres klar ist, daß allen (möglichen) Werten von α dieselbe Wahrscheinlichkeit (Häufigkeit) zukommt, daß also die Größe A von α unabhängig ist.

Es liege nun ein zweites physikalisches System vor, das sich von dem soeben betrachteten einzig darin unterscheide, daß auf das System eine nur von α abhängige Kraft vom Potential $\Phi(\alpha)$ wirke. Ist E die Energie des vorhin betrachteten Systems, so ist $E + \Phi$ die Energie des jetzt betrachteten, so daß wir die der Gleichung (1) analoge Beziehung erhalten:

$$dw' = C' e^{-\frac{N}{RT}(E+\Phi)} dp_1 \ldots dp_n.$$

Hieraus folgt für die Wahrscheinlichkeit dW dafür, daß in einem beliebig herausgegriffenen Zeitpunkt der Wert von α zwischen α und $\alpha + d\alpha$ liegt, die der Gleichung (2) analoge Beziehung:

$$(\text{I}) \quad \begin{cases} dW = \int C' e^{-\frac{N}{RT}(E+\Phi)} dp_1 \ldots dp_n = \frac{C'}{C} e^{-\frac{N}{RT}\Phi} A\,d\alpha \\ \qquad\qquad = A' e^{-\frac{N}{RT}\Phi} d\alpha, \end{cases}$$

wobei A' von α unabhängig ist.

Diese Beziehung, welche dem von Bolzmann in seinen gastheoretischen Untersuchungen vielfach benutzten Exponentialgesetz genau entspricht, ist für die molekulare Theorie der Wärme charakteristisch. Sie gibt Aufschluß darüber, wieviel sich ein einer konstanten äußeren Kraft unterworfener Parameter eines Systems infolge der ungeordneten Molekularbewegung von dem Werte entfernt, welcher dem stabilen Gleichgewicht entspricht.

§ 2. Anwendungsbeispiele für die in § 1 abgeleitete Gleichung.

Wir betrachten einen Körper, dessen Schwerpunkt sich längs einer Geraden (X-Achse eines Koordinatensystems) bewegen kann. Der Körper sei von einem Gase umgeben und es herrsche thermisches und mechanisches Gleichgewicht. Nach der Molekulartheorie wird sich der Körper infolge der Ungleichheit der Molekularstöße längs der Geraden in unregelmäßiger Weise hin und her bewegen, derart, daß bei dieser Bewegung kein Punkt der Geraden bevorzugt ist — vorausgesetzt, daß auf den Körper in Richtung der Geraden keine anderen Kräfte wirken als die Stoßkräfte der Moleküle. Die Abszisse x des Schwerpunktes ist also ein Parameter des Systems, welcher die oben für den Parameter α vorausgesetzten Eigenschaften besitzt.

Wir wollen nun eine auf den Körper in Richtung der Geraden wirkende Kraft $K = -Mx$ einführen. Dann wird der Schwerpunkt des Körpers nach der Molekulartheorie ebenfalls ungeordnete Bewegungen ausführen, ohne sich jedoch viel vom Punkte $x = 0$ zu entfernen, während er nach der klassischen Thermodynamik im Punkte $x = 0$ ruhen müßte. Nach der Molekulartheorie ist (Formel I)

$$dW = A' e^{-\frac{N}{RT} M \frac{x^2}{2}} dx,$$

gleich der Wahrscheinlichkeit dafür, daß in einem zufällig gewählten Zeitpunkt der Wert der Abszisse x zwischen x und $x + dx$ liegt. Hieraus findet man den mittleren Abstand des Schwerpunktes vom Punkte $x = 0$:

$$\sqrt{\overline{x^2}} = \frac{\int_{-\infty}^{+\infty} x^2 A' e^{-\frac{N}{RT} \frac{Mx^2}{2}} dx}{\int_{-\infty}^{+\infty} A' e^{-\frac{N}{RT} \frac{Mx^2}{2}} dx} = \sqrt{\frac{RT}{NM}}.$$

Damit $\sqrt{\overline{x^2}}$ genügend groß sei, um der Beobachtung zugänglich zu sein, muß die die Gleichgewichtslage des Körpers bestimmende Kraft sehr klein sein. Setzen wir als untere Grenze des Beobachtbaren $\sqrt{x^2} = 10^{-4}$ cm, so erhalten wir für $T = 300$ $M =$ ca. $5 \cdot 10^{-6}$. Damit der Körper mit dem Mikroskop beobachtbare Schwankungen ausführe, darf also die auf ihn wirkende Kraft bei einer Elongation von 1 cm nicht mehr als 5 milliontel Dyn betragen.

Wir wollen noch eine theoretische Bemerkung an die abgeleitete Gleichung anknüpfen. Der betrachtete Körper trage eine über einen sehr kleinen Raum verteilte elektrische Ladung und es sei das den Körper umgebende Gas so verdünnt, daß der Körper eine durch das umgebende Gas nur schwach modifizierte Sinusschwingung ausführe. Der Körper strahlt dann elektrische Wellen in den Raum aus und empfängt Energie aus der Strahlung des umliegenden Raumes; er vermittelt also einen Energieaustausch zwischen Strahlung und Gas. Wir gelangen zu einer Ableitung des Grenzgesetzes der Temperaturstrahlung, welches für große Wellenlängen und für hohe Temperaturen zu gelten scheint, indem wir die Bedingung dafür aufstellen, daß der betrachtete Körper im Durchschnitt ebensoviel Strahlung emittiert als absorbiert. Man gelangt so[1]) zu der folgenden Formel für die der Schwingungszahl ν entsprechende Strahlungsdichte ϱ_ν:

$$\varrho_\nu = \frac{R}{N} \frac{8\pi \nu^2}{L^3} T,$$

wobei L die Lichtgeschwindigkeit bedeutet.

Die von Hrn. Planck gegebene Strahlungsformel[2]) geht für kleine Periodenzahlen und hohe Temperaturen in diese Formel über. Aus dem Koeffizienten des Grenzgesetzes läßt sich die Größe N bestimmen, und man erhält so die Plancksche Bestimmung der Elementarquanta. Die Tatsache, daß man auf dem angedeuteten Wege nicht zu dem wahren Gesetz der Strahlung, sondern nur zu einem Grenzgesetz gelangt, scheint mir in einer elementaren Unvollkommenheit unserer physikalischen Anschauungen ihren Grund zu haben.

Wir wollen nun die Formel (I) noch dazu verwenden, zu entscheiden, wie klein ein suspendiertes Teilchen sein muß,

---

1) Vgl. Ann. d. Phys. **17**. p. 549. 1905. § 1 und 2.
2) M. Planck, Ann. d. Phys. **1**. p. 99. 1900.

damit es trotz der Wirkung der Schwere dauernd suspendiert bleibe. Wir können uns dabei auf den Fall beschränken, daß das Teilchen spezifisch schwerer ist als die Flüssigkeit, da der entgegengesetzte Fall vollkommen analog ist.

Ist v das Volumen des Teilchens, ϱ dessen Dichte, ϱ_0 die Dichte der Flüssigkeit, g die Beschleunigung der Schwere und x der vertikale Abstand eines Punktes vom Boden des Gefäßes, so ergibt Gleichung (I)

$$dW = \text{konst.} \, e^{-\frac{N}{RT} v (\varrho - \varrho_0) g x} dx.$$

Man wird also dann finden, daß suspendierte Teilchen in einer Flüssigkeit zu schweben vermögen, wenn für Werte von x, die nicht wegen ihrer Kleinheit sich der Beobachtung entziehen, die Größe

$$\frac{N}{RT} v (\varrho - \varrho_0) g x$$

keinen allzu großen Wert besitzt — vorausgesetzt, daß an den Gefäßboden gelangende Teilchen nicht durch irgendwelche Umstände an demselben festgehalten werden.

§ 3. Über die von der Wärmebewegung verursachten Veränderungen des Parameters α.

Wir kehren wieder zu dem in § 1 behandelten allgemeinen Falle zurück, für den wir Gleichung (I) abgeleitet haben. Der einfacheren Ausdrucksweise und Vorstellung halber wollen wir aber nun annehmen, daß eine sehr große Zahl (n) identischer Systeme von der dort charakterisierten Art vorliege; wir haben es dann mit Anzahlen statt mit Wahrscheinlichkeiten zu tun. Gleichung (I) sagt dann aus:

Von N Systemen liegt bei

(Ia) $$dn = \varphi \, e^{-\frac{N}{RT} \Phi} d\alpha = F(\alpha) d\alpha.$$

Systemen der Wert des Parameters α in einem zufällig herausgegriffenen Zeitpunkt zwischen α und $\alpha + d\alpha$.

Diese Beziehung wollen wir dazu benutzen, die Größe der durch die ungeordneten Wärmevorgänge erzeugten unregelmäßigen Veränderungen des Parameters α zu ermitteln. Zu diesem Zweck drücken wir in Zeichen aus, daß die Funktion $F(\alpha)$

sich unter der vereinten Wirkung der dem Potential Φ entsprechenden Kraft und des ungeordneten Wärmeprozesses sich innerhalb der Zeitspanne t nicht ändert; t bedeute hierbei eine so kleine Zeit, daß die zugehörigen Änderungen der Größen α der einzelnen Systeme als unendlich kleine Argumentänderungen der Funktion $F(\alpha)$ betrachtet werden können.

Trägt man auf einer Geraden von einem bestimmten Nullpunkte aus den Größen α numerisch gleiche Strecken ab, so entspricht jedem System ein Punkt (α) auf dieser Geraden. $F(\alpha)$ ist die Lagerungsdichte der Systempunkte (α) auf der Geraden. Durch einen beliebigen Punkt (α_0) der Geraden müssen nun während der Zeit t genau soviele Systempunkte in dem einen Sinne hindurchwandern, wie in dem anderen Sinne.

Die dem Potential Φ entsprechende Kraft bewirke eine Änderung von α von der Größe

$$\Delta_1 = -B\frac{\partial \Phi}{\partial \alpha} t,$$

wobei B von α unabhängig sei, d. h. die Änderungsgeschwindigkeit von α sei proportional der wirkenden Kraft und unabhängig vom Werte des Parameters. Den Faktor B nennen wir die „Beweglichkeit des Systems in bezug auf α".

Würde also die äußere Kraft wirken, ohne daß der unregelmäßige molekulare Wärmeprozeß die Größen α änderte, so gingen durch den Punkt (α_0) während der Zeit t

$$n_1 = B\left(\frac{\partial \Phi}{\partial \alpha}\right)_{\alpha=\alpha_0} \cdot t \cdot F(\alpha_0)$$

Systempunkte nach der negativen Seite hindurch.

Es sei ferner die Wahrscheinlichkeit dafür, daß der Parameter α eines Systems infolge des ungeordneten Wärmeprozesses innerhalb der Zeit t eine Änderung erfahre, deren Wert zwischen Δ und $\Delta + d\Delta$ liegt, gleich $\psi(\Delta)$, wobei $\psi(\Delta) = \psi(-\Delta)$ und ψ von α unabhängig sei. Die Anzahl der infolge des ungeordneten Wärmeprozesses durch den Punkt (α_0) während der Zeit t nach der positiven Seite hin wandernden Systempunkte ist dann:

$$n_2 = \int_{\Delta=0}^{\Delta=\infty} F(\alpha_0 - \Delta)\chi(\Delta)\, d\Delta,$$

wenn
$$\int_\Delta^\infty \psi(\Delta)\, d\Delta = \chi(\Delta)$$

gesetzt wird. Die Anzahl der nach der negativen Seite infolge des ungeordneten Wärmeprozesses wandernden Systempunkte ist:

$$n_3 = \int_\Delta^\infty F(\alpha_0 + \Delta)\chi(\Delta)\, d\Delta.$$

Der mathematische Ausdruck für die Unveränderlichkeit der Funktion F ist also:

$$-n_1 + n_2 - n_3 = 0.$$

Setzt man die für n_1, n_2, n_3 gefundenen Ausdrücke ein und berücksichtigt, daß Δ unendlich klein ist bez. daß $\psi(\Delta)$ nur für unendlich kleine Werte von Δ von 0 verschieden ist, so erhält man hieraus nach einfacher Rechnung:

$$B\left(\frac{\partial \Phi}{\partial \alpha}\right)_{\alpha=\alpha_0} F(\alpha_0)\, t + \tfrac{1}{2} F'(\alpha_0)\overline{\Delta^2} = 0.$$

Hierbei bedeutet

$$\overline{\Delta^2} = \int_{-\infty}^{+\infty} \Delta^2 \psi(\Delta)\, d\Delta$$

den Mittelwert der Quadrate der durch den unregelmäßigen Wärmeprozeß während der Zeit t hervorgerufenen Änderungen der Größen α. Aus dieser Beziehung erhält man unter Berücksichtigung von Gleichung (Ia):

II) $$\sqrt{\overline{\Delta^2}} = \sqrt{\frac{2R}{N}} \cdot \sqrt{BTt}.$$

Hierbei bedeutet R die Konstante der Gasgleichung ($8{,}31 \cdot 10^7$), N die Anzahl der wirklichen Moleküle in einem Grammolekül (ca. $4 \cdot 10^{23}$), B die „Beweglichkeit des Systems in bezug auf den Parameter α", T die absolute Temperatur, t die Zeit, innerhalb welcher die durch den ungeordneten Wärmeprozeß hervorgerufenen Änderungen von α stattfinden.

§ 4. Anwendung der abgeleiteten Gleichung auf die Brownsche Bewegung.

Wir berechnen nun mit Hilfe der Gleichungen (II) zunächst die mittlere Verschiebung, die ein kugelförmiger, in einer Flüssigkeit suspendierter Körper während der Zeit t in einer bestimmten Richtung (X-Richtung eines Koordinatensystems) erleidet. Zu diesem Zweck haben wir in jene Gleichung den entsprechenden Wert für B einzusetzen.

Wirkt auf eine Kugel vom Radius P, die in einer Flüssigkeit vom Reibungskoeffizienten k suspendiert ist, eine Kraft K, so bewegt sie sich mit der Geschwindigkeit[1]) $K/6\pi k P$. Es ist also zu setzen

$$B = \frac{1}{6 \pi k P},$$

so daß man — in Übereinstimmung mit der oben zitierten Arbeit — für die mittlere Verschiebung der suspendierten Kugel in Richtung der X-Achse den Wert erhält:

$$\sqrt{\overline{\Delta_x^2}} = \sqrt{t}\sqrt{\frac{RT}{N}\frac{1}{3\pi k P}}.$$

Wir behandeln zweitens den Fall, daß die betrachtete Kugel in der Flüssigkeit um einen ihrer Durchmesser (ohne Lagerreibung) frei drehbar gelagert sei und fragen nach der mittleren Drehung $\sqrt{\overline{\Delta_r^2}}$ der Kugel während der Zeit t infolge des ungeordneten Wärmeprozesses.

Wirkt auf eine Kugel vom Radius P, die in einer Flüssigkeit vom Reibungskoeffizienten k drehbar gelagert ist, das Drehmoment D, so dreht sie sich mit der Winkelgeschwindigkeit[1])

$$\psi = \frac{D}{8 \pi k P^3}$$

Es ist also zu setzen:

$$B = \frac{1}{8 \pi k P^3}.$$

Man erhält also:

$$\sqrt{\overline{\Delta_r^2}} = \sqrt{t}\sqrt{\frac{RT}{N}\frac{1}{4\pi k P^3}}.$$

Die durch die Molekularbewegung erzeugte Drehbewegung sinkt also mit wachsendem P viel rascher als die fortschreitende Bewegung.

Für $P = 0{,}5$ mm und Wasser von 17^0 liefert die Formel für den im Mittel in einer Sekunde zurückgelegten Winkel etwa 11 Bogensekunden, in der Stunde ca. 11 Bogenminuten.

---

1) Vgl. G. Kirchhoff, Vorles. über Mechanik. 26. Vorl.

Für $P = 0{,}5$ Mikron und Wasser von $17°$ erhält man für $t = 1$ Sekunde ca. 100 Winkelgrade.

Bei einem frei schwebenden suspendierten Teilchen finden drei voneinander unabhängige derartige Drehbewegungen statt.

Die für $\sqrt{\overline{\varDelta^2}}$ entwickelte Formel ließe sich noch auf andere Fälle anwenden. Setzt man z. B. für B den reziproken elektrischen Widerstand eines geschlossenen Stromkreises ein, so gibt sie an, wieviel Elektrizität im Durchschnitt während der Zeit t durch irgend einen Leiterquerschnitt geht, welche Beziehung abermals mit dem Grenzgesetz der Strahlung des schwarzen Körpers für große Wellenlängen und hohe Temperaturen zusammenhängt. Da ich jedoch keine durch das Experiment kontrollierbare Konsequenz mehr habe auffinden können, scheint mir die Behandlung weiterer Spezialfälle unnütz.

§ 5. Über die Gültigkeitsgrenze der Formel für $\sqrt{\overline{\varDelta^2}}$.

Es ist klar, daß die Formel (II) nicht für beliebig kleine Zeiten gültig sein kann. Die mittlere Veränderungsgeschwindigkeit von α infolge des Wärmeprozesses

$$\frac{\sqrt{\overline{\varDelta^2}}}{t} = \sqrt{\frac{2RTB}{N}} \cdot \frac{1}{\sqrt{t}}$$

wird nämlich für unendlich kleine Zeitdauer t unendlich groß, was offenbar unmöglich ist, denn es müßte sich ja sonst jeder suspendierte Körper mit unendlich großer Momentangeschwindigkeit bewegen. Der Grund liegt daran, daß wir in unserer Entwickelung implizite angenommen haben, daß der Vorgang während der Zeit t als von dem Vorgange in den unmittelbar vorangehenden Zeiten unabhängiges Ereignis aufzufassen sei. Diese Annahme trifft aber um so weniger zu, je kleiner die Zeiten t gewählt werden. Wäre nämlich zur Zeit $z = 0$

$$\frac{d\alpha}{dt} = \beta_0$$

der Momentanwert der Änderungsgeschwindigkeit, und würde die Änderungsgeschwindigkeit β in einem gewissen darauf folgenden Zeitintervall durch den ungeordneten thermischen Prozeß nicht beeinflußt, sondern die Änderung von β lediglich durch den passiven Widerstand $(1/B)$ bestimmt, so würde für $d\beta/dz$ die Beziehung gelten:

$$-\mu \frac{d\beta}{dz} = \frac{\beta}{B}.$$

μ ist hierbei durch die Festsetzung definiert, daß $\mu(\beta^2/2)$ die der Änderungsgeschwindigkeit β entsprechende Energie sein soll. In dem Falle der Translationsbewegung der suspendierten Kugel wäre also z. B. $\mu(\beta^2/2)$ die kinetische Energie der Kugel samt der kinetischen Energie der mitbewegten Flüssigkeit. Durch Integration folgt:

$$\beta = \beta_0 \, e^{-\frac{z}{\mu B}}.$$

Aus diesem Resultat folgert man, daß die Formel (II) nur für Zeitintervalle gilt, welche groß sind gegen μB.

Für Körperchen von 1 Mikron Durchmesser und von der Dichte $\varrho = 1$ in Wasser von Zimmertemperatur ist die untere Grenze der Gültigkeit der Formel (II) ca. 10^{-7} Sekunden; diese untere Grenze für die Zeitintervalle wächst proportional dem Quadrat des Radius des Körperchens. Beides gilt sowohl für die fortschreitende wie für die Rotationsbewegung der Teilchen.

Bern, Dezember 1905.

(Eingegangen 19. Dezember 1905.)

ANNALEN
DER
PHYSIK.

BEGRÜNDET UND FORTGEFÜHRT DURCH

F. A. C. GREN, L. W. GILBERT, J. C. POGGENDORFF, G. UND E. WIEDEMANN.

VIERTE FOLGE.

BAND 20.

DER GANZEN REIHE 325. BAND.

KURATORIUM:

F. KOHLRAUSCH, M. PLANCK, G. QUINCKE,
W. C. RÖNTGEN, E. WARBURG.

UNTER MITWIRKUNG

DER DEUTSCHEN PHYSIKALISCHEN GESELLSCHAFT

UND INSBESONDERE VON

M. PLANCK

HERAUSGEGEBEN VON

PAUL DRUDE.

MIT EINEM PORTRÄT UND SECHS FIGURENTAFELN.

LEIPZIG, 1906.

VERLAG VON JOHANN AMBROSIUS BARTH.

12. *Zur Theorie der Lichterzeugung und Lichtabsorption;* von *A. Einstein.*

In einer letztes Jahr erschienenen Arbeit[1]) habe ich gezeigt, daß die Maxwellsche Theorie der Elektrizität in Verbindung mit der Elektronentheorie zu Ergebnissen führt, die mit den Erfahrungen über die Strahlung des schwarzen Körpers im Widerspruch sind. Auf einem dort dargelegten Wege wurde ich zu der Ansicht geführt, daß Licht von der Frequenz ν lediglich in Quanten von der Energie $(R/N)\beta\nu$ absorbiert und emittiert werden könne, wobei R die absolute Konstante der auf das Grammolekül angewendeten Gasgleichung, N die Anzahl der wirklichen Moleküle in einem Grammolekül, β den Exponentialkoeffizienten der Wienschen (bez. der Planckschen) Strahlungsformel und ν die Frequenz des betreffenden Lichtes bedeutet. Diese Beziehung wurde entwickelt für einen Bereich, der dem Bereich der Gültigkeit der Wienschen Strahlungsformel entspricht.

Damals schien es mir, als ob die Plancksche Theorie der Strahlung[2]) in gewisser Beziehung ein Gegenstück bildete zu meiner Arbeit. Neue Überlegungen, welche im § 1 dieser Arbeit mitgeteilt sind, zeigten mir aber, daß die theoretische Grundlage, auf welcher die Strahlungstheorie von Hrn. Planck ruht, sich von der Grundlage, die sich aus der Maxwellschen Theorie und Elektronentheorie ergeben würde, unterscheidet, und zwar gerade dadurch, daß die Plancksche Theorie implizite von der eben erwähnten Lichtquantenhypothese Gebrauch macht.

In § 2 der vorliegenden Arbeit wird mit Hilfe der Lichtquantenhypothese eine Beziehung zwischen Voltaeffekt und lichtelektrischer Zerstreuung hergeleitet.

---

1) A. Einstein, Ann. d. Phys. **17**. p. 132. 1905.
2) M. Planck, Ann. d. Phys. **4**. p. 561. 1901.

200   *A. Einstein.*

§ 1. Die Plancksche Theorie der Strahlung und die Lichtquanten.

In § 1 meiner oben zitierten Arbeit habe ich gezeigt, daß die Molekulartheorie der Wärme zusammen mit der Maxwellschen Theorie der Elektrizität und Elektronentheorie zu der mit der Erfahrung im Widerspruch stehenden Formel für die Strahlung des schwarzen Körpers führt:

(1) $$\varrho_\nu = \frac{R}{N} \frac{8\pi\nu^2}{L^3} T.$$

Hierbei bedeutet ϱ_ν die Dichte der Strahlung bei der Temperatur T, deren Frequenz zwischen ν und $\nu + 1$ liegt.

Woher kommt es, daß Hr. Planck nicht zu der gleichen Formel, sondern zu dem Ausdruck

(2) $$\varrho_\nu = \frac{\alpha \nu^3}{e^{\frac{\beta \nu}{T}} - 1}$$

gelangt ist?

Hr. Planck hat abgeleitet[1]), daß die mittlere Energie $\bar{E}_\nu$ eines Resonators von der Eigenfrequenz ν, der sich in einem mit ungeordneter Strahlung erfüllten Raume befindet, durch die Gleichung

(3) $$\bar{E}_\nu = \frac{L^3}{8\pi\nu^2} \varrho_\nu$$

gegeben ist. Damit war das Problem der Strahlung des schwarzen Körpers reduziert auf die Aufgabe, $\bar{E}_\nu$ als Funktion der Temperatur zu bestimmen. Die letztere Aufgabe aber ist gelöst, wenn es gelingt, die Entropie eines aus einer großen Anzahl im dynamischen Gleichgewicht sich befindender, miteinander in Wechselwirkung stehender, gleich beschaffener Resonatoren von der Eigenfrequenz ν zu berechnen.

Die Resonatoren denken wir uns als Ionen, welche um eine Gleichgewichtslage geradlinige Sinusschwingungen auszuführen vermögen. Bei der Berechnung dieser Entropie spielt die Tatsache, daß die Ionen elektrische Ladungen besitzen, keine Rolle; wir haben diese Ionen einfach als Massenpunkte (Atome) aufzufassen, deren Momentanzustand durch ihre momentane Abweichung x von der Gleichgewichtslage und

---

1) M. Planck, Ann. d. Phys. **1**. p. 99. 1900.

durch ihre Momentangeschwindigkeit $dx/dt = \xi$ vollkommen bestimmt ist.

Damit bei thermodynamischem Gleichgewicht die Zustandsverteilung dieser Resonatoren eine eindeutig bestimmte sei, hat man anzunehmen, daß außer den Resonatoren frei bewegliche Moleküle in beliebig kleiner Zahl vorhanden seien, welche dadurch, daß sie mit den Ionen zusammenstoßen, Energie von Resonator zu Resonator übertragen können; die letzteren Moleküle werden wir bei Berechnung der Entropie nicht berücksichtigen.

Wir könnten $\bar{E}_\nu$ als Funktion der Temperatur aus dem Maxwell-Boltzmannschen Verteilungsgesetz ermitteln und würden dadurch zu der ungültigen Strahlungsformel (1) gelangen. Zu dem von Hrn. Planck eingeschlagenen Wege wird man in folgender Weise geführt.

Es seien $p_1 \ldots p_n$ geeignet gewählte Zustandsvariable[1]), welche den Zustand eines physikalischen Systems vollkommen bestimmen (z. B. in unserem Falle die Größen x und ξ sämtlicher Resonatoren). Die Entropie S dieses Systems bei der absoluten Temperatur T ist dargestellt durch die Gleichung[2]):

$$(4) \qquad S = \frac{\bar{H}}{T} + \frac{R}{N} \lg \int e^{-\frac{N}{RT}H} dp_1 \ldots dp_n,$$

wobei $\bar{H}$ die Energie des Systems bei der Temperatur T, H die Energie als Funktion der $p_1 \ldots p_n$ bedeutet, und das Integral über alle möglichen Wertkombinationen der $p_1 \ldots p_n$ zu erstrecken ist.

Besteht das System aus sehr vielen molekularen Gebilden — und nur in diesem Falle hat die Formel Bedeutung und Gültigkeit, so tragen nur solche Wertkombinationen der $p_1 \ldots p_n$ merklich zu dem Werte des in S auftretenden Integrales bei, deren H sehr wenig von $\bar{H}$ abweicht.[3]) Berücksichtigt man dies, so ersieht man leicht, daß bis auf Vernachlässigbares gesetzt werden kann:

$$S = \frac{R}{N} \lg \int_{H}^{H + \Delta H} dp_1 \ldots dp_n,$$

---

1) A. Einstein, Ann. d. Phys. 11. p. 170. 1903.
2) l. c. § 6.
3) Folgt aus § 3 und § 4 l. c.

wobei ΔH zwar sehr klein, aber doch so groß gewählt sei, daß $R \lg(\Delta H)/N$ eine vernachlässigbare Größe ist. S ist dann von der Größe von ΔH unabhängig.

Setzt man nun die Variabeln x_α und ξ_α der Resonatoren an Stelle der $dp_1 \ldots dp_n$ in die Gleichung ein und berücksichtigt man, daß für den α^ten Resonator die Gleichung

$$\int_{E_\alpha}^{E_\alpha + dE_\alpha} dx_\alpha \, d\xi_\alpha = \text{konst.} \, dE_\alpha$$

gilt (da E_α eine quadratische, homogene Funktion von x_α und ξ_α ist), so erhält man für S den Ausdruck:

(5) $$S = \frac{R}{N} \lg W,$$

wobei

(5a) $$W = \int_H^{H + \Delta H} dE_1 \ldots dE_n$$

gesetzt ist.

Würde man S nach dieser Formel berechnen, so würde man wieder zu der ungültigen Strahlungsformel (1) gelangen. Zur Planckschen Formel aber gelangt man, indem man voraussetzt, daß die Energie E_α eines Resonators nicht jeden beliebigen Wert annehmen kann, sondern nur Werte, welche ganzzahlige Vielfache von ε sind, wobei

$$\varepsilon = \frac{R}{N} \beta \nu.$$

Setzt man nämlich $\Delta H = \varepsilon$, so ersieht man sofort aus Gleichung (5a), daß nun W bis auf einen belanglosen Faktor gerade in diejenige Größe übergeht, welche Hr. Planck „Anzahl der Komplexionen" genannt hat.

Wir müssen daher folgenden Satz als der Planckschen Theorie der Strahlung zugrunde liegend ansehen:

Die Energie eines Elementarresonators kann nur Werte annehmen, die ganzzahlige Vielfache von $(R/N)\beta\nu$ sind; die Energie eines Resonators ändert sich durch Absorption und Emission sprungweise, und zwar um ein ganzzahliges Vielfache von $(R/N)\beta\nu$.

Diese Voraussetzung involviert aber noch eine zweite, indem sie im Widerspruch steht mit der theoretischen Grundlage, aus der heraus Gleichung (3) entwickelt ist. Wenn die Energie eines Resonators sich nur sprungweise ändern kann, so kann nämlich zur Ermittelung der mittleren Energie eines in einem Strahlungsraum befindlichen Resonators die übliche Theorie der Elektrizität nicht Anwendung finden, da diese keine *ausgezeichneten* Energiewerte eines Resonators kennt. Es liegt also der Planckschen Theorie die Annahme zugrunde:

Obwohl die Maxwellsche Theorie auf Elementarresonatoren nicht anwendbar ist, so ist doch die *mittlere* Energie eines in einem Strahlungsraume befindlichen Elementarresonators gleich derjenigen, welche man mittels der Maxwellschen Theorie der Elektrizität berechnet.

Der letztere Satz wäre ohne weiteres plausibel, wenn in allen Teilen des Spektrums, die für die Beobachtung in Betracht kommen, $\varepsilon = (R/N)\beta\nu$ klein wäre gegen die mittlere Energie $\bar{E}_\nu$ eines Resonators; dies ist aber durchaus nicht der Fall. Innerhalb des Gültigkeitsbereiches der Wienschen Strahlungsformel ist nämlich $e^{\beta\nu/T}$ groß gegen 1. Man beweist nun leicht, daß nach der Planckschen Strahlungstheorie $\bar{E}_\nu/\varepsilon$ innerhalb des Gültigkeitsbereiches der Wienschen Strahlungsformel den Wert $e^{-\beta\nu/T}$ hat; $\bar{E}_\nu$ ist also weit kleiner als ε. Es kommt also überhaupt nur wenigen Resonatoren ein von Null verschiedener Wert der Energie zu.

Die vorstehenden Überlegungen widerlegen nach meiner Meinung durchaus nicht die Plancksche Theorie der Strahlung; sie scheinen mir vielmehr zu zeigen, daß Hr. Planck in seiner Strahlungstheorie ein neues hypothetisches Element — die Lichtquantenhypothese — in die Physik eingeführt hat.

§ 2. Eine zu erwartende quantitative Beziehung zwischen lichtelektrischer Zerstreuung und Voltaeffekt.

Ordnet man die Metalle nach ihrer lichtelektrischen Empfindlichkeit in eine Reihe, so erhält man bekanntlich die Voltasche Spannungsreihe, wobei die Metalle desto lichtempfindlicher sind, je näher sie dem elektropositiven Ende der Spannungsreihe liegen.

Man begreift diese Tatsache bis zu einem gewissen Grade unter alleiniger Zugrundelegung der Annahme, daß die die wirksamen Doppelschichten erzeugenden, hier nicht zu untersuchenden Kräfte nicht an der Berührungsfläche zwischen Metall und Metall, sondern an der Berührungsfläche zwischen Metall und Gas ihren Sitz haben.

Jene Kräfte mögen an der Oberfläche eines an ein Gas angrenzenden Metallstückes M eine elektrische Doppelschicht erzeugen, welcher eine Potentialdifferenz V zwischen Metall und Gas entspreche — positiv gerechnet, wenn das Metall das höhere Potential besitzt.

Es seien V_1 und V_2 die Spannungsdifferenzen zweier Metalle M_1 und M_2 bei elektrostatischem Gleichgewichte, falls die Metalle gegeneinander isoliert sind. Bringt man die beiden Metalle zur Berührung, so wird das elektrische Gleichgewicht gestört und es findet ein vollständiger[1]) Spannungsausgleich zwischen den Metallen statt. Dabei werden sich über die vorerwähnten Doppelschichten an den Grenzflächen Metall–Gas einfache Schichten superponieren; diesen entspricht ein elektrostatisches Feld im Luftraume, dessen Linienintegral gleich der Voltadifferenz ist.

Nennt man V_{l_1} bez. V_{l_2} die elektrischen Potentiale in Punkten des Gasraumes, welche den einander berührenden Metallen unmittelbar benachbart sind, und V' das Potential im Innern der Metalle, so ist

$$V' - V_{l_1} = V_1,$$
$$V' - V_{l_2} = V_2,$$

also

$$V_{l_2} - V_{l_1} = V_1 - V_2.$$

Die elektrostatisch meßbare Voltadifferenz ist also numerisch gleich der Differenz der Potentiale, welche die Metalle im Gase annehmen, falls sie voneinander isoliert sind.

Ionisiert man das Gas, so findet im Gasraum eine durch die daselbst vorhandenen elektrischen Kräfte hervorgerufene Wanderung der Ionen statt, welcher Wanderung in den Metallen ein Strom entspricht, der an der Berührungsstelle der Metalle

---

1) Von der Wirkung der thermoelektrischen Kräfte sehen wir ab.

vom Metall mit größerem V (schwächer elektropositiv) nach dem Metall mit kleinerem V (stärker elektropositiv) gerichtet ist.

Es befinde sich nun ein Metall M isoliert in einem Gase. Seine der Doppelschicht entsprechende Potentialdifferenz gegen das Gas sei V. Um die Einheit negativer Elektrizität aus dem Metall in das Gas zu befördern, muß eine dem Potential V numerisch gleiche Arbeit geleistet werden. Je größer V, d. h. je weniger elektropositiv das Metall ist, desto mehr Energie ist also für die lichtelektrische Zerstreuung nötig, desto weniger lichtelektrisch empfindlich wird also das Metall sein.

Soweit übersieht man die Tatsachen, ohne über die Natur der lichtelektrischen Zerstreuung Annahmen zu machen. Die Lichtquantenhypothese liefert aber außerdem eine quantitative Beziehung zwischen Voltaeffekt und lichtelektrischer Zerstreuung. Es wird nämlich einem negativen Elementarquantum (Ladung ε) mindestens die Energie $V\varepsilon$ zugeführt werden müssen, um es aus dem Metall in das Gas zu bewegen. Es wird also eine Lichtart nur dann negative Elektrizität aus dem Metall entfernen können, wenn das „Lichtquant" der betreffenden Lichtart mindestens den Wert $V\varepsilon$ besitzt. Wir erhalten also:

$$V\varepsilon \leq \frac{R}{N}\beta\nu,$$

oder

$$V \leq \frac{R}{A}\beta\nu,$$

wobei A die Ladung eines Grammoleküls eines einwertigen Ions ist.

Nehmen wir nun an, daß ein Teil der absorbierenden Elektronen das Metall zu verlassen befähigt ist, sobald die Energie der Lichtquanten $V\varepsilon$ übertrifft[1]) — welche Annahme sehr plausibel ist —, so erhalten wir

$$V = \frac{R}{A}\beta\nu,$$

wobei ν die kleinste lichtelektrisch wirksame Frequenz bedeutet.

Sind also ν_1 und ν_2 die kleinsten Lichtfrequenzen, welche auf die Metalle M_1 und M_2 wirken, so soll für die Voltasche

---

1) Von der thermischen Energie der Elektronen ist dabei abgesehen.

Spannungsdifferenz V_{12} der beiden Metalle die Gleichung gelten:

$$-V_{12} = V_1 - V_2 = \frac{R}{A} \beta (\nu_1 - \nu_2),$$

oder, wenn V_{12} in Volt gemessen wird

$$V_{12} = 4{,}2 \cdot 10^{-15} (\nu_2 - \nu_1).$$

In dieser Formel ist folgender, im großen ganzen jedenfalls gültige Satz enthalten: Je stärker elektropositiv ein Metall ist, desto kleiner ist die unterste wirksame Lichtfrequenz für das betreffende Metall. Es wäre von hohem Interesse zu wissen, ob die Formel auch in quantitativer Beziehung als Ausdruck der Tatsachen zu betrachten ist.

Bern, März 1906.

(Eingegangen 13. März 1906.)

13. *Das Prinzip von der Erhaltung der Schwerpunktsbewegung und die Trägheit der Energie;* von *A. Einstein.*

In einer voriges Jahr publizierten Arbeit[1]) habe ich gezeigt, daß die Maxwellschen elektromagnetischen Gleichungen in Verbindung mit dem Relativitätsprinzip und Energieprinzip zu der Folgerung führen, daß die Masse eines Körpers bei Änderung von dessen Energieinhalt sich ändere, welcher Art auch jene Energieänderung sein möge. Es zeigte sich, daß einer Energieänderung von der Größe ΔE eine gleichsinnige Änderung der Masse von der Größe $\Delta E/V^2$ entsprechen müsse, wobei V die Lichtgeschwindigkeit bedeutet.

In dieser Arbeit will ich nun zeigen, daß jener Satz die notwendige und hinreichende Bedingung dafür ist, daß das Gesetz von der Erhaltung der Bewegung des Schwerpunktes (wenigstens in erster Annäherung) auch für Systeme gelte, in welchen außer mechanische auch elektromagnetische Prozesse vorkommen. Trotzdem die einfachen formalen Betrachtungen, die zum Nachweis dieser Behauptung durchgeführt werden müssen, in der Hauptsache bereits in einer Arbeit von H. Poincaré enthalten sind[2]), werde ich mich doch der Übersichtlichkeit halber nicht auf jene Arbeit stützen.

§ 1. Ein Spezialfall.

K sei ein im Raume frei schwebender, ruhender starrer Hohlzylinder. In A sei eine Einrichtung, um eine bestimmte Menge S strahlender Energie durch den Hohlraum nach B zu senden. Während der Aussendung jener Strahlungsmenge wirkt ein Strahlungsdruck auf die linke Innenwand des Hohlzylinders K, der letzterem eine gewisse nach links gerichtete Geschwindigkeit verleiht. Besitzt der Hohlzylinder die Masse M,

---

1) A. Einstein, Ann. d. Phys. **18**. p. 639. 1905.
2) H. Poincaré, Lorentz-Festschrift p. 252. 1900.

so ist diese Geschwindigkeit, wie aus den Gesetzen des Strahlungsdruckes leicht zu beweisen, gleich $\frac{1}{V}\frac{S}{M}$, wobei V die Lichtgeschwindigkeit bedeutet. Diese Geschwindigkeit behält K so lange, bis der Strahlenkomplex, dessen räumliche Ausdehnung

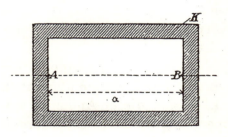

im Verhältnis zu der des Hohlraumes von K sehr klein sei, in B absorbiert ist. Die Dauer der Bewegung des Hohlzylinders ist (bis auf Glieder höherer Ordnung) gleich α/V, wenn α die Entfernung zwischen A und B bedeutet. Nach Absorption des Strahlenkomplexes in B ruht der Körper K wieder. Bei dem betrachteten Strahlungsvorgang hat sich K um die Strecke

$$\delta = \frac{1}{V}\frac{S}{M}\cdot\frac{\alpha}{V}$$

nach links verschoben.

Im Hohlraum von K sei ein der Einfachheit halber masselos gedachter Körper k vorhanden nebst einem (ebenfalls masselosen) Mechanismus, um den Körper k, der sich zunächst in B befinden möge, zwischen B und A hin und her zu bewegen. Nachdem die Strahlungsmenge S in B aufgenommen ist, werde diese Energiemenge auf k übertragen, und hierauf k nach A bewegt. Endlich werde die Energiemenge S in A wieder vom Hohlzylinder K aufgenommen und k wieder nach B zurückbewegt. Das ganze System hat nun einen vollständigen Kreisprozeß durchgemacht, den man sich beliebig oft wiederholt denken kann.

Nimmt man an, daß der Transportkörper k auch dann masselos ist, wenn er die Energiemenge S aufgenommen hat, so muß man auch annehmen, daß der Rücktransport der Energiemenge S nicht mit einer Lagenänderung des Hohlzylinders K verbunden sei. Der Erfolg des ganzen geschilderten Kreisprozesses besteht also einzig in einer Verschiebung δ des ganzen Systems nach links, welche Verschiebung durch Wiederholung des Kreisprozesses beliebig groß gemacht werden kann. Wir erhalten also das Resultat, daß ein ursprünglich ruhendes System, ohne daß äußere Kräfte auf dasselbe wirken, die Lage seines Schwerpunktes beliebig viel verändern kann, und zwar ohne daß das System irgend eine dauernde Veränderung erlitte.

Es ist klar, daß das erlangte Resultat keinen inneren Widerspruch enthält; wohl aber widerstreitet es den Grundgesetzen der Mechanik, nach denen ein ursprünglich ruhender Körper, auf welchen andere Körper nicht einwirken, keine Translationsbewegung ausführen kann.

Setzt man jedoch voraus, daß jeglicher Energie E die Trägheit E/V^2 zukomme, so verschwindet der Widerspruch mit den Elementen der Mechanik. Nach dieser Annahme besitzt nämlich der Transportkörper, während er die Energiemenge S von B nach A transportiert, die Masse S/V^2; und da der Schwerpunkt *des ganzen Systems* während dieses Vorganges nach dem Schwerpunktssatz ruhen muß, so erfährt der Hohlzylinder K während desselben im ganzen eine Verschiebung S' nach rechts von der Größe

$$\delta' = \alpha \cdot \frac{S}{V^2} \cdot \frac{1}{M}.$$

Ein Vergleich mit dem oben gefundenen Resultat zeigt, daß (wenigstens in erster Annäherung) $\delta = \delta'$ ist, daß also die Lage des Systems vor und nach dem Kreisprozeß dieselbe ist. Damit ist der Widerspruch mit den Elementen der Mechanik beseitigt.

§ 2. Über den Satz von der Erhaltung der Bewegung des Schwerpunktes.

Wir betrachten ein System von n diskreten materiellen Punkten mit den Massen $m_1, m_2 \ldots m_n$ und den Schwerpunktskoordinaten $x_1 \ldots z_n$. Diese materiellen Punkte seien in thermischer und elektrischer Beziehung nicht als Elementargebilde (Atome, Moleküle), sondern als Körper im gewöhnlichen Sinne von geringen Dimensionen aufzufassen, deren Energie durch die Schwerpunktsgeschwindigkeit nicht bestimmt sei. Diese Massen mögen sowohl durch elektromagnetische Vorgänge als auch durch konservative Kräfte (z. B. Schwerkraft, starre Verbindungen) aufeinander einwirken; wir wollen jedoch annehmen, daß sowohl die potentielle Energie der konservativen Kräfte als auch die kinetische Energie der Schwerpunktsbewegung der Massen stets als unendlich klein relativ zu der „inneren" Energie der Massen $m_1 \ldots m_n$ aufzufassen seien.

Es mögen im ganzen Raume die Maxwell-Lorenzschen Gleichungen

(1)
$$\begin{cases} \dfrac{u}{V}\varrho + \dfrac{1}{V}\dfrac{dX}{dt} = \dfrac{\partial N}{\partial y} - \dfrac{\partial M}{\partial z}, \\ \dfrac{u}{V}\varrho + \dfrac{1}{V}\dfrac{dY}{dt} = \dfrac{\partial L}{\partial z} - \dfrac{\partial N}{\partial x}, \\ \dfrac{u}{V}\varrho + \dfrac{1}{V}\dfrac{dZ}{dt} = \dfrac{\partial M}{\partial x} - \dfrac{\partial L}{\partial y}, \\ \dfrac{1}{V}\dfrac{dL}{dt} = \dfrac{\partial Y}{\partial z} - \dfrac{\partial Z}{\partial y}, \\ \dfrac{1}{V}\dfrac{dM}{dt} = \dfrac{\partial Z}{\partial x} - \dfrac{\partial X}{\partial z}, \\ \dfrac{1}{V}\dfrac{dN}{dt} = \dfrac{dX}{\partial y} - \dfrac{\partial Y}{\partial x} \end{cases}$$

gelten, wobei

$$\varrho = \frac{\partial X}{\partial x} + \frac{\partial Y}{\partial y} + \frac{\partial Z}{\partial z}$$

die 4π-fache Dichte der Elektrizität bedeutet.

Addiert man die der Reihe nach mit

$$\frac{V}{4\pi}Xx, \quad \frac{V}{4\pi}Yx \ldots \frac{V}{4\pi}Nx$$

multiplizierten Gleichungen (1) und integriert man dieselben über den ganzen Raum, so erhält man nach einigen partiellen Integrationen die Gleichung

(2)
$$\begin{cases} \int \dfrac{\varrho}{4\pi} x(uX + vY + wZ)\,d\tau \\ + \dfrac{d}{dt}\left\{ \int x \cdot \dfrac{1}{8\pi}(X^2 + Y^2 \ldots + N^2)\,d\tau \right\} \\ - \dfrac{V}{8\pi}\int (YN - ZM)\,d\tau = 0. \end{cases}$$

Das erste Glied dieser Gleichung stellt die von dem elektromagnetischen Felde den Körpern $m_1 \ldots m_n$ zugeführte Energie dar. Nach unserer Hypothese von der Abhängigkeit der Massen von der Energie hat man daher das erste Glied der Summe dem Ausdruck

$$V^2 \sum x_\nu \frac{dm_\nu}{dt}$$

gleichzusetzen, da wir nach dem Obigen annehmen, daß die einzelnen materiellen Punkte m_ν ihre Energie und daher auch ihre Masse *nur* durch Aufnahme von elektromagnetischer Energie ändern.

Schreiben wir ferner auch dem elektromagnetischen Felde eine Massendichte (ϱ_e) zu, die sich von der Energiedichte durch den Faktor $1/V^2$ unterscheidet, so nimmt das zweite Glied der Gleichung die Form an:

$$V^2 \frac{d}{dt}\left\{\int x \varrho_e \, d\tau\right\}.$$

Bezeichnet man mit J das im dritten Gliede der Gleichung (2) auftretende Integral, so geht letztere über in:

$$(2\mathrm{a}) \quad \sum\left(x_\nu \frac{d m_\nu}{dt}\right) + \frac{d}{dt}\left\{\int x \varrho_e \, d\tau\right\} - \frac{1}{4\pi V} J = 0.$$

Wir haben nun die Bedeutung des Integrales J aufzusuchen. Multipliziert man die zweite, dritte, fünfte und sechste der Gleichungen (1) der Reihe nach mit den Faktoren NV, $-MV$, $-ZV$, YV, addiert und integriert über den Raum, so erhält man nach einigen partiellen Integrationen

$$(3) \quad \frac{dJ}{dt} = -4\pi V \int \frac{\varrho}{4\pi}\left(X + \frac{v}{V} N - \frac{w}{V} M\right) d\tau = -4 V R_x,$$

wobei R_x die algebraische Summe der X-Komponenten aller vom elektromagnetischen Felde auf die Massen $m_1 \ldots m_n$ ausgeübten Kräfte bedeutet. Da die entsprechende Summe aller von den konservativen Wechselwirkungen herrührenden Kräfte verschwindet, so ist R_x gleichzeitig die Summe der X-Komponenten *aller* auf die Msssen m_ν ausgeübten Kräfte.

Wir wollen uns nun zunächst mit Gleichung (3) befassen, welche von der Hypothese, daß die Masse von der Energie abhängig sei, unabhängig ist. Sehen wir zunächst von der Abhängigkeit der Massen von der Energie ab und bezeichnen wir mit $\mathfrak{X}_\nu$ die Resultierende aller X-Komponenten der auf m_ν wirkenden Kräfte, so haben wir für die Masse m_ν die Bewegungsgleichung aufzustellen:

$$(4) \quad m_\nu \frac{d^2 x_\nu}{dt^2} = \frac{d}{dt}\left\{m_\nu \frac{d x_\nu}{dt}\right\} = \mathfrak{X}_\nu,$$

folglich erhalten wir auch:

(5) $$\frac{d}{dt}\sum\left(m_\nu \frac{dx_\nu}{dt}\right) = \sum \mathfrak{X}_\nu = R_x.$$

Aus Gleichung (5) und Gleichung (3) erhält man

(6) $$\frac{J}{4\pi V} + \sum m_\nu \frac{dx_\nu}{dt} = \text{konst.}$$

Führen wir nun die Hypothese wieder ein, daß die Größen m_ν von der Energie also auch von der Zeit abhängen, so stellt sich uns die Schwierigkeit entgegen, daß für diesen Fall die mechanischen Gleichungen nicht mehr bekannt sind; das erste Gleichheitszeichen der Gleichung (4) gilt nun nicht mehr. Es ist jedoch zu beachten, daß die Differenz

$$\frac{d}{dt}\left\{m_\nu \frac{dx_\nu}{dt}\right\} - m_\nu \frac{d^2 x_\nu}{dt^2} = \frac{dm_\nu}{dt}\frac{dx_\nu}{dt}$$

$$= \frac{1}{V^2}\int \frac{\varrho}{4\pi}\frac{dx_\nu}{dt}(uX + vY + wZ)d\tau$$

in den Geschwindigkeiten vom zweiten Grade ist. Sind daher alle Geschwindigkeiten so klein, daß Glieder zweiten Grades vernachlässigt werden dürfen, so gilt auch bei Veränderlichkeit der Masse m_ν die Gleichung

$$\frac{d}{dt}\left(m_\nu \frac{dx_\nu}{dt}\right) = \mathfrak{X}_\nu$$

sicher mit der in Betracht kommenden Genauigkeit. Es gelten dann auch die Gleichungen (5) und (6), und man erhält aus den Gleichungen (6) und (2a):

(2b) $$\frac{d}{dt}\left[\sum(m_\nu x_\nu) + \int x \varrho_e \, d\tau\right] = \text{konst.}$$

Bezeichnet ξ die X-Koordinate des Schwerpunktes der ponderabelen Massen und der Energiemasse des elektromagnetischen Feldes, so ist

$$\xi = \frac{\sum(m_\nu x_\nu) + \int x \varrho_e \, d\tau}{\sum m_\nu + \int \varrho_e \, d\tau},$$

wobei nach dem Energieprinzip der Wert des Nenners der

rechten Seite von der Zeit unabhängig ist.[1]) Wir können daher Gleichung (2b) auch in der Form schreiben:

(2c) $$\frac{d\xi}{dt} = \text{konst.}$$

Schreibt man also jeglicher Energie E die träge Masse E/V^2 zu, so gilt — wenigstens in erster Annäherung — das Prinzip von der Erhaltung der Bewegung des Schwerpunktes auch für Systeme, in denen elektromagnetische Prozesse vorkommen.

Aus der vorstehenden Untersuchung folgt, daß man entweder auf den Grundsatz der Mechanik, nach welchem ein ursprünglich ruhender, äußeren Kräften nicht unterworfener Körper keine Translationsbewegung ausführen kann, verzichten oder annehmen muß, daß die Trägheit eines Körpers nach dem angegebenen Gesetze von dessen Energieinhalt abhänge.

Bern, Mai 1906.

---

[1]) Nach der in dieser Arbeit entwickelten Auffassung ist der Satz von der Konstanz der Masse ein Spezialfall des Energieprinzipes.

(Eingegangen 17. Mai 1906.)

ANNALEN
DER
PHYSIK.

BEGRÜNDET UND FORTGEFÜHRT DURCH

F. A. C. GREN, L. W. GILBERT, J. C. POGGENDORFF, G. u. E. WIEDEMANN, P. DRUDE.

VIERTE FOLGE.

BAND 21.

DER GANZEN REIHE 326. BAND.

KURATORIUM:

F. KOHLRAUSCH, M. PLANCK, G. QUINCKE,
W. C. RÖNTGEN, E. WARBURG.

UNTER MITWIRKUNG

DER DEUTSCHEN PHYSIKALISCHEN GESELLSCHAFT

UND INSBESONDERE VON

M. PLANCK

HERAUSGEGEBEN VON

W. WIEN UND M. PLANCK.

MIT NEUN FIGURENTAFELN.

LEIPZIG, 1906.

VERLAG VON JOHANN AMBROSIUS BARTH.

9. *Über eine Methode zur Bestimmung des Verhältnisses der transversalen und longitudinalen Masse des Elektrons; von A. Einstein.*

Drei die Kathodenstrahlen betreffende Größen gibt es, welche einer präzisen Beobachtung zugänglich sind, nämlich die Spannung, welche den Strahlen ihre Geschwindigkeit verleiht (Erzeugungsspannung), die elektrostatische Ablenkbarkeit und die magnetische Ablenkbarkeit. Zwischen diesen drei Größen gibt es zwei voneinander unabhängige Beziehungen, deren Kenntnis für bedeutende Strahlengeschwindigkeiten von hervorragendem theoretischen Interesse ist. Eine dieser Beziehungen wurde für β-Strahlen von Hrn. Kaufmann untersucht, nämlich der Zusammenhang zwischen magnetischer und elektrostatischer Ablenkbarkeit.

Im folgenden soll darauf aufmerksam gemacht werden, daß eine zweite Beziehung zwischen diesen Größen mit hinreichender Genauigkeit bestimmt werden kann, nämlich die Beziehung zwischen Erzeugungsspannung und elektrostatischer Ablenkbarkeit der Kathodenstrahlen oder — was dasselbe bedeutet — das Verhältnis der transversalen zur longitudinalen Masse des Elektrons in Funktion der Erzeugungsspannung.

Wenn das Quadrat der Geschwindigkeit der Elektronen sehr klein ist gegenüber dem Quadrat der Lichtgeschwindigkeit, so gelten für die Bewegung des Elektrons die Gleichungen

$$\frac{d^2 x}{dt^2} = -\frac{\varepsilon}{\mu_0} X \text{ etc.,}$$

wobei ε/μ_0 das Verhältnis der Ladung zur Masse des Elektrons, x, y, z die Koordinaten des Elektrons und X, Y, Z die Komponenten der elektrischen Kraft des Feldes bedeuten, falls andere Kräfte als elektrostatische nicht auf das Elektron wirken. Wir nehmen an, die Elektronen bewegen sich mit der Anfangsgeschwindigkeit Null von einem gewissen Punkte x_0, y_0, z_0 (Kathode) aus. Die Bewegung ist dann eindeutig bestimmt durch obige Gleichungen; sie sei gegeben durch die Gleichungen

$$x = \varphi_1(t),$$
$$y = \varphi_2(t),$$
$$z = \varphi_3(t).$$

Denkt man sich alle elektrostatischen Kraftkomponenten überall mit n^2 multipliziert, so bewegt sich nunmehr — wie leicht aus den obigen Bewegungsgleichungen zu ersehen ist — das Elektron gemäß den Gleichungen

$$x = \varphi_1(n\,t),$$
$$y = \varphi_2(n\,t),$$
$$z = \varphi_3(n\,t).$$

Hieraus folgt, daß bei Proportionaländerung des Feldes wohl die Geschwindigkeit, nicht aber die Bahn der Elektronen sich ändert.

Eine Änderung der Bahn tritt bei Proportionaländerung des Feldes offenbar erst bei solchen Elektrongeschwindigkeiten ein, bei welchen das Verhältnis von transversaler und longitudinaler Masse merklich von der Einheit abweicht. Wählt man das elektrostatische Feld derart, daß die Kathodenstrahlen eine stark gekrümmte Bahn durchlaufen, so werden bereits geringe Verschiedenheiten der transversalen und longitudinalen

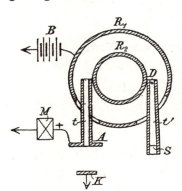

Masse einen beobachtbaren Einfluß auf die Bahnkurve haben. Nebenstehende schematische Skizze zeigt eine Anordnung, mittels welcher man das Verhältnis der transversalen zur longitudinalen Masse des Elektrons nach dem angedeuteten Prinzip bestimmen könnte. Die Kathodenstrahlen erlangen zwischen der geerdeten Kathode K und der an die positive Klemme der Stromquelle M angeschlossenen, zugleich als Blende dienenden Anode A ihre Geschwindigkeit, werden hierauf durch das mit A verbundene Röhrchen t in den Raum zwischen den Metallzylindern R_1 und R_2 eingeführt. R_1 ist geerdet, R_2 mit t, also mit dem positiven Pol der Stromquelle leitend verbunden, deren negativer Pol geerdet ist. Die Dimensionen seien so gewählt, daß sich langsame Kathodenstrahlen annähernd in einem Kreise bewegen, und zwar in geringer Entfernung von R_2. Die Strahlen gelangen hierauf in die mit R_2 metallisch verbundene, etwas konische Metallröhre t', in welcher sich der phosphoreszierende Schirm S befindet. Auf letzteren falle der Schatten des am inneren Ende von t' angeordneten vertikalen Drahtes D.

Bei Anwendung langsamer Kathodenstrahlen erhält der Schatten von D auf S eine ganz bestimmte Lage (Nullage). Erhöht man die Erzeugungsspannung der Strahlen, so wandert der Drahtschatten. Durch Einschalten einer Batterie B in die Erdungsleitung von R_1 werde jedoch der Schatten wieder in die Nullage zurückgeführt.

Bezeichnet man mit Π das Potential, bei welchem die Ablenkung der schattenbildenden Strahlen erfolgt, so ist Π auch diejenige Spannung, welcher die in Ablenkung begriffenen Strahlen ihre kinetische Energie verdanken. Bezeichnet ferner ϱ den Krümmungsradius der schattenbildenden Strahlen, so ist

$$\frac{\mu_t}{\mu_l} = \frac{\varrho}{2} \frac{X}{\Pi}.$$

Hierbei bedeutet μ_t die „transversale Masse" des Elektrons, μ_l diejenige longitudinale Masse, welche durch die Gleichung

$$\text{Kinetische Energie} = \mu_l \frac{v^2}{2}$$

definiert ist und X die ablenkende elektrische Kraft.

Nennt man P das Potential von R_2 (Potential des positiven Poles der Stromquelle M), p das Potential von R_1, bei welchem sich der Schatten in der Nullage befindet, so ist

$$\Pi = P - \alpha(P - p),$$

wobei α eine von den Apparatdimensionen abhängige, gegen 1 kleine Konstante bedeutet. Ferner ist die Größe X der Spannung $P-p$ proportional. Man erhält also aus obiger Gleichung

$$\frac{\mu_t}{\mu_l} = \text{konst.} \frac{P-p}{P - \alpha(P-p)},$$

oder (mit einigen erlaubten Vernachlässigungen)

$$\frac{\mu_t}{\mu_l} = \text{konst.} \left[1 - (1+\alpha)\frac{p}{P}\right].$$

Da α offenbar mit genügender Genauigkeit ermittelt werden kann und P und p bis auf wenige Prozent genau meßbar sind, so ist die Genauigkeit, mit welcher die Abweichung der Größe μ_t/μ_l von der Einheit ermittelt werden kann, im wesentlichen bestimmt durch die Genauigkeit, mit welcher auf die Nullage des Drahtschattens eingestellt werden kann. Man überzeugt sich leicht, daß letztere Genauigkeit so groß gemacht werden kann, daß eine Abweichung der Größe μ_t/μ_l von der Einheit um 0,3 Proz. (entsprechend einer Schattenverschiebung von ca. 1 mm, wenn $\overline{DS} = 10$ cm) noch bemerkt werden kann. Zu erwähnen ist insbesondere, daß die unvermeidlichen Schwankungen, denen beim Experiment das Potential P unterworfen ist, nur von unbedeutendem Einfluß auf die Genauigkeit der Messung sein können.

Wir wollen noch die Beziehung zwischen μ_t/μ_l und Π in erster Annäherung angeben, wie sie sich aus den verschiedenen Theorien ergibt. Wird Π in Volt ausgedrückt, so gilt

nach der Theorie von Bucherer:
$$\frac{\mu_t}{\mu_l} = 1 - 0{,}0070 \cdot \frac{\Pi}{10\,000},$$

nach der Theorie von Abraham:
$$\frac{\mu_t}{\mu_l} = 1 - 0{,}0084 \cdot \frac{\Pi}{10\,000},$$

nach der Theorie von Lorentz und Einstein:
$$\frac{\mu_t}{\mu_l} = 1 - 0{,}0104 \cdot \frac{\Pi}{10\,000}.$$

Da ich nicht in der Lage bin, selbst experimentell zu arbeiten, würde es mich freuen, wenn sich ein Physiker für die dargelegte Methode interessierte.

Bern, August 1906.

(Eingegangen 4. August 1906.)

ANNALEN
DER
PHYSIK.

BEGRÜNDET UND FORTGEFÜHRT DURCH

F. A. C. GREN, L. W. GILBERT, J. C. POGGENDORFF, G. u. E. WIEDEMANN, P. DRUDE.

VIERTE FOLGE.

BAND 22.

DER GANZEN REIHE 327. BAND.

KURATORIUM:

F. KOHLRAUSCH, M. PLANCK, G. QUINCKE,
W. C. RÖNTGEN, E. WARBURG.

UNTER MITWIRKUNG

DER DEUTSCHEN PHYSIKALISCHEN GESELLSCHAFT

UND INSBESONDERE VON

M. PLANCK

HERAUSGEGEBEN VON

W. WIEN UND M. PLANCK.

MIT EINEM PORTRÄT UND VIER FIGURENTAFELN.

LEIPZIG, 1907.

VERLAG VON JOHANN AMBROSIUS BARTH.

180

9. *Die Plancksche Theorie der Strahlung und die Theorie der spezifischen Wärme; von A. Einstein.*

In zwei früheren Arbeiten[1]) habe ich gezeigt, daß die Interpretation des Energieverteilungsgesetzes der schwarzen Strahlung im Sinne der Boltzmannschen Theorie des zweiten Hauptsatzes uns zu einer neuen Auffassung der Phänomene der Lichtemission und Lichtabsorption führt, die zwar noch keineswegs den Charakter einer vollständigen Theorie besitzt, die aber insofern bemerkenswert ist, als sie das Verständnis einer Reihe von Gesetzmäßigkeiten erleichtert. In der vorliegenden Arbeit soll nun dargetan werden, daß die Theorie der Strahlung — und zwar speziell die Plancksche Theorie — zu einer Modifikation der molekular-kinetischen Theorie der Wärme führt, durch welche einige Schwierigkeiten beseitigt werden, die bisher der Durchführung jener Theorie im Wege standen. Auch wird sich ein gewisser Zusammenhang zwischen dem thermischen und optischen Verhalten fester Körper ergeben.

Wir wollen zuerst eine Herleitung der mittleren Energie des Planckschen Resonators geben, die dessen Beziehung zur Molekularmechanik klar erkennen läßt.

Wir benutzen hierzu einige Resultate der allgemeinen molekularen Theorie der Wärme.[1]) Es sei der Zustand eines Systems im Sinne der molekularen Theorie vollkommen bestimmt durch die (sehr vielen) Variabeln $P_1, P_2 \ldots P_n$. Der Verlauf der molekularen Prozesse geschehe nach den Gleichungen

$$\frac{dP_\nu}{dt} = \Phi_\nu(P_1, P_2 \ldots P_n), \quad (\nu = 1, 2 \ldots n)$$

und es gelte für alle Werte der P_ν die Beziehung

(1) $$\sum \frac{\partial \Phi_\nu}{\partial P_\nu} = 0.$$

---

1) A. Einstein, Ann. d. Phys. **17**. p. 132. 1905 u. **20**. p. 199. 1905.

Es sei ferner ein Teilsystem des Systemes der P_ν, bestimmt durch die Variabeln $p_1 \ldots p_m$ (welche zu den P_ν gehören), und es sei angenommen, daß sich die Energie des ganzen Systems mit großer Annäherung aus zwei Teilen zusammengesetzt denken lasse, von denen einer (E) *nur* von den $p_1 \ldots p_m$ abhänge, während der andere von $p_1 \ldots p_m$ unabhängig sei. E sei ferner unendlich klein gegen die Gesamtenergie des Systems.

Die Wahrscheinlichkeit dW dafür, daß die p_ν in einem zufällig herausgegriffenen Zeitpunkt in einem unendlich kleinen Gebiete $(dp_1, dp_2 \ldots dp_m)$ liegen, ist dann durch die Gleichung gegeben[1])

$$(2) \qquad dW = C e^{-\frac{N}{RT}E} dp_1 \ldots dp_m.$$

Hierbei ist C eine Funktion der absoluten Temperatur (T), N die Anzahl der Moleküle in einem Grammäquivalent, R die Konstante der auf das Grammolekül bezogenen Gasgleichung.

Setzt man

$$\int_{dE} dp_1 \ldots dp_m = \omega(E) dE,$$

wobei das Integral über alle Kombinationen der p_ν zu erstrecken ist, welchen Energiewerte zwischen E und $E + dE$ entsprechen, so erhält man

$$(3) \qquad dW = C e^{-\frac{N}{RT}E} \omega(E) dE.$$

Setzt man als Variable P_ν die Schwerpunktskoordinaten und Geschwindigkeitskomponenten von Massenpunkten (Atomen, Elektronen), und nimmt man an, daß die Beschleunigungen nur von den Koordinaten, nicht aber von den Geschwindigkeiten abhängen, so gelangt man zur molekular-kinetischen Theorie der Wärme. Die Relation (1) ist hier erfüllt, so daß auch Gleichung (2) gilt.

Denkt man sich speziell als System der p_ν ein elementares Massenteilchen gewählt, welches längs einer Geraden Sinusschwingungen auszuführen vermag, und bezeichnet man mit x bez. ξ momentane Distanz von der Gleichgewichtslage bez. Geschwindigkeit desselben, so erhält man

$$(2\,\mathrm{a}) \qquad dW = C e^{-\frac{N}{RT}E} dx\, d\xi$$

---

[1]) A. Einstein, Ann. d. Phys. **11**. p. 170 u. f. 1903.

und, da $\int dx\, d\xi = \text{konst.}\, dE$, also $\omega = \text{konst.}$ zu setzen ist[1]):

$$(3\,\text{a}) \qquad dW = \text{konst.}\, e^{-\frac{N}{RT}E} dE.$$

Der Mittelwert der Energie des Massenteilchens ist also:

$$(4) \qquad \bar{E} = \frac{\int E e^{-\frac{N}{RT}E} dE}{\int e^{-\frac{N}{RT}E} dE} = \frac{RT}{N}.$$

Formel (4) kann offenbar auch auf ein geradlinig schwingendes Ion angewendet werden. Tut man dies, und berücksichtigt man, daß zwischen dessen mittlerer Energie $\bar{E}$ und der Dichte ϱ_ν der schwarzen Strahlung für die betreffende Frequenz nach einer Planckschen Untersuchung[2]) die Beziehung

$$(5) \qquad \bar{E}_\nu = \frac{L^3}{8\pi \nu^2} \varrho_\nu$$

gelten muß, so gelangt man durch Elimination von $\bar{E}$ aus (4) und (5) zu der Reileighschen Formel

$$(6) \qquad \varrho_\nu = \frac{R}{N} \frac{8\pi \nu^2}{L^3} T,$$

welcher bekanntlich nur die Bedeutung eines Grenzgesetzes für große Werte von T/ν zukommt.

Um zur Planckschen Theorie der schwarzen Strahlung zu gelangen, kann man wie folgt verfahren.[3]) Man behält Gleichung (5) bei, nimmt also an, daß durch die Maxwellsche Theorie der Elektrizität der Zusammenhang zwischen Strahlungsdichte und $\bar{E}$ richtig ermittelt sei. Dagegen verläßt man Gleichung (4), d. h. man nimmt an, daß die Anwendung der molekular-kinetischen Theorie den Widerspruch mit der Erfahrung bedinge. Hingegen halten wir an den Formeln (2) und (3) der allgemeinen molekularen Theorie der Wärme fest. Statt daß wir indessen gemäß der molekular-kinetischen Theorie

$$\omega = \text{konst.}$$

setzen, setzen wir $\omega = 0$ für alle Werte von E, welche den Werten $0, \varepsilon, 2\varepsilon, 3\varepsilon$ etc. nicht außerordentlich nahe liegen. Nur

---

1) Weil $E = a x^2 + b \xi^2$ zu setzen ist.
2) M. Planck, Ann. d. Phys. 1. p. 99. 1900.
3) Vgl. M. Planck, Vorlesungen über die Theorie der Wärmestrahlung. J. Ambr. Barth. 1906. §§ 149, 150, 154, 160, 166.

zwischen 0 und $0 + \alpha$, ε und $\varepsilon + \alpha$, 2ε und $2\varepsilon + \alpha$ etc. (wobei α unendlich klein sei gegen ε) sei ω von Null verschieden, derart, daß

$$\int_0^\alpha \omega \, dE = \int_\varepsilon^{\varepsilon+\alpha} \omega \, dE = \int_{2\varepsilon}^{2\varepsilon+\alpha} \omega \, dE = \ldots = A$$

sei. Diese Festsetzung involviert, wie man aus Gleichung (3) sieht, die Annahme, daß die Energie des betrachteten Elementargebildes lediglich solche Werte annehme, die den Werten 0, ε, 2ε etc. unendlich nahe liegen.

Unter Benutzung der eben dargelegten Festsetzung für ω erhält man mit Hilfe von (3):

$$\bar{E} = \frac{\int E e^{-\frac{N}{RT}E} \omega(E) \, dE}{\int e^{-\frac{N}{RT}E} \omega(E) \, dE} = \frac{0 + A\varepsilon e^{-\frac{N}{RT}\varepsilon} + A \cdot 2\varepsilon e^{-\frac{N}{RT}2\varepsilon} \ldots}{A + A e^{-\frac{N}{RT}\varepsilon} + A e^{-\frac{N}{RT}2\varepsilon} + \ldots}$$

$$= \frac{\varepsilon}{e^{\frac{N}{RT}\varepsilon} - 1};$$

Setzt man noch $\varepsilon = (R/N)\beta\nu$ (gemäß der Quantenhypothese), so erhält man hieraus:

(7) $$\bar{E} = \frac{\frac{R}{N}\beta\nu}{e^{\frac{\beta\nu}{T}} - 1},$$

sowie mit Hilfe von (5) die Plancksche Strahlungsformel:

$$\varrho_\nu = \frac{8\pi}{L^3} \cdot \frac{R\beta}{N} \frac{\nu^3}{e^{\frac{\beta\nu}{T}} - 1}.$$

Gleichung (7) gibt die Abhängigkeit der mittleren Energie des Planckschen Resonators von der Temperatur an.

Aus dem Vorhergehenden geht klar hervor, in welchem Sinne die molekular-kinetische Theorie der Wärme modifiziert werden muß, um mit dem Verteilungsgesetz der schwarzen Strahlung in Einklang gebracht zu werden. Während man sich nämlich bisher die molekularen Bewegungen genau denselben Gesetzmäßigkeiten unterworfen dachte, welche für die Bewegungen der Körper unserer Sinnenwelt gelten (wir fügen wesentlich nur das Postulat vollständiger Umkehrbarkeit hinzu), sind wir nun genötigt, für schwingungsfähige Ionen bestimmter Frequenz, die einen Energieaustausch zwischen Materie und Strahlung vermitteln können, die Annahme zu machen, daß die Mannigfaltigkeit der Zustände, welche sie anzunehmen vermögen, eine geringere sei als bei den Körpern unserer Erfahrung. Wir mußten ja annehmen, daß der Mechanismus der Energieübertragung ein solcher sei, daß die Energie des Elementargebildes ausschließlich die Werte 0, $(R/N)\beta\nu$, $2(R/N)\beta\nu$ etc. annehmen könne.[1])

Ich glaube nun, daß wir uns mit diesem Resultat nicht zufrieden geben dürfen. Es drängt sich nämlich die Frage auf: Wenn sich die in der Theorie des Energieaustausches zwischen Strahlung und Materie anzunehmenden Elementargebilde nicht im Sinne der gegenwärtigen molekular-kinetischen Theorie auffassen lassen, müssen wir dann nicht auch die Theorie modifizieren für die anderen periodisch schwingenden Gebilde, welche die molekulare Theorie der Wärme heranzieht? Die Antwort ist nach meiner Meinung nicht zweifelhaft. Wenn die Plancksche Theorie der Strahlung den Kern der Sache trifft, so müssen wir erwarten, auch auf anderen Gebieten der Wärmetheorie Widersprüche zwischen der gegenwärtigen molekular-kinetischen Theorie und der Erfahrung zu finden, die sich auf dem eingeschlagenen Wege heben lassen. Nach meiner Meinung trifft dies tatsächlich zu, wie ich im folgenden zu zeigen versuche.

Die einfachste Vorstellung, die man sich über die Wärmebewegung in festen Körpern bilden kann, ist die, daß die einzelnen in denselben enthaltenen Atome Sinusschwingungen um Gleichgewichtslagen ausführen. Unter dieser Voraussetzung erhält man durch Anwendung der molekular-kinetischen Theorie (Gleichung (4)) unter Berücksichtigung des Umstandes, daß jedem Atom drei Bewegungsfreiheiten zuzuschreiben sind,

---

[1]) Es ist übrigens klar, daß diese Voraussetzung auch auf schwingungsfähige Körper auszudehnen ist, die aus beliebig vielen Elementargebilden bestehen.

für die auf das Grammäquivalent bezogene spezifische Wärme des Stoffes
$$c = 3Rn$$
oder — in Grammkalorien ausgedrückt —
$$c = 5{,}94\,n,$$
wenn n die Anzahl der Atome im Molekül bedeutet. Es ist bekannt, daß diese Beziehung für die meisten Elemente und für viele Verbindungen im festen Aggregatzustand mit bemerkenswerter Annäherung erfüllt ist (Doulong-Petitsches Gesetz, Regel von F. Neumann und Kopp).

Betrachtet man jedoch die Tatsachen etwas genauer, so begegnet man zwei Schwierigkeiten, die der Anwendbarkeit der Molekulartheorie enge Grenzen zu ziehen scheinen.

1. Es gibt Elemente (Kohlenstoff, Bor und Silizium), welche im festen Zustande bei gewöhnlicher Temperatur eine bedeutend kleinere spezifische Atomwärme besitzen als 5,94. Es haben ferner alle festen Verbindungen, in denen Sauerstoff, Wasserstoff oder mindestens eines der eben genannten Elemente vorkommen, eine kleinere spezifische Wärme pro Grammmolekül als $n.\,5{,}94$.

2. Hr. Drude hat gezeigt[1]), daß die optischen Erscheinungen (Dispersion) dazu führen, jedem Atom einer Verbindung mehrere unabhängig voneinander bewegliche Elementarmassen zuzuschreiben, indem er mit Erfolg die ultraroten Eigenfrequenzen auf Schwingungen der Atome (Atomionen), die ultravioletten Eigenfrequenzen auf Schwingungen von Elektronen zurückführte. Hieraus ergibt sich für die molekularkinetische Theorie der Wärme eine zweite bedeutende Schwierigkeit, indem die spezifische Wärme — da die Zahl der beweglichen Massenpunkte pro Molekül größer ist als dessen Atomzahl — den Wert $5{,}94\,n$ beträchtlich übersteigen müßte.

Nach dem Obigen ist hierzu folgendes zu bemerken. Wenn wir die Träger der Wärme in festen Körpern als periodisch schwingende Gebilde ansehen, deren Frequenz von ihrer Schwingungsenergie unabhängig ist, dürfen wir nach der Planckschen Theorie der Strahlung nicht erwarten, daß die

---

1) P. Drude, Ann. d. Phys. **14**. p. 677. 1904.

186   *A. Einstein.*

spezifische Wärme stets den Wert 5,94 n besitze. Wir haben vielmehr zu setzen (7)

$$\bar{E} = \frac{3R}{N} \frac{\beta \nu}{e^{\frac{\beta \nu}{T}} - 1}.$$

Die Energie von N solchen Elementargebilden, in Grammkalorien gemessen, hat daher den Wert

$$5{,}94 \frac{\beta \nu}{e^{\frac{\beta \nu}{T}} - 1},$$

so daß jedes derartige schwingende Elementargebilde zur spezifischen Wärme pro Grammäquivalent den Wert

(8) $$5{,}94 \frac{e^{\frac{\beta \nu}{T}} \cdot \left(\frac{\beta \nu}{T}\right)^2}{\left(e^{\frac{\beta \nu}{T}} - 1\right)^2}.$$

beiträgt. Wir bekommen also, indem wir über alle Gattungen von schwingenden Elementargebilden summieren, welche in dem

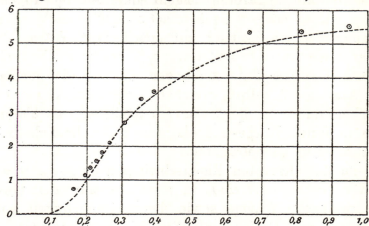

betreffenden festen Stoffe vorkommen, für die spezifische Wärme pro Grammäquivalent den Ausdruck[1])

(8a) $$c = 5{,}94 \sum \frac{e^{\frac{\beta \nu}{T}} \left(\frac{\beta \nu}{T}\right)^2}{\left(e^{\frac{\beta \nu}{T}} - 1\right)^2}.$$

Die vorstehende Figur[2]) zeigt den Wert des Ausdruckes (8) in Funktion von $x = (T/\beta \nu)$. Wenn $(T/\beta \nu) > 0{,}9$, unterscheidet

---

1) Die Betrachtung läßt sich leicht auf anisotrope Körper ausdehnen.
2) Vgl. deren gestrichelte Kurve.

sich der Beitrag des Gebildes zur molekularen spezifischen Wärme nicht beträchtlich vom Werte 5,94, der auch aus der bisher akzeptierten molekular-kinetischen Theorie sich ergibt; je kleiner v ist, bei um so tieferen Temperaturen wird dies bereits der Fall sein. Wenn dagegen $(T/\beta v) < 0,1$, so trägt das betreffende Elementargebilde nicht merklich zur spezifischen Wärme bei. Dazwischen findet ein anfänglich rascheres, dann langsameres Wachsen des Ausdruckes (8) statt.

Aus dem Gesagten folgt zunächst, daß die zur Erklärung der ultravioletten Eigenfrequenzen anzunehmenden schwingungsfähige Elektronen bei gewöhnlicher Temperatur ($T = 300$) zur spezifischen Wärme nicht merklich beitragen können; denn die Ungleichung $(T/\beta v) < 0,1$ geht für $T = 300$ über in die Ungleichung $\lambda < 4,8\,\mu$. Wenn dagegen ein Elementargebilde die Bedingung $\lambda > 48\,\mu$ erfüllt, so muß es nach dem Obigen bei gewöhnlicher Temperatur zur spezifischen Wärme pro Grammäquivalent nahezu den Beitrag 5,94 liefern.

Da für die ultraroten Eigenfrequenzen im allgemeinen $\lambda > 4,8\,\mu$ ist, so müssen nach unserer Auffassung jene Eigenschwingungen einen Beitrag zur spezifischen Wärme liefern, und zwar einen um so bedeutenderen, je größer das betreffende λ ist. Nach Drudes Untersuchungen sind es die ponderablen Atome (Atomionen) selbst, welchen diese Eigenfrequenzen zuzuschreiben sind. Es liegt also am nächsten, als Träger der Wärme in festen Körpern (Isolatoren) ausschließlich die positiven Atomionen zu betrachten.

Wenn die ultraroten Eigenschwingungsfrequenzen v eines festen Körpers bekannt sind, so wäre also nach dem Gesagten dessen spezifische Wärme sowie deren Abhängigkeit von der Temperatur durch Gleichung (8a) vollkommen bestimmt. Deutliche Abweichungen von der Beziehung $c = 5,94\,n$ wären bei gewöhnlicher Temperatur zu erwarten, wenn der betreffende Stoff eine optische ultrarote Eigenfrequenz aufweist, für welche $\lambda < 48\,\mu$; bei genügend tiefen Temperaturen sollen die spezifischen Wärmen aller festen Körper mit sinkender Temperatur bedeutend abnehmen. Ferner muß das Doulong-Petitsche Gesetz sowie das allgemeinere Gesetz $c = 5,94\,n$ für alle Körper bei genügend hohen Temperaturen gelten, falls sich bei letzteren keine neuen Bewegungsfreiheiten (Elektronionen) bemerkbar machen.

Die beiden oben genannten Schwierigkeiten werden durch die neue Auffassung beseitigt, und ich halte es für wahrscheinlich, daß letztere sich im Prinzip bewähren wird. Daran, daß sie den Tatsachen exakt entspreche, ist natürlich nicht zu denken. Die festen Körper erfahren beim Erwärmen Änderungen der molekularen Anordnung (z. B. Volumänderungen), die mit Änderungen des Energieinhaltes verbunden sind; alle festen Körper, die elektrisch leiten, enthalten frei bewegliche Elementarmassen, die zur spezifischen Wärme einen Beitrag liefern; die ungeordneten Wärmeschwingungen sind vielleicht von etwas anderer Frequenz als die Eigenschwingungen der nämlichen Elementargebilde bei optischen Prozessen. Endlich aber ist die Annahme, daß die in Betracht kommenden Elementargebilde eine von der Energie (Temperatur) unabhängige Schwingungsfrequenz besitzen, ohne Zweifel unzulässig.

Immerhin ist es interessant, unsere Konsequenzen mit der Erfahrung zu vergleichen. Da es sich nur um rohe Annäherung handelt, nehmen wir gemäß der F. Neumann-Koppschen Regel an, daß jedes Element, auch wenn dasselbe abnorm kleine spezifische Wärme besitzt, in allen seinen festen Verbindungen den gleichen Beitrag zur molekularen spezifischen Wärme liefere. Die in nachstehender Tabelle angegebenen Zahlen sind dem Lehrbuche der Chemie von Roskoe entnommen. Wir bemerken, daß alle Elemente von abnorm kleiner Atomwärme kleines Atomgewicht besitzen; dies ist nach unserer

Element	Spezifische Atomwärme	$\lambda_{\text{ber.}}$
S und P	5,4	42
Fl	5	33
O	4	21
Si	3,8	20
B	2,7	15
H	2,3	13
C	1,8	12

Auffassung zu erwarten, da ceteris paribus kleinen Atomgewichten große Schwingungsfrequenzen entsprechen. In der letzten Spalte der Tabelle sind die Werte von λ in Mikron angegeben, wie sie sich aus diesen Zahlen unter der Annahme, daß letztere für $T = 300$ gelten, mit Hilfe der dargestellten Beziehung zwischen x und c ergeben.

Wir entnehmen ferner den Tabellen von Landolt und Börnstein einige Angaben über ultrarote Eigenschwingungen (metallische Reflexion, Reststrahlen) einiger durchsichtiger fester Körper; die beobachteten λ sind in nachstehender Tabelle unter „$\lambda_{beob.}$" angegeben; die Zahlen unter „$\lambda_{ber.}$" sind obiger Tabelle entnommen, soweit sie sich auf Atome von abnorm kleiner spezifischer Wärme beziehen; für die übrigen soll $\lambda > 48\,\mu$ sein.

Körper	$\lambda_{beob.}$	$\lambda_{ber.}$
CaFl	24; 31,6	33; > 48
NaCl	51,2	> 48
KCl	61,2	> 48
CaCO$_3$	6,7; 11,4; 29,4	12; 21; > 48
SiO$_2$	8,5; 9,0; 20,7	20; 21

In der Tabelle enthalten NaCl und KCl nur Atome von normaler spezifischer Wärme; in der Tat sind die Wellenlängen ihrer ultraroten Eigenschwingungen größer als $48\,\mu$. Die übrigen Stoffe enthalten lauter Atome mit abnorm kleiner spezifischer Wärme (ausgenommen Ca); in der Tat liegen die Eigenfrequenzen dieser Stoffe zwischen $4,8\,\mu$ und $48\,\mu$. Im allgemeinen sind die aus den spezifischen Wärmen theoretisch ermittelten λ erheblich größer als die beobachteten. Diese Abweichungen können vielleicht in einer starken Veränderlichkeit der Frequenz des Elementargebildes mit der Energie desselben ihre Erklärung finden. Wie dem auch sein mag, jedenfalls ist die Übereinstimmung der beobachteten und berechneten λ hinsichtlich der Reihenfolge, sowie hinsichtlich der Größenordnung sehr bemerkenswert.

Wir wollen nun die Theorie noch auf den Diamanten anwenden. Die ultrarote Eigenfrequenz desselben ist nicht bekannt, läßt sich jedoch unter Zugrundelegung der dargelegten Theorie berechnen, wenn für einen Wert von T die molekulare spezifische Wärme c bekannt ist; das zu c gehörige x läßt sich aus der Kurve unmittelbar entnehmen, und man bestimmt hieraus λ nach der Beziehung $(TL/\beta\lambda) = x$.

Ich benutze die Beobachtungsresultate von H. F. Weber, die ich den Tabellen von Landolt und Börnstein entnahm (vgl. nachstehende Tabelle). Für $T = 331{,}3$ ist $c = 1{,}838$; hieraus folgt nach der angegebenen Methode $\lambda = 11{,}0\,\mu$. Unter Zugrundelegung dieses Wertes sind die in der dritten Spalte der Tabelle nach der Formel $x = (TL/\beta\lambda)$ berechnet ($\beta = 4{,}86 \cdot 10^{-11}$).

T	c	x
222,4	0,762	0,1679
262,4	1,146	0,1980
283,7	1,354	0,2141
306,4	1,582	0,2312
331,3	1,838	0,2500
358,5	2,118	0,2705
413,0	2,661	0,3117
479,2	3,280	0,3615
520,0	3,631	0,3924
879,7	5,290	0,6638
1079,7	5,387	0,8147
1258,0	5,507	0,9493

Die Punkte, deren Abszissen diese Werte von x, deren Ordinaten die in der Tabelle angegebenen, aus Beobachtungen Webers ermittelten Werte von c sind, sollen auf der oben dargestellten x, c-Kurve liegen. Wir haben diese Punkte — mit Ringen bezeichnet — in die obige Figur eingetragen; sie liegen tatsächlich nahezu auf der Kurve. Wir haben also anzunehmen, daß die elementaren Träger der Wärme beim Diamanten nahezu monochromatische Gebilde sind.

Es ist also nach der Theorie zu erwarten, daß der Diamant bei $\lambda = 11\,\mu$ ein Absorptionsmaximum aufweist.

Bern, November 1906.

(Eingegangen 9. November 1906.)

10. Über die Gültigkeitsgrenze des Satzes vom thermodynamischen Gleichgewicht und über die Möglichkeit einer neuen Bestimmung der Elementarquanta; von A. Einstein.

Der Zustand eines physikalischen Systems sei im Sinne der Thermodynamik bestimmt durch die Parameter λ, μ etc. (z. B. Anzeige eines Thermometers, Länge oder Volumen eines Körpers, Substanzmenge einer gewissen Art in einer Phase). Ist das System mit anderen Systemen nicht in Wechselwirkung, was wir annehmen, so wird nach der Tkermodynamik Gleichgewicht bei bestimmten Werten λ_0, μ_0 etc. der Parameter statthaben, für welche Werte die Entropie S des Systems ein Maximum ist. Nach der molekularen Theorie der Wärme jedoch ist dies nicht genau, sondern nur angenähert richtig; nach dieser Theorie besitzt der Parameter λ auch bei Temperaturgleichgewicht keinen konstanten Wert, sondern einen unregelmäßig schwankenden, der sich von λ_0 allerdings nur äußerst selten beträchtlich entfernt.

Die theoretische Untersuchung des statistischen Gesetzes, welchem diese Schwankungen unterworfen sind, scheint auf den ersten Blick bestimmte Festsetzungen in betreff des anzuwendenden molekularen Bildes zu erfordern. Dies ist jedoch nicht der Fall. Es genügt vielmehr im wesentlichen, die bekannte Boltzmannsche Beziehung anzuwenden, welche die Entropie S mit der statistischen Wahrscheinlichkeit eines Zustandes verbindet. Diese Beziehung lautet bekanntlich

$$S = \frac{R}{N} \lg W,$$

wobei R die Konstante der Gasgleichung und N die Anzahl der Moleküle in einem Grammäquivalent bedeutet.

Wir fassen einen Zustand des Systems ins Auge, in welchem der Parameter λ den von λ_0 sehr wenig abweichenden Wert $\lambda_0 + \varepsilon$ besitzt. Um den Parameter λ auf umkehrbarem Wege vom Werte λ_0 zum Werte λ bei konstanter Energie E

zu bringen, wird man eine Arbeit A dem System zuführen und die entsprechende Wärmemenge dem System entziehen müssen. Nach thermodynamischen Beziehungen ist:

$$A = \int dE - \int T dS,$$

oder, da die betrachtete Änderung unendlich klein und $\int dE = 0$ ist:

$$A = -T(S - S_0).$$

Andererseits ist aber nach dem Zusammenhang zwischen Entropie und Zustandswahrscheinlichkeit:

$$S - S_0 = \frac{R}{N} \lg\left(\frac{W}{W_0}\right).$$

Aus den beiden letzten Gleichungen folgt:

$$A = -\frac{RT}{N} \lg \frac{W}{W_0}$$

oder

$$W = W_0 e^{-\frac{N}{RT}A}.$$

Dies Resultat insolviert eine gewisse Ungenauigkeit, indem man ja eigentlich nicht von der Wahrscheinlichkeit eines *Zustandes*, sondern nur von der Wahrscheinlichkeit eines *Zustandsgebietes* reden kann. Schreiben wir statt der gefundenen Gleichung

$$dW = \text{konst.}\, e^{-\frac{N}{RT}A} d\lambda,$$

so ist das letztere Gesetz ein exaktes. Die Willkür, welche darin liegt, daß wir das Differential von λ und nicht das Differential irgendeiner Funktion von λ in die Gleichung eingesetzt haben, wird auf unser Resultat nicht von Einfluß sein.

Wir setzen nun $\lambda = \lambda_0 + \varepsilon$ und beschränken uns auf den Fall, daß A nach positiven Potenzen von ε entwickelbar ist, und daß nur das erste nicht verschwindende Glied dieser Entwickelung zum Werte des Exponenten merklich beiträgt bei solchen Werten von ε, für welche die Exponentialfunktion noch merklich von Null verschieden ist. Wir setzen also $A = \alpha \varepsilon^2$ und erhalten:

$$dW = \text{konst.}\, e^{-\frac{N}{RT}\alpha \varepsilon^2} d\varepsilon.$$

Es gilt also in diesem Falle für die Abweichungen ε das Gesetz der zufälligen Fehler. Für den Mittelwert der Arbeit A erhält man den Wert:
$$\bar{A} = \frac{1}{2} \frac{R}{N} T.$$

Das Quadrat der Schwankung ε eines Parameters λ ist also im Mittel so groß, daß die äußere Arbeit A, welche man bei strenger Gültigkeit der Thermodynamik anwenden müßte, um den Parameter λ bei konstanter Energie des Systems von λ_0 auf $\lambda^0 + \sqrt{\overline{\varepsilon^2}}$ zu verändern, gleich $\frac{1}{2} \frac{R}{N} T$ ist (also gleich dem dritten Teil der mittleren kinetischen Energie eines Atoms).

Führt man für R und N die Zahlenwerte ein, so erhält man angenähert:
$$\bar{A} = 10^{-16} T.$$

Wir wollen nun das gefundene Resultat auf einen kurz geschlossenen Kondensator von der (elektrostatisch gemessenen) Kapazität c anwenden. Ist $\sqrt{\overline{p^2}}$ die Spannung (elektrostatisch), welche der Kondensator im Mittel infolge der molekularen Unordnung annimmt, so ist
$$\bar{A} = \tfrac{1}{2} c \, \overline{p^2} = 10^{-16} T.$$

Wir nehmen an, der Kondensator sei ein Luftkondensator und er bestehe aus zwei ineinandergeschobenen Plattensystemen von je 30 Platten. Jede Platte habe von den benachbarten des anderen Systems im Mittel den Abstand 1 mm. Die Größe der Platten sei 100 cm². Die Kapazität c ist dann ca. 5000. Für gewöhnliche Temperatur erhält man dann
$$\sqrt{\overline{p^2_{\text{stat.}}}} = 3{,}4 \cdot 10^{-9}.$$

In Volt gemessen erhält man
$$\sqrt{\overline{p^2_{\text{Volt}}}} = 10^{-6}.$$

Denkt man sich die beiden Plattensysteme relativ zueinander beweglich, so daß sie vollständig auseinander geschoben werden können, so kann man erzielen, daß die Kapazität nach dem Auseinanderschieben von der Größenordnung 10 ist.

572 *A. Einstein. Thermodynamisches Gleichgewicht.*

Nennt man π die Potentialdifferenz, welche durch das Auseinanderschieben aus p entsteht, so hat man

$$\sqrt{\overline{\pi^2}} = 10^{-6} \cdot \frac{5000}{10} = 0{,}0005 \text{ Volt}.$$

Schließt man also den Kondensator bei zusammengeschobenen Plattensystemen kurz, und schiebt man dann, nachdem die Verbindung unterbrochen ist, die Plattensysteme auseinander, so erhält man zwischen den Plattensystemen Spannungsdifferenzen von der Größenordnung eines halben Millivolt.

Es scheint mir nicht ausgeschlossen zu sein, daß diese Spannungsdifferenzen der Messung zugänglich sind. Falls man nämlich Metallteile elektrisch verbinden und trennen kann, ohne daß hierbei noch andere *unregelmäßige* Potentialdifferenzen von gleicher Größenordnung wie die soeben berechneten auftreten, so muß man durch Kombination des obigen Plattenkondensators mit einem Multiplikator zum Ziele gelangen können. Es wäre dann ein der Brownschen Bewegung verwandtes Phänomen auf dem Gebiete der Elektrizität gegeben, daß zur Ermittelung der Größe N benutzt werden könnte.

Bern, Dezember 1906.

(Eingegangen 12. Dezember 1906.)

14. Berichtigung zu meiner Arbeit: „Die Plancksche Theorie der Strahlung etc."; von A. Einstein.

In der genannten, im Januarheft dieses Jahres erschienenen Arbeit habe ich geschrieben: „Nach Drudes Untersuchungen sind es die ponderabeln Atome (Atomionen) selbst, welchen diese Eigenfrequenzen zuzuschreiben sind. Es liegt also am nächsten, als Träger der Wärme in festen Körpern (Isolatoren) ausschließlich die positiven Atomionen zu betrachten."

Dieser Satz ist in zwei Beziehungen nicht aufrecht zu erhalten. Erstens sind nicht nur positiv, sondern auch negativ geladene Atomionen anzunehmen. Zweitens aber — und dies ist das Wesentliche — wird durch Drudes Untersuchungen nicht die Annahme gerechtfertigt, daß jedes schwingungsfähige Elementargebilde, welches als Träger von Wärme auftritt, stets eine elektrische Ladung besitze. Man kann also wohl aus der Existenz eines Absorptionsgebietes (unter den angegebenen Einschränkungen) auf die Existenz einer Gattung von Elementargebilden schließen, welche zur spezifischen Wärme einen Beitrag von charakteristischer Temperaturabhängigkeit liefert; der umgekehrte Schluß ist aber nicht statthaft, da es sehr wohl ungeladene Wärmeträger geben kann, d. h. solche, die sich optisch nicht bemerkbar machen. Letzteres ist besonders zu erwarten bei chemisch nicht gebundenen Atomen.

Der im letzten Satz der Abhandlung aus den Eigenschaften der spezifischen Wärme des Diamanten gezogene Schluß ist daher ebenfalls unstatthaft. Es sollte heißen:

„Es ist also nach der Theorie zu erwarten, daß der Diamant entweder bei $\lambda = 11\,\mu$ ein Absorptionsmaximum aufweist, oder daß derselbe überhaupt keine optisch nachweisbare ultrarote Eigenfrequenz besitzt."

(Eingegangen 3. März 1907.)

Berichtigung.

Bd. 22, p. 287 ist Zeile 4 von unten in Gleichung (2) der Buchstabe π zu streichen.

ANNALEN
DER
PHYSIK.

BEGRÜNDET UND FORTGEFÜHRT DURCH

F. A. C. GREN, L. W. GILBERT, J. C. POGGENDORFF, G. u. E. WIEDEMANN, P. DRUDE.

VIERTE FOLGE.

BAND 23.

DER GANZEN REIHE 328. BAND.

KURATORIUM:

F. KOHLRAUSCH, M. PLANCK, G. QUINCKE,
W. C. RÖNTGEN, E. WARBURG.

UNTER MITWIRKUNG

DER DEUTSCHEN PHYSIKALISCHEN GESELLSCHAFT

UND INSBESONDERE VON

M. PLANCK

HERAUSGEGEBEN VON

W. WIEN UND M. PLANCK.

MIT VIER FIGURENTAFELN.

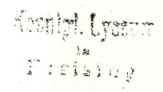

LEIPZIG, 1907.

VERLAG VON JOHANN AMBROSIUS BARTH.

12. Über die Möglichkeit einer neuen Prüfung des Relativitätsprinzips; von A. Einstein.

In einer letztes Jahr erschienenen wichtigen Arbeit[1]) hat Hr. J. Stark dargetan, daß die bewegten positiven Ionen der Kanalstrahlen Linienspektra emittieren, indem er den Doppler-Effekt nachwies und messend verfolgte. Er stellte auch Untersuchungen an in der Absicht, einen Effekt zweiter Ordnung (proportional $(v/V)^2$) nachzuweisen und zu messen; die nicht speziell für diesen Zweck eingerichtete Versuchsanordnung genügte jedoch nicht zur Erlangung eines sicheren Resultates.

Ich will im nachfolgenden kurz zeigen, daß das Relativitätsprinzip in Verbindung mit dem Prinzip der Konstanz der Geschwindigkeit des Lichtes jenen Effekt vorauszubestimmen gestattet. Wie ich in einer früheren Arbeit[2]) gezeigt habe, geht aus jenen Prinzipien hervor, daß eine gleichförmig bewegte Uhr, vom „ruhenden" System aus beurteilt, langsamer läuft als von einem mitbewegten Beobachter aus beurteilt. Bezeichnet ν die Anzahl der Schläge der Uhr pro Zeiteinheit für den ruhenden, ν_0 die entsprechende Anzahl für den mitbewegten Beobachter, so ist

$$\frac{\nu}{\nu_0} = \sqrt{1 - \left(\frac{v}{V}\right)^2}$$

oder in erster Annäherung

$$\frac{\nu - \nu_0}{\nu_0} = -\frac{1}{2}\left(\frac{v}{V}\right)^2.$$

Das Strahlung von bestimmten Frequenzen aussendende und absorbierende Atomion der Kanalstrahlen ist nun als eine rasch bewegte Uhr aufzufassen, und es ist daher die soeben angegebene Beziehung auf dasselbe anwendbar.

---

1) J. Stark, Ann. d. Phys. **21.** p. 401. 1906.
2) A. Einstein, Ann. d. Phys. **17.** p. 903. 1905.

Es ist aber zu beachten, daß die Frequenz ν_0 (für den mitbewegten Beobachter) unbekannt ist, so daß die obige Beziehung der experimentellen Prüfung nicht direkt zugänglich ist. Es ist aber anzunehmen, daß ν_0 auch gleich ist der Frequenz, welche dasselbe Ion im ruhenden Zustand emittiert bez. absorbiert, und zwar aus folgendem Grunde. Aus der Tatsache, daß dasselbe Linienspektrum unter sehr verschiedenen Bedingungen entsteht, entnehmen wir, daß die Frequenz ν_0 nicht abhängig ist von Wechselwirkungen zwischen bewegten Ionen und ruhendem Gas, sondern daß sie dem Ion allein eigentümlich ist; hieraus folgert man direkt mit Hilfe des Relativitätsprinzips, daß ν_0 gleich sein muß der Frequenz der von einem ruhenden Ion emittierten bez. absorbierten Strahlung.

Die Gleichung

$$\frac{\nu - \nu_0}{\nu_0} = -\frac{1}{2}\left(\frac{v}{V}\right)^2$$

gibt also direkt den gesuchten Effekt zweiter Ordnung.

Die von Hrn. Stark für den Effekt angegebenen Zahlenwerte sind mehr als zehnmal so groß als die aus der angegebenen Formel hervorgehenden. Es erscheint mir wahrscheinlich, daß sichere Resultate in der vorliegenden Frage erst dann zu erwarten sind, wenn es gelungen ist, (nichtleuchtende?) Kanalstrahlen im völlig gasfreien Raume zu erzielen.

Bern, März 1907.

(Eingegangen 17. März 1907.)

15. Bemerkungen zu der Notiz von Hrn. Paul Ehrenfest: „Die Translation deformierbarer Elektronen und der Flächensatz"; von A. Einstein.

In der genannten Abhandlung sind folgende Bemerkungen enthalten:

„Die Lorentzsche Relativitätselektrodynamik wird in der Formulierung, in der sie Hr. Einstein publiziert hat, ziemlich allgemein als abgeschlossenes System angesehen. Dementsprechend muß sich aus ihr rein deduktiv eine Antwort auf die Frage ergeben, die man durch Übertragung des Abrahamschen Problems vom starren auf das deformierbare Elektron erhält: Angenommen, es existiere ein deformierbares Elektron, das in der Ruhe irgend eine nicht-kugelförmige und nicht ellipsoidische Gestalt besitzt. Bei gleichförmiger Translation erfährt dieses Elektron nach Hrn. Einstein die bekannte Lorentz-Kontraktion. Ist nun für dieses Elektron gleichförmige Translation nach jeder Richtung hin kräftefrei möglich oder nicht?"

Hierzu habe ich folgendes zu bemerken:

1. Das Relativitätsprinzip oder — genauer ausgedrückt — das Relativitätsprinzip zusammen mit dem Prinzip von der Konstanz der Lichtgeschwindigkeit ist nicht als ein „abgeschlossenes System", ja überhaupt nicht als System aufzufassen, sondern lediglich als ein heuristisches Prinzip, welches für sich allein betrachtet nur Aussagen über starre Körper, Uhren und Lichtsignale enthält. Weiteres liefert die Relativitätstheorie nur dadurch, daß sie Beziehungen zwischen

sonst voneinander unabhängig erscheinenden Gesetzmäßigkeiten fordert.

Die Theorie der Bewegung des Elektrons beispielsweise kommt folgendermaßen zustande. Man setzt die Maxwellschen Gleichungen für das Vakuum für ein Koordinatenzeitsystem voraus. Durch Anwendung der vermittelst des Relativitätssystems hergeleiten Ort-Zeit-Transformation findet man die Transformationsgleichungen für die elektrischen und magnetischen Kräfte. Unter Benutzung der letzteren findet man durch abermalige Anwendung der Ort-Zeit-Transformation aus dem Gesetz für die Beschleunigung des langsam bewegten Elektrons (welches angenommen bez. der Erfahrung entnommen wurde) das Gesetz für die Beschleunigung des beliebig rasch bewegten Elektrons. Es handelt sich hier also keineswegs um ein „System", in welchem implizite die einzelnen Gesetze enthalten wären, und nur durch Deduktion daraus gefunden werden könnten, sondern nur um ein Prinzip, das (ähnlich wie der zweite Hauptsatz der Wärmetheorie) gewisse Gesetze auf andere zurückzuführen gestattet.

2. Als man sich noch nicht auf das Relativitätsprinzip stützte, sondern die Bewegungsgesetze des Elektrons auf elektrodynamischem Wege zu ermitteln strebte, sah man sich genötigt, über die Verteilung der Elektrizität bestimmtere Annahmen zu machen, damit das Problem kein unbestimmtes sei. Man dachte sich dabei die Elektrizität auf einem (starren) Gerüst verteilt. Es ist wohl zu beachten, daß die Gesetze, nach welchen ein solches Gebilde sich bewegt, nicht aus der Elektrodynamik allein hergeleitet werden können. Das Gerüst ist ja nichts anderes als die Einführung von Kräften, welche den elektrodynamischen das Gleichgewicht leisten. Wenn wir das Gerüst als einen starren (d. h. durch äußere Kräfte nicht deformierbaren) Körper ansehen, so kann das Problem der Bewegung des Elektrons dann und nur dann auf deduktivem Wege ohne Willkür gelöst werden, wenn die Dynamik des starren Körpers hinreichend genau bekannt ist.

Falls die Relativitätstheorie zutrifft, sind wir von letzterem Ziele noch weit entfernt. Wir besitzen erst eine Kinematik der Paralleltranslation und einen Ausdruck für die

208. *A. Einstein. Bemerkungen zu der Notiz von P. Ehrenfest.*

kinetische Energie eines in Paralleltranslation begriffenen Körpers, falls letzterer mit anderen Körpern nicht in Wechselwirkung steht[1]); im übrigen ist sowohl die Dynamik als auch die Kinematik des starren Körpers für den vorliegenden Fall noch als unbekannt zu betrachten.

Bern, den 14. April 1907.

---

1) Daß letztere Einschränkung wesentlich ist, werde ich demnächst in einer Arbeit zeigen.

(Eingegangen 16. April 1907.)

12. Über die vom Relativitätsprinzip geforderte Trägheit der Energie; von *A. Einstein.*

Das Relativitätsprinzip führt in Verbindung mit den Maxwellschen Gleichungen zu der Folgerung, daß die Trägheit eines Körpers mit dessen Energieinhalt in ganz bestimmter Weise wachse bez. abnehme. Betrachtet man nämlich einen Körper, der gleichzeitig nach zwei entgegengesetzten Richtungen eine bestimmte Strahlungsenergie aussendet, und untersucht man diesen Vorgang von zwei relativ zueinander gleichförmig bewegten Koordinatensystemen aus [1]), von denen das eine relativ zu dem Körper ruht, und wendet man auf den Vorgang — von beiden Koordinatensystemen aus — das Energieprinzip an, so gelangt man zu dem Resultat, daß einem Energiezuwachs ΔE des betrachteten Körpers stets ein Massenzuwachs $\Delta E/V^2$ entsprechen müsse, wobei V die Lichtgeschwindigkeit bedeutet.

Der Umstand, daß der dort behandelte spezielle Fall eine Annahme von so außerordentlicher Allgemeinheit (über die Abhängigkeit der Trägheit von der Energie) notwendig macht, fordert dazu auf, in allgemeinerer Weise die Notwendigkeit bez. Berechtigung der genannten Annahme zu prüfen. Insbesondere erhebt sich die Frage: Führen nicht andere spezielle Fälle zu mit der genannten Annahme unvereinbaren Folgerungen? Einen ersten Schritt in dieser Hinsicht habe ich letztes Jahr unternommen [2]), indem ich zeigte, daß jene Annahme den Widerspruch der Elektrodynamik mit dem Prinzip von der Konstanz der Schwerpunktsbewegung (mindestens was die Glieder erster Ordnung anbelangt) aufhebt.

Die *allgemeine* Beantwortung der aufgeworfenen Frage ist darum vorläufig nicht möglich, weil wir ein vollständiges, dem

---

1) A. Einstein, Ann. d. Phys. **18**. p. 639. 1905.
2) A. Einstein, Ann. d. Phys. **20**. p. 627. 1906.

Relativitätsprinzip entsprechendes Weltbild einstweilen nicht besitzen. Wir müssen uns vielmehr auf die speziellen Fälle beschränken, welche wir ohne Willkür vom Standpunkt der Relativitätselektrodynamik gegenwärtig behandeln können. Zwei solche Fälle werden wir im folgenden betrachten; bei dem ersten derselben besteht das System, dessen träge Masse untersucht werden soll, in einem starren, starr elektrisierten Körper, bei dem zweiten Fall aus einer Anzahl gleichförmig bewegter Massenpunkte, welche aufeinander keine Kräfte ausüben.

Bevor ich mit der Untersuchung beginne, muß ich hier noch eine Bemerkung über den mutmaßlichen Gültigkeitsbereich der Maxwellschen Gleichungen für den leeren Raum einschieben, um einem naheliegenden Einwand zu begegnen. In früheren Arbeiten habe ich gezeigt, daß unser heutiges elektromechanisches Weltbild nicht geeignet ist, die Entropieeigenschaften der Strahlung sowie die Gesetzmäßigkeiten der Emission und Absorption der Strahlung und die der spezifischen Wärme zu erklären; es ist vielmehr nach meiner Meinung nötig anzunehmen, daß die Beschaffenheit eines jeglichen periodischen Prozesses eine derartige ist, daß eine Umsetzung der Energie nur in bestimmten Quanten von endlicher Größe (Lichtquanten) vor sich gehen kann, daß also die Mannigfaltigkeit der in Wirklichkeit möglichen Prozesse eine kleinere ist als die Mannigfaltigkeit der im Sinne unserer heutigen theoretischen Anschauungen möglichen Prozesse.[1]) Einen Strahlungsvorgang im besonderen hätten wir uns so zu denken, daß der momentane elektromagnetische Zustand in einem Raumteile durch eine *endliche* Zahl von Größen vollständig bestimmt sei — im Gegensatze zur Vektorentheorie der Strahlung. Solange wir jedoch nicht im Besitz eines Bildes sind, welches den genannten Forderungen entspricht, werden wir uns naturgemäß in allen Fragen, welche nicht Entropieverhältnisse sowie Umwandlungen elementar kleiner Energiemengen betreffen, der gegenwärtigen Theorie bedienen, ohne fürchten zu müssen, dadurch zu unrichtigen Resultaten zu gelangen. Wie ich mir die heutige Sachlage in diesen Fragen denke, kann

---

1) A. Einstein, Ann. d. Phys. **17**. p. 132. 1905; **20**. p. 199. 1906 und **22**. p. 180. 1907.

ich am anschaulichsten durch folgenden fingierten Fall illustrieren.

Man denke sich, daß die molekularkinetische Theorie der Wärme noch nicht aufgestellt, daß aber mit voller Sicherheit nachgewiesen sei, daß die Brownsche Bewegung (Bewegung von in Flüssigkeiten suspendierten Teilchen) nicht auf äußerer Energiezufuhr beruhe, sondern daß klar erkannt sei, daß jene Bewegungen mit Hilfe der Mechanik und Thermodynamik nicht erklärt werden können. Man würde bei dieser Sachlage mit Recht zu dem Schlusse geführt, daß eine tiefgreifende Änderung der theoretischen Grundlagen Platz greifen müsse. Trotzdem würde sich aber niemand scheuen, bei Behandlung aller Fragen, welche sich nicht auf Momentanzustände in kleinen Raumteilen beziehen, die Grundgleichungen der Mechanik und Thermodynamik anzuwenden. In diesem Sinne können wir nach meiner Meinung mit Zuversicht unsere Betrachtungen auf die Maxwellschen Gleichungen stützen.

Es scheint mir in der Natur der Sache zu liegen, daß das Nachfolgende zum Teil bereits von anderen Autoren klargestellt sein dürfte. Mit Rücksicht darauf jedoch, daß hier die betreffenden Fragen von einem neuen Gesichtspunkt aus behandelt sind, glaubte ich, von einer für mich sehr umständlichen Durchmusterung der Literatur absehen zu dürfen, zumal zu hoffen ist, daß diese Lücke von anderen Autoren noch ausgefüllt werden wird, wie dies in dankenswerter Weise bei meiner ersten Arbeit über das Relativitätsprinzip durch Hrn. Planck und Hrn. Kaufmann bereits geschehen ist.

§ 1. Über die kinetische Energie eines in gleichförmiger Translation begriffenen, äußeren Kräften unterworfenen starren Körpers.

Wir betrachten einen in gleichförmiger Translationsbewegung (Geschwindigkeit v) in Richtung der wachsenden x-Koordinate eines ruhend gedachten Koordinatensystems (x, y, z) befindlichen starren Körper. Wirken äußere Kräfte nicht auf ihn, so ist nach der Relativitätstheorie seine kinetische Energie K_0 gegeben durch die Gleichung[1])

---

1) A. Einstein, Ann. d. Phys. 17. p. 917 ff. 1905.

$$K_0 = \mu V^2 \left\{ \frac{1}{\sqrt{1 - \left(\frac{v}{V}\right)^2}} - 1 \right\},$$

wobei μ seine Masse (im gewöhnlichen Sinne) und V die Lichtgeschwindigkeit im Vakuum bedeutet. Wir wollen nun zeigen, daß nach der Relativitätstheorie dieser Ausdruck nicht mehr gilt, falls äußere Kräfte auf den Körper wirken, welche einander das Gleichgewicht halten. Um den Fall behandeln zu können, müssen wir voraussetzen, daß jene Kräfte elektrodynamische seien. Wir denken uns daher den Körper starr elektrisiert (mit kontinuierlich verteilter Elektrizität), und es wirke auf ihn ein elektromagnetisches Kraftfeld. Die elektrische Dichte denken wir uns allenthalben als sehr gering und das Kraftfeld als intensiv, derart, daß die den Wechselwirkungen zwischen den elektrischen Massen des Körpers entsprechenden Kräfte gegenüber den vom äußeren Kraftfelde auf die elektrischen Ladungen des Körpers ausgeübten Kräfte vernachlässigt werden können.[1]) Die von dem Kraftfeld auf den Körper zwischen den Zeiten t_0 und t_1 übertragene Energie ΔE ist gegeben durch den Ausdruck:

$$\Delta E = \int_{t_0}^{t_1} dt \int v X \frac{\varrho}{4\pi} \, dx \, dy \, dz,$$

wobei das Raumintegral über den Körper zu erstrecken und

$$\varrho = \frac{\partial X}{\partial x} + \frac{\partial Y}{\partial y} + \frac{\partial Z}{\partial z}$$

gesetzt ist. Diesen Ausdruck transformieren wir nach den in der oben zitierten Abhandlung angegebenen Transformationsgleichungen[2]) auf dasjenige Ort-Zeitsystem (ξ, η, ζ, τ), welches einem relativ zu dem Körper ruhenden, zu (x, y, z) parallelachsigen Koordinatensystem entspricht. Man erhält so in einer Bezeichnung, welche der in jener Abhandlung benutzten genau entspricht, nach einfacher Rechnung

$$\Delta E = \iint \beta v X' \frac{\varrho'}{4\pi} d\xi \, d\eta \, d\zeta \, d\tau,$$

---

1) Wir führen diese Annahme ein, um annehmen zu können, daß die wirkenden Kräfte vermöge der Art, wie sie erzeugt sind, keinen beschränkenden Bedingungen unterworfen seien.

2) A. Einstein, Ann. d. Phys. **17**. §§ 3 u. 6. 1905.

wobei β wie dort den Ausdruck

$$\frac{1}{\sqrt{1-\left(\frac{v}{V}\right)^2}}$$

bedeutet. Es ist zu beachten, daß gemäß unseren Voraussetzungen die Kräfte X' keine beliebigen sein dürfen. Sie müssen vielmehr zu jeder Zeit so beschaffen sein, daß der betrachtete Körper keine Beschleunigung erfährt. Hierfür erhält man nach einem Satze der Statik die notwendige (aber nicht hinreichende) Bedingung, daß von einem mit dem Körper bewegten Koordinatensystem aus betrachtet die Summe der X-Komponenten der auf den Körper wirkenden Kräfte stets verschwindet. Man hat also für jedes τ:

$$\int X'\varrho'\,d\xi\,d\eta\,d\zeta = 0\,.$$

Wären also die Grenzen für τ in dem obigen Integralausdruck für $\varDelta E$ von ξ, η, ζ unabhängig, so wäre $\varDelta E = 0$. Dies ist jedoch nicht der Fall. Aus der Transformationsgleichung

$$t = \beta\left(\tau + \frac{v}{V^2}\xi\right)$$

folgt nämlich unmittelbar, daß die Zeitgrenzen im bewegten System sind:

$$\tau = \frac{t_0}{\beta} - \frac{v}{V^2}\xi \quad \text{und} \quad \tau = \frac{t_1}{\beta} - \frac{v}{V^2}\xi\,.$$

Wir denken uns das Integral im Ausdruck für $\varDelta E$ in drei Teile zerlegt.

Der erste Teil umfasse die Zeiten τ zwischen

$$\frac{t_0}{\beta} - \frac{v}{V^2}\xi \quad \text{und} \quad \frac{t_0}{\beta},$$

der zweite Teil zwischen

$$\frac{t_0}{\beta} \quad \text{und} \quad \frac{t_1}{\beta},$$

der dritte zwischen

$$\frac{t_1}{\beta} \quad \text{und} \quad \frac{t_1}{\beta} - \frac{v}{V^2}\xi\,.$$

Der zweite Teil verschwindet, weil er von ξ, η, ζ unabhängige Zeitgrenzen hat. Der erste und dritte Teil hat überhaupt nur dann einen bestimmten Wert, wenn die Annahme gemacht wird, daß in der Nähe der Zeiten $t = t_0$ und $t = t_1$ die auf den Körper wirkenden Kräfte von der Zeit unabhängig seien, derart, daß für alle Punkte des starren Körpers zwischen den Zeiten

$$\tau = \frac{t_0}{\beta} - \frac{v}{V^2}\xi \quad \text{und} \quad \tau = \frac{t_0}{\beta}$$

bez. zwischen

$$\tau = \frac{t_1}{\beta} \quad \text{und} \quad \tau = \frac{t_1}{\beta} - \frac{v}{V^2}\xi$$

die elektrische Kraft X' von der Zeit unabhängig ist. Nennt man X_0' bez. X_1' die in diesen beiden Zeiträumen vorhandenen X', so erhält man:

$$\Delta E = -\int \frac{v^2}{V^2}\beta \frac{\xi X_1' \varrho'}{4\pi} d\xi\, d\eta\, d\zeta + \int \frac{v^2}{V^2}\beta \frac{\xi X_0' \varrho'}{4\pi} d\xi\, d\eta\, d\zeta.$$

Nimmt man ferner an, daß am Anfang ($t = t_0$) keine Kräfte auf den Körper wirken, so verschwindet das zweite dieser Integrale. Mit Rücksicht darauf, daß

$$\frac{X_1' \varrho'}{4\pi} d\xi\, d\eta\, d\zeta$$

die ξ-Komponente K_ξ der auf das Raumelement wirkenden ponderomotorischen Kraft ist, erhält man

$$\Delta E = -\frac{\left(\frac{v}{V}\right)^2}{\sqrt{1 - \left(\frac{v}{V}\right)^2}} \sum (\xi K_\xi),$$

wobei die Summe über alle Massenelemente des Körpers zu erstrecken ist.

Wir haben also folgendes merkwürdige Resultat erhalten. Setzt man einen starren Körper, auf den ursprünglich keine Kräfte wirken, dem Einflusse von Kräften aus, welche dem Körper keine Beschleunigung erteilen, so leisten diese Kräfte — von einem relativ zu dem Körper bewegten Koordinatensystem aus betrachtet — eine Arbeit ΔE auf den Körper, welche lediglich abhängt von der endgültigen Kräfteverteilung und der Translationsgeschwindigkeit. Nach dem Energieprinzip folgt hieraus unmittelbar, daß die kinetische Energie eines Kräften unterworfenen starren Körpers um ΔE größer ist als die kinetische Energie desselben, ebenso rasch bewegten Körpers, falls keine Kräfte auf denselben wirken.

§ 2. Über die Trägheit eines elektrisch geladenen starren Körpers.

Wir betrachten abermals einen starren, starr elektrisierten Körper, welcher eine gleichförmige Translationsbewegung im Sinne der wachsenden x-Koordinaten eines „ruhenden" Koordinatensystems ausführt (Geschwindigkeit v). Ein äußeres elektromagnetisches Kraftfeld sei nicht vorhanden. Wir wollen indessen jetzt das von den elektrischen Massen des Körpers erzeugte elektromagnetische Feld berücksichtigen. Wir berechnen zunächst die elektromagnetische Energie

$$E_e = \frac{1}{8\pi} \int (X^2 + Y^2 + Z^2 + L^2 + M^2 + N^2)\, dx\, dy\, dz.$$

Zu diesem Zweck transformieren wir diesen Ausdruck unter Benutzung der in der mehrfach zitierten Abhandlung enthaltenen Transformationsgleichungen, indem wir unter dem Integral Größen einführen, welche sich auf ein mit dem Körper bewegtes Koordinatensystem beziehen. Wir erhalten so:

$$E_e = \frac{1}{8\pi} \int \frac{1}{\beta} \left[X'^2 + \frac{1 + \left(\frac{v}{V}\right)^2}{1 - \left(\frac{v}{V}\right)^2} (Y'^2 + Z'^2) \right] d\xi\, d\eta\, d\zeta.$$

Es ist zu beachten, daß der Wert dieses Ausdruckes abhängt von der Orientierung des starren Körpers relativ zur Bewegungsrichtung. Wenn sich daher die gesamte kinetische Energie des elektrisierten Körpers ausschließlich zusammensetzte aus der kinetischen Energie K_0, welche dem Körper wegen seiner ponderabeln Masse zukommt, und dem Überschuß der elektromagnetischen Energie des bewegten Körpers über die elektrostatische Energie des Körpers für den Fall der Ruhe, so wären wir damit zu einem Widerspruche gelangt, wie leicht aus folgendem zu ersehen ist.

Wir denken uns, der betrachtete Körper sei relativ zu dem mitbewegten Koordinatensystem in unendlich langsamer Drehung begriffen, ohne daß äußere Einwirkungen während dieser Bewegung auf ihn stattfinden. Es ist klar, daß diese

Bewegung kräftefrei möglich sein muß, da ja nach dem Relativitätsprinzip die Bewegungsgesetze des Körpers relativ zu dem mitbewegten System dieselben sind wie die Bewegungsgesetze in bezug auf ein „ruhendes" System. Wir betrachten nun den gleichförmig bewegten und unendlich langsam sich drehenden Körper vom „ruhenden" System aus. Da die Drehung unendlich langsam sein soll, trägt sie zur kinetischen Energie nichts bei. Der Ausdruck der kinetischen Energie ist daher in dem betrachteten Fall derselbe wie wenn keine Drehung, sondern ausschließlich gleichförmige Paralleltranslation stattfände. Da nun der Körper relativ zur Bewegungsrichtung im Laufe der Bewegung verschiedene (beliebige) Lagen annimmt, und während der ganzen Bewegung das Energieprinzip gelten muß, so ist klar, daß eine Abhängigkeit der kinetischen Energie eines in Translationsbewegung begriffenen elektrisierten Körpers von der Orientierung unmöglich ist.

Dieser Widerspruch wird durch die Resultate des vorigen Paragraphen beseitigt. Die kinetische Energie des betrachteten Körpers kann nämlich nicht berechnet werden wie die eines starren Körpers, auf den keine Kräfte wirken. Wir haben vielmehr gemäß § 1 zu berücksichtigen, daß unser starrer Körper Kräften unterworfen ist, welche ihre Ursache in der Wechselwirkung zwischen den elektrischen Massen haben. Bezeichnen wir also mit K_0 die kinetische Energie für den Fall, daß keine elektrischen Ladungen vorhanden sind, so erhalten wir für die gesamte kinetische Energie K des Körpers den Ausdruck

$$K = K_0 + \Delta E + (E_e - E_s),$$

wobei E_s die elektrostatische Energie des betrachteten Körpers im Zustand der Ruhe bedeutet. In unserem Falle hat man

$$\Delta E = - \frac{v^2}{V^2} \beta \frac{1}{4\pi} \int \xi X' \left(\frac{\partial X'}{\partial \xi} + \frac{\partial Y'}{\partial \eta} + \frac{\partial Z'}{\partial \zeta} \right) d\xi\, d\eta\, d\zeta,$$

woraus man durch partielle Integration mit Berücksichtigung des Umstandes, daß X', Y', Z' von einem Potential ableitbar sind, erhält

$$\Delta E = \frac{v^2}{V^2} \beta \frac{1}{8\pi} \int (X'^2 - Y'^2 - Z'^2)\, d\xi\, d\eta\, d\zeta.$$

Berücksichtigt man die im § 1 angegebenen Ausdrücke für K_0 und β, so erhält man für die kinetische Energie des elektrisierten starren Körpers den Ausdruck

$$K = \left(\mu + \frac{E_s}{V^2}\right) \cdot V^2 \left(\frac{1}{\sqrt{1 - \left(\frac{v}{V}\right)^2}} - 1\right).$$

Dieser Ausdruck ist, wie es sein muß, von der Orientierung des Körpers relativ zur Translationsrichtung unabhängig. Vergleicht man den Ausdruck für K mit dem für die Energie K_0 eines nicht elektrisch geladenen Körpers

$$K_0 = \mu V^2 \left(\frac{1}{\sqrt{1 - \left(\frac{v}{V}\right)^2}} - 1\right),$$

so erkennt man, daß der elektrostatisch geladene Körper eine träge Masse besitzt, welche die des nicht geladenen Körpers um die durch das Quadrat der Lichtgeschwindigkeit dividierte elektrostatische Energie übertrifft. Der Satz von der Trägheit der Energie wird also durch unser Resultat in dem behandelten speziellen Fall bestätigt.

§ 3. Bemerkungen betreffend die Dynamik des starren Körpers.

Nach dem Vorangehenden könnte es scheinen, als ob wir von dem Ziele, eine dem Relativitätsprinzip entsprechende Dynamik der Paralleltranslation des starren Körpers zu schaffen, nicht mehr weit entfernt wären. Man muß sich indessen daran erinnern, daß die im § 1 ausgeführte Untersuchung die Energie des Kräften unterworfenen starren Körpers nur für den Fall lieferte, daß jene Kräfte zeitlich konstant sind. Wenn zur Zeit t_1 die Kräfte X' von der Zeit abhängen, so erweist sich die Arbeit ΔE, also auch die Energie des starren Körpers, nicht nur als abhängig von denjenigen Kräften, welche zu *einer* bestimmten Zeit herrschen.

Um die hier vorliegende Schwierigkeit möglichst drastisch zu beleuchten, denken wir uns folgenden einfachen Spezialfall. Wir betrachten einen starren Stab AB, welcher relativ zu einem Koordinatensystem (ξ, η, ζ) ruhe, wobei die Stabachse in der ζ-Achse ruhe. Zu einer bestimmten Zeit τ_0 mögen auf die Stabenden für ganz kurze Zeit entgegengesetzt gleiche Kräfte P wirken, während der Stab in allen übrigen Zeiten Kräften nicht unterworfen sei. Es ist klar, daß die genannte, zur Zeit τ_0 auf den Stab ausgeübte Wirkung eine Bewegung des Stabes *nicht* erzeugt. Wir betrachten nun genau denselben Vorgang von einem zum vorher benutzten parallelachsigen Koordinatensystem aus, relativ zu welchem

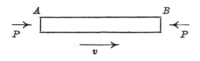

sich unser Stab in der Richtung A—B mit der Geschwindigkeit v bewegt. Von dem letztgenannten Koordinatensystem aus beurteilt, wirken nun aber die Kraftimpulse in A und B nicht gleichzeitig; der Impuls in B ist vielmehr gegen den Impuls in A verspätet um $l\beta(v/V^2)$ Zeiteinheiten, wobei l die (ruhend gemessene) Stablänge bedeutet. Wir sind also zu dem folgenden sonderbar aussehenden Resultat gekommen. Auf den bewegten Stab AB wirkt zuerst in A ein Kraftimpuls und darauf nach einiger Zeit ein entgegengesetzter in B. Diese beiden Kraftimpulse kompensieren einander derart, daß die Bewegung des Stabes durch sie nicht modifiziert wird. Noch merkwürdiger erscheint der Fall, wenn wir nach der Energie des Stabes fragen zu einer Zeit, in welcher der Impuls in A bereits vorbei ist, während der Impuls in B noch nicht zu wirken begonnen hat. Der Impuls in A hat auf den Stab Arbeit übertragen (weil der Stab bewegt ist); um diese Arbeit muß sich also die Energie des Stabes vermehrt haben. Gleichwohl hat sich weder die Geschwindigkeit des Stabes noch sonst eine auf ihn Bezug habende Größe, von der wir die Energiefunktion des Stabes abhängen lassen könnten, geändert. Es scheint also eine Verletzung des Energieprinzipes vorzuliegen.

Die *prinzipielle* Lösung dieser Schwierigkeit liegt auf der Hand. Indem wir implizite annehmen, durch die auf den Stab wirkenden Kräfte und durch die in demselben Augenblick herrschende Stabgeschwindigkeit den Momentanzustand des Stabes vollständig bestimmen zu können, nehmen wir an, daß ein Geschwindigkeitszuwachs des Körpers durch die ihn erzeugende, irgendwo am Körper angreifende Kraft *momentan* erzeugt werde, daß also die Ausbreitung der auf einen Punkt des Körpers ausgeübten Kraft über den ganzen Körper keine

Zeit erfordere. Eine derartige Annahme ist, wie nachher gezeigt wird, mit dem Relativitätsprinzip nicht vereinbar. Wir sind also in unserem Falle offenbar genötigt, bei Einwirkung des Impulses in A eine Zustandsänderung unbekannter Qualität im Körper anzunehmen, welche sich mit endlicher Geschwindigkeit in demselben ausbreitet und in kurzer Zeit eine Beschleunigung des Körpers bewirkt, falls innerhalb dieser Zeit nicht noch andere Kräfte auf den Körper wirken, deren Wirkungen die der erstgenannten kompensieren. Wenn also die Relativitätselektrodynamik richtig ist, sind wir noch weit davon entfernt, eine Dynamik der Paralleltranslation des starren Körpers zu besitzen.

Wir wollen nun zeigen, daß nicht nur die Annahme *momentaner* Ausbreitung irgend einer Wirkung, sondern allgemeiner jede Annahme von der Ausbreitung einer Wirkung mit Überlichtgeschwindigkeit mit der Relativitätstheorie nicht vereinbar ist.

Längs der x-Achse eines Koordinatensystems (x, y, z) erstrecke sich ein Materialstreifen, relativ zu welchem sich eine gewisse Wirkung mit der Geschwindigkeit W fortzupflanzen vermöge, und es möge sowohl in $x = 0$ (Punkt A) als auch in $x = +l$ (Punkt B) sich je ein relativ zum Koordinatensystem (x, y, z) ruhender Beobachter befinden. Der Beobachter in A sende vermittelst der oben genannten Wirkung Zeichen zu dem Beobachter in B durch den Materialstreifen, welch letzterer nicht ruhe, sondern sich mit der Geschwindigkeit $v(< V)$ in der negativen x-Richtung bewege. Das Zeichen wird dann, wie aus § 5 (l. c.) hervorgeht, mit der Geschwindigkeit

$$\frac{W - v}{1 - \frac{W v}{V^2}}$$

von A nach B übertragen. Die Zeit T, welche zwischen Zeichengebung in A und Zeichenempfang in B verstreicht, ist also

$$T = l \frac{1 - \frac{W v}{V^2}}{W - v}.$$

Die Geschwindigkeit v kann jeglichen Wert annehmen, der kleiner ist als V. Wenn also $W > V$ ist, wie wir angenommen haben, so kann man v stets so wählen, daß $T < 0$ ist. Dies Resultat besagt, daß wir einen Übertragungsmechanismus für möglich halten müßten, bei dessen Benutzung die erzielte Wirkung der (etwa von einem Willensakt begleiteten) Ursache *vorangeht*. Wenn dies Resultat auch, meiner Meinung nach, rein logisch genommen keinen Widerspruch enthält, so widerstreitet es doch so unbedingt dem Charakter unserer gesamten Erfahrung, daß durch dasselbe die Unmöglichkeit der Annahme $W > V$ zur Genüge erwiesen ist.

§ 4. Über die Energie eines Systems, welches aus einer Anzahl kräftefrei bewegter Massenpunkte besteht.

Betrachtet man den Ausdruck für die kinetische Energie k eines mit der Geschwindigkeit v bewegten Massenpunktes (μ)

$$k = \mu V^2 \left\{ \frac{1}{\sqrt{1 - \left(\frac{v}{V}\right)^2}} - 1 \right\},$$

so fällt auf, daß dieser Ausdruck die Gestalt einer Differenz besitzt. Es ist nämlich

$$k = \left| \mu V^2 \frac{1}{\sqrt{1 - \left(\frac{v}{V}\right)^2}} \right|_{v=0}^{v=v}.$$

Frägt man nicht speziell nach der kinetischen Energie, sondern nach der Energie ε des bewegten Massenpunktes schlechtweg, so ist $\varepsilon = k + $ konst. Während man nun in der klassischen Mechanik die willkürliche Konstante in dieser Gleichung am bequemsten verschwinden läßt, erhält man in der Relativitätsmechanik den einfachsten Ausdruck für ε, indem man den Nullpunkt der Energie so wählt, daß die Energie ε_0 für den ruhenden Massenpunkt μV^2 gesetzt wird.[1]) Man erhält dann

$$\varepsilon = \mu V^2 \frac{1}{\sqrt{1 - \left(\frac{v}{V}\right)^2}}.$$

An dieser Wahl des Nullpunktes der Energie werden wir im folgenden festhalten.

---

1) Es ist zu beachten, daß die vereinfachende Festsetzung $\mu V^2 = \varepsilon_0$ zugleich der Ausdruck des Prinzipes der Äquivalenz von Masse und Energie ist, und daß im Falle des masselosen elektrisierten Körpers ε_0 nichts anderes ist als seine elektrostatische Energie.

Trägheit der Energie.

Wir führen nun wieder die zwei stets relativ zueinander bewegten Koordinatensysteme (x, y, z) und (ξ, η, ζ) ein. Relativ zu (ξ, η, ζ) sei ein Massenpunkt μ mit der Geschwindigkeit w bewegt in einer Richtung, welche mit der positiven ξ-Achse den Winkel φ bilde. Unter Benutzung der in § 5 (l. c.) hergeleiteten Beziehungen läßt sich leicht die Energie ε des Massenpunktes, bezogen auf das System (x, y, z) bestimmen. Man erhält

$$\varepsilon = \mu V^2 \frac{1 + \frac{v w \cos \varphi}{V^2}}{\sqrt{1 - \frac{v^2}{V^2}} \sqrt{1 - \frac{w^2}{V^2}}}.$$

Sind mehrere Massenpunkte vorhanden, denen verschiedene Massen, Geschwindigkeiten und Bewegungsrichtungen zukommen, so erhalten wir für deren Gesamtenergie E den Ausdruck

$$E = \frac{1}{\sqrt{1 - \left(\frac{v}{V}\right)^2}} \left\{ \sum \mu V^2 \cdot \frac{1}{\sqrt{1 - \left(\frac{w}{V}\right)^2}} \right\}$$
$$+ \frac{v}{\sqrt{1 - \left(\frac{v}{V}\right)^2}} \left\{ \sum \frac{\mu w \cos \varphi}{\sqrt{1 - \left(\frac{w}{V}\right)^2}} \right\}.$$

Bis jetzt haben wir über den Bewegungszustand des Systems (ξ, η, ζ) relativ zu den bewegten Massen nichts festgesetzt. Wir können und wollen hierüber nun folgende, den Bewegungszustand von (ξ, η, ζ) eindeutig bestimmende Bedingungen festsetzen:

$$\sum \frac{\mu w_\xi}{\sqrt{1 - \left(\frac{w}{V}\right)^2}} = 0, \quad \sum \frac{\mu w_\eta}{\sqrt{1 - \left(\frac{w}{V}\right)^2}} = 0,$$
$$\sum \frac{\mu w_\zeta}{\sqrt{1 - \left(\frac{w}{V}\right)^2}} = 0,$$

wobei w_ξ, w_η, w_ζ die Komponenten von w bezeichnen. Dieser Festsetzung entspricht in der klassischen Mechanik die Bedingung, daß das Bewegungsmoment des Massensystems in bezug auf (ξ, η, ζ) verschwinde. Dann erhalten wir

$$E = \left(\sum \mu V^2 \cdot \frac{1}{\sqrt{1 - \left(\frac{w}{V}\right)^2}} \right) \cdot \frac{1}{\sqrt{1 - \left(\frac{v}{V}\right)^2}},$$

oder, indem man die Energie E_0 des Systems relativ zum System (ξ, η, ζ) einführt:

$$E = \frac{E_0}{V^2} \cdot V^2 \frac{1}{\sqrt{1 - \left(\frac{v}{V}\right)^2}}.$$

Vergleicht man diesen Ausdruck mit dem für die Energie eines mit der Geschwindigkeit v bewegten Massenpunktes

$$\varepsilon = \mu V^2 \frac{1}{\sqrt{1 - \left(\frac{v}{V}\right)^2}},$$

so erhält man folgendes Resultat: In bezug auf die Abhängigkeit der Energie vom Bewegungszustand des Koordinatensystems, auf welches die Vorgänge bezogen werden, läßt sich ein System gleichförmig bewegter Massenpunkte ersetzen durch einen einzigen Massenpunkt von der Masse $\mu = E_0/V^2$.

Ein System bewegter Massenpunkte besitzt also — als Ganzes genommen — desto mehr Trägheit, je rascher die Massenpunkte relativ zueinander bewegt sind. Die Abhängigkeit ist wieder gegeben durch das in der Einleitung angegebene Gesetz.

Bern, Mai 1907.

(Eingegangen 14. Mai 1907.)

ANNALEN DER PHYSIK.

BEGRÜNDET UND FORTGEFÜHRT DURCH
F. A. C. GREN, L. W. GILBERT, J. C. POGGENDORFF, G. u. E. WIEDEMANN, P. DRUDE.

VIERTE FOLGE.

BAND 26.
DER GANZEN REIHE 331. BAND.

KURATORIUM:
F. KOHLRAUSCH, M. PLANCK, G. QUINCKE,
W. C. RÖNTGEN, E. WARBURG.

UNTER MITWIRKUNG
DER DEUTSCHEN PHYSIKALISCHEN GESELLSCHAFT
UND INSBESONDERE VON
M. PLANCK

HERAUSGEGEBEN VON
W. WIEN UND M. PLANCK.

MIT ACHT FIGURENTAFELN.

LEIPZIG, 1908.
VERLAG VON JOHANN AMBROSIUS BARTH.

5. Über die elektromagnetischen Grundgleichungen für bewegte Körper; von A. Einstein und J. Laub.

In einer kürzlich veröffentlichten Abhandlung[1]) hat Hr. Minkowski die Grundgleichungen für die elektromagnetischen Vorgänge in bewegten Körpern angegeben. In Anbetracht des Umstandes, daß diese Arbeit in mathematischer Beziehung an den Leser ziemlich große Anforderungen stellt, halten wir es nicht für überflüssig, jene wichtigen Gleichungen im folgenden auf elementarem Wege, der übrigens mit dem Minkowskischen im wesentlichen übereinstimmt, abzuleiten.

§ 1. Ableitung der Grundgleichungen für bewegte Körper.

Der einzuschlagende Weg ist folgender: Wir führen zwei Koordinatensysteme K und K' ein, welche beide beschleunigungsfrei, jedoch relativ zueinander bewegt sind. Ist im Raume Materie vorhanden, die relativ zu K' ruht, gelten in bezug auf K' die Gesetze der Elektrodynamik ruhender Körper, welche durch die Maxwell-Hertzschen Gleichungen dargestellt sind. Transformieren wir diese Gleichungen auf das System K, so erhalten wir unmittelbar die elektrodynamischen Gleichungen bewegter Körper für den Fall, daß die Geschwindigkeit der Materie räumlich und zeitlich konstant ist. Die so erhaltenen Gleichungen gelten offenbar mindestens in erster Annäherung auch dann, wenn die Geschwindigkeitsverteilung der Materie eine beliebige ist. Diese Annahme rechtfertigt sich zum Teil auch dadurch, daß das auf diese Weise erhaltene Resultat streng gilt in dem Falle, daß eine Anzahl von mit verschiedenen Geschwindigkeiten gleichförmig bewegten Körpern vorhanden ist, welche voneinander durch Vakuumzwischenräume getrennt sind.

---

1) H. Minkowski, Göttinger Nachr. 1908.

Elektromagnetische Grundgleichungen für bewegte Körper. 533

Wir wollen mit Bezug auf das System K' den Vektor der elektrischen Kraft $\mathfrak{E}'$, der magnetischen Kraft $\mathfrak{H}'$, der dielektrischen Verschiebung $\mathfrak{D}'$, der magnetischen Induktion $\mathfrak{B}'$, den des elektrischen Stromes $\mathfrak{F}'$ nennen; ferner bezeichne ϱ' die elektrische Dichte. Es mögen für das Bezugssystem K' die Maxwell-Hertzschen Gleichungen gelten:

(1) $$\mathrm{curl}' \mathfrak{H}' = \frac{1}{c}\left(\frac{\partial \mathfrak{D}'}{\partial t'} + \mathfrak{F}'\right),$$

(2) $$\mathrm{curl}' \mathfrak{E}' = -\frac{1}{c}\frac{\partial \mathfrak{B}'}{\partial t'},$$

(3) $$\mathrm{div}' \mathfrak{D}' = \varrho',$$

(4) $$\mathrm{div}' \mathfrak{B}' = 0.$$

Wir betrachten ein zweites rechtwinkliges Bezugssystem K, dessen Achsen dauernd parallel sind denen von K'. Der Anfangspunkt von K' soll sich mit der konstanten Geschwindigkeit v in der positiven Richtung der x-Achse von K bewegen. Dann gelten bekanntlich bei passend gewähltem Anfangspunkt der Zeit nach der Relativitätstheorie für jedes Punktereignis folgende Transformationsgleichungen[1]):

(5) $$\begin{cases} x' = \beta(x - vt), \\ y' = y, \\ z' = z, \\ t' = \beta\left(t - \frac{v}{c^2}x\right), \end{cases} \left(\beta = \frac{1}{\sqrt{1-\frac{v^2}{c^2}}}\right),$$

wobei x, y, z, t die Raum- und Zeitkoordinaten im System K bedeuten. Führt man die Transformationen aus, so erhält man die Gleichungen:

(1 a) $$\mathrm{curl}\, \mathfrak{H} = \frac{1}{c}\left(\frac{\partial \mathfrak{D}}{\partial t} + \mathfrak{F}\right),$$

(2 a) $$\mathrm{curl}\, \mathfrak{E} = -\frac{1}{c}\frac{\partial \mathfrak{B}}{\partial t},$$

(3 a) $$\mathrm{div}\, \mathfrak{D} = \varrho,$$

(4 a) $$\mathrm{div}\, \mathfrak{B} = 0,$$

---

1) A. Einstein, Ann. d. Phys. **17**. p. 902. 1905.

wobei gesetzt ist:

(6)
$$\begin{cases} \mathfrak{E}_x = \mathfrak{E}_x', \\ \mathfrak{E}_y = \beta\left(\mathfrak{E}_y' + \frac{v}{c}\mathfrak{B}_z'\right), \\ \mathfrak{E}_z = \beta\left(\mathfrak{E}_z' - \frac{v}{c}\mathfrak{B}_y'\right), \\ \mathfrak{D}_x = \mathfrak{D}_x', \\ \mathfrak{D}_y = \beta\left(\mathfrak{D}_y' + \frac{v}{c}\mathfrak{H}_z'\right), \\ \mathfrak{D}_z = \beta\left(\mathfrak{D}_z' - \frac{v}{c}\mathfrak{H}_y'\right); \end{cases}$$

(7)
$$\begin{cases} \mathfrak{H}_x = \mathfrak{H}_x', \\ \mathfrak{H}_y = \beta\left(\mathfrak{H}_y' - \frac{v}{c}\mathfrak{D}_z'\right), \\ \mathfrak{H}_z = \beta\left(\mathfrak{H}_z' + \frac{v}{c}\mathfrak{D}_y'\right), \\ \mathfrak{B}_x = \mathfrak{B}_x', \\ \mathfrak{B}_y = \beta\left(\mathfrak{B}_y' - \frac{v}{c}\mathfrak{E}_z'\right), \\ \mathfrak{B}_z = \beta\left(\mathfrak{B}_z' + \frac{v}{c}\mathfrak{E}_y'\right) \end{cases}$$

und

(8) $$\varrho = \beta\left(\varrho' + \frac{v}{c}\mathfrak{z}_x'\right),$$

(9)
$$\begin{cases} \mathfrak{z}_x = \beta\left(\mathfrak{z}_x' + \frac{v}{c}\varrho'\right), \\ \mathfrak{z}_y = \mathfrak{z}_y', \\ \mathfrak{z}_z = \mathfrak{z}_z'. \end{cases}$$

Will man die Ausdrücke für die gestrichenen Größen als Funktion der ungestrichenen haben, so vertauscht man die gestrichenen und ungestrichenen Größen und ersetzt v durch $-v$.

Die Gleichungen (1a) bis (4a), welche die elektromagnetischen Vorgänge relativ zum System K beschreiben, haben dieselbe Gestalt, wie die Gleichungen (1) bis (4). *Wir wollen daher die Größen*

$$\mathfrak{E}, \mathfrak{D}, \mathfrak{H}, \mathfrak{B}, \varrho, \mathfrak{z}$$

analog benennen, wie die entsprechenden Größen relativ zum System K'. Es sind also $\mathfrak{E}, \mathfrak{D}, \mathfrak{H}, \mathfrak{B}, \varrho, \mathfrak{z}$ *die elektrische Kraft, die dielektrische Verschiebung, die magnetische Kraft, die magne-*

Elektromagnetische Grundgleichungen für bewegte Körper. 535

tische Induktion, die elektrische Dichte, der elektrische Strom in bezug auf K.

Die Transformationsgleichungen (6) und (7) reduzieren sich für das Vakuum auf die früher gefundenen[1]) Gleichungen für elektrische und magnetische Kräfte.

Es ist klar, daß man durch wiederholte Anwendung solcher Transformationen, wie die soeben durchgeführte, stets auf Gleichungen von derselben Gestalt wie die ursprünglichen (1) bis (4) kommen muß, und daß für solche Transformationen die Gleichungen (6) bis (9) maßgebend sind. Denn es wurde bei der ausgeführten Transformation in formaler Beziehung nicht davon Gebrauch gemacht, daß die Materie relativ zu dem ursprünglichen System K' ruhte.

Die Gültigkeit der transformierten Gleichungen (1a) bis (4a) nehmen wir an auch für den Fall, daß die Geschwindigkeit der Materie räumlich und zeitlich variabel ist, was in erster Annäherung richtig sein wird.

Es ist bemerkenswert, daß die Grenzbedingungen für die Vektoren $\mathfrak{E}$, $\mathfrak{D}$, $\mathfrak{H}$, $\mathfrak{B}$ an der Grenze zweier Medien dieselben sind, wie für ruhende Körper. Es folgt dies direkt aus den Gleichungen (1a) bis (4a).

Die Gleichungen (1a) bis (4a) gelten genau wie die Gleichungen (1) bis (4) ganz allgemein für inhomogene und anisotrope Körper. Dieselben bestimmen die elektromagnetischen Vorgänge noch nicht vollständig. Es müssen vielmehr noch Beziehungen gegeben sein, welche die Vektoren $\mathfrak{D}$, $\mathfrak{B}$ und $\mathfrak{J}$ als Funktion von $\mathfrak{E}$ und $\mathfrak{H}$ ausdrücken. Solche Gleichungen wollen wir nun für den Fall angeben, daß die *Materie isotrop* ist. Betrachten wir zunächst wieder den Fall, daß alle Materie relativ zu K' ruht, so gelten in bezug auf K' die Gleichungen:

(10) $$\mathfrak{D}' = \varepsilon\,\mathfrak{E}',$$

(11) $$\mathfrak{B}' = \mu\,\mathfrak{H}',$$

(12) $$\mathfrak{J}' = \sigma\,\mathfrak{E}',$$

wobei ε = Dielektrizitätskonstante, μ = Permeabilität, σ = elektrische Leitfähigkeit als bekannte Funktionen von x', y', z', t' anzusehen sind. Durch die Transformation von (10) bis (12)

---

[1] A. Einstein, l. c. p. 909.

auf K mittels der Umkehrung unserer Transformationsgleichungen (6) bis (9) erhält man die für das System K geltenden Beziehungen:

$$(10\,\mathrm{a})\quad\begin{cases}\mathfrak{D}_x = \varepsilon\,\mathfrak{E}_x,\\[2pt]\mathfrak{D}_y - \dfrac{v}{c}\mathfrak{H}_z = \varepsilon\left(\mathfrak{E}_y - \dfrac{v}{c}\mathfrak{B}_z\right),\\[2pt]\mathfrak{D}_z + \dfrac{v}{c}\mathfrak{H}_y = \varepsilon\left(\mathfrak{E}_z + \dfrac{v}{c}\mathfrak{B}_y\right),\end{cases}$$

$$(11\,\mathrm{a})\quad\begin{cases}\mathfrak{B}_x = \mu\,\mathfrak{H}_x,\\[2pt]\mathfrak{B}_y + \dfrac{v}{c}\mathfrak{E}_z = \mu\left(\mathfrak{H}_y + \dfrac{v}{c}\mathfrak{D}_z\right),\\[2pt]\mathfrak{B}_z - \dfrac{v}{c}\mathfrak{E}_y = \mu\left(\mathfrak{H}_z - \dfrac{v}{c}\mathfrak{D}_y\right),\end{cases}$$

$$(12\,\mathrm{a})\quad\begin{cases}\beta\left(\mathfrak{z}_x - \dfrac{v}{c}\varrho\right) = \sigma\,\mathfrak{E}_x,\\[2pt]\mathfrak{z}_y = \sigma\beta\left(\mathfrak{E}_y - \dfrac{v}{c}\mathfrak{B}_z\right),\\[2pt]\mathfrak{z}_z = \sigma\beta\left(\mathfrak{E}_z + \dfrac{v}{c}\mathfrak{B}_y\right),\end{cases}$$

Ist die Geschwindigkeit der Materie nicht der X-Achse parallel, sondern ist diese Geschwindigkeit durch den Vektor $\mathfrak{v}$ bestimmt, so erhält man die mit den Gleichungen (10a) bis (12a) gleichartigen vektoriellen Beziehungen:

$$(13)\quad\begin{cases}\mathfrak{D} + \dfrac{1}{c}[\mathfrak{v}\,\mathfrak{H}] = \varepsilon\left\{\mathfrak{E} + \dfrac{1}{c}[\mathfrak{v}\,\mathfrak{B}]\right\},\\[2pt]\mathfrak{B} - \dfrac{1}{c}[\mathfrak{v}\,\mathfrak{E}] = \mu\left\{\mathfrak{H} - \dfrac{1}{c}[\mathfrak{v}\,\mathfrak{D}]\right\},\\[2pt]\beta\left(\mathfrak{z}_\mathfrak{v} - \dfrac{|\mathfrak{v}|}{c}\varrho\right) = \sigma\left\{\mathfrak{E} + \dfrac{1}{c}[\mathfrak{v}\,\mathfrak{B}]\right\}_\mathfrak{v},\\[2pt]\mathfrak{z}_{\bar{\mathfrak{v}}} = \sigma\beta\left\{\mathfrak{E} + \dfrac{1}{c}[\mathfrak{v}\,\mathfrak{B}]\right\}_{\bar{\mathfrak{v}}},\end{cases}$$

wobei der Index $\mathfrak{v}$ bedeutet, daß die Komponente nach der Richtung von $\mathfrak{v}$, der Index $\bar{\mathfrak{v}}$, daß die Komponenten nach den auf $\mathfrak{v}$ senkrechten Richtungen $\bar{\mathfrak{v}}$ zu nehmen ist.

§ 2. Über das elektromagnetische Verhalten bewegter Dielektrika. Versuch von Wilson.

Im folgenden Abschnitt wollen wir noch an einem einfachen Spezialfall zeigen, wie sich bewegte Dielektrika nach

Elektromagnetische Grundgleichungen für bewegte Körper.

der Relativitätstheorie verhalten, und worin sich die Resultate von den durch die Lorentzsche Theorie gelieferten, unterscheiden.

Es sei S ein im Querschnitt angedeuteter, prismatischer Streifen (vgl. Figur) aus einem homogenen, isotropen Nichtleiter, der sich senkrecht zur Papierebene in beiderlei Sinn ins Unendliche erstreckt und sich vom Beschauer nach der Papierebene zu mit der konstanten Geschwindigkeit v zwischen den beiden Kondensatorplatten A_1 und A_2 hindurchbewegt. Die Ausdehnung des Streifens S senkrecht zu den Platten A sei unendlich klein relativ zu dessen Ausdehnung parallel den Platten und zu beiden Ausdehnungen der Platten A; der Zwischenraum zwischen S und den Platten A (im folgenden kurz Zwischenraum genannt)

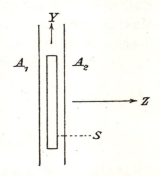

sei außerdem gegenüber der Dicke von S zu vernachlässigen. Das betrachtete Körpersystem beziehen wir auf ein relativ zu den Platten A ruhendes Koordinatensystem, dessen positive X-Richtung in die Bewegungsrichtung falle, und dessen Y- und Z-Achsen parallel bzw. senkrecht zu den Platten A sind. Wir wollen das elektromagnetische Verhalten des zwischen den Platten A sich befindenden Streifenstückes untersuchen, falls der elektromagnetische Zustand stationär ist.

Wir denken uns eine geschlossene Fläche, welche gerade den wirksamen Teil der Kondensatorplatten nebst dem des dazwischen liegenden Streifenstückes einschließt. Da sich innerhalb dieser Fläche weder bewegte wahre Ladungen, noch elektrische Leitungsströme befinden, gelten die Gleichungen (vgl. Gleichungen (1a) bis (4a)):

$$\operatorname{curl} \mathfrak{H} = 0,$$
$$\operatorname{curl} \mathfrak{E} = 0.$$

Innerhalb dieses Raumes sind also sowohl die elektrische, wie auch die magnetische Kraft von einem Potential ableitbar. Wir können daher sofort die Verteilung der Vektoren $\mathfrak{E}$ und $\mathfrak{H}$, falls die Verteilung der freien elektrischen bzw. magnetischen Dichte bekannt ist. Wir beschränken uns auf die Betrachtung des Falles, daß die magnetische Kraft $\mathfrak{H}$ parallel der Y-Achse ist, die elektrische $\mathfrak{E}$ parallel der Z-Achse. Dazu, sowie zu der Voraussetzung, daß die in Betracht kommenden Felder innerhalb des Streifens, sowie innerhalb des Zwischenraumes homogen sind, berechtigen uns die oben erwähnten Größenordnungsbedingungen für die Abmessungen des betrachteten Systems. Ebenso schließen wir unmittelbar, daß die an den Enden des Streifenquerschnittes sich befindenden magnetischen Massen nur einen verschwindend kleinen Beitrag zum magnetischen Feld liefern.[1]) Die Gleichungen (13) geben dann für das Innere des Streifens folgende Beziehungen:

$$\mathfrak{D}_z + \frac{v}{c}\mathfrak{H}_y = \varepsilon\left(\mathfrak{E}_z + \frac{v}{c}\mathfrak{B}_y\right),$$

$$\mathfrak{B}_y + \frac{v}{c}\mathfrak{E}_z = \mu\left(\mathfrak{H}_y + \frac{v}{c}\mathfrak{D}_z\right).$$

Diese Gleichungen lassen sich auch in folgender Form schreiben:

$$(1) \quad \begin{cases} \left(1 - \varepsilon\mu\frac{v^2}{c^2}\right)\mathfrak{B}_y = \frac{v}{c}(\varepsilon\mu - 1)\mathfrak{E}_z + \mu\left(1 - \frac{v^2}{c^2}\right)\mathfrak{H}_y, \\ \left(1 - \varepsilon\mu\frac{v^2}{c^2}\right)\mathfrak{D}_z = \varepsilon\left(1 - \frac{v^2}{c^2}\right)\mathfrak{E}_z + \frac{v}{c}(\varepsilon\mu - 1)\mathfrak{H}_y. \end{cases}$$

Zur Deutung von (1) bemerken wir folgendes: An der Oberfläche des Streifens erfährt die dielektrische Verschiebung $\mathfrak{D}_z$ keinen Sprung, also ist $\mathfrak{D}_z$ die Ladung der Kondensatorplatten (genauer der Platte A_1) pro Flächeneinheit. Ferner ist $\mathfrak{E}_z \times \delta$ gleich der Potentialdifferenz zwischen den Kondensatorplatten A_1 und A_2, falls δ den Abstand der Platten bezeichnet, denn denkt man sich den Streifen durch einen parallel der XZ-Ebene verlaufenden unendlich engen Spalt getrennt, so ist $\mathfrak{E}$, nach den für diesen Vektor geltenden Grenzbedingungen, gleich der elektrischen Kraft in dem Spalt.

Wir betrachten nun zunächst den Fall, daß ein von außen erregtes Magnetfeld nicht vorhanden ist, d. h. nach dem obigen, daß in dem betrachteten Raume die magnetische Feldstärke $\mathfrak{H}_y$

---

1) Es erhellt dies auch daraus, daß wir ohne wesentliche Änderung der Verhältnisse den Kondensatorplatten und dem Streifen Kreiszylinderform geben könnten, in welchem Falle freie magnetische Massen aus Symmetriegründen überhaupt nicht auftreten könnten.

Elektromagnetische Grundgleichungen für bewegte Körper.

überhaupt verschwindet. Dann haben die Gleichungen (1) folgende Gestalt:

$$\left(1 - \varepsilon\mu\frac{v^2}{c^2}\right)\mathfrak{B}_y = \frac{v}{c}(\varepsilon\mu - 1)\mathfrak{E}_z,$$

$$\left(1 - \varepsilon\mu\frac{v^2}{c^2}\right)\mathfrak{D}_z = \varepsilon\left(1 - \frac{v^2}{c^2}\right)\mathfrak{E}_z.$$

Da $v < c$ sein muß, so sind, falls $\varepsilon\mu - 1 > 0$ ist, die Koeffizienten von $\mathfrak{E}_z$ in den beiden letzten Gleichungen positiv. Die Koeffizienten von $\mathfrak{B}_y$ und $\mathfrak{D}_z$ sind dagegen größer, gleich bzw. kleiner als Null, je nachdem die Streifengeschwindigkeit kleiner, gleich oder größer als $c/\sqrt{\varepsilon\mu}$, d. h. als die Geschwindigkeit elektromagnetischer Wellen in dem Streifenmedium, ist. Hat also $\mathfrak{E}_z$ einen bestimmten Wert, d. h. legt man an die Kondensatorplatten eine bestimmte Spannung an und variiert man die Streifengeschwindigkeit von kleineren zu größeren Werten, so wächst zunächst sowohl die dem Vektor $\mathfrak{D}$ proportionale Ladung der Kondensatorplatten, wie die magnetische Induktion $\mathfrak{B}$ im Streifen. Erreicht v den Wert $c/\sqrt{\varepsilon\mu}$, so wird sowohl die Ladung des Kondensators, wie auch die magnetische Induktion unendlich groß. Es würde also in diesem Falle eine Zerstörung des Streifens durch beliebig kleine angelegte Potentialdifferenzen stattfinden. Für alle $v > c/\sqrt{\varepsilon\mu}$ resultiert ein negativer Wert für $\mathfrak{D}$ und $\mathfrak{B}$. In dem letzten Falle würde also eine an die Kondensatorplatten gelegte Spannung eine Ladung des Kondensators in dem der Spannungsdifferenz entgegengesetzten Sinne bewirken.

Wir betrachten jetzt noch den Fall, daß ein von außen erregtes magnetisches Feld $\mathfrak{H}_y$ vorhanden ist. Dann hat man die Gleichung:

$$\left(1 - \varepsilon\mu\frac{v^2}{c^2}\right)\mathfrak{D}_z = \varepsilon\left(1 - \frac{v^2}{c^2}\right)\mathfrak{E}_z + \frac{v}{c}(\varepsilon\mu - 1)\mathfrak{H}_y,$$

welche bei gegebenem $\mathfrak{H}_y$ eine Beziehung zwischen $\mathfrak{E}_z$ und $\mathfrak{D}_z$ gibt. Beschränkt man sich nur auf Größen erster Ordnung in v/c, so hat man:

(2) $$\mathfrak{D}_z = \varepsilon\mathfrak{E}_z + \frac{v}{c}(\varepsilon\mu - 1)\mathfrak{H}_y,$$

während die Lorentzsche Theorie auf den Ausdruck:

(3) $$\mathfrak{D}_z = \varepsilon\mathfrak{E}_z + \frac{v}{c}(\varepsilon - 1)\mu\mathfrak{H}_y$$

führt.

Die letzte Gleichung wurde bekanntlich von H. A. Wilson (Wilsoneffekt) experimentell geprüft. Man sieht, daß sich (2) und (3) in Gliedern erster Ordnung unterscheiden. Hätte man einen dielektrischen Körper von beträchtlicher Permeabilität, so könnte man eine experimentelle Entscheidung zwischen den Gleichungen (2) und (3) treffen.

Verbindet man die Platten A_1 und A_2 durch einen Leiter, so tritt auf den Kondensatorplatten eine Ladung von der Größe $\mathfrak{D}_z$ pro Flächeneinheit auf; man erhält sie aus der Gleichung (2), indem man berücksichtigt, daß bei verbundenen Kondensatorplatten $\mathfrak{E}_z = 0$ ist. Es ergibt sich:

$$\mathfrak{D}_z = \frac{v}{c}(\varepsilon \mu - 1)\mathfrak{H}_y.$$

Verbindet man die Kondensatorplatten A_1 und A_2 mit einem Elektrometer von unendlich kleiner Kapazität, so ist $\mathfrak{D}_z = 0$, und man bekommt für die Spannung ($\mathfrak{E}_z \cdot \delta$) die Gleichung:

$$0 = \varepsilon \mathfrak{E}_z + \frac{v}{c}(\varepsilon \mu - 1)\mathfrak{H}_y.$$

Bern, 29. April 1908.

(Eingegangen 2. Mai 1908.)

6. Über die im elektromagnetischen Felde auf ruhende Körper ausgeübten ponderomotorischen Kräfte; *von A. Einstein und J. Laub.*

In einer kürzlich erschienenen Abhandlung[1]) hat Hr. Minkowski einen Ausdruck für die auf beliebig bewegte Körper wirkenden ponderomotorischen Kräfte elektromagnetischen Ursprunges angegeben. Spezialisiert man die Minkowskischen Ausdrücke auf ruhende, isotrope und homogene Körper, so erhält man für die X-Komponente der auf die Volumeneinheit wirkenden Kraft:

$$(1) \qquad K_x = \varrho\, \mathfrak{E}_x + \mathfrak{s}_y \mathfrak{B}_z - \mathfrak{s}_z \mathfrak{B}_y,$$

wobei ϱ die elektrische Dichte, $\mathfrak{s}$ den elektrischen Leitungsstrom, $\mathfrak{E}$ die elektrische Feldstärke, $\mathfrak{B}$ die magnetische Induktion bedeuten. Dieser Ausdruck scheint uns aus folgenden Gründen mit dem elektronentheoretischen Bild nicht in Einklang zu stehen: Während nämlich ein von einem elektrischen Strom (Leitungsstrom) durchflossener Körper im Magnetfeld eine Kraft erleidet, wäre dies nach Gleichung (1) nicht der Fall, wenn der im Magnetfeld befindliche Körper statt von einem Leitungsstrom von einem Polarisationsstrom ($\partial \mathfrak{D}/\partial t$) durchsetzt wird. Nach Minkowski besteht also hier ein prinzipieller Unterschied zwischen einem Verschiebungsstrom und einem Leitungsstrom derart, daß ein Leiter nicht betrachtet werden kann als ein Dielektrikum von unendlich großer Dielektrizitätskonstante.

Angesichts dieser Sachlage schien es uns von Interesse zu sein, die ponderomotorischen Kräfte für beliebige magnetisierbare Körper auf elektronentheoretischem Wege abzuleiten. Wir geben im folgenden eine solche Ableitung, wobei wir uns aber auf ruhende Körper beschränken.

---

1) H. Minkowski, Gött. Nachr. 1908. p. 45.

§ 1. Kräfte, welche nicht von Geschwindigkeiten der Elementarteilchen abhängen.

Wir wollen uns bei der Ableitung konsequent auf den Standpunkt der Elektronentheorie stellen[1]); wir setzen also:

$$(2) \qquad \mathfrak{D} = \mathfrak{E} + \mathfrak{P},$$

$$(3) \qquad \mathfrak{B} = \mathfrak{H} + \mathfrak{Q},$$

wobei $\mathfrak{P}$ den elektrischen, $\mathfrak{Q}$ den magnetischen Polarisationsvektor bedeutet. Die elektrische bzw. die magnetische Polarisation denken wir uns bestehend in räumlichen Verschiebungen von an Gleichgewichtslagen gebundenen, elektrischen bzw. magnetischen Massenteilchen von Dipolen. Außerdem nehmen wir noch das Vorhandensein von nicht an Dipole gebundenen, beweglichen elektrischen Teilchen (Leitungselektronen) an. In dem Raume zwischen den genannten Teilchen mögen die Maxwellschen Gleichungen für den leeren Raum gelten, und es seien, wie bei Lorentz, *die Wechselwirkungen zwischen Materie und elektromagnetischem Felde ausschließlich durch diese Teilchen bedingt.* Dementsprechend nehmen wir an, daß die vom elektromagnetischen Felde auf das Volumenelement der Materie ausgeübten Kräfte gleich sind der Resultierenden der ponderomotorischen Kräfte, welche von diesem Felde auf alle in dem betreffenden Volumenelement befindlichen elektrischen und magnetischen Elementarteilchen ausgeübt werden. Unter Volumenelement der Materie verstehen wir stets einen so großen Raum, daß er eine sehr große Zahl von elektrischen und magnetischen Teilchen enthält. Die Grenzen eines betrachteten Volumenelementes muß man sich ferner stets so genommen denken, daß die Grenzfläche keine elektrische bzw. magnetische Dipole schneidet.

Wir berechnen zunächst diejenige auf einen elektrischen Dipol wirkende Kraft, welche daher herrührt, daß die Feldstärke $\mathfrak{E}$ an den Orten, an welchen sich die Elementarmassen des Dipols befinden, nicht genau dieselbe ist. Bezeichnet man

---

[1]) Der einfacheren Darstellung halber halten wir aber an der dualen Behandlung der elektrischen und magnetischen Erscheinungen fest.

mit $\mathfrak{p}$ den Vektor des Dipolmomentes, so erhält man für die X-Komponente der gesuchten Kraft den Ausdruck:

$$\mathfrak{f}_x = \mathfrak{p}_x \frac{\partial \mathfrak{E}_x}{\partial x} + \mathfrak{p}_y \frac{\partial \mathfrak{E}_x}{\partial y} + \mathfrak{p}_z \frac{\partial \mathfrak{E}_x}{\partial z}.$$

Denkt man sich den letzten Ausdruck für alle Dipole in der Volumeneinheit gebildet und summiert, so erhält man unter Berücksichtigung der Beziehung:

$$\sum \mathfrak{p} = \mathfrak{P}$$

die Gleichung:

(4) $$\mathfrak{F}_{1x} = \left\{ \mathfrak{P}_x \frac{\partial \mathfrak{E}_x}{\partial x} + \mathfrak{P}_y \frac{\partial \mathfrak{E}_x}{\partial y} + \mathfrak{P}_z \frac{\partial \mathfrak{E}_x}{\partial z} \right\}.$$

Wenn die algebraische Summe der positiven und negativen Leitungselektronen nicht verschwindet, dann kommt zum Ausdruck (4) noch ein Term hinzu, den wir nun berechnen wollen. Die X-Komponente der auf ein Leitungselektron von der elektrischen Masse e wirkenden ponderomotorischen Kraft ist $e\mathfrak{E}_x$. Summiert man über alle Leitungselektronen der Volumeneinheit, so erhält man:

(5) $$\mathfrak{F}_{2x} = \mathfrak{E}_x \sum e.$$

Denkt man sich die betrachtete in der Volumeneinheit befindliche Materie von einer Fläche umschlossen, welche keine Dipole schneidet, so erhält man nach dem Gaussschen Satz und nach der Definition des Verschiebungsvektors $\mathfrak{D}$:

$$\sum e = \operatorname{div} \mathfrak{D},$$

so daß

(5a) $$\mathfrak{F}_{2x} = \mathfrak{E}_x \operatorname{div} \mathfrak{D}$$

wird. Die X-Komponente der von der elektrischen Feldstärke auf die Volumeneinheit der Materie ausgeübten Kraft ist daher gleich:

(6) $$\mathfrak{F}_{e_x} = \mathfrak{F}_{1x} + \mathfrak{F}_{2x} = \mathfrak{P}_x \frac{\partial \mathfrak{E}_x}{\partial x} + \mathfrak{P}_y \frac{\partial \mathfrak{E}_x}{\partial y} + \mathfrak{P}_z \frac{\partial \mathfrak{E}_x}{\partial z} + \mathfrak{E}_x \operatorname{div} \mathfrak{D}.$$

Analog erhalten wir unter Berücksichtigung der Beziehung

$$\operatorname{div} \mathfrak{B} = 0$$

für die X-Komponente der von der magnetischen Feldstärke gelieferten Kraft:

(7) $$\mathfrak{F}_{mx} = \left\{ \mathfrak{D}_x \frac{\partial \mathfrak{H}_x}{\partial x} + \mathfrak{D}_y \frac{\partial \mathfrak{H}_x}{\partial y} + \mathfrak{D}_z \frac{\partial \mathfrak{H}_x}{\partial z} \right\}.$$

Es ist zu bemerken, daß für die Herleitung der Ausdrücke (6) und (7) keinerlei Voraussetzung gemacht werden muß über die Beziehungen, welche die Feldstärken $\mathfrak{E}$ und $\mathfrak{H}$ mit den Polarisationsvektoren $\mathfrak{P}$ und $\mathfrak{Q}$ verbinden.

Hat man es mit anisotropen Körpern zu tun, so liefern die elektrische bzw. die magnetische Feldstärke nicht nur eine Kraft, sondern auch Kräftepaare, welche sich auf die Materie übertragen. Das gesuchte Drehmoment ergibt sich leicht für die einzelnen Dipole und Summation über alle elektrischen und magnetischen Dipole in der Volumeneinheit. Man erhält:

$$(8) \qquad \mathfrak{L} = \{[\mathfrak{P}\,\mathfrak{E}] + [\mathfrak{Q}\,\mathfrak{H}]\}.$$

Die Formel (6) liefert diejenigen ponderomotorischen Kräfte, welche bei elektrostatischen Problemen eine Rolle spielen. Wir wollen diese Gleichung für den Fall, daß es sich um isotrope Körper handelt, so umformen, daß sie einen Vergleich gestattet mit demjenigen Ausdrucke für die ponderomotorischen Kräfte, wie er in der Elektrostatik angegeben wird. Setzen wir

$$\mathfrak{P} = (\varepsilon - 1)\,\mathfrak{E},$$

so geht die Gleichung (6) über in:

$$\mathfrak{F}_{e_x} = \mathfrak{E}_x \operatorname{div} \mathfrak{D} - \frac{1}{2}\mathfrak{E}^2 \frac{\partial \varepsilon}{\partial x} + \frac{1}{2}\frac{\partial}{\partial x}(\varepsilon - 1)\mathfrak{E}^2.$$

Die ersten beiden Glieder dieses Ausdruckes sind identisch mit den aus der Elektrostatik bekannten. Das dritte Glied ist, wie man sieht, von einem Potential ableitbar. Handelt es sich um Kräfte, die auf einen im Vakuum befindlichen Körper wirken, so liefert das Glied bei Integration über den Körper keinen Beitrag. Handelt es sich aber um die ponderomotorische Wirkung auf Flüssigkeiten, so wird der dem dritten Gliede entsprechende Anteil der Kraft bei Gleichgewicht durch eine Druckverteilung in der Flüssigkeit kompensiert.

§ 2. Kräfte, welche von den Geschwindigkeiten der Elementarteilchen abhängen.

Wir gehen jetzt über zu demjenigen Anteile der ponderomotorischen Kraft, welcher durch die Bewegungsgeschwindigkeiten der Elementarladungen geliefert wird.

Ponderomotorische Kräfte.

Wir gehen aus vom Biot-Savartschen Gesetz. Auf ein stromdurchflossenes Volumenelement, welches sich in einem magnetischen Felde befindet, wirkt erfahrungsgemäß pro Volumeneinheit die Kraft:

$$\frac{1}{c}\,[\mathfrak{J}\,\mathfrak{H}],$$

falls die betrachtete, stromdurchflossene Materie nicht magnetisch polarisierbar ist. Für das Innere von magnetisch polarisierbaren Körpern wurde, soviel uns bekannt ist, bis jetzt jene Kraft gleich[1])

$$\frac{1}{c}\,[\mathfrak{J}\,\mathfrak{B}]$$

gesetzt, wobei $\mathfrak{B}$ die magnetische Induktion bedeutet. Wir wollen nun zeigen, daß *auch* im Falle, daß das stromdurchflossene Material *magnetisch polarisierbar ist*, die auf das stromdurchflossene Volumenelement wirkende Kraft erhalten wird, wenn man zu der durch die Gleichung (7) ausgedrückten Kraft noch die Volumenkraft:

(9) $$\mathfrak{F}_s = \frac{1}{c}\,[\mathfrak{J}\,\mathfrak{H}]$$

hinzufügt. Wir wollen dies zuerst an einem einfachen Beispiel anschaulich machen.

Der unendlich dünne im Querschnitt gezeichnete Streifen S erstrecke sich senkrecht zur Papierebene nach beiden Seiten ins Unendliche. Er bestehe aus magnetisch polarisierbarem Material und befinde sich in einem homogenen Magnetfelde $\mathfrak{H}_a$, dessen Richtung durch die Pfeile (vgl. Figur) angedeutet ist. Wir fragen

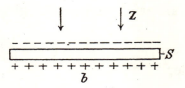

nach der auf den Materialstreifen wirkenden Kraft, falls derselbe von einem Strome i durchflossen ist.

Die Erfahrung lehrt, daß diese Kraft von der magnetischen Permeabilität des Leitermateriales unabhängig ist, und man schloß daraus, daß es nicht die Feldstärke $\mathfrak{H}$, sondern die magnetische Induktion $\mathfrak{B}_i$ sein müsse, welche für die pondero-

---

[1]) Vgl. z. B. auch M. Abraham, Theorie der Elektrizität 2. p. 319. 1905.

motorische Kraft maßgebend ist, denn im Innern des Streifens ist die magnetische Induktion $\mathfrak{B}_i$ gleich der außerhalb des Streifens wirkenden Kraft $\mathfrak{H}_a$, unabhängig von dem Werte der Permeabilität des Streifens, während die im Innern des Streifens herrschende Kraft $\mathfrak{H}_i$ bei gegebenem äußeren Felde von μ abhängt. Dieser Schluß ist aber nicht stichhaltig, weil die ins Auge gefaßte ponderomotorische Kraft nicht die einzige ist, welche auf unseren Materialstreifen wirkt. Das äußere Feld $\mathfrak{H}_a$ induziert nämlich auf der Oberseite und Unterseite des Materialstreifens magnetische Belegungen von der Dichte[1]: $\mathfrak{H}_a(1 - 1/\mu)$, und zwar auf der Oberseite eine negative, auf der Unterseite eine positive Belegung. Auf jede dieser Belegungen wirkt eine von dem im Streifen fließenden Strom erzeugte Kraft von der Stärke $i/2b$ pro Längeneinheit des Streifens[2]), welche magnetische Kraft an der Oberseite und Unterseite verschieden gerichtet ist. Die so resultierenden ponderomotorischen Kräfte addieren sich, so daß wir die ponderomotorische Kraft erhalten: $(1 - 1/\mu)\mathfrak{H}_a i$. Diese Kraft scheint bis jetzt nicht berücksichtigt worden zu sein.

Die auf die Längeneinheit unseres Streifens im ganzen ausgeübte Kraft ist nun gleich der Summe der soeben berechneten und der auf die Volumenelemente des Streifens infolge des Stromdurchganges im Magnetfeld wirkenden Kraft R. Da die gesamte auf die Längeneinheit wirkende ponderomotorische Kraft erfahrungsgemäß gleich $i\mathfrak{H}_a$ ist, so besteht die Gleichung:

$$\left(1 - \frac{1}{\mu}\right) i \mathfrak{H}_a + R = i \mathfrak{H}_a$$

oder

$$R = \frac{i \mathfrak{H}_a}{\mu} = i \mathfrak{H}_i.$$

Man sieht also, daß für die Berechnung der ponderomotorischen Kraft R, welche auf stromdurchflossene Volumenelemente

---

[1] Die Dichte ist nämlich gleich:
$$\mathfrak{D}_i = \mathfrak{B}_i - \mathfrak{H}_i = \mathfrak{H}_a\left(1 - \frac{1}{\mu}\right).$$

[2] Statt dieser auf die Belegungen wirkenden Kräfte hätten wir streng genommen nach den Resultaten des vorigen Paragraphen allerdings Volumenkräfte einführen müssen, was jedoch ohne Belang ist.

wirkt, nicht die Induktion $\mathfrak{B}_i$, sondern die Feldstärke $\mathfrak{H}_i$ maßgebend ist.

Um jeden Zweifel zu beseitigen, wollen wir noch ein Beispiel behandeln, aus welchem man ersieht, daß das Prinzip der Gleichheit von Wirkung und Gegenwirkung den von uns gewählten Ansatz fordert.

Wir denken uns einen zylindrischen, von leerem Raum umgebenen und vom Strom $\mathfrak{z}$ durchflossenen Leiter, welcher sich längs der X-Achse eines Koordinatensystems beiderseits ins Unendliche erstreckt. Die Materialkonstanten des Leiters, sowie die im folgenden auftretenden Feldvektoren seien von x unabhängig, aber Funktionen von y und z. Der Leiter sei ein magnetisch harter Körper und besitze eine Magnetisierung quer zur X-Achse. Wir nehmen an, daß ein äußeres Feld auf den Leiter nicht wirkt, daß also die magnetische Kraft $\mathfrak{H}$ in großen Entfernungen vom Leiter verschwindet.

Es ist klar, daß auf den Leiter als Ganzes keine ponderomotorische Kraft wirkt, denn es würde zu dieser Wirkung keine Gegenwirkung angebbar sein. Wir wollen nun zeigen, daß bei Wahl unseres Ansatzes jene Kraft in der Tat verschwindet. Die gesamte auf die Längeneinheit unseres Leiters in der Richtung der Z-Achse wirkende Kraft läßt sich darstellen gemäß den Gleichungen (7) und (9) in der Form:

$$(10) \quad R = \int \left(\mathfrak{D}_y \frac{\partial \mathfrak{H}_z}{\partial y} + \mathfrak{D}_z \frac{\partial \mathfrak{H}_z}{\partial z} \right) df + \int \frac{1}{c} \mathfrak{z}_x \mathfrak{H}_y \, df;$$

wobei df ein Flächenelement der YZ-Ebene bedeutet. Wir nehmen an, daß sämtliche in Betracht kommende Größen an der Oberfläche des Leiters stetig sind. Wir behandeln zuerst das erste Integral der Gleichung (10). Es ist:

$$\mathfrak{D}_y \frac{\partial \mathfrak{H}_z}{\partial y} + \mathfrak{D}_z \frac{\partial \mathfrak{H}_z}{\partial z} = \frac{\partial \mathfrak{D}_y \mathfrak{H}_z}{\partial y} + \frac{\partial \mathfrak{D}_z \mathfrak{H}_z}{\partial z} - \mathfrak{H}_z \left(\frac{\partial \mathfrak{D}_y}{\partial y} + \frac{\partial \mathfrak{D}_z}{\partial z} \right).$$

Setzt man die rechte Seite dieser Gleichung in unser Integral ein, so verschwinden bei Integration über die YZ-Ebene die beiden ersten Glieder, da die Kräfte im Unendlichen verschwinden. Das dritte Glied kann unter Berücksichtigung:

$$\operatorname{div} \mathfrak{B} = 0$$

umgeformt werden, so daß unser Integral die Form annimmt:
$$\int \mathfrak{H}_z \left(\frac{\partial \mathfrak{H}_y}{\partial y} + \frac{\partial \mathfrak{H}_z}{\partial z} \right) df.$$
Nun ist:
$$\mathfrak{H}_z \left(\frac{\partial \mathfrak{H}_y}{\partial y} + \frac{\partial \mathfrak{H}_z}{\partial z} \right) = \frac{\partial \mathfrak{H}_y \mathfrak{H}_z}{\partial y} + \frac{1}{2} \frac{\partial \mathfrak{H}_z^2}{\partial z} - \mathfrak{H}_y \frac{\partial \mathfrak{H}_z}{\partial y}.$$

Bei der Integration verschwinden aber die beiden Glieder $\frac{\partial \mathfrak{H}_y \mathfrak{H}_z}{\partial y} + \frac{1}{2} \frac{\partial \mathfrak{H}_z^2}{\partial z}$. Das Glied $- \mathfrak{H}_y \frac{\partial \mathfrak{H}_z}{\partial y}$ läßt sich umformen mittels der Maxwellschen Gleichungen in:
$$- \frac{1}{c} \mathfrak{H}_y \left\{ \mathfrak{z}_x + \frac{\partial \mathfrak{H}_y}{\partial z} \right\},$$
so daß wir endlich die Gleichung (10) schreiben können:
$$R = - \frac{1}{c} \int \mathfrak{H}_y \left\{ \mathfrak{z}_x + \frac{\partial \mathfrak{H}_y}{\partial z} \right\} df + \frac{1}{c} \int \mathfrak{z}_x \mathfrak{H}_y \, df$$
$$= - \frac{1}{c} \int \mathfrak{H}_y \frac{\partial \mathfrak{H}_y}{\partial z} df = - \frac{1}{2c} \int \frac{\partial \mathfrak{H}_y^2}{\partial z} df.$$

Das letzte Integral wird Null, weil im Unendlichen die Kräfte verschwinden. —

Nachdem wir so die Kraft festgestellt haben, welche auf von einem Leitungsstrom durchflossene Materie wirkt, erhalten wir die Kraft, die auf einen von einem Polarisationsstrom durchsetzten Körper wirkt, indem wir beachten, daß Polarisationsstrom und Leitungsstrom in bezug auf elektrodynamische Wirkung vom Standpunkt der Elektronentheorie durchaus äquivalent sein müssen.

Durch Berücksichtigung der Dualität von magnetischen und elektrischen Erscheinungen erhält man auch noch die Kraft, welche auf einen von einem magnetischen Polarisationsstrom durchsetzten Körper im elektrischen Felde ausgeübt wird. Als Gesamtausdruck für diejenigen Kräfte, welche von der Geschwindigkeit der Elementarteilchen abhängen, erhalten wir auf diese Weise die Gleichungen:

(11) $$\mathfrak{F}_a = \frac{1}{c} [\mathfrak{z} \, \mathfrak{H}] + \frac{1}{c} \left[\frac{\partial \mathfrak{P}}{\partial t} \, \mathfrak{H} \right] + \frac{1}{c} \left[\mathfrak{E} \, \frac{\partial \mathfrak{M}}{\partial t} \right].$$

§ 3. Gleichheit von actio und reactio.

Addiert man die Gleichungen (6), (7) und (11), so erhält man den Gesamtausdruck für die X-Komponente der pro Volumeneinheit auf die Materie wirkenden ponderomotorischen Kraft in der Form:

$$\mathfrak{F}_x = \mathfrak{E}_x \operatorname{div} \mathfrak{D} + \mathfrak{P}_x \frac{\partial \mathfrak{E}_x}{\partial x} + \mathfrak{P}_y \frac{\partial \mathfrak{E}_x}{\partial y} + \mathfrak{P}_z \frac{\partial \mathfrak{E}_x}{\partial z}$$

$$+ \mathfrak{Q}_x \frac{\partial \mathfrak{H}_x}{\partial x} + \mathfrak{Q}_y \frac{\partial \mathfrak{H}_x}{\partial y} + \mathfrak{Q}_z \frac{\partial \mathfrak{H}_x}{\partial z}$$

$$+ \frac{1}{c}[\mathfrak{z}\,\mathfrak{H}]_x + \frac{1}{c}\left[\frac{\partial \mathfrak{P}}{\partial t} \mathfrak{H}\right]_x + \frac{1}{c}\left[\mathfrak{E}\,\frac{\partial \mathfrak{Q}}{\partial t}\right]_x.$$

Die Gleichung kann man auch schreiben:

$$\mathfrak{F}_x = \mathfrak{E}_x \operatorname{div} \mathfrak{E} + \frac{1}{c}[\mathfrak{z}\,\mathfrak{H}]_x + \frac{1}{c}\left[\frac{\partial \mathfrak{D}}{\partial t}\mathfrak{H}\right]_x + \mathfrak{H}_x \operatorname{div} \mathfrak{H} + \frac{1}{c}\left[\mathfrak{E}\,\frac{\partial \mathfrak{B}}{\partial t}\right]_x$$

$$+ \frac{\partial (\mathfrak{P}_x \mathfrak{E}_x)}{\partial x} + \frac{\partial (\mathfrak{P}_y \mathfrak{E}_x)}{\partial y} + \frac{\partial (\mathfrak{P}_z \mathfrak{E}_x)}{\partial z}$$

$$+ \frac{\partial (\mathfrak{Q}_x \mathfrak{H}_x)}{\partial x} + \frac{\partial (\mathfrak{Q}_y \mathfrak{H}_x)}{\partial y} + \frac{\partial (\mathfrak{Q}_z \mathfrak{H}_x)}{\partial z} - \frac{1}{c}\frac{\partial}{\partial t}[\mathfrak{E}\,\mathfrak{H}]_x.$$

Ersetzt man

$$\frac{1}{c}\left(\sigma + \frac{\partial \mathfrak{D}}{\partial t}\right) \quad \text{und} \quad \frac{1}{c}\frac{\partial \mathfrak{B}}{\partial t}$$

mittels der Maxwellschen Gleichungen durch $\operatorname{curl} \mathfrak{H}$ bzw. durch $\operatorname{curl} \mathfrak{E}$, so erhält man durch eine einfache Umformung:

(12) $$\mathfrak{F}_x = \frac{\partial X_x}{\partial x} + \frac{\partial X_y}{\partial y} + \frac{\partial X_z}{\partial z} - \frac{1}{c^2}\frac{\partial \mathfrak{S}_x}{\partial t},$$

wobei gesetzt ist[1]):

(13) $$\begin{cases} X_x = -\tfrac{1}{2}(\mathfrak{E}^2 + \mathfrak{H}^2) + \mathfrak{E}_x \mathfrak{D}_x + \mathfrak{H}_x \mathfrak{B}_x, \\ X_y = \phantom{-\tfrac{1}{2}(\mathfrak{E}^2 + \mathfrak{H}^2)} \mathfrak{E}_x \mathfrak{D}_y + \mathfrak{H}_x \mathfrak{B}_y, \\ X_z = \phantom{-\tfrac{1}{2}(\mathfrak{E}^2 + \mathfrak{H}^2)} \mathfrak{E}_x \mathfrak{D}_z + \mathfrak{H}_x \mathfrak{B}_z, \\ \mathfrak{S}_x = \phantom{-\tfrac{1}{2}(\mathfrak{E}^2 + \mathfrak{H}^2)} c\,[\mathfrak{E}\,\mathfrak{H}]_x. \end{cases}$$

---

1) Hr. Geheimrat Wien hatte die Güte, uns darauf aufmerksam zu machen, daß bereits H. A. Lorentz die ponderomotorischen Kräfte für nicht magnetisierbare Körper in dieser Form angegeben hat. Enzykl. d. mathem. W. 5. p. 247.

Entsprechende Gleichungen gelten für die beiden anderen Komponenten der ponderomotorischen Kraft.

Integriert man (12) über den unendlichen Raum, so erhält man, falls im Unendlichen die Feldvektoren verschwinden, die Gleichung:

$$(14) \qquad \int \mathfrak{F}_x \, d\tau = -\frac{1}{c^2} \int d\tau \cdot \frac{d\mathfrak{S}_x}{dt}.$$

Sie sagt aus, daß unsere ponderomotorischen Kräfte bei Einführung der elektromagnetischen Bewegungsgröße dem Satz von der Gleichheit von actio und reactio genügen.

Bern, 7. Mai 1908.

(Eingegangen 13. Mai 1908.)

ANNALEN
DER
PHYSIK.

BEGRÜNDET UND FORTGEFÜHRT DURCH
F. A. C. GREN, L. W. GILBERT, J. C. POGGENDORFF, G. u. E. WIEDEMANN, P. DRUDE.

VIERTE FOLGE.

BAND 27.

DER GANZEN REIHE 332. BAND.

KURATORIUM:
F. KOHLRAUSCH, M. PLANCK, G. QUINCKE,
W. C. RÖNTGEN, E. WARBURG.

UNTER MITWIRKUNG
DER DEUTSCHEN PHYSIKALISCHEN GESELLSCHAFT
UND INSBESONDERE VON
M. PLANCK

HERAUSGEGEBEN VON
W. WIEN UND M. PLANCK.

MIT ELF FIGURENTAFELN.

LEIPZIG, 1908.
VERLAG VON JOHANN AMBROSIUS BARTH.

**14. Berichtigung zur Abhandlung:
„Über die elektromagnetischen Grundgleichungen
für bewegte Körper";
von A. Einstein und J. Laub.**

In der genannten Abhandlung dieser Zeitschrift **26**. p. 532. 1908 sind zwei Fehler unterlaufen:

p. 534 Formel (8) muß heißen:

$$\varrho = \beta\left(\varrho' + \frac{v}{c^2}\mathfrak{z}_x'\right)$$

statt:

$$\varrho = \beta\left(\varrho' + \frac{v}{c}\mathfrak{z}_x'\right),$$

ferner die erste der Formeln (9):

$$\mathfrak{z}_x = \beta\left(\mathfrak{z}_x' + v\varrho'\right)$$

statt:

$$\mathfrak{z}_x = \beta\left(\mathfrak{z}_x' + \frac{v}{c}\varrho'\right).$$

Die erste der Formeln (12a) sowie die dritte der Formeln (13) müssen ebenfalls heißen:

$$\beta(\mathfrak{z}_x - v\varrho) = \sigma\mathfrak{E}_x$$

und

$$\beta(\mathfrak{z}_\mathfrak{v} - |\mathfrak{v}|\varrho) = \sigma\left\{\mathfrak{E} + \frac{1}{c}[\mathfrak{v}\mathfrak{B}]\right\}_\mathfrak{v}.$$

(Eingegangen 24. August 1908.)

ANNALEN DER PHYSIK.

BEGRÜNDET UND FORTGEFÜHRT DURCH
F. A. C. GREN, L. W. GILBERT, J. C. POGGENDORFF, G. u. E. WIEDEMANN, P. DRUDE.

VIERTE FOLGE.

BAND 28.
DER GANZEN REIHE 333. BAND.

KURATORIUM:
F. KOHLRAUSCH, M. PLANCK, G. QUINCKE,
W. C. RÖNTGEN, E. WARBURG.

UNTER MITWIRKUNG
DER DEUTSCHEN PHYSIKALISCHEN GESELLSCHAFT
UND INSBESONDERE VON
M. PLANCK

HERAUSGEGEBEN VON
W. WIEN UND M. PLANCK.

MIT ACHT FIGURENTAFELN.

LEIPZIG, 1909.
VERLAG VON JOHANN AMBROSIUS BARTH.

11. *Bemerkungen zu unserer Arbeit: „Über die elektromagnetischen Grundgleichungen für bewegte Körper";* von *A. Einstein und J. Laub.*

Hr. Laue war so freundlich, uns auf eine in unserer im Titel genannten Arbeit enthaltene Unrichtigkeit hinzuweisen.[1]) Wir sagen dort (Ann. d. Phys. **26**. p. 535. 1908):

„Es ist bemerkenswert, daß die Grenzbedingungen für die Vektoren $\mathfrak{E}, \mathfrak{D}, \mathfrak{H}, \mathfrak{B}$ an der Grenze zweier Medien dieselben sind, wie für ruhende Körper. Es folgt dies direkt aus den Gleichungen (1a) bis (4a)."

Abgesehen davon, daß für die Herleitung der Grenzbedingungen die Gleichungen (3a) und (4a) nicht in Betracht kommen, ist diese Behauptung nur dann richtig, wenn die Bewegungskomponente normal zur Grenzfläche verschwindet, was bei der im § 2 der genannten Arbeit behandelten Aufgabe tatsächlich zutrifft. Die allgemein gültigen Grenzbedingungen findet man wohl am leichtesten auf folgendem Wege, der dem von Heinrich Hertz eingeschlagenen entspricht.

Ist die Grenzfläche, oder besser gesagt, die unendlich dünne Grenzübergangsschale, beliebig bewegt, so werden sich in einem momentan in ihr gelegenen ruhenden Punkt die das elektromagnetische Feld bestimmenden Größen im allgemeinen unstetig bzw. unendlich rasch mit der Zeit ändern; diese Änderungen werden aber stetig sein für einen Punkt, der sich *mit der Materie bewegt*. Es wird also die Anwendung des Operators

$$\frac{\partial}{\partial t} + (\mathfrak{v} \nabla)$$

an einem Skalar oder einem Vektor auch in der Grenzfläche

---

1) Hr. Laue hat uns in seinem Briefe bereits die Grenzbedingungen richtig angegeben und uns eine andere Ableitung derselben mitgeteilt.

nicht zu unendlich großen Werten führen. Schreiben wir nun die Gleichung (1 a)[1]) in der Form:

$$\frac{1}{c}\left\{\frac{\partial \mathfrak{D}}{\partial t} + (\mathfrak{v}\nabla)\mathfrak{D}\right\} + \mathfrak{z} = \operatorname{curl} \mathfrak{H} + \frac{1}{c}(\mathfrak{v}\nabla)\mathfrak{D}$$

und nehmen wir an, daß die Stromdichte $\mathfrak{z}$ auch in der Grenzschicht endlich sei, so ist die linke Seite dieser Gleichung in der Grenzschicht endlich. Dasselbe gilt also auch für die rechte Seite der Gleichung.

Zur leichten Interpretation dieses Resultates denken wir uns das Koordinatensystem so gelegt, daß ein bestimmtes, unendlich kleines Stück der Grenzfläche, das wir nun betrachten wollen, der YZ-Ebene parallel sei. Dann ist klar, daß die Ableitungen aller Größen nach y und z in dem betrachteten Stück der Grenzfläche endlich bleiben. Es muß also auch der Inbegriff derjenigen Glieder der rechten Seite obiger Gleichung, die Differentiationen nach x enthalten, etwas Endliches liefern. Durch einfaches Entwickeln der rechten Seite und Weglassen der nach y und z differenzierten Glieder gelangt man zu dem Resultate, daß in der Grenzschicht die Ausdrücke:

$$\frac{\mathfrak{v}_x}{c}\frac{\partial \mathfrak{D}_x}{\partial x},$$

$$\frac{\partial \mathfrak{H}_z}{\partial x} - \frac{\mathfrak{v}_x}{c}\frac{\partial \mathfrak{D}_y}{\partial x},$$

$$\frac{\partial \mathfrak{H}_y}{\partial x} + \frac{\mathfrak{v}_x}{c}\frac{\partial \mathfrak{D}_z}{\partial x}$$

endlich bleiben. Setzen wir noch voraus, daß die Geschwindigkeitskomponenten an der Grenzfläche keinen Sprung erleiden, so folgt daraus, daß die Ausdrücke:

$$\mathfrak{D}_x,$$

$$\mathfrak{H}_y + \frac{\mathfrak{v}_x}{c}\mathfrak{D}_z,$$

$$\mathfrak{H}_z - \frac{\mathfrak{v}_x}{c}\mathfrak{D}_y$$

auf beiden Seiten der Grenzfläche (YZ-Ebene) denselben Wert

---

[1] l. c.

haben. Da $\mathfrak{D}_x$ und die Komponenten von $\mathfrak{v}$ stetig sind, können wir die beiden letzten Ausdrücke auch ersetzen durch:

$$\mathfrak{H}_y - \frac{1}{c}(\mathfrak{v}_z \mathfrak{D}_x - \mathfrak{v}_x \mathfrak{D}_z),$$

$$\mathfrak{H}_z - \frac{1}{c}(\mathfrak{v}_x \mathfrak{D}_y - \mathfrak{v}_y \mathfrak{D}_x).$$

Von der speziellen Wahl der Lage der Koordinatenachsen relativ zum betrachteten Element der Grenzfläche machen wir uns frei, indem wir das Resultat in den Bezeichnungen der Vektoranalysis schreiben. Bezeichnen wir durch die Indizes n bzw. $\bar{n}$ die Komponente des betreffenden Vektors im Sinne bzw. senkrecht zur Normale der Unstetigkeitsfläche, so folgt, daß

$$\mathfrak{D}_n,$$
$$\left\{\mathfrak{H} - \frac{1}{c}[\mathfrak{v}\,\mathfrak{D}]\right\}_{\bar{n}}$$

an der Grenzfläche stetig sein müssen.

In gleicher Weise schließt man aus der Gleichung (2a)[1]) die Stetigkeit der Komponenten:

$$\mathfrak{B}_n,$$
$$\left\{\mathfrak{E} + \frac{1}{c}[\mathfrak{v}\,\mathfrak{B}]\right\}_{\bar{n}}.$$

Bern und Würzburg, November 1908.

---

1) l. c.

(Eingegangen 6. Dezember 1908.)

Nachtrag. Wenn an der betrachteten Grenzfläche eine Schicht wahrer Elektrizität ($\int \varrho\, d\tau$) von der Flächendichte η sich befindet, so wird $\mathfrak{s}$ unendlich. Es ist dann

$$\operatorname{curl} \mathfrak{H} + \frac{1}{c}(\mathfrak{v}\,\triangle)\mathfrak{D} - \mathfrak{s}$$

in der Grenzschicht endlich, wobei $\mathfrak{s}$ durch $(\mathfrak{v}/c)\varrho$ ersetzt werden kann. Für diesen Fall findet man ebenfalls die obigen Grenzbedingungen, mit dem Unterschiede, daß die erste derselben durch

$$\mathfrak{D}_{n2} - \mathfrak{D}_{n1} = \eta$$

zu ersetzen ist.

(Eingegangen 19. Januar 1909.)

5. Bemerkung zu der Arbeit von D. Mirimanoff „Über die Grundgleichungen..."; von A. Einstein.

1. Das in dieser Arbeit[1]) angegebene System von Differentialgleichungen und Transformationsgleichungen unterscheidet sich von dem Minkowskis in keiner Weise bzw. *nur* dadurch, daß derjenige Vektor, welcher gewöhnlich mit $\mathfrak{H}$ bezeichnet wird (magnetische Kraft), vom Verfasser mit

$$\mathfrak{O} = \mathfrak{H} - \frac{1}{c}[\mathfrak{P}\mathfrak{w}]$$

bezeichnet wurde.

Differentialgleichung (I) ist nämlich bei Einführung von $\mathfrak{O}$, wie der Verfasser selbst zeigt, identisch mit der betreffenden Gleichung Minkowskis, während die übrigen drei Differentialgleichungen $\mathfrak{H}$ nicht enthalten und bereits die Form der entsprechenden Gleichungen Minkowskis haben. Der Verfasser sagt auch selbst, daß sich seine Vektoren $\mathfrak{E}, \mathfrak{D}, \mathfrak{O}, \mathfrak{B}$ transformieren, wie die gewöhnlich mit $\mathfrak{E}, \mathfrak{D}, \mathfrak{H}, \mathfrak{B}$ bezeichneten Vektoren.

2. Auch die Beziehungen zwischen den Vektoren, welche Materialkonstanten (ε, μ und σ) enthalten, unterscheiden sich nicht von den entsprechenden Minkowskis. Der Verfasser geht nämlich davon aus, daß für ein relativ zu dem betrachteten Systempunkt momentan ruhendes Koordinatensystem die Gleichungen

$$\mathfrak{D} = \varepsilon \mathfrak{E}, \quad \mathfrak{H} = \frac{1}{\mu}\mathfrak{B}, \quad \mathfrak{J} = \sigma \mathfrak{E}$$

gelten sollen; bedenkt man nun, daß der Vektor $\mathfrak{H}$ (des Verfassers) für $\mathfrak{w} = 0$ mit dem Vektor $\mathfrak{O}$ identisch ist, und daß $\mathfrak{O}$ in den Differentialgleichungen des Verfassers und in dessen Transformationsgleichungen genau dieselbe Rolle spielt, wie $\mathfrak{m}$ in Minkowskis Gleichungen (gewöhnlich mit $\mathfrak{H}$ bezeichnet),

---

[1]) D. Mirimanoff, Ann. d. Phys. **28**. p. 192. 1909.

so ersieht man, daß auch diese Gleichungen mit den entsprechenden Minkowskis übereinstimmen, bis auf den Umstand, daß die Bezeichnung $\mathfrak{H}$ durch die Bezeichnung $\mathfrak{Q}$ ersetzt ist.

3. Es ist also gezeigt, daß die Größe $\mathfrak{Q}$ Mirimanoffs in dessen sämtlichen Gleichungen dieselbe Rolle spielt wie diejenige Größe, welche man gewöhnlich mit $\mathfrak{H}$ bezeichnet und „magnetische Kraft" oder „magnetische Feldstärke" nennt. Trotzdem hätten die Gleichungen Mirimanoffs einen anderen Inhalt als die Gleichungen Minkowskis, wenn die Größe $\mathfrak{Q}$ Mirimanoffs definitionsgemäß eine andere physikalische Bedeutung hätte als die gewöhnlich mit $\mathfrak{H}$ bezeichnete Größe.

Um hierüber ein Urteil zu gewinnen, fragen wir uns zunächst, was in den Minkowskischen Gleichungen

(A)
$$\begin{cases} \operatorname{curl} \mathfrak{H} = \frac{1}{c}\frac{\partial \mathfrak{D}}{\partial t} + \mathfrak{i}, \\ \operatorname{curl} \mathfrak{E} = -\frac{1}{c}\frac{\partial \mathfrak{B}}{\partial t}, \\ \operatorname{div} \mathfrak{D} = \varrho, \\ \operatorname{div} \mathfrak{B} = 0 \end{cases}$$

die Vektoren $\mathfrak{E}, \mathfrak{D}, \mathfrak{H}, \mathfrak{B}$ für eine Bedeutung haben. Man muß zugeben, daß diese Vektoren für den Fall, daß die Geschwindigkeit $\mathfrak{w}$ der Materie von Null abweicht, bisher nicht eigens definiert worden sind; Definitionen, auf welchen (ideale) Messungen dieser Größen basiert werden könnten, besitzen wir nur für den Fall, daß $\mathfrak{w}$ verschwindet, und zwar denke ich an jene Definitionen, welche aus der Elektrodynamik ruhender Körper wohlbekannt sind. Wenn daher unter Benutzung der Minkowskischen Gleichungen gefunden ist, daß in einem bestimmten, mit der Geschwindigkeit $\mathfrak{w}$ bewegten Volumelement des Körpers die Feldvektoren zu einer gewissen Zeit die bestimmten (Vektor-)Werte $\mathfrak{E}, \mathfrak{D}, \mathfrak{H}, \mathfrak{B}$ haben, so müssen wir diese Feldvektoren erst auf ein mit Bezug auf das betreffende Volumelement ruhendes Bezugssystem transformieren. Die so erhaltenen Vektoren $\mathfrak{E}', \mathfrak{D}', \mathfrak{H}', \mathfrak{B}'$ haben erst eine bestimmte physikalische Bedeutung, die aus der Elektrodynamik ruhender Körper bekannt ist.

Bemerkung zu der Arbeit von D. Mirimanoff.

Die Minkowskischen Differentialgleichungen sagen also für Punkte, in denen $\mathfrak{w} \neq 0$ ist, für sich allein noch gar nichts aus, wohl aber die Minkowskischen Differentialgleichungen zusammen mit den Minkowskischen Transformationsgleichungen und mit der Bestimmung, daß für den Fall $\mathfrak{w} = 0$ die Definitionen der Elektrodynamik ruhender Körper für die Feldvektoren gelten sollen.

Wir haben nun zu fragen: Ist der Vektor $\mathfrak{H}$ Mirimanoffs in anderer Weise definiert als der von uns soeben mit $\mathfrak{H}$ bezeichnete Vektor? Dies ist nicht der Fall, und zwar aus folgenden Gründen:

1. Für die Feldvektoren $\mathfrak{E}, \mathfrak{D}, \mathfrak{H}, \mathfrak{B}$ Mirimanoffs gelten dieselben Differentialgleichungen und Transformationsgleichungen wie für die Vektoren $\mathfrak{E}, \mathfrak{D}, \mathfrak{H}, \mathfrak{B}$ der Minkowskischen Gleichungen (A).

2. Sowohl Mirimanoffs Vektor $\mathfrak{H}$ als auch der Vektor $\mathfrak{H}$ von (A) sind nur für den Fall $\mathfrak{w} = 0$ definiert. In diesem Falle ist aber wegen Mirimanoffs Gleichung

$$\mathfrak{H} = \mathfrak{H} - \frac{1}{c}[\mathfrak{B}\,\mathfrak{w}]$$

$\mathfrak{H} = \mathfrak{H} =$ Feldstärke zu setzen; für den Vektor $\mathfrak{H}$ der Gleichungen (A) gilt genau in gleicher Weise, daß er im Falle $\mathfrak{w} = 0$ mit der Feldstärke im Sinne der Elektrodynamik ruhender Körper gleichbedeutend ist.

Aus diesen beiden Argumenten folgt, daß der Vektor $\mathfrak{H}$ Mirimanoffs und der Vektor $\mathfrak{H}$ von (A) durchaus gleichwertig sind.

4. Um seine Resultate bezüglich der Wilsonschen Anordnung mit den von Hrn. Laub und mir erhaltenen zu vergleichen, hätte der Verfasser die Betrachtung so weit durchführen müssen, daß er zu Beziehungen zwischen definierten, d. h. wenigstens prinzipiell der Erfahrung zugänglichen Größen gelangt wäre. Er hätte zu diesem Zwecke nur die seinem Gleichungssystem entsprechenden Grenzbedingungen anzuwenden gehabt. Nach dem Vorigen hätte er so zu genau denselben Folgerungen gelangen müssen wie wir, da seine Theorie mit der von Minkowski identisch ist.

Schließlich möchte ich noch hinweisen auf die Bedeutung

888. *A. Einstein. Bemerkung zu der Arbeit von D. Mirimanoff.*

der neulich erschienenen Arbeit von Ph. Frank[1]), welche die Übereinstimmung zwischen der Lorentzschen elektronentheoretischen und der Minkowskischen Behandlung der Elektrodynamik bewegter Körper durch Berücksichtigung der Lorentzkontraktion wiederherstellt. Der Vorzug der elektronentheoretischen Behandlungsweise liegt einerseits darin, daß sie eine anschauliche Deutung der Feldvektoren liefert, andererseits darin, daß sie auskommt ohne die willkürliche Voraussetzung, daß die Differentialquotienten der Geschwindigkeit der Materie in den Differentialgleichungen nicht auftreten.

Bern, Januar 1909.

---

1) Ph. Frank, Ann. d. Phys. 27. p. 1059. 1908.

(Eingegangen 22. Januar 1909.)

ANNALEN DER PHYSIK.

BEGRÜNDET UND FORTGEFÜHRT DURCH
F. A. C. GREN, L. W. GILBERT, J. C. POGGENDORFF, G. u. E. WIEDEMANN, P. DRUDE.

VIERTE FOLGE.

BAND 33.

DER GANZEN REIHE 338. BAND.

KURATORIUM:

M. PLANCK, G. QUINCKE,
W. C. RÖNTGEN, W. VOIGT, E. WARBURG.

UNTER MITWIRKUNG

DER DEUTSCHEN PHYSIKALISCHEN GESELLSCHAFT

UND INSBESONDERE VON

M. PLANCK

HERAUSGEGEBEN VON

W. WIEN UND M. PLANCK.

MIT SECHS FIGURENTAFELN.

LEIPZIG, 1910.

VERLAG VON JOHANN AMBROSIUS BARTH.

2. Über einen Satz der Wahrscheinlichkeitsrechnung und seine Anwendung in der Strahlungstheorie; *von A. Einstein und L. Hopf.*

§ 1. Das physikalische Problem als Ausgangspunkt.

Will man in der Theorie der Temperaturstrahlung irgend eine Wirkung der Strahlung berechnen, etwa die auf einen Oszillator wirkende Kraft, so verwendet man dazu stets als analytischen Ausdruck für die elektrische oder magnetische Kraft Fouriersche Reihen der allgemeinen Gestalt

$$\sum_n A_n \sin 2\pi n \frac{t}{T} + B_n \cos 2\pi n \frac{t}{T}.$$

Hierbei ist das Problem gleich auf einen bestimmten Raumpunkt spezialisiert, was für das Folgende ohne Bedeutung ist, t bedeutet die variable Zeit, T die sehr große Zeitdauer, für welche die Entwickelung gilt. Bei der Berechnung irgendwelcher Mittelwerte — und nur solche kommen in der Strahlungstheorie überhaupt vor — nimmt man die einzelnen Koeffizienten A_n, B_n als unabhängig voneinander an, man setzt voraus, daß jeder Koeffizient unabhängig von den Zahlenwerten der anderen das Gaußsche Fehlergesetz befolge, so daß die Wahrscheinlichkeit[1]) dW einer Kombination von Werten A_n, B_n sich aus den Wahrscheinlichkeiten der einzelnen Koeffizienten einfach als Produkt darstellen müsse.

(1) $\quad dW = W_{A_1} \cdot W_{A_2} \ldots W_{B_1} \cdot W_{B_2} \ldots dA_1 \ldots dB_1 \ldots$

Da bekanntlich die Strahlungslehre, so wie sie exakt aus den allgemein anerkannten Fundamenten der Elektrizitäts-

---

1) Unter „Wahrscheinlichkeit eines Koeffizienten" ist offenbar folgendes zu verstehen: Wir denken uns die elektrische Kraft in sehr vielen Zeitmomenten in Fouriersche Reihen entwickelt. Derjenige Bruchteil dieser Entwickelungen, bei welchem ein Koeffizient in einem bestimmten Wertbereich liegt, ist die Wahrscheinlichkeit dieses Wertbereiches des betreffenden Koeffizienten.

theorie und der statistischen Mechanik folgt, in unlösbare Widersprüche mit der Erfahrung führt, liegt es nahe, dieser einfachen Annahme der Unabhängigkeit zu mißtrauen und ihr die Schuld an den Mißerfolgen der Strahlungstheorie zuzuschreiben.

Im folgenden soll nun gezeigt werden, daß dieser Ausweg unmöglich ist, daß sich vielmehr das physikalische Problem auf ein rein mathematisches zurückführen läßt, das zum statistischen Gesetze (1) führt.

Betrachten wir nämlich die aus einer bestimmten Richtung herkommende[1]) Strahlung, so hat diese gewiß einen höheren Grad von Ordnung, als die gesamte in einem Punkte wirkende Strahlung. Die Strahlung aus einer bestimmten Richtung können wir aber immer noch auffassen als von sehr vielen Emissionszentren herrührend, d. h. wir können die Fläche, welche die Strahlung aussendet, noch in sehr viele unabhängig voneinander ausstrahlende Flächenelemente zerlegen; denn der Entfernung dieser Fläche vom Aufpunkt sind ja keine Grenzen gesteckt, also auch nicht ihrer gesamten Ausdehnung. In diese von den einzelnen Flächenelementen herrührenden Strahlungselemente führen wir wieder ein höheres Ordnungsprinzip ein, indem wir diese Strahlungselemente alle als von gleicher Form und nur durch eine zeitliche Phase verschieden auffassen; mathematisch gesprochen: die Koeffizienten der Fourierschen Reihen, welche die Strahlung der einzelnen Flächenelemente darstellen, seien für alle Flächenelemente dieselben, nur der Anfangspunkt der Zeit von Element zu Element verschieden. Können wir Gleichung (1) unter Zugrundelegung dieser Ordnungsprinzipien beweisen, so gilt sie a fortiori für den Fall, daß man diese Ordnungsprinzipien fallen läßt. Bezeichnet der Index s das einzelne Flächenelement, so erhält die dort ausgesandte Strahlung die Form:

$$\sum_{(n)} a_n \sin 2\pi n \frac{t - t_s}{T}.$$

Die gesamte von uns betrachtete Strahlung wird also dargestellt durch die Doppelsummen:

$$(2) \quad \sum_s \sum_n a_n \left(\sin 2\pi n \frac{t}{T} \cos 2\pi n \frac{t_s}{T} - \cos 2\pi n \frac{t}{T} \sin 2\pi n \frac{t_s}{T} \right).$$

---

1) genauer: „einem bestimmten Elementarwinkel $d\varkappa$ entsprechende"

Vergleichung von (2) und (1) führt also zu den Ausdrücken:

$$(3) \quad \begin{cases} A_n = a_n \sum_s \cos 2\pi n \frac{t_s}{T}, \\ B_n = a_n \sum_s \sin 2\pi n \frac{t_s}{T}, \end{cases}$$

n ist eine sehr große Zahl, t_s kann jeden Wert zwischen 0 und T annehmen, die einzelnen Summanden

$$\cos 2\pi n \frac{t_s}{T} \quad \text{bzw.} \quad \sin 2\pi n \frac{t_s}{T}$$

liegen also regellos zwischen -1 und $+1$ verteilt und sind gleich wahrscheinlich positiv wie negativ. Können wir für eine Kombination von Summen solcher Größen allgemein die Gültigkeit unserer Gleichung (1) nachweisen, so ist damit auch die Unmöglichkeit erwiesen, irgend ein Ordnungsprinzip in die im leeren Raum sich ausbreitende Strahlung einzuführen.

§ 2. Formulierung des allgemeinen mathematischen Problems.

Wir stellen uns also folgendes mathematische Problem: Gegeben ist eine sehr große Anzahl von Elementen, deren Zahlenwerte α ein bekanntes statistisches Gesetz befolgen (entsprechend den t_s). Von jedem dieser Zahlenwerte werden gewisse Funktionen $f_1(\alpha) f_2(\alpha) \ldots$ gebildet (entsprechend $\sin 2\pi n \frac{t_s}{T} n . \cos 2\pi n \frac{t_s}{T}$). Diese Funktionen müssen wir noch einer Einschränkung unterwerfen: Es ergibt sich nämlich aus der Wahrscheinlichkeit, daß eine der Größen α zwischen $\alpha + d\alpha$ liegt, ein statistisches Gesetz für die f; die Wahrscheinlichkeit $\varphi(f) df$, daß f einen Zahlenwert zwischen f und $f + df$ habe, sei nun stets eine solche Funktion, daß der Mittelwert

$$\bar{f} = \int_{-\infty}^{+\infty} f \varphi(f) df = 0.$$

(Es ist leicht einzusehen, daß unsere Funktionen sin und cos wirklich diese Voraussetzung erfüllen; denn wenn jeder Wert von t_s zwischen $\underline{0 \text{ und } T}$ gleich wahrscheinlich ist, verschwinden die Mittelwerte $\overline{\sin 2\pi n \frac{t_s}{T}}$ und $\overline{\cos 2\pi n \frac{t_s}{T}}$.)

Wir fassen nun eine (sehr große) Anzahl Z solcher Elemente α zu einem System zusammen. Zu einem derartigen System gehören bestimmte Summen

$$\Sigma_{(Z)} f_1(\alpha), \quad \Sigma_{(Z)} f_2(\alpha) \ldots$$

(entsprechend den Koeffizienten A_n/a_n, B_n/a_n). Wir stellen uns die Aufgabe, das statistische Gesetz zu ermitteln, welches eine Kombination dieser Summen befolgt.

Zunächst müssen wir über einen prinzipiellen Punkt Klarheit schaffen:

Das statistische Gesetz, das die Summen Σ selbst befolgen, wird gar nicht von der Anzahl Z der Elemente unabhängig sein. Das können wir leicht an dem einfachen Spezialfall sehen, daß $f(\alpha)$ nur die Werte $+1$ und -1 annehmen könne. Dann ist offenbar:

$$\Sigma_{(Z+1)} = \Sigma_{(Z)} \pm 1$$

und

$$\overline{\Sigma_{(Z+1)}^2} = \overline{\Sigma_{(Z)}^2} + 1.$$

Der quadratische Mittelwert der Summe wächst also proportional mit der Anzahl der Elemente. Wollen wir also zu einem von Z unabhängigen statistischen Gesetze gelangen, so dürfen wir nicht die Σ betrachten, sondern, da $\overline{\Sigma^2}/Z$ konstant bleibt, die Größen

$$S = \frac{\Sigma}{\sqrt{Z}}.$$

§ 3. Statistisches Gesetz der einzelnen S.

Ehe wir nun eine Kombination aller Größen

$$S^{(n)} = \frac{\Sigma_{(Z)} f_n(\alpha)}{\sqrt{Z}}$$

untersuchen, wollen wir das Wahrscheinlichkeitsgesetz einer einzelnen solchen Größe aufstellen.

Wir betrachten eine Vielheit von N-Systemen der oben definierten Art. Zu jedem System gehört ein Zahlenwert S. Diese Größen befolgen wegen der statistischen Verteilung der α ein gewisses Wahrscheinlichkeitsgesetz, so daß die Anzahl der Systeme, deren Zahlenwert zwischen S und $S+dS$ liegt:

(4) $$dN = F(S)\, dS.$$

Fügen wir nun zu den aus Z-Elementen bestehenden Systemen noch je ein weiteres Element, d. h. gehen wir von S_Z zu S_{Z+1} über, so werden die einzelnen Glieder unserer Vielheit ihren Zahlenwert ändern und in ein anderes Gebiet dS einrücken. Wenn es trotzdem möglich sein soll, zu einem von Z unabhängigen statistischen Gesetz zu gelangen, so darf sich bei diesem Übergang die Anzahl dN nicht ändern. Es muß also in ein bestimmtes (in unserem einfachsten Fall eindimensionales) Gebiet dS die gleiche Anzahl von Systemen ein- wie austreten. Bezeichnet Φ die Zahl der Systeme, welche vom Übergang von Z zu $Z+1$ Elementen einen gewissen Zahlenwert S_0 durchschreiten und zwar sowohl der Größe wie der Richtung nach, so muß:

(5) $$\operatorname{div} \Phi = 0,$$

also

$$\frac{d\Phi}{dS} = 0$$

und, da ja Φ für $S = \infty$ jedenfalls gleich 0 sein muß, auch

(6) $$\Phi = 0.$$

Nun ist:

$$S_{(Z+1)} = \frac{\Sigma_{(Z+1)} f(\alpha)}{\sqrt{Z+1}} = S_{(Z)} \sqrt{\frac{Z}{Z+1}} + \frac{f(\alpha)}{\sqrt{Z+1}},$$

oder, da Z eine sehr große Zahl sein soll:

(7) $$S_{(Z+1)} = S_{(Z)} - \frac{S_{(Z)}}{2Z} + \frac{f(\alpha)}{\sqrt{Z}}.$$

Die Anzahl Φ setzt sich also aus zwei Teilen zusammen, einem Φ_1, der vom Summanden $-S/2Z$ und einem Φ_2, der von $f(\alpha)/\sqrt{Z}$ herrührt.

Φ_1 enthält alle diejenigen S, welche in einem positiven Abstand $\leq S_0/2Z$ vom Werte S_0 gelegen waren; und zwar durchschreiten diese Glieder S_0 in negativer Richtung. Ihre Anzahl ist, da $S_0/2Z$ eine sehr kleine Zahl ist, bis auf unendlich kleine Größen höherer Ordnung:

(8) $$\Phi_1 = -\frac{S_0}{2Z} F(S_0).$$

Zur Anzahl Φ_2 kommt ein Beitrag aus jeder beliebigen positiven und negativen Entfernung Δ von S_0, und zwar ein positiver oder negativer Beitrag, je nachdem Δ negativ oder positiv ist. In der Entfernung Δ ist die Anzahl dN gegeben durch

$$F(S_0 + \Delta)\, dS = F(S_0 + \Delta)\, d\Delta,$$

oder, da doch nur kleine Werte von Δ ins Gewicht fallen, durch

$$\left\{F(S_0) + \Delta \left(\frac{dF}{d\Delta}\right)_{S_0}\right\} d\Delta.$$

Von dieser Anzahl durchqueren alle diejenigen den Wert S_0 in positiver Richtung, die, von einem negativen Δ herkommend, ein so großes $f(\alpha)$ haben, daß

$$\frac{f(\alpha)}{\sqrt{Z}} \geqq |\Delta|,$$

also die Anzahl

$$\int_{-\Delta\sqrt{Z}}^{+\infty} \varphi(f)\, df.$$

In der negativen Richtung geht analog die Anzahl

$$\int_{-\infty}^{-\Delta\sqrt{Z}} \varphi(f)\, df.$$

So wird:

$$\Phi_2 = \int_{-\infty}^{0} d\Delta \left\{F(S_0) + \Delta \left(\frac{dF}{d\Delta}\right)_{S_0}\right\} \int_{-\Delta\sqrt{Z}}^{+\infty} \varphi(f)\, df$$

$$- \int_{0}^{\infty} d\Delta \left\{F(S_0) + \Delta \left(\frac{dF}{d\Delta}\right)_{S_0}\right\} \int_{-\infty}^{-\Delta\sqrt{Z}} \varphi(f)\, df.$$

Durch partielle Integration geht dies über in:

$$\Phi_2 = -\int_{-\infty}^{0} d\Delta \left\{\Delta \cdot F(S_0) + \frac{\Delta^2}{2} \left(\frac{dF}{d\Delta}\right)_{S_0}\right\} \varphi(-\Delta\sqrt{Z}) \cdot \sqrt{Z}$$

$$- \int_{0}^{\infty} d\Delta \left\{\Delta \cdot F(S_0) + \frac{\Delta^2}{2} \left(\frac{dF}{d\Delta}\right)_{S_0}\right\} \varphi(-\Delta\sqrt{Z}) \cdot \sqrt{Z}.$$

Da nun nach Voraussetzung

$$\int_{-\infty}^{+\infty} f\varphi(f)\, df = 0$$

wird, wenn wir $\Delta \sqrt{Z} = f$ als Variable einführen:

(9) $$\begin{cases} \Phi_2 = -\frac{1}{2Z}\left(\frac{dF}{d\Delta}\right)_{S_0} \int_{-\infty}^{+\infty} f^2 \varphi(f) df \\ \quad = -\frac{1}{2Z}\left(\frac{dF}{d\Delta}\right)_{S_0} \cdot \overline{f^2}. \end{cases}$$

(8) und (9) in (6) eingesetzt, ergeben die Differentialgleichung:

$$S F + \overline{f^2} \frac{dF}{dS} = 0,$$

deren Lösung:

(10) $$F = \text{const.}\, e^{-\frac{S^2}{2\overline{f^2}}},$$

das Gausssche Fehlergesetz ausspricht.

§ 4. Statistisches Gesetz einer Kombination aller $S^{(n)}$.

Wir dehnen nun die Betrachtungen des vorigen Paragraphen vom eindimensionalen Fall auf den beliebig vieler Dimensionen aus. Wir haben diesmal eine Kombination von vielen Größen $S^{(n)}$ zu betrachten. Die Anzahl der in einem unendlich kleinen Gebiete $dS^{(1)} dS^{(2)} \ldots$ liegenden Systeme sei:

(11) $$dN = F(S^{(1)}, S^{(2)} \ldots) dS^{(1)} dS^{(2)} \ldots$$

Wieder fordern wir, daß dN sich nicht ändern soll, wenn wir von $S^{(n)}_{(Z)}$ zu $S^{(n)}_{(Z+1)}$ übergehen, wieder führt dies zu der Differentialgleichung (5)

$$\text{div } \Phi = 0.$$

Nur hat die Anzahl Φ in unserem jetzigen Fall Komponenten in jeder Richtung $S^{(1)}, S^{(2)} \ldots$, die wir mit $\Phi^{(1)}, \Phi^{(2)} \ldots$ bezeichnen wollen. (5) nimmt also die Gestalt an

$$\sum_n \frac{\partial \Phi^{(n)}}{\partial S^{(n)}} = 0.$$

Zwischen $S^{(n)}_{(Z)}$ und $S^{(n)}_{(Z+1)}$ besteht, wie früher Gleichung (7), daher bleiben die Betrachtungen des vorigen Paragraphen vollkommen gültig zur Berechnung der einzelnen $\Phi^{(n)}$. Es wird also

$$\Phi^{(n)} = S^{(n)} F + \overline{f_n^2} \frac{\partial F}{\partial S^{(n)}}.$$

Wir können diesen Ausdruck noch vereinfachen, indem wir alle $\overline{f_n^2}$ als gleich annehmen. Dies kommt ersichtlich nur darauf hinaus, daß wir die einzelnen Funktionen f_n mit passenden Konstanten multipliziert denken. (Im speziellen Fall unserer sin und cos ist diese vereinfachende Annahme von selbst erfüllt.)

So erhalten wir schließlich für die Funktion F die Differentialgleichung:

$$(12) \qquad \sum_n \frac{\partial}{\partial S^{(n)}}\left(S^{(n)} F + \overline{f^2}\frac{\partial F}{\partial S^{(n)}}\right) = 0.$$

Zur Lösung dieser Differentialgleichung führt uns die Betrachtung des über den ganzen Raum erstreckten Integrals:

$$(13) \begin{cases} \int \frac{1}{F} \sum_{n}^{n_1} \left\{\left(S^{(n)} F + \overline{f^2}\frac{\partial F}{\partial S^{(n)}}\right)^2\right\} dS^{(1)} \ldots dS^{(n_1)} \\ = \int \sum_{n}^{n_1} \left\{\left(S^{(n)} F + \overline{f^2}\frac{\partial F}{\partial S^{(n)}}\right)\left(S^{(n)} + \overline{f^2}\frac{\partial \log F}{\partial S^{(n)}}\right)\right\} dS^{(1)} \ldots dS^{(n_1)}. \end{cases}$$

Nun ist aber:

$$\int \sum_{n}^{n_1}\left\{\left(S^{(n)} F + \overline{f^2}\frac{\partial F}{\partial S^{(n)}}\right) S^{(n)}\right\} dS^{(1)} \ldots dS^{(n_1)}$$

$$= \int \left(F \sum_{n}^{n_1} S^{(n)\,2} + \overline{f^2} \sum_{n}^{n_1} S^{(n)}\frac{\partial F}{\partial S^{(n)}}\right) dS^{(1)} \ldots dS^{(n_1)},$$

oder wenn wir den zweiten Summanden partiell integrieren und bedenken, daß im Unendlichen $F = 0$ sein muß,

$$= \int F \left(\sum_{n}^{n_1} S^{(n)\,2} - \overline{f^2} \cdot n_1\right) dS^{(1)} \ldots dS^{(n_1)}.$$

Dieser Ausdruck verschwindet aber, weil

$$\int F S^{(n)\,2} dS^{(1)} \ldots dS^{(n_1)}$$

nichts anderes ist, als der im letzten Paragraphen abgeleitete Mittelwert $\overline{S^{(n)\,2}}$, falls nur ein einziges S betrachtet wird; für diesen folgt aus Gleichung (10)

$$\overline{S^2} = \overline{f^2}.$$

Andererseits wird durch partielle Integration:

$$\int \sum \left\{ \left(S^{(n)} F + \overline{f^2} \frac{\partial F}{\partial S^{(n)}} \right) \overline{f^2} \frac{\partial \log F}{\partial S^{(n)}} \right\} dS^{(1)} \ldots dS^{(n_1)}$$
$$= \int \overline{f^2} \log F \sum \left(\frac{\partial}{\partial S^{(n)}} \left(S^{(n)} F + \overline{f^2} \frac{\partial F}{\partial S^{(n)}} \right) \right) dS^{(1)} \ldots dS^{(n_1)},$$

was nach Gleichung (12) ebenfalls verschwindet.

Somit ist erwiesen, daß das Integral (13) verschwindet; dies ist aber wegen des quadratischen Charakters des Integranden nur möglich, wenn überall für jedes n gilt:

(14) $$S^{(n)} F + \overline{f^2} \frac{\partial F}{\partial S^{(n)}} = 0.$$

So gelangen wir also für F zu einem statistischen Gesetz, welches in bezug auf jedes $S^{(n)}$ mit dem Gaussschen Fehlergesetz identisch ist:

(15) $$F = \text{const.}\, e^{-\frac{S^{(1)2}}{2\overline{f^2}}} \cdot e^{-\frac{S^{(2)2}}{2\overline{f^2}}} \cdot$$

Die Wahrscheinlichkeit einer Kombination von Werten $S^{(n)}$ setzt sich also einfach als Produkt aus den Wahrscheinlichkeiten der einzelnen $S^{(n)}$ zusammen.

Es ist klar, daß, wenn für $S^{(1)}, S^{(2)} \ldots$ die Gleichung (15) gilt, dieselbe Gleichung für eine Kombination von Größen

$$S^{(n)'} = a_n S^{(n)}$$

erfüllt ist. In diesem Falle tritt statt $\overline{f^2}$ die Größe $a_n^2 \overline{f^2}$ in die Exponenten ein. Von der Art der $S^{(n)'}$ sind aber die Koeffizienten A_n, B_n unseres physikalischen Problems; und zwar ist

$$S^{(n)} = \frac{A_n}{a_n \sqrt{Z}},$$

also

$$\alpha_n = a_n \sqrt{Z}$$

zu setzen.

Somit ist auch die Gültigkeit der Gleichung (1) und die Unmöglichkeit erwiesen, eine wahrscheinlichkeits-theoretische Beziehung zwischen den Koeffizienten der die Temperaturstrahlung darstellenden Fourierreihe aufzustellen.

(Eingegangen 29. August 1910.)

3. Statistische Untersuchung der Bewegung eines Resonators in einem Strahlungsfeld; *von A. Einstein und L. Hopf.*

§ 1. Gedankengang.

Es ist bereits auf verschiedenen Wegen gezeigt worden und heute wohl allgemein anerkannt, daß unsere gegenwärtigen Anschauungen von der Verteilung und Ausbreitung der elektromagnetischen Energie einerseits, von der statistischen Energieverteilung anderseits, bei richtiger Anwendung in der Strahlentheorie zu keinem anderen als dem sogenannten Rayleighschen (Jeansschen) Strahlungsgesetz führen können. Da dieses mit der Erfahrung in vollkommenem Widerspruch steht, ist es nötig, an den Grundlagen der zur Ableitung verwendeten Theorien eine Änderung vorzunehmen, und man hat vielfach vermutet, daß die Anwendung der statistischen Energieverteilungsgesetze auf die Strahlung oder auf rasch oszillierende Bewegungen (Resonatoren) nicht einwandfrei sei. Die folgende Untersuchung soll nun zeigen, daß es einer derartigen zweifelhaften Anwendung gar nicht bedarf, und daß es genügt, den Satz der Äquipartition der Energie nur auf die *fortschreitende* Bewegung der Moleküle und Oszillatoren anzuwenden, um zum Rayleighschen Strahlungsgesetz zu gelangen. Die Anwendungsfähigkeit des Satzes auf die fortschreitende Bewegung ist durch die Erfolge der kinetischen Gastheorie genügend erwiesen; wir werden daher schließen dürfen, daß erst eine prinzipiellere und tiefer gehende Änderung der grundlegenden Anschauungen zu einem der Erfahrung besser entsprechenden Strahlungsgesetz führen kann.

Wir betrachten einen beweglichen elektromagnetischen Oszillator[1]), der einesteils den Wirkungen eines Strahlungsfeldes unterliegt, andernteils mit einer Masse m behaftet ist und mit den im Strahlungsraum vorhandenen Molekülen in Wechselwirkung

---

1) Der Einfachheit halber werden wir annehmen, der Oszillator schwinge nur in der z-Richtung und sei nur in der x Richtung beweglich.

tritt. Betände diese letztere Wechselwirkung allein, so wäre der quadratische Mittelwert der Bewegungsgröße der fortschreitenden Bewegung des Oszillators durch die statistische Mechanik vollkommen bestimmt. In unserem Falle besteht außerdem die Wechselwirkung des Oszillators mit dem Strahlungsfelde. Damit statistisches Gleichgewicht möglich sei, darf diese letztere Wechselwirkung an jenem Mittelwerte nichts ändern. Mit anderen Worten: der quadratische Mittelwert der Bewegungsgröße der fortschreitenden Bewegung, welchen der Oszillator unter der Einwirkung *der Strahlung allein* annimmt, muß derselbe sein wie derjenige, welchen er nach der statistischen Mechanik unter der mechanischen Einwirkung der Moleküle allein annähme. Damit reduziert sich das Problem auf dasjenige, den quadratischen Mittelwert $\overline{(mv)^2}$ der Bewegungsgröße zu ermitteln, den der Oszillator unter der Einwirkung des Strahlungsfeldes allein annimmt.

Dieser Mittelwert muß zur Zeit $t = 0$ derselbe sein wie zur Zeit $t = \tau$, so daß man hat:

$$\overline{(mv)^2_{t=0}} = \overline{(mv)^2_{t=\tau}}.$$

Für das folgende ist es zweckmäßig, zweierlei Kraftwirkungen zu unterscheiden, durch welche das Strahlungsfeld den Oszillator beeinflußt, nämlich

1. Die Widerstandskraft K, welche der Strahlungsdruck einer geradlinigen Bewegung des Oszillators entgegenstellt. Diese ist bei Vernachlässigung der Glieder von Größenordnung $(v/c)^2$ (c = Lichtgeschwindigkeit) proportional der Geschwindigkeit v, wir können also schreiben: $K = -Pv$. Nehmen wir ferner an, daß während der Zeit τ die Geschwindigkeit v sich nicht merklich ändert, so wird der von dieser Kraft herrührende Impuls $= -Pv\tau$.

2. Die Schwankungen Δ des elektromagnetischen Impulses, die infolge der Bewegung elektrischer Massen im ungeordneten Strahlungsfelde auftreten. Diese können ebensowohl positiv, wie negativ sein und sind von dem Umstande, daß der Oszillator bewegt ist, in erster Annäherung unabhängig.

Diese Impulse superponieren sich während der Zeit τ auf den Impuls $(mv)_{t=0}$ und unsere Gleichung wird:

(1) $$\overline{(mv)^2_{t=0}} = \overline{(mv_{t=0} + \Delta - Pv\tau)^2}.$$

Bewegung eines Resonators in einem Strahlungsfeld. **1107**

Durch Vergrößerung der Masse m können wir jederzeit erreichen, daß das mit τ^2 multiplizierte Glied, welches auf der rechten Seite von Gleichung (1) erscheint, vernachlässigt werden darf. Ferner verschwindet das mit $\overline{v\varDelta}$ multiplizierte Glied, da v und $\varDelta$ voneinander ganz unabhängig sowohl negativ wie positiv werden können. Ersetzen wir noch $m\overline{v^2}$ durch die Temperatur Θ mittels der aus der Gastheorie bekannten Gleichung:

$$m\overline{v^2} = \frac{R}{N}\Theta$$

(R = absolute Gaskonstante, N = Loschmidtsche Zahl), so erhält Gleichung (1) die Form:

(2) $$\overline{\varDelta^2} = 2\frac{R}{N}P\Theta\tau.$$

Wir haben also nur $\overline{\varDelta^2}$ und P (bzw. $\overline{K}$) durch elektromagnetische Betrachtungen zu ermitteln, dann liefert Gleichung (2) das Strahlungsgesetz.

§ 2. Berechnung der Kraft $\overline{K}$.[1])

Um die Kraft zu berechnen, welche die Strahlung einem bewegten Oszillator entgegenstellt, berechnen wir zuerst die Kraft auf einen ruhenden Oszillator und transformieren diese dann mit Hilfe der aus der Relativitätstheorie folgenden Formeln.

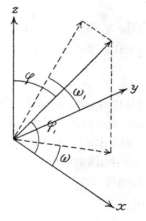

Der Oszillator mit Eigenschwingung v_0 schwinge frei in der z-Richtung eines rechtwinkeligen Koordinatensystems x, y, z. Bezeichnen dann $\mathfrak{E}$ und $\mathfrak{H}$ die elektrische bzw. magnetische Kraft des äußeren Feldes, so gehorcht das Moment f des Oszillators nach Planck[2]) der Differentialgleichung:

(3) $$16\pi^4 v_0^3 f + 4\pi^2 v_0 \ddot{f} - 2\sigma \dddot{f} = 3\sigma c^3 \mathfrak{E}_z.$$

Hierbei ist noch σ eine für die Dämpfung des Oszillators durch Ausstrahlung charakteristische Konstante.

---

1) Vgl. auch M. Abraham, Ann. d. Phys. **14.** p. 273 ff.. 1904.
2) M. Planck, Vorl. über die Theorie der Wärmestrahlung p. 113.

Es falle nun eine ebene Welle auf den Oszillator; der Strahl schließe mit der z-Achse den Winkel φ ein, seine Projektion auf die xy-Ebene mit der x-Achse den Winkel ω. Zerlegen wir diese Welle in zwei senkrecht zueinander polarisierte, davon die elektrische Kraft der einen in der Strahloszillatorebene liege, die der anderen senkrecht dazu, so ist klar, daß nur die erstere dem Oszillator ein gewisses Moment erteilt. Schreiben wir die elektrische Kraft dieser ersteren Wellen als Fouriersche Reihe

$$(4) \quad \mathfrak{E} = \sum_n A_n \cos\left\{\frac{2\pi n}{T}\left(t - \frac{\alpha x + \beta y + \gamma z}{c}\right) - \vartheta_n\right\},$$

wobei T eine sehr große Zeit bedeute, so drücken sich die Richtungskosinus α, β, γ des Strahles durch φ und ω in folgender Weise aus:

$$\alpha = \sin\varphi \cos\omega, \quad \beta = \sin\varphi \sin\omega, \quad \gamma = \cos\varphi$$

und die für unsere weitere Rechnung in Betracht kommenden Komponenten der elektrischen und der magnetischen Kraft sind:

$$(5) \quad \begin{cases} \mathfrak{E}_x = \mathfrak{E} \cos\varphi \cos\omega, \\ \mathfrak{E}_z = -\mathfrak{E} \sin\varphi, \\ \mathfrak{H}_y = \mathfrak{E} \cos\varphi \sin\omega. \end{cases}$$

Die ponderomotorische Kraft, welche auf den Oszillator ausgeübt wird, ist

$$k = f\frac{\partial \mathfrak{E}}{\partial x} + \frac{1}{c}\left[\frac{df}{dt}\mathfrak{H}\right].$$

Damit diese Gleichung, sowie Gleichung (3) gültig sei, muß angenommen werden, daß die Abmessungen des Oszillators stets klein seien gegen die in Betracht kommenden Strahlungswellenlängen. Die x-Komponente k_x der ponderomotorischen Kraft ist

$$(6) \quad k_x = \frac{\partial \mathfrak{E}_x}{\partial x}f - \frac{1}{c}\mathfrak{H}_y\frac{df}{dt}.$$

Durch Auflösung von (3)[1]) erhalten wir mit Berücksichtigung von (4) und (5):

$$f = -\frac{3c^3}{16\pi^3}T^3 \sin\varphi \sum_n A_n \frac{\sin\gamma_n}{n^3}\cos(\tau_n - \gamma_n),$$

$$\dot f = \frac{3c^3}{8\pi^2}T^2 \sin\varphi \sum_n A_n \frac{\sin\gamma_n}{n^2}\sin(\tau_n - \gamma_n),$$

---

1) M. Planck, l. c. p. 114.

wobei zur Abkürzung
$$\tau_n = 2\pi n \frac{t}{T} - \vartheta_n$$
gesetzt ist und γ_n durch die Gleichung gegeben ist:
$$\cotg \gamma_n = \frac{\pi \nu_0 \left(\nu_0^2 - \frac{n^2}{T^2}\right)}{\sigma \frac{n^3}{T^3}}.$$

Da ferner:
$$\frac{\partial \mathfrak{E}_x}{\partial z} = \frac{2\pi}{cT} \cos^2 \varphi \cos \omega \sum_n n A_n \sin^2 \tau_n{}^1),$$
erscheint k_x als Doppelsumme:
$$k_x = -\frac{3 c^2}{8\pi} T^2 \cos^2 \varphi \sin \varphi \cos \omega \sum_n \sum_m A_n \frac{\sin \gamma_n}{n^3}$$
$$A_m m \cos(\tau_n - \gamma_n) \sin \tau_m,$$
$$-\frac{3 c^2}{8\pi} T^2 \sin \varphi \cos \omega \sum_n \sum_m A_n \frac{\sin \gamma_n}{n^2}$$
$$A_m \sin(\tau_n - \gamma_n) \cos \tau_m.$$

Bei der Mittelwertbildung kommen wegen der Unabhängigkeit der Phasenwinkel ϑ voneinander nur die Glieder $n = m$ in Betracht[2]) und es wird:

(7) $\begin{cases} \overline{k_x} = \dfrac{3 c^2}{16 \pi^2} T^2 \sin^3 \varphi \cos \omega \sum_n A_n^2 \dfrac{\sin \gamma_n}{n^2} \\ \quad = \dfrac{3 c^2}{16 \pi^2} \overline{A_{\nu_0 T}^2} T \dfrac{\sigma}{2\nu_0} \sin^3 \varphi \cos \omega.{}^3) \end{cases}$

Dies ist der Mittelwert der x-Komponente der Kraft, welche eine in Richtung φ, ω einfallende Welle auf den ruhenden Oszillator ausübt.

Bewegt sich der Oszillator in der x-Richtung mit der Geschwindigkeit v, so ersetzen wir die Winkel φ, ω praktischer durch den Winkel φ_1 zwischen Strahl und x-Achse und den

---

1) Eigentlich wäre dieser Ausdruck für $\partial \mathfrak{E}_x/\partial z$ ebenso wie der für $\mathfrak{H}_y$ durch die Komponenten der Welle zu ergänzen, die senkrecht zu der den Oszillator erregenden polarisiert ist; doch ist klar, daß diese Ausdrücke wegen der Unabhängigkeit ihrer Phasen von denjenigen des Oszillators nichts zum Mittelwert der Kraft beitragen.

2) Diese Unabhängigkeit folgt aus dem Endergebnis der vorhergehenden Abhandlung.

3) M. Planck, l. c. p. 122.

Winkel ω_1 zwischen der Projektion des Strahles auf die yz-Ebene und der y-Achse. Es gelten dann die Beziehungen:
$$\cos \varphi_1 = \sin \varphi \cos \omega,$$
$$\sin \varphi_1 \cos \omega_1 = \sin \varphi \sin \omega,$$
$$\sin \varphi_1 \sin \omega_1 = \cos \varphi.$$

Zum Werte der Kraft $\overline{k_x'}$, welche auf den bewegten Oszillator wirkt, führen uns die Transformationsformeln der Relativitätstheorie[1])
$$A' = A\left(1 - \frac{v}{c}\cos\varphi_1\right),$$
$$T' = T\left(1 + \frac{v}{c}\cos\varphi_1\right),$$
$$\nu' = \nu\left(1 - \frac{v}{c}\cos\varphi_1\right),$$
$$\cos\varphi_1' = \frac{\cos\varphi_1 - \dfrac{v}{c}}{1 - \dfrac{v}{c}\cos\varphi_1}, \quad \omega_1' = \omega_1.$$

Es wird:
$$\overline{k_x'} = \frac{3c^2}{16\pi^2}\overline{A_{\nu_0'}^{'2}T'}\cdot T'\frac{\sigma}{2\nu_0'}(1 - \sin^2\varphi_1' \sin^2\omega_1')\cos\varphi_1'.$$

Nun ist, wenn Glieder mit $(v/c)^2$ vernachlässigt werden:
$$\overline{A_{\nu_0'}^{'2}T'} = \overline{A_{\nu_0}^2 T}\left(1 - 2\frac{v}{c}\cos\varphi_1\right),$$

oder, da wir alles auf die Eigenschwingung ν_0' des bewegten Oszillators zu beziehen haben:
$$\overline{A_{\nu_0'}^{'2}T'} = \overline{A^2}_{\nu_0'\left(1 + \frac{v}{c}\cos\varphi_1\right)T_\bullet}\left(1 - 2\frac{v}{c}\cos\varphi_1\right)$$
$$= \left\{\overline{A_{\nu_0'}^2 T} + \nu_0'\frac{v}{c}\cos\varphi_1\left(\frac{d\overline{A^2}}{d\nu}\right)_{\nu_0 T}\right\}\cdot\left(1 - 2\frac{v}{c}\cos\varphi_1\right).$$

Wir drücken weiterhin die Größe $\overline{A^2}T$ durch die mittlere Strahlungsdichte ϱ aus. Die mittlere Energie einer ebenen Welle, welche aus einer bestimmten Richtung kommt, setzen wir gleich der Energiedichte in einem Kegel vom Öffnungswinkel $d\varkappa$. Nehmen wir noch Rücksicht auf die Gleichheit der elektrischen und magnetischen Kraft und auf die beiden Polarisationsebenen, so gelangen wir zu der Beziehung:
$$\varrho\frac{d\varkappa}{4\pi} = \frac{1}{8\pi}\frac{\overline{A^2}T}{2}\cdot 2\cdot 2.$$

---

[1]) A. Einstein, Ann. d. Phys. 17. p. 914. 1905.

Unser Kraftausdruck wird:

$$(8) \quad \begin{cases} \overline{k_x'} = \dfrac{3\,c^2}{16\,\pi^2} \cdot \dfrac{\sigma}{2\,\nu_0'} \left\{ \varrho_{\nu_0'} + \nu_0' \dfrac{v}{c} \cos\varphi_1 \left(\dfrac{d\varrho}{d\nu}\right)_{\nu_0'} \right\} \left(\cos\varphi_1 - \dfrac{v}{c}\right) \\ \qquad\qquad \left(1 - \dfrac{\sin^2\varphi_1}{1 - 2\dfrac{v}{c}\cos\varphi_1} \sin^2\omega_1 \right) dx. \end{cases}$$

Integrieren wir schließlich noch über alle Öffnungswinkel, so erhalten wir die gesuchte Gesamtkraft:

$$(9) \quad \overline{K} = -\dfrac{3\,c\,\sigma}{10\,\pi\,\nu_0'} v \left\{ \varrho_{\nu_0'} - \dfrac{\nu_0'}{3}\left(\dfrac{d\varrho}{d\nu}\right)_{\nu_0'} \right\}.$$

§ 3. Berechnung der Impulsschwankungen $\overline{\varDelta^2}$.

Die Berechnung der Impulsschwankungen läßt sich gegenüber der Kraftberechnung bedeutend vereinfachen, da eine Transformation nach der Relativitätstheorie unnötig ist.[1]) Es genügt, die elektrische und magnetische Kraft im Anfangspunkt, als nur von der Zeit abhängig, in eine Fourierreihe zu entwickeln, wenn man nur den Beweis führen kann, daß die einzelnen in diesem Ausdruck auftretenden Kraftkomponenten voneinander unabhängig sind.

Der Impuls, welchen der Oszillator in der Zeit τ in der x-Richtung erfährt, ist:

$$J = \int_0^\tau k_x\, dt = \int_0^\tau \left(\dfrac{\partial \mathfrak{E}_x}{\partial x} f - \dfrac{1}{c} \mathfrak{H}_y \dfrac{df}{dt} \right) dt.$$

Partielle Integration ergibt:

$$\int_0^\tau \mathfrak{H}_y \dfrac{df}{dt}\, dt = [\mathfrak{H}_y f]_0^\tau - \int_0^\tau \dfrac{\partial \mathfrak{H}_y}{\partial t} f\, dt.$$

Der erste Summand verschwindet, wenn man τ passend wählt, bzw. wenn τ groß genug ist. Setzt man noch — nach der Maxwellschen Gleichung

$$\dfrac{1}{c} \dfrac{\partial \mathfrak{H}_y}{\partial t} = \dfrac{\partial \mathfrak{E}_z}{\partial x} - \dfrac{\partial \mathfrak{E}_x}{\partial z},$$

so gelangt man zu dem einfachen Ausdruck:

$$(10) \quad J = \int_0^\tau \dfrac{\partial \mathfrak{E}_z}{\partial x} f\, dt.$$

---

[1]) Die von den Unregelmäßigkeiten des Strahlungsvorganges herrührenden Impulse wechselnden Vorzeichens können nämlich für einen *ruhenden* Resonator ermittelt werden.

Nun treten in unserem Ausdruck nur die Komponente E_z und ihre Ableitung $\partial \mathfrak{E}_z/\partial x$ auf. Deren Unabhängigkeit läßt sich aber leicht nachweisen. Denn betrachten wir nur zwei sich entgegenkommende Wellenzüge (vom gleichen Öffnungswinkel), so können wir schreiben:

$$E_z = \sum \left\{ a_n \sin \frac{2\pi n}{T} \left(t - \frac{\alpha x + \beta y + \gamma z}{c} \right) \right.$$
$$+ b_n \cos \frac{2\pi n}{T} \left(t - \frac{\alpha x + \beta y + \gamma z}{c} \right)$$
$$+ a_n' \sin \frac{2\pi n}{T} \left(t + \frac{\alpha x + \beta y + \gamma z}{c} \right)$$
$$\left. + b_n' \cos \frac{2\pi n}{T} \left(t + \frac{\alpha x + \beta y + \gamma z}{c} \right) \right\}$$

und

$$\frac{\partial \mathfrak{E}_z}{\partial x} = \sum \left\{ \frac{2\pi n \alpha}{Tc} \left[-a_n \cos \frac{2\pi n}{T}(\cdots) + b_n \sin \frac{2\pi n}{T}(\cdots) \right.\right.$$
$$\left.\left. + a_n' \cos \frac{2\pi n}{T}(\cdots) - b_n' \sin \frac{2\pi n}{T}(\cdots) \right] \right\}.$$

Die Größen $a_n + a_n'$, $a_n - a_n' \cdots$ sind aber voneinander unabhängig und vom selben Charakter, wie die in der vorangehenden Abhandlung mit S bezeichneten; für solche ist dort nachgewiesen, daß sich das Wahrscheinlichkeitsgesetz einer Kombination darstellt als Produkt von Gaussschen Fehlerfunktionen der einzelnen Größen. Aus dem Gesagten schließt man leicht, daß zwischen den Koeffizienten der Entwickelungen von $\mathfrak{E}_z$ und $\partial \mathfrak{E}_z/\partial x$ keinerlei Wahrscheinlichkeitsbeziehung bestehen kann.

Wir setzen nun $\mathfrak{E}_z$ und $\partial \mathfrak{E}_z/\partial x$ als Fourierreihen an:

$$\mathfrak{E}_z = \sum{}_m B_n \cos\left(2\pi n \frac{t}{T} - \vartheta_n\right),$$
$$\frac{\partial \mathfrak{E}_z}{\partial x_z} = \sum{}_n C_m \cos\left(2\pi m \frac{t}{T} - \xi_m\right).$$

Dann wird:

$$f = \frac{3c^3}{16\pi^3} T^3 \sum{}_n B_n \frac{\sin \gamma_n}{n^3} \cos\left(2\pi n \frac{t}{T} - \vartheta_n - \gamma_n\right)$$

und

$$J = \frac{3c^3}{16\pi^3} T^3 \int_0^\tau dt \sum{}_m \overline{\sum{}_n} C_m B_n \frac{\sin \gamma_n}{n^3}$$
$$\left[\cos\left\{2\pi(n+m)\frac{t}{T} - \xi_m - \vartheta_n - \gamma_n\right\} \right.$$
$$\left. - \cos\left\{2\pi(n-m)t + \xi_m - \vartheta_n - \gamma_n\right\} \right].$$

Bei der Integration über t ergeben sich zwei Summanden mit den Faktoren $1/n+m$ und $1/n-m$; da n und m sehr große Zahlen sind, ist der erstere sehr klein, kann also vernachlässigt werden. Man gelangt so zu dem Ausdruck:

$$(11) \quad J = -\frac{3\,c^3}{32\,\pi^4} T^4 \sum_m \sum_n C_m B_n \cdot \frac{\sin \gamma_n}{n^3} \frac{1}{n-m} \cos \delta_{mn} \cdot \sin \pi(n-m)\frac{\tau}{T}$$

mit der Abkürzung:

$$\delta_{mn} = \pi(n-m)\frac{\tau}{T} + \xi_m - \vartheta_n - \gamma_n.$$

J^2 erscheint dann als vierfache Summe über n, m und zwei weitere Variable n' und m'. Bilden wir den Mittelwert $\overline{J^2}$, so haben wir darauf zu achten, daß die Winkel δ_{mn} und $\delta_{m'n'}$ vollkommen voneinander unabhängig sind, daß also bei der Mittelwertbildung nur die Terme in Betracht kommen, bei denen diese Unabhängigkeit aufgehoben ist. Ersichtlich ist dies nur der Fall, wenn

$$m = m' \quad \text{und} \quad n = n',$$

gelangen wir zu dem gesuchten Mittelwert:

$$\overline{J^2} = \left(\frac{3\,c^3\,T^4}{32\,\pi^4}\right)^2 \sum_m \sum_n \frac{1}{2} C_m^{\,2} B_n^{\,2} \left(\frac{\sin\gamma_n}{n^3}\right)^2 \frac{1}{(n-m)^2} \sin^2 \pi(n-m)\frac{t}{T},$$

da

$$\sum_m \frac{1}{(n-m)^2} \sin^2 \pi(n-m)\frac{t}{T}$$
$$= \frac{1}{T} \int_0^\infty \frac{1}{(\nu-\mu)^2} \sin^2(\nu-\mu)\pi\tau \cdot d\mu = \frac{\pi^2\,\tau}{T}.$$

und

$$\sum_n \frac{\sin^2 \gamma_n}{n^6} = \frac{1}{T^5} \int_0^\infty \frac{\sin\gamma_n}{\nu^6} d\nu = \frac{1}{T^5} \cdot \frac{\sigma}{2\,\nu_0^{\,5}},$$

wird:

$$(12) \quad \overline{J^2} = \left(\frac{3\,c^3}{32\,\pi^3}\right)^2 \frac{\sigma\,\tau}{4\,\nu_0^{\,5}} \overline{B_{\nu_0}^{\,2} T} \cdot \overline{C_{\nu_0}^{\,2} T} \cdot T^2.$$

Nun ist:

$$\overline{J^2} = \overline{(\overline{J}+\Delta)^2} = \overline{J}^2 + 2\overline{J}\overline{\Delta} + \overline{\Delta^2},$$

und da die Mittelwerte $\overline{J}$ und $\overline{\Delta}$ verschwinden, gibt Ausdruck (12) den Wert der Impulsschwankungen $\overline{\Delta^2}$ selbst an.

Es erübrigt noch die Mittelwerte der Amplituden $\overline{B^2_{\nu_0 T}}$ und $\overline{C^2_{\nu_0 T}}$ durch die Strahlungsdichte ϱ_{ν_0} auszudrücken.

Zu diesem Zweck müssen wir wieder die von den verschiedenen Richtungen herkommende Strahlung betrachten und, wie oben, die Amplitude der aus einer bestimmten Richtung kommenden Strahlung mit der Energiedichte in Beziehung setzen durch die Gleichung:
$$\overline{A^2_{\nu_0 T}}\, T = \varrho_{\nu_0}\, d\varkappa.$$

Die Amplitude:
$$B_{\nu_0 T} = \sum A_{\nu_0 T} \sin \varphi$$

über alle Einfallswinkel, also

(13) $\qquad \overline{B^2_{\nu_0 T}} \cdot T = \overline{A^2_{\nu_0 T}} \cdot T \sum \sin^2 \varphi = \tfrac{8}{3}\pi \varrho_{\nu_0}.$

Analog ergibt sich:

(14) $\qquad \overline{C^2_{\nu_0 T}}\, T = \left(\dfrac{2\pi\nu}{c}\right)^2 \overline{A^2_{\nu_0 T}}\, T \sum \sin^4 \varphi \cos^2 \omega = \dfrac{64}{15}\dfrac{\pi^3 \nu_0^2}{c^2} \varrho_{\nu_0}.$

So erhalten wir schließlich durch Einsetzen von (13) und (14) in (12):

(15) $\qquad \overline{\varDelta^2} = \dfrac{c^4 \sigma \tau}{40\, \pi^2 \nu_0^3} \varrho_{\nu_0}^2.$

§ 5. Das Strahlungsgesetz.

Wir haben jetzt nur noch die gefundenen Werte (9) und (15) in unsere Gleichung (2) einzusetzen, so gelangen wir zu der das Strahlungsgesetz enthaltenden Differentialgleichung:
$$\dfrac{c^3 N}{24\,\pi R\,\Theta\,\nu^2}\varrho^2 = \varrho - \dfrac{\nu}{3}\dfrac{d\varrho}{d\nu},$$

welche integriert ergibt:

(16) $\qquad \varrho = \dfrac{8\pi R\,\Theta\,\nu^2}{c^3 N}.$

Dies ist das wohlbekannte Rayleighsche Strahlungsgesetz, welches mit der Erfahrung im grellsten Widerspruche steht. In den Grundlagen unserer Ableitung muß also eine Aussage stecken, welche sich mit den wirklichen Erscheinungen bei der Temperaturstrahlung nicht im Einklang befindet.

Betrachten wir darum diese Grundlagen kritisch näher:

Man hat den Grund dafür, daß alle exakten statistischen Betrachtungen im Gebiete der Strahlungslehre zum Rayleighschen Gesetze führen, in der Anwendung dieser Betrachtungsweise auf die Strahlung selbst finden wollen. Planck[1]) hält dies Argument mit einem gewissen Recht der Jeansschen Ableitung entgegen. Bei der obigen Ableitung war aber von einer irgendwie willkürlichen Übertragung statistischer Betrachtungen auf die Strahlung gar nicht die Rede; der Satz von der Äquipartition der Energie wurde nur auf die fortschreitende Bewegung der Oszillatoren angewandt. Die Erfolge der kinetischen Gastheorie zeigen aber, daß für die fortschreitende Bewegung dieser Satz als durchaus bewiesen angesehen werden kann.

Das bei unserer Ableitung benutzte theoretische Fundament, das eine unzutreffende Annahme enthalten muß, ist also kein anderes, als das der Dispersionstheorie des Lichtes bei vollkommen durchsichtigen Körpern zugrunde liegende. Die wirklichen Erscheinungen unterscheiden sich von den aus diesem Fundament zu erschließenden Resultaten dadurch, daß bei ersteren noch Impulsschwankungen anderer Art sich bemerkbar machen, die bei kurzwelliger Strahlung von geringer Dichte die von der Theorie gelieferten ungeheuer überwiegen.[2])

Zürich, August 1910.

---

1) M. Planck, l. c. p. 178.
2) Vgl. A. Einstein, Phys. Zeitschr. **10**. p. 185 ff. Das wesentlich Neue der vorliegenden Arbeit besteht darin, daß die Impulsschwankungen zum erstenmal exakt ausgerechnet wurden.

(Eingegangen 29. August 1910.)

11. *Theorie der Opaleszenz von homogenen Flüssigkeiten und Flüssigkeitsgemischen in der Nähe des kritischen Zustandes;* von *A. Einstein.*

Smoluchowski hat in einer wichtigen theoretischen Arbeit[1]) gezeigt, daß die Opaleszenz bei Flüssigkeiten in der Nähe des kritischen Zustandes sowie die Opaleszenz bei Flüssigkeitsgemischen in der Nähe des kritischen Mischungsverhältnisses und der kritischen Temperatur vom Standpunkte der Molekulartheorie der Wärme aus in einfacher Weise erklärt werden kann. Jene Erklärung beruht auf folgender allgemeiner Folgerung aus Boltzmanns Entropie — Wahrscheinlichkeitsprinzip: Ein nach außen abgeschlossenes physikalisches System durchläuft im Laufe unendlich langer Zeit alle Zustände, welche mit dem (konstanten) Wert seiner Energie vereinbar sind. Die statistische Wahrscheinlichkeit eines Zustandes ist hierbei aber nur dann merklich von Null verschieden, wenn die Arbeit, die man nach der Thermodynamik zur Erzeugung des Zustandes aus dem Zustande idealen thermodynamischen Gleichgewichtes aufwenden müßte, von derselben Größenordnung ist, wie die kinetische Energie eines einatomigen Gasmoleküls bei der betreffenden Temperatur.

Wenn eine derart kleine Arbeit genügt, um in Flüssigkeitsräumen von der Größenordnung eines Wellenlängenkubus eine von der mittleren Dichte der Flüssigkeit merklich abweichende Dichte bzw. ein von dem mittleren merklich abweichendes Mischungsverhältnis herbeizuführen, so muß slso offenbar die Erscheinung der Opaleszenz (Tyndallphänomen) auftreten. Smoluchowski zeigte, daß diese Bedingung in der Nähe der kritischen Zustände tatsächlich erfüllt ist; er hat aber keine exakte Berechnung der Menge des durch Opaleszenz seitlich abgegebenen Lichtes gegeben. Diese Lücke soll im folgenden ausgefüllt werden.

---

1) M. v. Smoluchowski, Ann. d. Phys. 25. p. 205—226. 1908.

§ 1. Allgemeines über das Boltzmannsche Prinzip.

Das Boltzmannsche Prinzip kann durch die Gleichung

(1) $$S = \frac{R}{N} \lg W + \text{konst.}$$

formuliert werden. Hierbei bedeutet

R die Gaskonstante,
N die Zahl der Moleküle in einem Grammolekül,
S die Entropie,
W ist die Größe, welche als die „Wahrscheinlichkeit" desjenigen Zustandes bezeichnet zu werden pflegt, welchem der Entropiewert S zukommt.

Gewöhnlich wird W gleichgesetzt der Anzahl der möglichen verschiedenen Arten (Kompexionen), in welchen der ins Auge gefaßte, durch die beobachtbaren Parameter eines Systems im Sinne einer Molekulartheorie unvollständig definierte Zustand realisiert gedacht werden kann. Um W berechnen zu können, braucht man eine *vollständige* Theorie (etwa eine vollständige molekular-mechanische Theorie) des ins Auge gefaßten Systems. Deshalb erscheint es fraglich, ob bei dieser Art der Auffassung dem Boltzmannschen Prinzip *allein*, d. h. ohne eine *vollständige* molekular-mechanische oder sonstige die Elementarvorgänge vollständig darstellende Theorie (Elementartheorie) irgend ein Sinn zukommt. Gleichung (1) erscheint ohne Beigabe einer Elementartheorie oder — wie man es auch wohl ausdrücken kann — vom phänomenologischen Standpunkt aus betrachtet inhaltlos.

Das Boltzmannsche Prinzip erhält jedoch einen Inhalt unabhängig von jeder Elementartheorie, wenn man aus der Molekularkinetik den Satz annimmt und verallgemeinert, daß die Nichtumkehrbarkeit der physikalischen Vorgänge nur eine scheinbare sei.

Es sei nämlich der Zustand eines Systems in phänomenologischem Sinne bestimmt durch die prinzipiell beobachtbaren Variabeln $\lambda_1 \ldots \lambda_n$. Jedem Zustand Z entspricht eine Kombination von Werten dieser Variabeln. Ist das System nach außen abgeschlossen, so ist die Energie — und zwar im allgemeinen außer dieser keine andere Funktion der Variabeln — unveränderlich. Wir denken uns alle mit dem Energie-wert des Systems vereinbarten Zustände des Systems und bezeichnen sie mit $Z_1 \ldots Z_l$. Wenn die Nichtumkehrbarkeit der Vorgänge keine prinzipielle ist, so werden diese Zustände $Z_1 \ldots Z_l$ im Laufe der Zeit immer wieder vom System durchlaufen werden. Unter dieser Annahme kann man in folgendem Sinne von der Wahrscheinlichkeit der einzelnen Zustände sprechen. Denkt man sich das System eine ungeheuer lange Zeit Θ hindurch beobachtet und den Bruchteil τ_1 der Zeit Θ ermittelt, in welchem das System den Zustand Z_1 hat, so ist τ_1/Θ die Wahrscheinlichkeit des Zustandes Z_1. Analoges gilt für die Wahrscheinlichkeit der übrigen Zustände Z. Wir haben nach Boltzmann die scheinbare Nichtumkehrbarkeit darauf zurückzuführen, daß die Zustände von verschiedener Wahrscheinlichkeit sind, und daß das System wahrscheinlich Zustände größerer Wahrscheinlichkeit annimmt, wenn es sich gerade in einem Zustande relativ geringer Wahrscheinlichkeit befindet. Das scheinbar vollkommen Gesetzmäßige nichtumkehrbarer Vorgänge ist darauf zurückzuführen, daß die Wahrscheinlichkeiten der einzelnen Zustände Z von *verschiedener Größenordnung* sind, so daß von allen an einen bestimmten Zustand Z angrenzenden Zuständen *einer* wegen seiner gegenüber den anderen ungeheuren Wahrscheinlichkeit praktisch immer auf den erstgenannten Zustand folgen wird.

Die soeben fortgesetzte Wahrscheinlichkeit, zu deren Defination es keiner Elementartheorie bedarf, ist es, welche mit der Entropie in der durch Gleichung (1) ausgedrückten Beziehung steht. Daß Gleichung (1) für die so definierte Wahrscheinlichkeit wirklich gelten muß, ist leicht einzusehen. Die Entropie ist nämlich eine Funktion, welche (innerhalb des Gültigkeitsbereiches der Thermodynamik) bei keinem Vorgange abnimmt, bei welchem das System ein isoliertes ist. Es gibt noch andere Funktionen, welche diese Eigenschaft haben; alle aber sind, falls die Energie E die einzige zeitlich invariante Funktion des Systems ist, von der Form $\varphi(S, E)$, wobei $\partial \varphi/\partial S$ stets positiv ist. Da die Wahrscheinlichkeit W ebenfalls eine bei keinem Prozesse abnehmende Funktion ist, so ist auch W eine Funktion von S und E allein, oder — wenn nur Zustände derselben Energie verglichen werden — eine Funktion von S allein. Daß die zwischen S und W in Gleichung (1) gegebene Beziehung die einzig mögliche ist, kann bekanntlich aus dem Satze abgeleitet werden, daß die Entropie eines aus Teilsystemen bestehenden Gesamtsystems gleich ist der Summe der Entropien der Teilsysteme. So kann Gleichung (1) für alle Zustände Z bewiesen werden, die zu demselben Wert der Energie gehören.

Dieser Auffassung des Boltzmannschen Prinzipes steht zunächst folgender Einwand entgegen. Man kann nicht von der statistischen Wahrscheinlichkeit eines *Zustandes*, sondern nur von der eines *Zustandsgebietes* reden. Ein solches ist definiert durch einen Teil g der „Energiefläche" $E(\lambda_1 \ldots \lambda_n) = 0$. W sinkt offenbar mit der Größe des gewählten Teiles der Energiefläche zu Null herab. Hierdurch würde Gleichung (1) durchaus bedeutungslos, wenn die Beziehung zwischen S und W nicht von ganz besonderer Art wäre. Es tritt nämlich in (1) $\lg W$ mit dem sehr kleinen Faktor R/N multipliziert auf. Denkt man sich W für ein so großes Gebiet G_w ermittelt, daß dessen Abmessungen etwa an der Grenze des Wahrnehmbaren liegen, so wird $\lg W$ einen bestimmten Wert haben. Wird das Gebiet etwa e^{10} mal verkleinert, so wird die rechte Seite nur um die verschwindend kleine Größe $10(R/N)$ wegen der Verminderung der Gebietsgröße verkleinert. Wenn daher die Abmessungen des Gebietes zwar klein gewählt werden gegenüber beobachtbaren Abmessungen, aber doch so groß, daß $R/N \lg G_w/G$ numerisch von vernachlässigbarer Größe ist, so hat Gleichung (1) einen genügend genauen Inhalt.

Es wurde bisher angenommen, daß $\lambda_1 \ldots \lambda_n$ den Zustand des betrachteten Systems im phänomenologischen Sinne *vollständig* bestimmen. Gleichung (1) behält ihre Bedeutung aber auch ungeschmälert bei, wenn wir nach der Wahrscheinlichkeit eines im phänomenologischen Sinne unvollständig bestimmten Zustandes fragen. Fragen wir nämlich nach der Wahrscheinlichkeit eines Zustandes, der durch bestimmte Werte von $\lambda_1 \ldots \lambda_\nu$ definiert ist (wobei $\nu < n$), während wir die Werte von $\lambda_\nu \ldots \lambda_n$ unbestimmt lassen. Unter allen Zuständen mit den Werten $\lambda_1 \ldots \lambda_\nu$ werden diejenigen Werte von $\lambda_\nu \ldots \lambda_n$ weitaus die häufigsten sein, welche die Entropie des Systems bei konstantem $\lambda_1 \ldots \lambda_\nu$ zu einem Maximum machen. Zwischen diesem Maximalwerte der Energie und der Wahrscheinlichkeit *dieses* Zustandes wird in diesem Falle Gleichung (1) bestehen.

§ 2. Über die Abweichungen von einem Zustande thermodynamischen Gleichgewichtes.

Wir wollen nun aus Gleichung (1) Schlüsse ziehen über den Zusammenhang zwischen den thermodynamischen Eigenschaften eines Systems und dessen statistischen Eigenschaften. Gleichung (1) liefert unmittelbar die Wahrscheinlichkeit eines Zustandes, wenn die Entropie desselben gegeben ist. Wir haben jedoch gesehen, daß diese Beziehung keine exakte ist; es kann vielmehr bei bekanntem S nur die Größenordnung der Wahrscheinlichkeit W des betreffenden Zustandes ermittelt werden. Trotzdem aber können aus (1) genaue Beziehungen über das statistische Verhalten eines Systems abgeleitet werden, und zwar in dem Falle, daß der Bereich der Zustandsvariabeln, für welchen W in Betracht kommende Werte hat, als unendlich klein angesehen werden kann.

Aus Gleichung (1) folgt
$$W = \text{konst.}\, e^{\frac{N}{R} S}.$$

Diese Gleichung gilt der Größenordnung nach, wenn man jedem Zustand Z ein kleines Gebiet, von der Größenordnung wahrnehmbarer Gebiete, zuordnet. Die Konstante bestimmt sich der Größenordnung nach durch die Erwägung, daß W für den Zustand des Entropiemaximums (Entropie S_0) von der Größenordnung Eins ist, so daß man der Größenordnung nach hat
$$W = e^{\frac{N}{R}(S - S_0)}.$$

Daraus ist zu folgern, daß die Wahrscheinlichkeit dW dafür, daß die Größen $\lambda_1 \ldots \lambda_n$ zwischen λ_1 und $\lambda_1 + d\lambda_1 \ldots \lambda_n$ und $\lambda_n + d\lambda_n$ liegen, der Größenordnung nach gegeben ist durch die Gleichung[1])
$$dW = e^{\frac{N}{R}(S - S_0)} \cdot d\lambda_1 \ldots d\lambda_n$$

---

1) Wir wollen annehmen, daß Gebiete von Ausdehnungen beobachtbarer Größe in den λ endlich ausgedehnt sind.

und zwar in dem Falle, daß das System durch die $\lambda_1 \ldots \lambda_n$ (in phänomenologischem Sinne) nur unvollständig bestimmt ist.[1]) Genau genommen unterscheidet sich dW von dem gegebenen Ausdruck noch durch einen Faktor f, so daß zu setzen ist

$$dW = e^{\frac{N}{R}(S-S_0)} \cdot f \cdot d\lambda_1 \ldots d\lambda_n.$$

Dabei wird f eine Funktion von $\lambda_1 \ldots \lambda_n$ und von solcher Größenordnung sein, daß es die Größenordnung des Faktors auf der rechten Seite nicht beeinträchtigt.[2])

Wir bilden nun dW für die unmittelbare Umgebung eines Entropiemaximums. Es ist, falls die Taylorsche Entwickelung in dem in Betracht kommenden Bereich konvergiert, zu setzen

$$S = S_0 - \tfrac{1}{2} \sum \sum s_{\mu\nu} \lambda_\mu \lambda_\nu + \ldots$$

$$f = f_0 + \sum \lambda_\nu \left(\frac{\partial f}{\partial \lambda_\nu}\right) + \ldots$$

falls für den Zustand des Entropiemaximums $\lambda_1 = \lambda_2 = \ldots \lambda_n = 0$ ist. Die Doppelsumme im Ausdruck für S ist, weil es sich um ein Entropiemaximum handelt, wesentlich positiv. Man kann daher statt der λ neue Variable einführen, so daß sich jene Doppelsumme in eine einfache Summe verwandelt, in der nur die Quadrate der wieder mit λ bezeichneten neuen Variabeln auftreten. Man erhält

$$dW = \text{konst.}\, e^{-\frac{N}{2R} \sum s_\nu \lambda_\nu^2 + \ldots} \cdot \left[f_0 + \sum \left(\frac{\partial f}{\partial \lambda_\nu} \lambda_\nu\right)\right] d\lambda_1 \ldots d\lambda_n.$$

Die im Exponenten auftretenden Glieder erscheinen mit der sehr großen Zahl N/R multipliziert. Deshalb wird der Exponentialfaktor im allgemeinen bereits für solche Werte der λ praktisch verschwinden, die wegen ihrer Kleinheit keinen vom Zustand thermodynamischen Gleichgewichtes irgendwie erheblich abweichenden Zuständen des Systems entsprechen. Für

---

1) Im anderen Falle wäre die Mannigfaltigkeit der möglichen Zustände wegen des Energieprinzipes nur $(n-1)$ dimensional.

2) Über die Größenordnung der Ableitungen der Funktion f nach den λ wissen wir nichts. Wir wollen aber im folgenden annehmen, daß die Ableitungen von f der Größenordnung nach der Funktion f selbst gleich sind.

derartig kleine Werte λ wird man stets den Faktor f durch denjenigen Wert f_0 ersetzen können, den er im Zustand des thermodynamischen Gleichgewichtes hat. In allen diesen Fällen, in denen die Variablen nur wenig von ihren dem idealen thermischen Gleichgewicht entsprechenden Werten abweichen, kann also die Formel durch

$$(2) \qquad dW = \text{konst.}\, e^{-\frac{N}{R}(S-S_0)} \cdot d\lambda_1 \ldots d\lambda_n$$

ersetzt werden.

Für derart kleine Abweichungen vom thermodynamischen Gleichgewicht, wie sie für unseren Fall in Betracht kommen, hat die Größe $S - S_0$ eine anschauliche Bedeutung. Denkt man sich die uns interessierenden Zustände in der Nähe des thermodynamischen Gleichgewichtes durch äußere Einwirkung in umkehrbarer Weise hergestellt, so gilt nach der Thermodynamik für jeden Elementarvorgang die Energiegleichung

$$dU = dA + T\, dS,$$

falls man mit U die Energie des Systems, mit dA die demselben zugeführte elementare Arbeit bezeichnet. Uns interessieren nur Zustände, welche ein nach außen abgeschlossenes System annehmen kann, also Zustände, die zu dem nämlichen Energiewerte gehören. Für den Übergang eines solchen Zustandes in einen benachbarten ist $dU = 0$. Es wird ferner nur einen vernachlässigbaren Fehler bedingen, wenn wir in obiger Gleichung T durch die Temperatur T_0 des thermodynamischen Gleichgewichtes ersetzen. Obige Gleichung geht dann über in

$$dA + T_0\, dS = 0$$

oder

$$(3) \qquad \int dS = S - S_0 = \frac{1}{T_0} A,$$

wobei A die Arbeit bedeutet, welche man nach der Thermodynamik aufwenden müßte, um das System aus dem Zustande thermodynamischen Gleichgewichtes in den betrachteten Zustand überzuführen. Wir können also Gleichung (2) in der Form schreiben

$$(2\text{a}) \qquad dW = \text{konst.}\, e^{\frac{N}{R T_0} A}\, d\lambda_1 \ldots d\lambda_n.$$

Die Parameter λ denken wir uns nun so gewählt, daß sie beim thermodynamischen Gleichgewicht gerade verschwinden. In einer gewissen Umgebung wird A nach den λ nach dem Taylorschen Satz entwickelbar sein, welche Entwickelung bei passender Wahl der λ die Gestalt haben wird

$A + \frac{1}{2} \sum a_\nu \lambda_\nu^2 +$ Glieder höheren als zweiten Grades in den λ,

wobei die a_ν sämtlich positiv sind. Da ferner im Exponenten der Gleichung (2a) die Größe A mit dem sehr großen Faktor N/RT_0 multipliziert erscheint, so wird der Exponentialfaktor im allgemeinen nur für sehr kleine Werte von A, also auch für sehr kleine Werte der λ merkbar von Null abweichen. Für derart kleine Werte der λ werden im allgemeinen die Glieder höheren als ersten Grades im Ausdruck von A gegenüber den Gliedern zweiten Grades nur vernachlässigbare Beiträge liefern. Ist dies der Fall, so können wir für Gleichung (2a) setzen

(2b)
$$dW = \text{konst.} \, e^{-\frac{N}{2RT_0} \sum a_\nu \lambda_\nu^2} d\lambda_1 \ldots d\lambda_n,$$

eine Gleichung, welche die Form des Gaussschen Fehlergesetzes hat.

Auf diesen wichtigsten Spezialfall wollen wir uns in dieser Arbeit beschränken. Aus (2b) folgt unmittelbar, daß der Mittelwert der auf den Parameter λ_ν entfallenden Abweichungsarbeit A_ν den Wert hat

(4)
$$\overline{A_\nu} = \overline{\tfrac{1}{2} a_\nu \lambda_\nu^2} = \frac{RT_0}{2N}.$$

Diese mittlere Arbeit ist also gleich dem dritten Teil der mittleren kinetischen Energie eines einatomigen Gasmoleküls.

§ 3. Über die Abweichungen der räumlichen Verteilung von Flüssigkeiten und Flüssigkeitsgemischen von der gleichmäßigen Verteilung.

Wir bezeichnen mit ϱ_0 die mittlere Dichte einer homogenen Substanz bzw. die mittlere Dichte der einen Komponente eines binären Flüssigkeitsgemisches. Wegen der Unregelmäßigkeit der Wärmebewegung wird die Dichte ϱ in einem Punkte der Flüssigkeit von ϱ_0 im allgemeinen verschieden sein. Ist die Flüssigkeit in einen Würfel eingeschlossen, welcher bezüglich eines Koordinatensystems durch

$$0 < x < L,$$
$$0 < y < L$$

und

$$0 < z < L$$

charakterisiert ist, so können wir für das Innere dieses Würfels setzen

$$(5) \quad \begin{cases} \varrho = \varrho_0 + \triangle, \\ \triangle = \sum_\varrho \sum_\sigma \sum_\tau B_{\varrho\sigma\tau} \cos 2\pi \varrho \frac{x}{2L} \cos 2\pi \sigma \frac{y}{2L} \cos 2\pi \tau \frac{z}{2L}. \end{cases}$$

Die Größen ϱ, σ, τ bedeuten die ganzen positiven Zahlen. Hierzu ist aber folgendes zu bemerken.

Streng genommen kann man nicht von der Dichte einer Flüssigkeit in einem Raumpunkte reden, sondern nur von der mittleren Dichte in einem Raume, dessen Abmessungen groß sind gegenüber der mittleren Distanz benachbarter Moleküle. Aus diesem Grunde werden die Glieder der Entwickelung, bei denen eine der Größen ϱ, σ, τ oberhalb gewisser Grenzen liegt, keine physikalische Bedeutung besitzen. Aus dem folgenden wird man aber ersehen, daß dieser Umstand für uns nicht von Bedeutung ist.

Die Größen $B_{\varrho, \sigma, \tau}$ werden sich mit der Zeit ändern, derart, daß sie im Mittel gleich Null sind. Wir fragen nach den statistischen Gesetzen, denen die Größen B unterliegen. Diese spielen die Rolle der Parameter λ des vorigen Paragraphen, welche den Zustand unseres Systems im phänomenologischen Sinne bestimmen.

Diese statistischen Gesetze erhalten wir nach dem vorigen Paragraphen, indem wir die Arbeit A in Funktion der Größen B ermitteln. Dies ist auf folgende Weise möglich. Bezeichnen wir mit $\varphi(\varrho)$ die Arbeit, die man aufwenden muß, um die Masseneinheit von der mittleren Dichte ϱ_0 isotherm auf die Dichte ϱ zu bringen, so hat diese Arbeit für die im Volumenelement $d\tau$ befindliche Masse $\varrho \, d\tau$ den Wert

$$\varrho \varphi \, d\tau,$$

also für den ganzen Flüssigkeitswürfel den Wert
$$A = \int \varrho \cdot \varphi \cdot d\tau.$$

Wir werden anzunehmen haben, daß die Abweichungen $\triangle$ der Dichte von der mittleren sehr klein sind und setzen
$$\varrho = \varrho_0 + \triangle,$$
$$\varphi = \varphi(\varrho_0) + \left(\frac{\partial \varphi}{\partial \varrho}\right)_0 \triangle + \tfrac{1}{2}\left(\frac{\partial^2 \varphi}{\partial \varrho^2}\right)_0 \triangle^2 + \cdots$$

Hieraus folgt, weil $\varphi(\varrho_0) = 0$ und $\int \triangle \, d\tau = 0$ ist,
$$A = \left(\frac{\partial \varphi}{\partial \varrho} + \tfrac{1}{2}\varrho \frac{\partial^2 \varphi}{\partial \varrho^2}\right)\int \triangle^2 d\tau,$$

wobei der Index „0" der Einfachheit halber fortgelassen ist. Dabei sind im Integranden die Glieder vierten und höheren Grades weggelassen, was offenbar nur dann erlaubt ist, wenn
$$\frac{\partial \varphi}{\partial \varrho} + \tfrac{1}{2}\varrho \frac{\partial^2 \varphi}{\partial \varrho^2}$$

nicht allzu klein und die mit $\triangle^4$ usw. multiplizierten Glieder nicht allzu groß sind. Nach (5) ist aber
$$\int \triangle^2 d\tau = \frac{L^3}{8}\sum_\varrho \sum_\sigma \sum_\tau B_{\varrho,\sigma,\tau}^2,$$

da die Raumintegrale der Doppelprodukte der Fourierschen Summenglieder verschwinden. Es ist also
$$A = \left(\frac{\partial \varphi}{\partial \varrho} + \tfrac{1}{2}\varrho \frac{\partial^2 \varphi}{\partial \varrho^2}\right)\frac{L^3}{8}\sum_\varrho \sum_\sigma \sum_\tau B_{\varrho\sigma\tau}^2.$$

Drücken wir die Arbeit, die pro Masseneinheit geleistet werden muß, um aus dem Zustande thermodynamischen Gleichgewichtes einen Zustand von bestimmtem ϱ zu erzielen, als Funktion des spezifischen Volumens $1/\varrho = v$ aus, setzt man also
$$\varphi(\varrho) = \psi(v),$$

so erhält man noch einfacher

(6) $$A = \frac{L^3}{16} v^3 \frac{\partial^2 \psi}{\partial v^2} \sum_\varrho \sum_\sigma \sum_\tau B_{\varrho\sigma\tau}^2,$$

wobei die Größen v und $\partial^2\psi/\partial v^2$ für den Zustand des idealen thermodynamischen Gleichgewichtes einzusetzen sind. Wir bemerken, daß die Koeffizienten B nur quadratisch, nicht aber als Doppelprodukte im Ausdrucke für A vorkommen. Es sind also die Größen B Parameter des Systems von der Art, wie sie in den Gleichungen (2b) und (4) des vorigen Paragraphen auftreten. Die Größen B befolgen daher (unabhängig voneinander) das Gausssche Fehlergesetz, und Gleichung (4) ergibt unmittelbar

$$(7) \qquad \frac{L^3}{8} v^3 \frac{\partial^2 \psi}{\partial v^2} \overline{B^2_{\varrho \sigma \tau}} = \frac{R T_0}{N}.$$

Die statistischen Eigenschaften unseres Systems sind also vollkommen bestimmt bzw. auf die thermodynamisch ermittelbare Funktion ψ zurückgeführt.

Wir bemerken, daß die Vernachlässigung der Glieder mit $\triangle^3$ usw. nur dann gestattet ist, wenn $\partial^2 \psi / \partial v^2$ für das ideale thermodynamische Gleichgewicht nicht allzu klein ist, oder gar verschwindet. Letzteres findet statt bei Flüssigkeiten und Flüssigkeitsgemischen, die sich genau im kritischen Zustande befinden. Innerhalb eines gewissen (sehr kleinen) Bereiches um den kritischen Zustand werden die Formeln (6) und (7) ungültig. Es besteht jedoch keine *prinzipielle* Schwierigkeit gegen eine Vervollständigung der Theorie durch Berücksichtigung der Glieder höheren Grades in den Koeffizienten.[1]

§ 4. Berechnung des von einem unendlich wenig inhomogenen absorptionsfreien Medium abgebeugten Lichtes.

Nachdem wir aus dem Boltzmannschen Prinzip das statistische Gesetz ermittelt haben, nach welchem die Dichte einer einheitlichen Substanz bzw. das Mischungsverhältnis einer Mischung mit dem Orte variiert, gehen wir dazu über, den Einfluß zu untersuchen, den das Medium auf einen hindurchgehenden Lichtstrahl ausübt.

$\varrho = \varrho_0 + \triangle$ sei wieder die Dichte in einem Punkte des Mediums, bzw. falls es sich um eine Mischung handelt, die räumliche Dichte der einen Komponente. Der betrachtete Lichtstrahl sei monochromatisch. In bezug auf ihn läßt sich das Medium durch den Brechungsindex g charakterisieren, oder durch die zu der betreffenden Frequenz gehörige schein-

---

[1] Vgl. M. v. Smoluchowski, l. c., p. 215.

bare Dielektrizitätskonstante ε, die durch die Beziehung $g = \sqrt{\varepsilon}$ mit dem Brechungsindex verknüpft ist. Wir setzen

$$(8) \quad \varepsilon = \varepsilon_0 + \left(\frac{\partial \varepsilon}{\partial \varrho}\right)_0 \triangle = \varepsilon_0 + \iota;$$

wobei ι ebenso wie $\triangle$ als unendlich kleine Größe zu behandeln ist.

In jedem Punkte des Mediums gelten die Maxwellschen Gleichungen, welche — da wir den Einfluß der Geschwindigkeit der zeitlichen Änderung von ε auf das Licht vernachlässigen können, die Form annehmen

$$\frac{\varepsilon}{c}\frac{\partial \mathfrak{E}}{\partial t} = \operatorname{curl} \mathfrak{H}, \qquad \operatorname{div} \mathfrak{H} = 0,$$

$$\frac{1}{c}\frac{\partial \mathfrak{H}}{\partial t} = -\operatorname{curl} \mathfrak{E}, \qquad \operatorname{div}(\varepsilon \mathfrak{E}) = 0,$$

Hierin bedeutet $\mathfrak{E}$ die elektrische, $\mathfrak{H}$ die magnetische Feldstärke, c die Vakuum-Lichtgeschwindigkeit. Durch Eliminieren von $\mathfrak{H}$ erhält man daraus

$$(9) \quad \frac{\varepsilon}{c^2}\frac{\partial^2 \mathfrak{E}}{\partial t^2} = \triangle \mathfrak{E} - \operatorname{grad} \operatorname{div} \mathfrak{E},$$

$$(10) \quad \operatorname{div}(\varepsilon \mathfrak{E}) = 0.$$

Es sei nun $\mathfrak{E}_0$ das elektrische Feld einer Lichtwelle, wie es verlaufen würde, wenn ε nicht mit dem Orte variierte, wir wollen sagen „das Feld der erregenden Lichtwelle". Das wirkliche Feld (Gesamtfeld) $\mathfrak{E}$ wird sich von $\mathfrak{E}_0$ unendlich wenig unterscheiden um das Opaleszenzfeld $\mathfrak{e}$, so daß zu setzen ist

$$(11) \quad \mathfrak{E} = \mathfrak{E}_0 + \mathfrak{e}.$$

Setzt man die Ausdrücke für ε und $\mathfrak{E}$ aus (8) und (11) in (9) und (10) ein, so erhält man bei Vernachlässigung von unendlich Kleinem zweiter Ordnung, indem man berücksichtigt, daß $\mathfrak{E}_0$ die Maxwellschen Gleichungen mit konstanter Dielektrizitätskonstante ε_0 befriedigt,

$$(9\,\mathrm{a}) \quad \frac{\varepsilon_0}{c^2}\frac{\partial^2 \mathfrak{e}}{\partial t^2} - \triangle \mathfrak{e} = -\frac{1}{c^2} \iota \frac{\partial^2 \mathfrak{E}_0}{\partial t^2} - \operatorname{grad} \operatorname{div} \mathfrak{e},$$

$$(10\,\mathrm{a}) \quad \operatorname{div}(\iota \mathfrak{E}_0) + \operatorname{div}(\varepsilon_0 \mathfrak{e}) = 0.$$

Entwickelt man (10a), und berücksichtigt man dabei, daß div $\mathfrak{E}_0 = 0$ und grad $\varepsilon_0 = 0$ ist, so erhält man

$$\text{div } \mathfrak{e} = -\frac{1}{\varepsilon_0} \mathfrak{E}_0 \text{ grad } \iota.$$

Setzt man dies in (9a) ein, so ergibt sich

(9b) $\qquad \dfrac{\varepsilon_0}{c^2} \dfrac{\partial^2 \mathfrak{e}}{\partial t^2} - \triangle \mathfrak{e} = -\dfrac{1}{c^2} \iota \dfrac{\partial^2 \mathfrak{E}_0}{\partial t^2} + \dfrac{1}{\varepsilon_0} \text{ grad } \{\mathfrak{E}_0 \text{ grad } \iota\} = \mathfrak{a},$

wobei die rechte Seite ein als bekannt anzusehender Vektor ist, der zur Abkürzung mit „$\mathfrak{a}$" bezeichnet ist. Zwischen dem Opaleszenzfelde $\mathfrak{e}$ und dem Vektor $\mathfrak{a}$ besteht also eine Beziehung von derselben Form wie zwischen dem Vektorpotential und der elektrischen Strömung. Die Lösung lautet bekanntlich

(12) $\qquad \mathfrak{e} = \dfrac{1}{4\pi} \displaystyle\int \dfrac{\{\mathfrak{a}\}_{t_0 - \frac{r}{V}}}{r} d\tau,$

wobei r die Entfernung von $d\tau$ vom Aufpunkt, $V = c/\sqrt{\varepsilon_0}$ die Fortpflanzungsgeschwindigkeit der Lichtwellen bedeutet. Das Raumintegral ist über den ganzen Raum auszudehnen, in welchem das erregende Lichtfeld $\mathfrak{E}_0$ von Null verschieden ist. Erstreckt man es nur über einen Teil dieses Raumes, so erhält man den Teil des Opaleszenzfeldes, welchen die erregende Lichtwelle dadurch erzeugt, daß sie den betreffenden Raumteil durchsetzt.

Wir stellen uns die Aufgabe, denjenigen Teil des Opaleszenzfeldes zu ermitteln, der von einer erregenden ebenen monochromatischen Lichtwelle im Innern des Würfels

$$0 < x < l,$$
$$0 < y < l,$$
$$0 < z < l$$

erzeugt wird. Dabei sei die Kantenlänge l dieses Würfels klein gegenüber der Kantenlänge L des früher betrachteten Würfels.

Die erregende ebene Lichtwelle sei gegeben durch

(13) $\qquad \mathfrak{E}_0 = \mathfrak{A} \cos 2\pi n \left(t - \dfrac{n \mathfrak{r}}{V}\right),$

wobei $\mathfrak{n}$ den Einheitsvektor der Wellennormale (Komponenten α, β, γ) und $\mathfrak{r}$ den vom Koordinatenursprung gezogenen Radiusvektor (Komponenten x, y, z) bedeute. Den Aufpunkt wählen wir der Einfachheit halber in einer gegen l unendlich großen Entfernung D auf der X-Achse unseres Koordinatensystems. Für einen solchen Aufpunkt nimmt Gleichung (12) die Form an:

(12a) $$e = \frac{1}{4\pi D} \int \{\mathfrak{a}\}_{t_1 + \frac{x}{V}} d\tau.$$

Es ist nämlich
$$t_0 - \frac{r}{V} = t_0 - \frac{D-x}{V}$$
zu setzen, wobei zur Abkürzung
$$t_0 - \frac{D}{V} = t_1$$
gesetzt ist, und man kann den Faktor $1/r$ des Integranden durch den bis auf relativ unendlich Kleines gleichen konstanten Faktor $1/D$ ersetzen.

Wir haben nun das über unsern Würfel von der Kantenlänge l erstreckte, in (12a) auftretende Raumintegral zu berechnen, indem wir den Ausdruck für $\mathfrak{a}$ aus (9b) einsetzen. Diese Rechnung erleichtern wir uns durch die Einführung des folgenden Symbols. Ist φ ein Skalar oder Vektor, der Funktion ist von x, y, z mit t, so setzen wir
$$\varphi\left(x, y, z, t_1 + \frac{x}{V}\right) = \varphi^*,$$
so daß also φ^* nur von x, y und z abhängig ist. Daraus folgt für einen Skalar φ sofort die Gleichung
$$\operatorname{grad} \varphi^* = (\operatorname{grad} \varphi)^* + \mathfrak{i}\frac{1}{V}\left(\frac{\partial \varphi}{\partial t}\right)^*,$$
woraus folgt
$$\int (\operatorname{grad} \varphi)^* d\tau = \int \operatorname{grad} \varphi^* d\tau - \mathfrak{i}\frac{1}{V}\int \left(\frac{\partial \varphi}{\partial t}\right)^* d\tau,$$
wobei $\mathfrak{i}$ den Einheitsvektor in Richtung der X-Achse bedeutet. Das erste der Integrale auf der rechten Seite läßt sich durch partielle Integration umformen. Bedeutet $\mathfrak{N}$ die äußere Einheitsnormale der Oberfläche des Integrationsraumes, ds das Oberflächenelement, so ist
$$\int \operatorname{grad} \varphi^* d\tau = \int \varphi^* \mathfrak{N} \, ds.$$

Man hat also

(14) $$\int (\operatorname{grad} \varphi)^* \, d\tau = \int \varphi^* \mathfrak{N} \, ds - \mathfrak{i} \frac{1}{V} \int \left(\frac{\partial \varphi}{\partial t}\right)^* dt.$$

Ist φ eine Funktion undulatorischen Charakters, so wird das Flächenintegral der rechten Seite unserer Gleichung keinen dem Volum des Integrationsraumes proportionalen, überhaupt keinen für uns in Betracht kommenden Beitrag leisten. In diesem Falle kann also ein Integral von der Gestalt

$$\int (\operatorname{grad} \varphi)^* \, d\tau$$

nur zur X-Komponente einen Beitrag liefern.

Bildet man nun die beiden Integrale, welche durch Einsetzen von $\mathfrak{a}$ (Gleichung (9b)) in das in (12a) auftretende Integral

$$\int \mathfrak{a}^* \, d\tau$$

entstehen, so ersieht man, daß das zweite dieser Integrale die Gestalt der linken Seite von (14) hat, wobei $\varphi = \mathfrak{E}_0 \operatorname{grad} \iota$ ist. Da dies tatsächlich eine Funktion undulatorischen Charakters ist, welche zudem verschwindet, wenn $\operatorname{grad} \iota$ an der Oberfläche verschwindet, so kann nach (14) dies zweite Integral nur zur X-Komponente von $\mathfrak{e}$ einen in Betracht kommenden Anteil liefern. Eine genauere Rechnung lehrt, daß dies zweite Integral gerade die X-Komponente des ersten Integrales kompensiert. Wir brauchen dies nicht eigens zu beweisen, weil $\mathfrak{e}_x$ wegen der Transversalität des Lichtes verschwinden muß. Vermöge des soeben Gesagten folgt aus (12a) und (9b)

(12b) $$\begin{cases} \mathfrak{e}_x = 0, \\ \mathfrak{e}_y = -\frac{1}{4\pi D c^2} \int \iota \left(\frac{\partial^2 \mathfrak{E}_{0y}}{\partial t^2}\right)^* d\tau, \\ \mathfrak{e}_z = -\frac{1}{4\pi D c^2} \int \iota \left(\frac{\partial^2 \mathfrak{E}_{0z}}{\partial t^2}\right)^* d\tau. \end{cases}$$

Wir berechnen nun $\mathfrak{e}_y$, indem wir in die zweite dieser Gleichungen aus Gleichung (13)

$$\left(\frac{\partial^2 \mathfrak{E}_{0y}}{\partial t^2}\right)^* = -\mathfrak{A}_y (2\pi n)^2 \cos 2\pi n \left(t_1 + \frac{x}{V} - \frac{\alpha x + \beta y + \gamma z}{V}\right)$$

einsetzen. Ferner ersetzen wir ι mittels der Gleichungen (8) und (5). Wir erhalten so, indem wir Summen- und Integrationszeichen vertauschen,

$$e_y = \frac{\mathfrak{A}_y (2\pi n)^2}{4\pi D c^2} \frac{\partial \varepsilon}{\partial \varrho} \sum_\varrho \sum_\sigma \sum_\tau B_{\varrho\sigma\tau} \iiint \cos 2\pi n \left(t_1 + \frac{(1-\alpha)x - \beta y - \gamma z}{V}\right)$$
$$\cdot \cos\left(2\pi \varrho \frac{x}{2L}\right) \cdot \cos\left(2\pi \sigma \frac{y}{2L}\right) \cdot \cos\left(2\pi \tau \frac{z}{2L}\right) dx\, dy\, dz\,,$$

wobei das Raumintegral über den Würfel von der Kantenlänge l zu erstrecken ist. Das Raumintegral ist von der Form

$$J_{\varrho\sigma\tau} = \iiint \cos(\lambda x + \mu y + \nu z) \cos \lambda' x \cos \mu' y \cos \nu' z\, dx\, dy\, dz\,,$$

wobei zu berücksichtigen ist, daß λ, μ, ν, λ', μ', ν' als sehr große Zahlen zu betrachten sind.[1] In diesem Falle ist zu setzen

$$(15) \quad \begin{cases} J_{\varrho\sigma\tau} = \left(\frac{1}{2}\right)^3 l^3 \dfrac{\sin(\lambda-\lambda')\frac{l}{2}}{\frac{(\lambda-\lambda')l}{2}} \cdot \dfrac{\sin(\mu-\mu')\frac{l}{2}}{\frac{(\mu-\mu')l}{2}} \cdot \dfrac{\sin(\nu-\nu')\frac{l}{2}}{\frac{(\nu-\nu')l}{2}} \\ \qquad\qquad \cos\left(2\pi n t_1 + \frac{(\lambda-\lambda')l}{2} + \frac{(\mu-\mu')l}{2} + \frac{(\nu-\nu')l}{2}\right). \end{cases}$$

Neben diesem Ausdruck sind bei der Integration solche Ausdrücke vernachlässigt, welche eine oder mehrere der sehr großen Größen $(\lambda + \lambda')$ usw. im Nenner haben. Man sieht, daß J nur für solche $\varrho\,\sigma\,\tau$ merklich von Null abweicht, für welche die Differenzen $(\lambda - \lambda')$ usw. nicht sehr groß sind. Wir merken an, daß hierbei gesetzt ist

$$(15\,\mathrm{a}) \quad \begin{cases} \lambda = 2\pi n \dfrac{1-\alpha}{V}, & \lambda' = \dfrac{\pi \varrho}{L}, \\ \mu = -2\pi n \dfrac{\beta}{V}, & \mu' = \dfrac{\pi \sigma}{L}, \\ \nu = -2\pi n \dfrac{\gamma}{V}, & \nu' = \dfrac{\pi \tau}{L}. \end{cases}$$

---

[1] Es ist im folgenden so gerechnet, wie wenn λ, μ, ν *positiv* wären. Ist dies nicht der Fall, so ändern sich ein oder mehrere Vorzeichen in (15). Das Endresultat ist aber stets das gleiche.

Setzen wir zur Abkürzung

$$\frac{\mathfrak{A}_y (2\pi n)^2}{4\pi D c^2} \frac{\partial \varepsilon}{\partial \varrho} = A,$$

so ist
(12c)
$$\mathfrak{e}_y = A \sum_\varrho \sum_\sigma \sum_\tau B_{\varrho\sigma\tau} J_{\varrho\sigma\tau}.$$

Diese Gleichung ergibt in Verbindung mit (15) und (15a) den Momentanwert des Opaleszenzfeldes für jeden Moment $t_0 = t_1 + D/V$ an der Stelle $x = D$, $y = z = 0$. Uns interessiert besonders die mittlere Intensität des Opaleszenzlichtes, wobei der Mittelwert zu nehmen ist sowohl hinsichtlich der Zeit als auch hinsichtlich der auftretenden opaleszenz-erregenden Dichteschwankungen. Als Maß für diese mittlere Intensität kann der Mittelwert von $\mathfrak{e}^2 = \mathfrak{e}_y^2 + \mathfrak{e}_z^2$ dienen. Es ist

$$\mathfrak{e}_y^2 = A^2 \sum_\varrho \sum_\sigma \sum_\tau \sum_{\varrho'} \sum_{\sigma'} \sum_{\tau'} B_{\varrho\sigma\tau} B_{\varrho'\sigma'\tau'} J_{\varrho\sigma\tau} J_{\varrho'\sigma'\tau'},$$

wobei die Summe über alle Kombinationen der Indizes ϱ, σ, τ, ϱ', σ', τ' zu erstrecken ist — stets für denselben Wert von t_1. Wir bilden nun den Mittelwert dieser Größe in bezug auf die verschiedenen Dichteverteilungen. Aus (15) ersieht man, daß die Größen $J_{\varrho\sigma\tau}$ von der Dichteverteilung nicht abhängen, ebensowenig die Größe A. Bezeichnen wir also den Mittelwert einer Größe durch einen darüber gesetzten Strich, so erhalten wir

$$\overline{\mathfrak{e}_y^2} = A^2 \sum \sum \sum \sum \sum \overline{B_{\varrho\sigma\tau} B_{\varrho'\sigma'\tau'}} J_{\varrho\sigma\tau} J_{\varrho'\sigma'\tau'}.$$

Da aber gemäß § 3 die Größen B voneinander unabhängig das Gausssche Fehlergesetz erfüllen (wenigstens soweit die von uns verfolgte Annäherung reicht), so ist, falls nicht $\varrho = \varrho'$, $\sigma = \sigma'$ und $\tau = \tau'$ ist

$$\overline{B_{\varrho\sigma\tau} B_{\varrho'\sigma'\tau'}} = 0.$$

Unser Ausdruck für $\overline{\mathfrak{e}_y^2}$ reduziert sich deshalb auf

$$\overline{\mathfrak{e}_y^2} = A^2 \sum \sum \sum \overline{B_{\varrho\sigma\tau}^2} J_{\varrho\sigma\tau}^2.$$

Dieser Mittelwert ist aber noch nicht der gesuchte. Es muß auch bezüglich der *Zeit* der Mittelwert genommen werden. Diese tritt lediglich auf im letzten Faktor des Ausdruckes für $J_{\varrho\sigma\tau}$. Berücksichtigt man, daß der zeitliche Mittelwert dieses Faktors den Wert $\frac{1}{2}$ hat und setzt man zur Abkürzung

(16)
$$\begin{cases} \frac{(\lambda - \lambda')l}{2} = \xi, \\ \frac{(\mu - \mu')l}{2} = \eta, \\ \frac{(\nu - \nu')l}{2} = \zeta, \end{cases}$$

so erhält man für den endgültigen Mittelwert $\overline{\overline{e_y^2}}$ den Ausdruck

$$\overline{\overline{e_y^2}} = \tfrac{1}{2} A^2 \cdot \left(\frac{l}{2}\right)^6 \sum\sum\sum \overline{B_{\varrho\sigma\tau}^2} \frac{\sin^2\xi}{\xi^2} \frac{\sin^2\eta}{\eta^2} \frac{\sin^2\zeta}{\zeta^2}.$$

Nach (7) ist ferner $\overline{B_{\varrho\sigma\tau}^2}$ von $\varrho\,\sigma\,\tau$ unabhängig, kann also vor die Summenzeichen gestellt werden. Es unterscheiden sich ferner die ξ, welche zu aufeinanderfolgenden Werten von ϱ gehören, nach (16) und (15a) um $\frac{\pi}{2}\frac{l}{L}$, also um eine unendlich kleine Größe. Deshalb kann man die auftretende dreifache Summe in ein dreifaches Integral verwandeln. Da nach dem Gesagten für das Intervall $\varDelta\xi$ zweier aufeinanderfolgender ξ-Werte in dreifacher Summe die Beziehung

$$\varDelta\xi \cdot \frac{2}{\pi}\frac{L}{l} = 1$$

ist, so ist

$$\sum\sum\sum \frac{\sin^2\xi}{\xi^2}\frac{\sin^2\eta}{\eta^2}\frac{\sin^2\zeta}{\zeta^2}$$
$$= \left(\frac{2}{\pi}\frac{L}{l}\right)^3 \sum\sum\sum \frac{\sin^2\xi}{\xi^2}\frac{\sin^2\eta}{\eta^2}\frac{\sin^2\zeta}{\zeta^2} \varDelta\xi\,\varDelta\eta\,\varDelta\zeta,$$

welche letztere Summe ohne weiteres als dreifaches Integral geschrieben werden kann. Aus (16) und (15a) schließt man, daß dies Integral praktisch zwischen den Grenzen $-\infty$ und $+\infty$ zu nehmen ist, so daß es in ein Produkt dreier Integrale zerfällt, deren jedes den Wert π hat. Berücksichtigt man dies, so erhält man endlich mit Hilfe von (7) und durch Einsetzen des Ausdruckes für A für $\overline{\overline{e_y^2}}$ den Ausdruck

$$\overline{\overline{e_y^2}} = \frac{R\,T_0}{N} \frac{\left(\frac{\partial\varepsilon}{\partial\varrho}\right)^2}{v^2 \frac{\partial^2\psi}{\partial v^2}} \left(\frac{2\pi n}{c}\right)^4 \frac{l^3}{(4\pi D)^2}\frac{\mathfrak{A}_y^2}{2}.$$

oder, wenn man konsequent das spezifische Volumen v einführt und c/n durch die Wellenlänge λ des erregenden Lichtes ersetzt:

$$(17) \quad \overline{e_y^2} = \frac{R T_0}{N} \frac{v \left(\frac{\partial \varepsilon}{\partial v}\right)^2}{\frac{\partial^2 \psi}{\partial v^2}} \left(\frac{2\pi}{\lambda}\right)^4 \frac{\Phi}{(4\pi D)^2} \frac{\mathfrak{A}_y^2}{2}.$$

Hierbei ist das durchstrahlte opaleszenzerregende Volumen, auf dessen Gestalt es nicht ankommt, mit Φ bezeichnet. Eine analoge Formel gilt bezüglich der z-Komponente, während seine x-Komponente von e verschwindet. Man sieht daraus, daß für Intensität und Polarisationszustand des nach einer bestimmten Richtung entsandten Opaleszenzlichtes die Projektion des elektrischen Vektors des erregenden Lichtes auf die Normalebene zum Opaleszenzstrahl maßgebend ist, welches auch die Fortpflanzungsrichtung des erregenden Lichtes sein mag.[1]) Bezeichnet J_e die Intensität des erregenden Lichtes, J_0 die des Opaleszenzlichtes in der Distanz D von der Erregerstelle in bestimmter Richtung, φ den Winkel zwischen elektrischem Vektor des Erregerlichtes und der Normalebene zum betrachteten Opaleszenzstrahl, so ist nach (17)

$$(17\,\mathrm{a}) \quad \frac{J_0}{J_e} = \frac{R T_0}{N} \frac{v \left(\frac{\partial \varepsilon}{\partial v}\right)^2}{\frac{\partial^2 \psi}{\partial v^2}} \left(\frac{2\pi}{\lambda}\right)^4 \frac{\Phi}{(4\pi D)^2} \cos^2 \varphi.$$

Wir berechnen noch die scheinbare Absorption infolge Opaleszenz durch Integration des Opaleszenzlichtes über alle Richtungen. Man erhält, wenn man mit δ die Dicke der durchstrahlten Schicht, mit α die Absorptionskonstante bezeichnet ($e^{-\alpha\delta}$ = Schwächungsfaktor der Intensität):

$$(18) \quad \alpha = \frac{1}{6\pi} \frac{R T_0}{N} \frac{v \left(\frac{\partial \varepsilon}{\partial v}\right)^2}{\frac{\partial^2 \psi}{\partial v^2}} \left(\frac{2\pi}{\lambda}\right)^4.$$

---

1) Daß unser Opaleszenzlicht diese Eigenschaft mit demjenigen Opaleszenzlicht gemein hat, das durch gegen die Wellenlänge des Lichtes kleine suspendierte Körper veranlaßt wird, kann nicht auffallen. Denn in beiden Fällen handelt es sich um unregelmäßige, örtlich rasch veränderliche Störungen der Homogenität der durchstrahlten Substanz.

Es ist von Bedeutung, daß das Hauptresultat unserer Untersuchung, das durch Formel (17a) gegeben ist, eine exakte Bestimmung der Konstante N, d. h. der absoluten Größe der Moleküle gestattet. Im folgenden soll dies Resultat auf den Spezialfall der homogenen Substanz sowie auf den flüssiger binärer Gemische in der Nähe des kritischen Zustandes angewendet werden.

§ 5. Homogene Substanz.

Im Falle einer homogenen Substanz haben wir zu setzen

$$\psi = - \int p \, dv,$$

also

$$\frac{\partial^2 \psi}{\partial v^2} = - \frac{\partial p}{\partial v}.$$

Ferner ist nach der Beziehung von Clausius-Mosotti-Lorentz

$$\frac{\varepsilon - 1}{\varepsilon + 2} v = \text{konst.},$$

also

$$\left(\frac{\partial \varepsilon}{\partial v}\right)^2 = \frac{(\varepsilon - 1)^2 (\varepsilon + 2)^2}{9 v^2}.$$

Setzt man diese Werte in (17a) ein, so erhält man

(17 b) $\quad \dfrac{J_0}{J_e} = \dfrac{R T_0}{N} \dfrac{(\varepsilon - 1)^2 (\varepsilon + 2)^2}{9 v \left(-\dfrac{\partial p}{\partial v}\right)} \left(\dfrac{2 \pi}{\lambda}\right)^4 \dfrac{\Phi}{(4 \pi D)^2} \cos^2 \varphi.$

In dieser Formel, welche das Verhältnis der Intensität des Opaleszenzlichtes zum erregenden Licht ergibt, falls letzteres in der Distanz D vom primär bestrahlten Volumen Φ gemessen wird, bedeutet:

R die Gaskonstante,
T die absolute Temperatur,
N die Zahl der Moleküle in einem Grammolekül,
ε das Quadrat des Brechungsexponenten für die Wellenlänge λ,
v das spezifische Volumen,
$\dfrac{\partial p}{\partial v}$ den isothermen Differentialquotienten des Druckes nach dem Volumen,
φ den Winkel zwischen dem elektrischen Feldvektor der erregenden Welle und der Normalebene zum betrachteten Opaleszenzstrahl.

Daß $\partial p/\partial v$ der isotherm und nicht etwa der adiabatisch genommene Differentialquotient ist, hängt damit zusammen, daß von allen Zuständen, die zu einer gegebenen Dichteverteilung gehören, der Zustand gleicher Temperatur bei gegebener Gesamtenergie der Zustand größter Entropie, also auch größter statistischer Wahrscheinlichkeit ist.

Ist die Substanz, um welche es sich handelt, ein ideales Gas, so ist nahe $\varepsilon + 2 = 3$ zu setzen. Man erhält für diesen Fall

$$(17\,\mathrm{c}) \qquad \frac{J_0}{J_e} = \frac{R\,T_0}{N} \frac{(\varepsilon - 1)^2}{p} \left(\frac{2\,\pi}{\lambda}\right)^4 \frac{\Phi}{(4\,\pi\,D)^2} \cos^2 \varphi\,.$$

Diese Formel vermag, wie eine Überschlagsrechnung zeigt, sehr wohl die Existenz des von dem bestrahlten Luftmeer ausgesandten vorwiegend blauen Lichtes zu erklären.[1]) Dabei ist bemerkenswert, daß unsere Theorie nicht *direkt* Gebrauch macht von der Annahme einer diskreten Verteilung der Materie.

§ 6. Flüssigkeitsgemisch.

Auch im Falle eines Flüssigkeitsgemisches gilt der Herleitung gemäß Gleichung (17a), wenn man setzt

$v =$ spezifisches Volumen der Masseneinheit der ersten Komponente,
$\psi =$ Arbeit, welche man braucht, um auf umkehrbarem Wege die Masseneinheit der ersten Komponente bei konstanter Temperatur auf umkehrbarem Wege vom spezifischen Volumen des Temperaturgleichgewichtes auf ein bestimmtes anderes spezifisches Volumen zu bringen.

Die Größe ψ läßt sich in dem Falle, daß der mit dem betrachteten Flüssigkeitsgemisch koexistierende Dampf als Gemisch idealer Gase betrachtet werden kann, und daß die Mischung als inkompressibel anzusehen ist, durch der Erfahrung zugängliche Größen ersetzen. Wir finden dann ψ durch folgende elementare Betrachtung.

Der Masseneinheit der ersten Komponente sei die Masse k der zweiten Komponente zugemischt. k ist dann ein Maß für die Zusammensetzung des Gemisches, dessen Gesamtmasse

---

1) Gleichung (17c) kann man auch erhalten, indem man die Ausstrahlungen der einzelnen Gasmoleküle summiert, wobei diese als vollkommen unregelmäßig verteilt angesehen werden. (Vgl. Rayleigh, Phil. Mag. 47. p. 375. 1899 und Papers 4. p. 400.)

$1 + k$ ist. Dies Gemisch besitze eine Dampfphase, und es sei p'' der Partialdruck, v'' das spezifische Volumen der zweiten Komponente in der Dampfphase. Dies System sei in eine Hülle eingeschlossen, welche einen semipermeabeln Wandteil besitzt, durch den die zweite Komponente, nicht aber die erste in Gasform aus- und eingeführt werden kann. In eine zweite, relativ unendlich große Hülle sei eine relativ unendlich große Menge des Gemisches eingeschlossen von derjenigen Zusammensetzung (charakterisiert durch k_0), für welche wir die Opaleszenz berechnen wollen. Dies zweite Gemisch besitze auch einen Dampfraum mit semipermeabler Wand, und es sei Partialdruck – spezifisches Volumen der zweiten Komponente im Dampfraum mit p_0'', v_0'' bezeichnet. Im Innern beider Hüllen möge die Temperatur T_0 herrschen. Wir berechnen nun die Arbeit $d\psi$, welche nötig ist, um durch Transportieren der Masse dk der zweiten Komponente von dem zweiten Behälter in den ersten in Gasform auf umkehrbarem Wege das Konzentrationsmaß k im ersten Behälter um dk zu erhöhen. Diese Arbeit setzt sich aus folgenden drei Teilen zusammen:

$-\dfrac{dk}{M''} p_0'' v_0''$ (Arbeit bei der Entnahme aus dem zweiten Behälter)

$\dfrac{dk}{M''} R\, T_0 \lg \dfrac{p''}{p_0''}$ (Isothermische Kompression bis auf den Partialdruck im ersten Behälter)

$+\dfrac{dk}{M''} p'' v''$ (Arbeit beim Einführen in den ersten Behälter).

Hierbei ist das Flüssigkeitsvolumen neben dem Gasvolumen vernachlässigt. M'' ist das Molekulargewicht der zweiten Komponente in der Dampfphase. Da sich das erste und dritte Glied nach dem Gesetz von Mariotte wegheben, erhalten wir

$$d\psi = \frac{R\, T_0}{M''} dk \lg \frac{p''}{p_0''}.$$

Die Funktion ψ ist also unmittelbar aus Konzentrationen und Partialdrucken berechenbar. Wir haben nun $\partial^2 \psi / \partial v^2$ zu ermitteln für denjenigen Zustand, den wir durch den Index „$_0$" bezeichnet haben. Es ist

$$\lg \left(\frac{p''}{p_0''}\right) = \lg \left(1 + \frac{p'' - p_0''}{p_0''}\right) = \lg(1 + \pi) = \pi - \frac{\pi^2}{2} \cdots,$$

wobei π die relative Druckänderung der zweiten Komponente gegenüber dem Ursprungszustande bezeichnet. Aus den beiden letzten Gleichungen folgt

$$\frac{\partial \psi}{\partial v} = \frac{R T_0}{M''} \frac{\pi - \frac{\pi^2}{2} \cdots}{\frac{\partial v}{\partial k}}.$$

Differenziert man noch einmal nach v und berücksichtigt, daß

$$\frac{\partial}{\partial v} = \frac{\frac{\partial}{\partial k}}{\frac{\partial v}{\partial k}}$$

ist, so erhält man, wenn man im Resultat $\pi = 0$ setzt:

$$\left(\frac{\partial^2 \psi}{\partial v^2}\right)_0 = \frac{R T_0}{M''} \frac{\frac{\partial \pi}{\partial k}}{\left(\frac{\partial v}{\partial k}\right)^2} = \frac{R T_0}{M''} \frac{\frac{1}{p''}\frac{\partial p''}{\partial k}}{\left(\frac{\partial v}{\partial k}\right)^2}.$$

Berücksichtigen wir dies, und ebenso, daß

$$\frac{\partial \varepsilon}{\partial v} = \frac{\frac{\partial \varepsilon}{\partial k}}{\frac{\partial v}{\partial k}},$$

so geht die Formel (17a) über in

$$(17\,\mathrm{d}) \qquad \frac{J_0}{J_e} = \frac{M''}{N} \frac{v \left(\frac{\partial \varepsilon}{\partial k}\right)^2}{\frac{\partial (\lg p'')}{\partial k}} \left(\frac{2\pi}{\lambda}\right)^4 \frac{\Phi}{(4\pi D)^2} \cos^2 \varphi.$$

Diese Formel, welche nur noch dem Experiment zugängliche Größen enthält, bestimmt die Opaleszenzeigenschaften von binären Flüssigkeitsgemischen, insoweit man deren gesättigte Dämpfe als ideale Gase behandeln darf, vollkommen bis auf ein kleines Gebiet in unmittelbarer Nähe des kritischen Punktes. Hier aber dürfte wegen der starken Lichtabsorption und deren großer Temperaturabhängigkeit eine quantitative Untersuchung ohnehin ausgeschlossen sein. Wir wiederholen hier die Bedeutungen der in der Formel auftretenden Zeichen, soweit sie nicht bei Formel (17b) angegeben sind; es ist

M'' das Molekulargewicht der zweiten Komponente in der Dampfphase,

v das Volumen des Flüssigkeitsgemisches, in welchem die Masseneinheit der ersten Komponente enthalten ist,

k die Masse zweiter Komponente, welche auf die Masseneinheit erster Komponente entfällt,

p'' der Dampfdruck der zweiten Komponente.

Damit es nicht wunderlich erscheine, daß in (17d) die beiden Komponenten eine verschiedene Rolle spielen, bemerke ich, daß die bekannte thermodynamische Beziehung

$$\frac{1}{M''} \frac{dp''}{p''} = -\frac{1}{M'} \cdot \frac{1}{k} \frac{dp'}{p'}$$

besteht. Aus dieser kann man schließen, daß es gleichgültig ist, welche Komponente man als erste bzw. zweite behandelt.

Eine quantitative experimentelle Untersuchung der hier behandelten Erscheinungen wäre von großem Interesse. Denn einerseits wäre es wertvoll, zu wissen, ob das Boltzmannsche Prinzip wirklich die hier in Betracht kommenden Erscheinungen richtig ergibt, andererseits könnte man durch solche Untersuchungen zu genauen Werten für die Zahl N gelangen.

Zürich, Oktober 1910.

(Eingegangen 8. Oktober 1910.)

ANNALEN DER PHYSIK.

BEGRÜNDET UND FORTGEFÜHRT DURCH

F. A. C. GREN, L. W. GILBERT, J. C. POGGENDORFF, G. u. E. WIEDEMANN, P. DRUDE.

VIERTE FOLGE.

BAND 34.

DER GANZEN REIHE 339. BAND.

KURATORIUM:

M. PLANCK, G. QUINCKE,
W. C. RÖNTGEN, W. VOIGT, E. WARBURG.

UNTER MITWIRKUNG

DER DEUTSCHEN PHYSIKALISCHEN GESELLSCHAFT

UND INSBESONDERE VON

M. PLANCK

HERAUSGEGEBEN VON

W. WIEN UND M. PLANCK.

MIT SECHS FIGURENTAFELN.

LEIPZIG, 1911.

VERLAG VON JOHANN AMBROSIUS BARTH.

8. *Bemerkung zu dem Gesetz von Eötvös;* von *A. Einstein.*

Eötvös hat empirisch folgende Gesetzmäßigkeit für Flüssigkeiten aufgestellt, die bekanntlich mit bemerkenswerter Annäherung sich bestätigt:

(1) $$\gamma v^{2/3} = k(\tau - T).$$

Hierbei ist γ die Oberflächenspannung, v das Molekularvolumen, k eine universelle Konstante, T die Temperatur, τ eine Temperatur, die von der kritischen nur wenig abweicht.

γ ist die freie Energie pro Oberflächeneinheit, also $\gamma - T\frac{d\gamma}{dT}$ die Energie pro Oberflächeneinheit. Berücksichtigt man, daß v im Vergleich zu γ wenig von der Temperatur abhängt, so kann man mit ähnlicher Annäherung setzen:

(1a) $$\left(\gamma - T\frac{d\gamma}{dT}\right) v^{2/3} = k\tau.$$

Nach der Regel von den übereinstimmenden Zuständen ist aber einerseits die Siedetemperatur bei Atmosphärendruck angenähert ein bestimmter Bruchteil der kritischen Temperatur, andererseits besteht zwischen Siedetemperatur und Verdampfungswärme Proportionalität (Regel von Trouton).

Hieraus ergibt sich, daß die Gleichung (1a) auch die angenäherte Gültigkeit der Gleichung:

(1b) $$\left(\gamma - T\frac{d\gamma}{dT}\right) v_s^{2/3} = k'(D_s - RT_s)$$

zur Folge hat. Da γ mit großer Annäherung eine lineare Funktion der Temperatur ist, braucht die Klammer der linken Seite nicht für die atmosphärische Siedetemperatur berechnet zu werden. Die linke Seite der Gleichung ist gleich derjenigen Energie Uf, welche notwendig ist, um eine Oberflächenvergrößerung der Substanz herbeizuführen, die gleich ist einer

Seitenfläche des Grammolekülwürfels. $D_s - RT_s$ ist die innere Energie U_i, die bei der Verdampfung eines Grammoleküls aufzuwenden ist. Gleichung (1b) kann daher geschrieben werden in der Form:

$$\text{(1c)} \qquad \frac{U_f}{U_i} = k'.$$

Wir wollen die letzte Gleichung nun interpretieren. Es sei S (vgl. die Figur) ein Schnitt durch einen Grammolekülwürfel parallel einer Seitenfläche. $2 U_f$ ist dann gleich der (negativ genommen) potentiellen Energie, welche der Gesamtheit der Wechselwirkungen zwischen den Molekülen auf einer Seite von S und den Molekülen auf der andern Seite von S entspricht; U_i ist die (negativ genommene) potentielle Energie, welche den Wechselwirkungen sämtlicher Moleküle des Würfels entspricht.[1]

Die nächstliegende Fundamentalhypothese über die Molekularkräfte, die zu einer einfachen Beziehung zwischen U_f und U_i führt, ist diese:

Der Radius der Wirkungssphäre der Moleküle ist groß gegen das Molekül, jedoch für Moleküle verschiedener Art gleich groß. Zwei Moleküle üben in der Entfernung r eine Kraft aufeinander aus, deren negative potentielle Energie durch $c^2 f(r)$ gegeben ist, wobei c eine für das Molekül charakteristische Konstante, $f(r)$ eine universelle Funktion von r, $f(\infty)$ gleich Null sei. Der Fall führt nur dann zu einfachen Beziehungen, wenn $f(r)$ derart beschaffen ist, daß die Summen, weche U_f und U_i darstellen, als Integrale geschrieben werden können; wir wollen auch dies (mit van der Waals) voraussetzen. Dann erhält man durch einfache Rechnung:

$$U_f = c^2 N^2 K_2 v^{-4/3},$$
$$U_i = c^2 N^2 K_1 v^{-1}.$$

---

[1] Hierin liegt insofern eine bemerkenswerte Ungenauigkeit, als sicherlich nicht die ganze Energie U_i als potentielle Energie im Sinne der Mechanik angesprochen werden darf; dies wäre nur dann zulässig, wenn die spezifische Wärme bei konstantem Volumen im flüssigen und im Gaszustand gleich groß wäre. Es wäre wohl richtiger, die auf den absoluten Nullpunkt extrapolierte Verdampfungswärme einzuführen.

Hierbei ist
$$K_1 = \int f(r)\,d\tau,$$
ausgedehnt über den ganzen Raum,
$$K_2 = \frac{1}{2} \int_0^\infty \psi(\Delta)\,d\Delta,$$
wobei
$$\psi(\Delta) = \int_\Delta^\infty dx \int_{-\infty}^{+\infty} \int_{-\infty}^{+\infty} f(r)\,dy\,dz.$$

K_1 und K_2 sind also universelle Konstante, die nur von dem Elementargesetz der Molekularkräfte abhängen. Man erhält hieraus:

(2) $$\frac{U_f}{U_i} = \frac{K_2}{K_1} v^{-1/3},$$

im Widerspruch mit der als Ausdruck der Erfahrung anzusehenden Gleichung (1c). Man sieht auch ohne alle Rechnung ein, daß sich abgesehen von universellen Faktoren U_f zu U_i verhalten muß wie der Radius der molekularen Wirkungssphäre zur Seite des Grammolekülwürfels ($v^{1/3}$). Wenn also der Radius der Wirkungssphäre universell ist, so kann man nicht zu Gleichung (1c) gelangen, sondern nur zu (2).

Man sieht leicht ein, daß es im Falle der Gültigkeit von Gleichung (2) unmöglich wäre, aus der Kapillaritätskonstante einen Rückschluß auf das Molekulargewicht einer Flüssigkeit zu ziehen.

Damit Gleichung (1c) herauskomme, muß man von der Annahme ausgehen, daß der Radius der molekularen Wirkungssphäre der Größe $v^{1/3}$, oder, was dasselbe bedeutet, dem Abstand benachbarter Moleküle der Flüssigkeit proportional sei. Diese Annahme erscheint zunächst recht ungereimt, denn was sollte der Radius der Wirkungssphäre eines Moleküls damit zu tun haben, in welcher Distanz sich die benachbarten Moleküle befinden? Vernünftig wird diese Supposition nur in dem Falle, daß sich *nur die benachbarten Moleküle*, nicht aber die weiter entfernten, im Wirkungsbereich eines Moleküls befinden. In diesem Falle muß nach dem Gesagten Gleichung (1a) herauskommen, und wir sind sogar in der Lage, die Größe der Konstanten k' abzuschätzen. Die Betrachtung, die ich im folgenden hierfür gebe, ließe sich wohl durch eine exaktere ersetzen; ich wähle sie aber, weil sie mit einem Minimum formaler Elemente auskommt.

Ich denke mir die Moleküle regelmäßig verteilt in einem quadratischen Gitter. In diesem betrachte ich einen Elementarkubus, dessen Kanten je drei Moleküle enthalten, so daß der ganze Kubus $3^3 = 27$ Moleküle enthält. Eines davon ist in der Mitte. Die übrigen 26, und nur diese, betrachte ich als dem in der Mitte befindlichen Molekül benachbart, und rechne so, wie wenn deren Abstände vom mittleren Molekül gleich groß wären. Bezeichnet man die negativ genommene potentielle Energie eines Moleküls gegenüber einem benachbarten mit φ, so ist dessen potentielle Energie gegenüber allen benachbarten Molekülen gleich 26φ, und deshalb

$$U_i = \frac{1}{2} N \cdot 26 \varphi.$$

Denken wir uns ferner, daß unser mittleres Molekül M unmittelbar unterhalb der Ebene S in der Figur liegt, und daß die Grenzebene des dort gezeichneten Grammolekülwürfels den Seitenflächen der Elementarwürfel des Molekülgitters parallel seien, so steht unser Molekül M mit 9 Molekülen der nächstoberen Schicht in Wechselwirkung. Da $N^{2/3}$ solcher Moleküle M unmittelbar unterhalb der Fläche S liegen, so ist die potentielle Energie, die wir oben mit $2 U_f$ bezeichnet haben, gegeben durch:

$$2 U_f = 9 \cdot N^{2/3} \cdot \varphi.$$

Es ergibt sich also:

$$\frac{U_f}{U_i} = \frac{9}{26} N^{-1/3},$$

oder, wenn man für N den Wert $7 \cdot 10^{23}$ einsetzt,

$$\frac{U_f}{U_i} = 3 \cdot 10^{-9}.$$

Ich habe andererseits mittels der Gleichung (1b) aus der Erfahrung die Konstante k', welche nach (1c) der soeben berechneten Größe gleich sein soll, für Quecksilber und Benzol aus Versuchsdaten berechnet, und die Werte

$$5 \cdot 18 \times 10^{-9}$$
$$5 \cdot 31 \times 10^{-9}$$

erhalten. Diese Übereinstimmung bezüglich der Größenordnung mit der durch jene rohe theoretische Betrachtung ermittelten Größe ist eine sehr bemerkenswerte.

Angeregt durch eine mündliche Bemerkung meines Kollegen G. Bredig überlegte ich mir noch, von welcher Größenordnung der theoretische ermittelte Wert U_f/U_i wird, wenn man annimmt, daß das Molekül nicht nur mit den unmittelbar benachbarten, sondern auch noch mit weiter entfernten in Wechselwirkung steht. Der Würfel, der die Moleküle enthält, welche mit einem Molekül in Wechselwirkung stehen, hat dann nicht 3^3, sondern n^3 Moleküle. Es ergibt sich dann, daß U_f/U_i nahe proportional n herauskommt. Es kommen also für $n = 5$ oder $n = 7$ auch noch Werte für U_f/U_i von der richtigen Größenordnung heraus. Trotzdem ist es höchst wahrscheinlich, daß ein Molekül nur mit dem nächstbenachbarten in Wechselwirkung steht, da es eben als sehr unwahrscheinlich betrachtet werden muß, daß der Radius der molekularen Wirkungssphäre der dritten Wurzel aus dem Molekularvolumen proportional, sonst aber von keiner physikalischen Konstante des Moleküls abhängig sei.

Noch eine Bemerkung drängt sich bei dieser Betrachtung auf. Es ist bekannt, daß Stoffe mit sehr kleinem Molekül vom Gesetze der übereinstimmenden Zustände erheblich abweichen; sollte dies nicht damit im Zusammenhang stehen, daß bei solchen Stoffen der Radius der molekularen Wirkungssphäre größer ist als der dreifache Molekülradius?

(Eingegangen 30. November 1910.)

9. *Eine Beziehung zwischen dem elastischen Verhalten und der spezifischen Wärme bei festen Körpern mit einatomigem Molekül;* von *A. Einstein.*

Mein Kollege, Hr. Prof. Zangger, machte mich auf eine wichtige Bemerkung aufmerksam, die Sutherland[1]) neulich publizierte. Dieser stellte sich die Frage, ob die elastischen Kräfte fester Körper Kräfte derselben Art seien wie diejenigen Kräfte, welche die Träger der ultraroten Eigenschwingungen in ihre Ruhelage zurücktreiben, also deren Eigenfrequenzen bedingen. Er fand, daß diese Frage mit großer Wahrscheinlichkeit zu bejahen sei auf Grund folgender Tatsache: die ultraroten Eigenfrequenzen sind von derselben Größenordnung wie diejenigen Frequenzen, welche man anwenden mußte, um elastische Transversalschwingungen durch den Körper zu senden, deren halbe Wellenlänge gleich ist dem Abstand benachbarter Moleküle des Körpers.

Bei aller Wichtigkeit der Sutherlandschen Betrachtung ist es aber klar, daß man auf diesem Wege nicht mehr erlangen kann als eine rohe Größenordnungsbeziehung, und zwar insbesondere aus dem Grunde, weil anzunehmen ist, daß die bekannten ultraroten Eigenschwingungen in der Hauptsache als Schwingungen der verschieden geladenen Ionen eines Moleküls gegeneinander, die elastischen Schwingungen aber als Schwingungen der ganzen Moleküle gegeneinander aufzufassen sind. Es scheint mir deshalb, daß eine genauere Prüfung der Sutherlandschen Idee nur bei Stoffen mit einatomigem Molekül möglich sei, denen nach der Erfahrung und nach dem theoretischen Bilde optisch nachweisbare Eigenschwingungen von der bekannten Art nicht zukommen. Nach der von mir auf die

---

1) W. Sutherland, Phil. Mag. (6) **20.** p. 657. 1910.

Planksche Theorie der Strahlung gegründete Theorie der spezifischen Wärme fester Körper[1]) ist es aber möglich, die Eigenfrequenzen der einatomigen Körper, welche Träger der Wärme sind, aus der Abhängigkeit der spezifischen Wärme von der Temperatur zu ermitteln. Diese Eigenfrequenzen kann man benutzen, um die Sutherlandsche Auffassung zu prüfen, indem man diese Eigenfrequenzen mit jenen vergleicht, die sich aus der Elastizität ergeben. Eine Art, wie dies geschehen kann, ist im folgenden gegeben, und es sei gleich hier bemerkt, daß sich beim Silber auf dem angedeuteten Wege Sutherlands Auffassung von der Wesensgleichheit der elastischen und der die Eigenfrequenz bestimmenden Kräfte befriedigend bestätigte.

An eine *exakte* Berechnung der Eigenschwingungsfrequenzen aus den elastischen Konstanten ist vorläufig nicht zu denken. Wir bedienen uns vielmehr hier einer rohen, der in der vorangehenden Arbeit benutzten ähnlichen Rechenmethode, die aber wohl im Wesentlichen das Richtige treffen dürfte.

Wir denken uns zunächst die Moleküle der Substanz nach einem quadratischen Raumgitter angeordnet. Es hat dann jedes Molekül 26 Nachbarmoleküle, die allerdings nicht gleich weit von demselben entfernt sind. Wir werden aber so rechnen, wie wenn diese 26 Nachbarmoleküle im Ruhestande alle gleich weit vom betrachteten Molekül entfernt wären.

Wir haben nun irgend eine plausible, möglichst einfache Darstellung der Molekularkräfte zu wählen. Da führen wir zuerst die für das folgende fundamentale, in der vorangehenden Mitteilung für Flüssigkeiten erwiesene Voraussetzung ein, daß jedes Molekül nur mit seinen Nachbarmolekülen, nicht aber mit entfernteren Molekülen in Wechselwirkung stehe. Zwei Nachbarmoleküle mögen eine Zentralkraft aufeinander ausüben, welche verschwindet, wenn der Abstand der Moleküle gleich d ist. Ist ihr Abstand gleich $d - \Delta$, so wirke eine Abstoßungskraft von der Größe $a\Delta$.

Nun berechnen wir die Kraft, welche die 26 Nachbarmoleküle der Verrückung eines Moleküls entgegensetzen. Dabei denken wir uns die 26 Nachbarmoleküle, statt auf einer Würfel-

---

[1]) A. Einstein, Ann. d. Phys. **22**. p. 180. 1907.

oberfläche, auf einer Kugelfläche von gleich großem räumlichem Inhalt verteilt, deren Radius gleich d zu wählen ist, so daß wir haben

(1) $$\frac{4}{3} d^3 \pi = 8 \frac{v}{N},$$

wenn v das Molekularvolumen der Substanz und N die Zahl der Moleküle in einem Grammolekül bedeutet. Wir denken uns das im Mittelpunkt der Kugel liegende Molekül in beliebiger Richtung um die gegen d kleine Länge x verschoben und berechnen die der Verschiebung entgegenwirkende Kraft so, wie wenn die Masse der 26 Moleküle gleichförmig über die Kugeloberfläche verteilt wäre. Auf dem vom Molekül aus gezogenen elementar kleinen körperlichen Winkel $d\varkappa$, dessen Achse mit der Richtung der Verschiebung x den Winkel ϑ bilde, liegen dann $26 \cdot (d\varkappa/4\pi)$ Moleküle, welche in Richtung der Verschiebung x die Kraft

$$-\frac{26}{4\pi} d\varkappa \cdot a \cdot x \cos\vartheta \cdot \cos\vartheta$$

liefern. Durch Integration bekommen wir für die auf das verschobene Molekül wirkende Kraft den Wert

$$-\frac{26}{3} a x.$$

Hieraus ergibt sich, wenn man hinzunimmt, daß M/N gleich ist der Masse eines Moleküls (M = Molekulargewicht der Substanz), die Eigenfrequenz ν und die dieser entsprechende Vakuumwellenlänge λ des Moleküls. Es ist

(2) $$\nu = \frac{1}{2\pi} \sqrt{\frac{26}{3} a \cdot \frac{N}{M}}$$

und

(2a) $$\lambda = 2\pi e \sqrt{\frac{3}{26} \frac{M}{aN}}.$$

Wir berechnen nun auf Grund derselben Näherungsannahmen den Kompressibilitätskoeffizienten der Substanz. Zu diesem Zwecke drücken wir die bei einer gleichmäßigen Kompression aufzuwendende Arbeit A auf zwei verschiedene Arten aus und setzen beide Ausdrücke einander gleich.

Es ist $(a/2)\varDelta^2$ die für die Verkleinerung des Abstandes zweier benachbarter Moleküle um $\varDelta$ aufzuwendende Arbeit.

Da jedes Molekül 26 benachbarte Moleküle hat, so ist die zur Verkleinerung seines Abstandes von den Nachbarmolekülen aufzuwendende Arbeit $26 \cdot (a/2) \Delta^2$. Da es in der Volumeneinheit N/v Moleküle gibt und jeder Term $(a/2)\Delta^2$ zu zwei Molekülen gehört, erhält man

$$A = \frac{26}{4} \cdot \frac{N}{v} a \Delta^2.$$

Ist $\varkappa$ andererseits die Kompressibilität, Θ die Kontraktion der Volumeneinheit, so ist $A = 1/2 \varkappa \cdot \Theta^2$, oder, da $\Theta = 3\Delta/d$ ist:

$$A = \frac{9}{2} \frac{\Delta^2}{\varkappa \cdot d^2}.$$

Durch gleichsetzen dieser beiden Werte für A erhält man

(3) $$\varkappa = \frac{18}{26} \frac{v}{a \cdot d^2 \cdot N}.$$

Durch Eliminieren von a und d aus den Gleichungen (1), 2a) und (3) erhält man

$$\lambda = \frac{2\pi}{\sqrt{6}} \left(\frac{6}{\pi}\right)^{1/3} \frac{C}{N^{1/3}} M^{1/3} \varrho^{1/6} \sqrt{\varkappa} = 1{,}08 \cdot 10^3 \cdot M^{1/3} \varrho^{1/6} \sqrt{\varkappa}$$

Die Formel setzt natürlich voraus, daß Polymerisation nicht stattfindet. Im folgenden sind die Eigenwellenlängen (als Maß für die Eigenfrequenzen) derjenigen Metalle nach dieser Formel berechnet, für welche Grüneisen[1]) die kubische Kompressibilität angegeben hat. Es ergibt sich[2]):

Stoff	$\lambda \cdot 10^4$	Stoff	$\lambda \cdot 10^4$
Aluminium . . .	45	Palladium	58
Kupfer	53	Platin	66
Silber	73	Kadmium	115
Gold	79	Zinn	102
Nickel	45	Blei	135
Eisen	46	Wismut	168

Nach der aus der Planckschen Strahlungstheorie abgeleiteten Theorie der spezifischen Wärme soll letztere gegen

---

1) E. Grüneisen, Ann. d. Phys. **25.** p. 848. 1908.
2) Die Temperaturabhängigkeit der kubischen Kompressibilität ist hierbei vernachlässigt.

den Nullwert der absoluten Temperatur abfallen nach folgendem Gesetz:

$$C = 3R \frac{e^{-\frac{a}{T}} \left(\frac{a}{T}\right)^2}{\left(e^{-\frac{a}{T}} - 1\right)^2},$$

wobei C die auf das Grammolekel bezogene spezifische Wärme bedeutet, und

$$\frac{h\nu}{k} = a = \frac{h \cdot c}{k \cdot \lambda}$$

gesetzt ist. Hierbei sind h und $\varkappa$ die Konstanten der Planckschen Strahlungsformel. Man kann daher aus dem Verlauf der spezifischen Wärme λ ein zweites Mal bestimmen. Der einzige, der oben angeführten Stoffe, dessen spezifische Wärme bei tiefen Temperaturen hinreichend genau bestimmt ist, ist das Silber. Für dieses fand Nernst[1]) $a = 162$, woraus sich $\lambda \cdot 10^4 = 90$ ergibt, während wir aus den elastischen Konstanten $\lambda \cdot 10^4 = 73$ berechnet haben. Diese nahe Übereinstimmung ist wahrhaft überraschend. Eine noch exaktere Prüfung der Sutherlandschen Auffassung wird sich wohl nur dadurch erzielen lassen, daß man die molekulare Theorie der festen Körper vervollkommnet.

---

1) Vgl. W. Nernst, Bulletin des Seances de la Société franç. de Phys. 1910. 1· fasc.

(Eingegangen 30. November 1910.)

175

10. Bemerkungen zu den P. Hertzschen Arbeiten: „Über die mechanischen Grundlagen der Thermodynamik"[1]; von A. Einstein.

———

Hr. P. Hertz hat in seinen soeben genannten vortrefflichen Arbeiten zwei Stellen, die sich in Arbeiten von mir über den gleichen Gegenstand vorfinden, angegriffen. Zu diesen Angriffen will ich im folgenden kurz Stellung nehmen, wobei ich bemerke, daß das hier Gesagte das Resultat einer mündlichen Besprechung mit Hrn. Hertz ist, in welcher wir uns über die beiden in Betracht kommenden Punkte vollkommen geeinigt haben.

1. Im vorletzten Absatz des § 13 seiner zweiten Arbeit kritisiert Hertz eine von mir gegebene Ableitung des Entropiesatzes für nicht umkehrbare Vorgänge. Ich halte diese Kritik für vollkommen zutreffend. Meine Ableitung hatte mich schon damals nicht befriedigt, weshalb ich kurz darauf eine zweite Ableitung gab, die auch von Hrn. Hertz zitiert ist.

2. Die in § 4 seiner ersten Abhandlung enthaltenen Bemerkungen gegen eine in meiner ersten einschlägigen Abhandlung enthaltene Betrachtung[2] über das Temperaturgleichgewicht beruht auf einem Mißverständnis, das durch eine allzu knappe und nicht genügend sorgfältige Formulierung jener Betrachtung hervorgerufen wurde.

Da jedoch der Gegenstand durch die Arbeiten anderer Autoren genügend klar gelegt worden ist, und zudem ein Eingehen auf diesen speziellen Punkt wenig Interesse be-

———

1) A. Einstein, Ann. d. Phys. **9.** p. 425 1902 und **11.** p. 176. 1903.
2) P. Hertz, Ann. d. Phys. **33.** p. 225 u. 537. 1910.

anspruchen dürfte, will ich an dieser Stelle nicht weiter darauf eingehen. Ich bemerke nur noch, daß der von Gibbs in seinem Buche eingeschlagene Weg, der darin besteht, daß man gleich von einer kanonischen Gesamtheit ausgeht, nach meiner Meinung, dem von mir eingeschlagenen vorzuziehen ist. Wenn mir das Gibbssche Buch damals bekannt gewesen wäre, hätte ich jene Arbeiten überhaupt nicht publiziert, sondern mich auf die Behandlung einiger weniger Punkte beschränkt.

Zürich, Oktober 1910.

(Eingegangen 30. November 1910.)

10. Bemerkung zu meiner Arbeit[1]: "Eine Beziehung zwischen dem elastischen Verhalten..."; von A. Einstein.

In der genannten Arbeit habe ich als den Entdecker des Zusammenhanges zwischen elastischem und optischem Verhalten fester Stoffe Sutherland angegeben. Es war mir entgangen, daß E. Madelung zuerst auf diesen fundamental wichtigen Zusammenhang aufmerksam gemacht hat.[2] Madelung hat einen quantitativen Zusammenhang zwischen Elastizität und (optischer) Eigenfrequenz zweiatomiger Verbindungen gefunden, welcher dem von mir für den Fall einatomiger Stoffe abgeleiteten genau entspricht und mit der Erfahrung recht befriedigend übereinstimmt. Besonders muß hervorgehoben werden, daß Madelung zu seiner Beziehung nur unter der Voraussetzung gelangen kann, daß die Kräfte, die zwischen den Atomen eines Moleküls wirken, von derselben Größenordnung sind wie die Kräfte, die zwischen gleichartigen Atomen benachbarter Moleküle wirken; m. a. W. der Molekülverband scheint bei den von Madelung untersuchten Stoffen im festen Zustand nicht zu bestehen; diese Stoffe scheinen vollkommen dissoziiert zu sein. Es entspricht dies ganz den Vorstellungen, zu welchen die Untersuchung geschmolzener Salze geführt hat.

Zürich, Januar 1911.

---

1) A. Einstein, Ann. d. Phys. **34**. p. 170 ff. 1911.
2) E. Madelung, Nachr. d. kgl. Ges. d. Wissensch. zu Göttingen. Math.-phys. Kl. 20. II. 1909 und 29. I. 1910; Physik. Zeitschr. **11**. p. 898 bis 905. 1910.

(Eingegangen 30. Januar 1911.)

11. *Berichtigung zu meiner Arbeit: „Eine neue Bestimmung der Moleküldimensionen"* [1]; *von A. Einstein.*

Vor einigen Wochen teilte mir Hr. Bacelin, der auf Veranlassung von Hrn. Perrin eine Experimentaluntersuchung über die Viskosität von Suspensionen ausführte, brieflich mit, daß der Viskositätskoeffizient von Suspensionen nach seinen Resultaten erheblich größer sei, als der in § 2 meiner Arbeit entwickelten Formel entspricht. Ich ersuchte deshalb Hrn. Hopf, meine Rechnungen nachzuprüfen, und er fand in der Tat einen Rechenfehler, der das Resultat erheblich fälscht. Diesen Fehler will ich im folgenden berichtigen.

Auf p. 296 der genannten Abhandlung stehen Ausdrücke für die Spannungskomponenten X_y und X_z, die durch einen Fehler im Differenzieren der Geschwindigkeitskomponenten u, v, w gefälscht sind. Es muß heißen:

$$X_x = -2kA + 10kP^3 \frac{A\xi^2}{\varrho^5} - 25kP^3 \frac{M\xi^2}{\varrho^7},$$

$$X_y = \phantom{-2kA+{}} 5kP^3 \frac{(A+B)\xi\eta}{\varrho^5} - 25kP^3 \frac{M\xi\eta}{\varrho^7},$$

$$X_z = \phantom{-2kA+{}} 5kP^3 \frac{(A+C)\xi\zeta}{\varrho^5} - 25kP^3 \frac{M\xi\zeta}{\varrho^7},$$

wobei gesetzt ist

$$M = A\xi^2 + B\eta^2 + C\zeta^2.$$

Berechnet man dann die pro Zeiteinheit auf die in der Kugel vom Radius R enthaltene Flüssigkeit durch die Druckkräfte übertragene Energie, so erhält man statt Gleichung (7) auf p. 296:

(7) $$W = 2\delta^2 k(V + \tfrac{1}{2}\Phi).$$

---

1) A. Einstein, Ann. d. Phys. **19**. p. 289 ff. 1906.

592 *A. Einstein. Berichtigung.*

Unter Benutzung dieser berichtigten Gleichung erhält man dann statt der in § 2 entwickelten Gleichung $k^* = k(1 + \varphi)$ die Gleichung

$$k^* = k(1 + 2{,}5\,\varphi).$$

Der Viskositätskoeffizient k^* der Suspension wird also durch das Gesamtvolumen φ der in der Volumeinheit suspendierten Kugeln 2,5 mal stärker beeinflußt als nach der dort gefundenen Formel.

Legt man die berichtigte Formel zugrunde, so erhält man für das Volumen von 1 g in Wasser gelöstem Zucker statt des in § 3 angegebenen Wertes 2,45 cm³ den Wert 0,98, also einen vom Volumen 0,61 von 1 g festem Zucker erheblich weniger abweichenden Wert. Endlich erhält man aus der inneren Reibung und Diffusion von verdünnten Zuckerlösungen statt des im Anhange jener Arbeit angegebenen Wertes $N = 4{,}15 \cdot 10^{23}$ für die Anzahl der Moleküle im Grammolekül den Wert $6{,}56 \cdot 10^{23}$.

Zürich, Januar 1911.

(Eingegangen 21. Januar 1911.)

ANNALEN DER PHYSIK.

BEGRÜNDET UND FORTGEFÜHRT DURCH

F. A. C. GREN, L. W. GILBERT, J. C. POGGENDORFF, G. u. E. WIEDEMANN, P. DRUDE.

VIERTE FOLGE.

BAND 35.

DER GANZEN REIHE 340. BAND.

KURATORIUM:

M. PLANCK, G. QUINCKE,
W. C. RÖNTGEN, W. VOIGT, E. WARBURG.

UNTER MITWIRKUNG

DER DEUTSCHEN PHYSIKALISCHEN GESELLSCHAFT

HERAUSGEGEBEN VON

W. WIEN UND M. PLANCK.

MIT SECHS FIGURENTAFELN.

LEIPZIG, 1911.

VERLAG VON JOHANN AMBROSIUS BARTH.

2. *Elementare Betrachtungen über die thermische Molekularbewegung in festen Körpern;*
von A. Einstein.

In einer früheren Arbeit[1]) habe ich dargelegt, daß zwischen dem Strahlungsgesetz und dem Gesetz der spezifischen Wärme fester Körper (Abweichung vom Dulong-Petitschen Gesetz) ein Zusammenhang existieren müsse[2]). Die Untersuchungen Nernsts und seiner Schüler haben nun ergeben, daß die spezifische Wärme zwar im ganzen das aus der Strahlungstheorie gefolgerte Verhalten zeigt, daß aber das wahre Gesetz der spezifischen Wärme von dem theoretisch gefundenen systematisch abweicht. Es ist ein erstes Ziel dieser Arbeit, zu zeigen, daß diese Abweichungen darin ihren Grund haben, daß die Schwingungen der Moleküle weit davon entfernt sind, *monochromatische* Schwingungen zu sein. Die thermische Kapazität eines Atoms eines festen Körpers ist nicht gleich der eines schwach gedämpften, sondern ähnlich der eines stark gedämpften Oszillators im Strahlungsfelde. Der Abfall der spezifischen Wärme nach Null hin bei abnehmender Temperatur erfolgt deshalb weniger rasch, als er nach der früheren Theorie erfolgen sollte; der Körper verhält sich ähnlich wie ein Gemisch von Resonatoren, deren Eigenfrequenzen über ein gewisses Gebiet verteilt sind. Des weiteren wird gezeigt, daß sowohl Lindemanns Formel, als auch meine Formel zur Berechnung der Eigenfrequenz v der Atome durch Dimensionalbetrachtung abgeleitet werden können, insbesondere auch die Größenordnung der in diesen Formeln auftretenden Zahlen-

---

1) A. Einstein, Ann. d. Phys. **22**. p. 184. 1907.
2) Die Wärmebewegung in festen Körpern wurde dabei aufgefaßt als in monochromatischen Schwingungen der Atome bestehend. Vgl. hierzu § 2 dieser Arbeit.

koeffizienten. Endlich wird gezeigt, daß die Gesetze der Wärmeleitung in kristallisierten Isolatoren mit der Molekularmechanik nicht im Einklang sind, daß man aber die Größenordnung der tatsächlich zu beobachtenden Wärmeleitfähigkeit durch eine Dimensionalbetrachtung ableiten kann, wobei sich gleichzeitig ergibt, wie die thermische Leitfähigkeit einatomiger Stoffe von deren Atomgewicht, Atomvolumen und Eigenfrequenz mutmaßlich abhängt.

§ 1. Über die Dämpfung der thermischen Atomschwingungen.

In einer kürzlich erschienenen Arbeit[1]) habe ich gezeigt, daß man zu angenähert richtigen Werten für die Eigenfrequenzen der thermischen Atomschwingungen gelangt, indem man von folgenden Annahmen ausgeht:

1. Die die Atome an ihre Ruhelage fesselnden Kräfte sind wesensgleich den elastischen Kräften der Mechanik.

2. Die elastischen Kräfte wirken nur zwischen unmittelbar benachbarten Atomen.

Durch diese beiden Annahmen ist zwar die Theorie noch nicht vollständig festgelegt, da man die Elementargesetze der Wechselwirkung zwischen unmittelbar benachbarten Atomen noch bis zu einem gewissen Grade frei wählen kann. Auch ist nicht a priori klar, wie viele Moleküle man noch als „unmittelbar benachbart" ansehen will. Die spezielle Wahl der hieher gehörigen Hypothesen ändert jedoch wenig an den Resultaten, so daß ich mich wieder an die einfachen Annahmen halten will, die ich in jener Arbeit eingeführt habe. Auch die dort eingeführte Bezeichnungsweise will ich hier wieder benutzen.

In der zitierten Arbeit denke ich mir, daß jedes Atom 26 mit ihm elastisch in Wechselwirkung stehende Nachbaratome habe, die rechnerisch in bezug auf ihre elastische Wirkung auf das betrachtete Atom alle als gleichwertig angesehen werden dürfen. Die Berechnung der Eigenfrequenz wurde folgendermaßen durchgeführt. Man denkt sich die 26 Nachbaratome festgehalten und nur das betrachtete Atom schwingend; dieses führt dann eine ungedämpfte Pendel-

---

1) A. Einstein, Ann. d. Phys. **34**. p. 170. 1911.

schwingung aus, deren Frequenz man berechnet (aus der kubischen Kompressibilität). In Wahrheit sind aber die 26 Nachbarmoleküle nicht festgehalten, sondern sie schwingen in ähnlicher Weise wie das betrachtete Atom um ihre Gleichgewichtslage. Durch ihre elastischen Verknüpfungen mit dem betrachteten Atom beeinflussen sie die Schwingungen dieses letzteren, so daß dessen Schwingungsamplituden in den Koordinatenrichtungen sich fortwährend ändern, oder — was auf dasselbe hinauskommt — die Schwingung weicht von einer monochromatischen Schwingung ab. Es ist unsere erste Aufgabe, den Betrag dieser Abweichung abzuschätzen.

Es sei M das betrachtete Molekül, dessen Schwingungen in der x-Richtung wir untersuchen; x sei die momentane Entfernung des Moleküls aus seiner Ruhelage; M_1' sei ein Nachbarmolekül von M in der Ruhelage, das sich aber momentan im Abstand $d + \xi_1$ von der Ruhelage von M befinde, dann übt M_1' auf M in der Richtung MM_1' eine Kraft aus von der Größe $a(\xi_1 - x \cos \varphi_1)$. Die X-Komponente dieser Kraft ist

$$a(\xi_1 - x \cos \varphi_1) \cos \varphi_1.$$

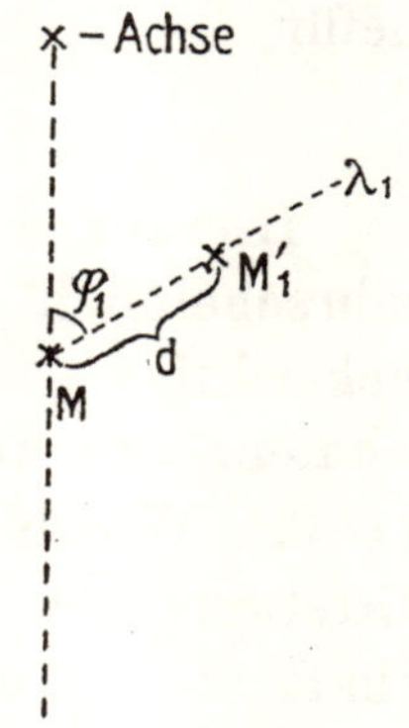

Fig. 1.

Ist m die Masse von M, so erhält man für M die Bewegungsgleichung

$$m \frac{d^2 x}{dt^2} = -x \cdot \sum a \cos^2 \varphi_1 + \sum a \, \xi_1 \cos \varphi_1,$$

wobei über alle 26 Nachbaratome zu summieren ist.

Nun berechnen wir die auf das Atom von den Nachbaratomen während einer halben Schwingung übertragene Energie. Dabei rechnen wir so, wie wenn die Oszillation sowohl des betrachteten Moleküls, als auch der Nachbarmoleküle während der Zeit einer halben Schwingung rein sinusartig erfolgte, d. h. wir setzen

$$x = A \sin 2 \pi \nu t,$$
$$\xi_1 = A_1' \sin (2 \pi \nu t + \alpha_1)$$

. . . .

Indem wir obige Gleichung mit $(dx/dt)\,dt$ multiplizieren und über die genannte Zeit integrieren, erhalten wir als Ausdruck für die Änderung der Energie

$$\int d\left\{m\frac{x^2}{2} + \sum (a\cos^2\varphi)\cdot\frac{x^2}{2}\right\} = \sum a\cos\varphi_1 \int \xi_1 \frac{dx}{dt}\,dt.$$

Bezeichnen wir mit $\varDelta$ die ganze Energiezunahme des Atoms, mit η_1, η_2 usw. die von den einzelnen Nachbaratomen während der Zeit einer halben Schwingung auf das Atom übertragenen Energiemengen, so können wir diese Gleichung in der Form

$$\varDelta = \sum \eta_n$$

schreiben, wobei

$$\eta_n = a\cos\varphi_n \int \xi_n \frac{dx}{dt}\,dt$$

gesetzt ist. Nach obigen Ansätzen für x, $\xi_1 \ldots$ ergibt sich hiefür

$$\eta_n = \frac{\pi}{2} a \cos\varphi_n \sin\alpha_n\, A\, A_n'.$$

Hieraus ergibt sich, daß die einzelnen Größen η_n gleich wahrscheinlich positiv wie negativ sind, wenn man berücksichtigt, daß die Winkel α_n jeden Wert gleich oft annehmen, und zwar unabhängig voneinander. Deshalb ist auch $\overline{\varDelta} = 0$. Wir bilden nun als Maß für die Energieänderung den Mittelwert $\overline{\varDelta^2}$. Wegen der angegebenen statistischen Eigenschaft von η_1 usw. ist

$$\overline{\varDelta^2} = \sum \overline{\eta_n^2}.$$

Da, wie leicht einzusehen ist,

$$\overline{\sin^2\alpha_n A^2 A_n'^2} = \tfrac{1}{2}\overline{A^2}^2,$$

so hat man

$$\overline{\eta_n^2} = \left(\frac{\pi}{2}a\right)^2 \cdot \tfrac{1}{2}\overline{A^2}^2 \cdot \cos^2\varphi_n$$

und

$$\overline{\varDelta^2} = \frac{\pi^2}{8} a^2 \overline{A^2}^2 \sum \cos^2\varphi_n.$$

Zur angenäherten Ausführung dieser Summe nehmen wir an, daß zwei der 26 Atome M' auf der x-Ache liegen, 16 derselben einen Winkel von nahezu 45° (bzw. 135°) gegen die x-Achse machen, die übrigen acht in der y-z-Ebene liegen. Wir erhalten dann $\sum \cos^2\varphi_n = 10$, so daß folgt:

$$\sqrt{\overline{\varDelta^2}} = \sqrt{\frac{10}{8}}\,\pi\, a\, \overline{A^2}.$$

Wir vergleichen nun mit diesem Mittelwert für die Energie-

Molekularbewegung in festen Körpern.

zunahme des Atoms die mittlere Energie des Atoms. Der Momentanwert für die potentielle Energie des Atoms ist

$$a \frac{x^2}{2} \sum \cos^2 \varphi = a \frac{x^2}{2} \cdot 10.$$

Der Mittelwert der potentiellen Energie ist also

$$5 a \overline{x^2} = \tfrac{5}{2} a \overline{A^2}.$$

Der Mittelwert der Gesamtenergie E ist also

$$\overline{E} = 5 a \overline{A^2}.$$

Der Vergleich von $\overline{E}$ mit $\sqrt{\overline{A^2}}$ zeigt, *daß die Energieänderung während der Zeit einer halben Schwingung von derselben Größenordnung ist wie die Energie selbst.*

Die von uns zugrunde gelegten Ansätze für x, ξ_1 usw. sind also eigentlich nicht einmal für die Zeit einer halben Schwingung angenähert richtig. Unser Resultat aber, daß sich die Schwingungsenergie bereits während einer halben Schwingung bedeutend ändert, wird hiervon nicht berührt.

§ 2. Spezifische Wärme einfacher fester Stoffe und Strahlungstheorie.

Bevor wir uns fragen, was für eine Konsequenz das soeben erlangte Resultat für die Theorie der spezifischen Wärme hat, müssen wir uns des Gedankenganges erinnern, der von der Strahlungstheorie zur Theorie der spezifischen Wärme führt. Planck hat bewiesen, daß ein durch Ausstrahlung schwach gedämpfter Oszillator von der Eigenfrequenz ν_0 in einem Strahlungsfelde von der Dichte $\mathfrak{u}$ ($\mathfrak{u}\, d\nu$ = Strahlungsenergie des Frequenzbereiches $d\nu$ pro Volumeneinheit) die mittlere Energie

$$\overline{E} = \frac{c^3 \mathfrak{u}_0}{8 \pi \nu_0^2}$$

annimmt, wenn c die Vakuumlichtgeschwindigkeit, ν_0 die Eigenfrequenz des Oszillators, $\mathfrak{u}_0$ die Strahlungsdichte für die Frequenz ν_0 bedeutet.

Der betrachtete Oszillator bestehe in einem Ion, das durch quasielastische Kräfte an eine Gleichgewichtslage gebunden sei. Es mögen sich im Strahlungsraum auch noch Gasmoleküle befinden, welche sich mit der Strahlung im statistischen (Temperatur-) Gleichgewichte befinden, und welche mit dem unseren Oszillator bildenden Ion Zusammenstöße erfahren können. Durch diese Zusammenstöße darf auf den Oszillator im Mittel keine Energie übertragen werden, da sonst der Oszillator das thermodynamische Gleichgewicht zwischen Gas und Strahlung stören würde. Es muß deshalb geschlossen werden, daß die mittlere Energie, welche die Gasmoleküle allein unserem Oszillator erteilen würden, genau gleich groß ist wie die mittlere Energie, welche die Strahlung allein dem Oszillator erteilt, also gleich $\bar{E}$. Da es ferner für die molekularen Zusammenstöße prinzipiell ohne Belang ist, ob das betreffende Gebilde eine elektrische Ladung trägt oder nicht, so gilt die obige Relation für jedes annähernd monochromatisch schwingende Gebilde. Seine mittlere Energie ist verknüpft mit der mittleren Dichte u der Strahlung von der gleichen Frequenz bei der betreffenden Temperatur. Faßt man die Atome fester Körper als nahezu monochromatisch schwingende Gebilde auf, so erhält man demnach aus der Strahlungsformel direkt die Formel für die spezifische Wärme, welche für ein Grammolekül den Wert $N(d\bar{E}/dT)$ haben müßte.

Man sieht, daß diese Überlegung, deren Resultat mit den Resultaten der statistischen Mechanik bekanntlich nicht im Einklang steht, unabhängig ist von der Quantentheorie, überhaupt unabhängig von jeder speziellen Theorie der Strahlung. Sie stützt sich nur

    1. auf das empirisch bekannte Strahlungsgesetz,
    2. auf die Plancksche Resonatorenbetrachtung, welche ihrerseits auf die Maxwellsche Elektromagnetik und Mechanik gegründet ist,
    3. auf die Auffassung, daß die Atomschwingungen mit großer Annäherung sinusförmig sind.

Zu 2. ist ausdrücklich zu bemerken, daß die von Planck benutzte Schwingungsgleichung des Oszillators nicht ohne Mechanik streng abgeleitet werden kann. Die Elektromagnetik bedient sich nämlich bei der Lösung von Bewegungsaufgaben der Voraussetzung, daß die Summe der am Gerüst eines Elektrons angreifenden elektrodynamischen und sonstigen Kräfte stets Null sei, oder — wenn man dem betreffenden Gebilde ponderable Masse zuschreibt — daß die Summe der elektrodynamischen und sonstigen Kräfte gleich sei der Masse multipliziert mit der Beschleunigung. Man hat also a priori wohl Grund, an der Richtigkeit des Resultates der Planckschen Betrachtung zu zweifeln, wenn man bedenkt, daß das Fundament unserer Mechanik, auf rasch periodische Vorgänge angewendet, zu der Erfahrung widersprechenden Resultaten führt[1]), daß also die Anwendung jenes Fundamentes auch hier Bedenken erregen muß. Trotzdem glaube ich, daß an der Planckschen Beziehung zwischen u_0 und $\bar{E}$ festzuhalten ist, schon deshalb, weil sie eben zu einer angenähert richtigen Darstellung der spezifischen Wärme bei tiefen Temperaturen geführt hat.

Dagegen haben wir im vorigen Paragraphen gezeigt, daß die Annahme 3. nicht aufrecht erhalten werden kann. Die Atomschwingungen sind nicht angenähert harmonische Schwingungen. Der Frequenzbereich eines Atoms ist so groß, daß sich die Schwingungsenergie während einer halben Schwingung um einen Betrag von der Größenordnung der Schwingungsenergie ändert. Wir haben also jedem Atom nicht eine bestimmte Frequenz, sondern einen Frequenzbereich $\Delta\nu$ zuzuschreiben, der von derselben Größenordnung wie die Frequenz selber ist. Um die Formel für die spezifische Wärme fester Körper exakt abzuleiten, müßte man für ein Atom eines festen Körpers unter Zugrundelegung eines mechanischen Modelles eine Betrachtung durchführen, die der von Planck für den unendlich wenig gedämpften Oszillator durchgeführten völlig analog ist. Man müßte berechnen, bei welcher mittleren Schwingungsenergie ein Atom, wenn es mit einer elektrischen Ladung versehen wird, in einem Temperaturstrahlungsfelde ebensoviel Energie emittiert wie absorbiert.

Während ich mich ziemlich resultatlos mit der Durchführung dieses Planes quälte, erhielt ich von Nernst den Korrekturbogen einer Arbeit zugesandt[2]), in welcher eine über-

---

1) Unsere Mechanik vermag nämlich die kleinen spezifischen Wärmen fester Körper bei tiefen Temperaturen nicht zu erklären.

2) W. Nernst u. F. A. Lindemann, Sitzungsber. d. preuß. Akad. d. Wiss. 22. 1911.

raschend brauchbare vorläufige Lösung der Aufgabe enthalten ist. Er findet, daß die Form

$$\frac{3}{2} R \left(\frac{\left(\frac{\beta \nu}{T}\right)^2 e^{\frac{\beta \nu}{T}}}{\left(e^{\frac{\beta \nu}{T}} - 1\right)^2} + \frac{\left(\frac{\beta \nu}{2T}\right)^2 e^{\frac{\beta \nu}{2T}}}{\left(e^{\frac{\beta \nu}{2T}} - 1\right)^2} \right)$$

die Temperaturabhängigkeit der Atomwärme vorzüglich darstellt. Daß diese Form sich der Erfahrung besser anschmiegt als die ursprünglich von mir gewählte, ist nach dem Vorangehenden leicht zu erklären. Man kommt ja zu derselben unter der Annahme, daß ein Atom in der halben Zeit mit der Frequenz ν, in der andern Hälfte der Zeit mit der Frequenz $\nu/2$ quasi ungedämpft sinusartig schwinge. Die bedeutende Abweichung des Gebildes vom monochromatischen Verhalten findet auf diese Weise ihren primitivsten Ausdruck.

Allerdings ist es dann nicht gerechtfertigt, ν als die Eigenfrequenz des Gebildes zu betrachten, sondern es wird als mittlere Eigenfrequenz ein zwischen ν und $\nu/2$ liegender Wert anzusehen sein. Es muß ferner bemerkt werden, daß an eine genaue Übereinstimmung der thermischen und optischen Eigenfrequenz nicht gedacht werden kann, auch wenn die Eigenfrequenzen der verschiedenen Atome der betreffenden Verbindung nahe übereinstimmen, weil bei der thermischen Schwingung das Atom gegenüber allen benachbarten Atomen schwingt, bei der optischen Schwingung aber nur gegenüber den benachbarten Atomen entgegengesetzten Vorzeichens.

§ 3. Dimensionalbetrachtung zu Lindemanns Formel und zu meiner Formel zur Berechnung der Eigenfrequenz.

Aus Dimensionalbetrachtungen kann man bekanntlich zunächst allgemeine funktionelle Zusammenhänge zwischen physikalischen Größen finden, wenn man alle physikalischen Größen kennt, welche in dem betreffenden Zusammenhang vorkommen. Wenn man z. B. weiß, daß die Schwingungszeit Θ eines mathematischen Pendels von der Pendellänge l, von der Beschleunigung g des freien Falles, von der Pendelmasse m, aber von keiner anderen Größe abhängen kann, so führt eine einfache

Dimensionalbetrachtung dazu, daß der Zusammenhang durch die Gleichung

$$\Theta = C \cdot \sqrt{\frac{l}{g}}$$

gegeben sein muß, wobei C eine dimensionslose Zahl ist. Man kann aber bekanntlich noch etwas mehr aus der Dimensionalbetrachtung entnehmen, wenn auch nicht mit voller Strenge. Es pflegen nämlich dimensionale Zahlenfaktoren (wie hier der Faktor C), deren Größe sich nur durch eine mehr oder weniger detaillierte mathematische Theorie deduzieren läßt, im allgemeinen von der Größenordnung Eins zu sein. Dies läßt sich zwar nicht streng fordern, denn warum sollte ein numerischer Faktor $(12\pi)^3$ nicht bei einer mathematisch-physikalischen Betrachtung auftreten können? Aber derartige Fälle gehören unstreitig zu den Seltenheiten. Gesetzt also, wir würden an einem einzigen mathematischen Pendel die Schwingungszeit Θ und die Pendellänge l messen, und wir würden aus obiger Formel für die Konstante C den Wert 10^{10} herausbekommen, so würden wir unserer Formel bereits mit berechtigtem Mißtrauen gegenüberstehen. Umgekehrt werden wir, falls wir aus unseren Versuchsdaten für C etwa 6,3 finden, an Vertrauen gewinnen; unsere Grundannahme, daß in der gesuchten Beziehung nur die Größen Θ, l und g, aber keine anderen Größen vorkommen, wird für uns an Wahrscheinlichkeit gewinnen.

Wir suchen nun die Eigenfrequenz ν eines Atoms eines festen Körpers durch eine Dimensionalbetrachtung zu ermitteln. Die einfachste Möglichkeit ist offenbar die, daß der Schwingungsmechanismus durch folgende Größen bestimmt ist:

1. durch die Masse m eines Atoms (Dimension m),

2. durch den Abstand d zweier benachbarter Atome (Dimension l),

3. durch die Kräfte, welche benachbarte Atome einer Veränderung ihres Abstandes entgegensetzen. Diese Kräfte äußern sich auch bei elastischen Deformationen; ihre Größe wird gemessen durch den Koeffizienten der Kompressibilität $\varkappa$ (Dimension $l\,t^2/m$).

Der einzige Ausdruck für ν aus diesen drei Größen, welcher die richtige Dimension hat, ist

$$\nu = C\sqrt{\frac{d}{m\,\varkappa}},$$

wobei C wieder ein dimensionsloser Zahlenfaktor ist. Führt man für d das Molekularvolumen v ein ($d = \sqrt[3]{v/N}$), statt m das sogenannte Atomgewicht M ($M = N\cdot m$), so erhält man daraus

$$\nu = C\, N^{1/3}\, v^{1/6}\, M^{-1/2}\, \varkappa^{-1/2} = C\cdot 1{,}9\cdot 10^7\, M^{-1/2}\, \varrho^{-1/6}\, \varkappa^{-1/2},$$

wobei ϱ die Dichte bezeichnet.

Die von mir durch molekularkinetische Betrachtung gefundene Formel

$$\lambda = 1{,}08\cdot 10^3 \cdot M^{1/3}\, \varrho^{1/6}\, \varkappa^{1/2}$$

oder

$$\nu = 2{,}8\cdot 10^7\, M^{-1/3}\, \varrho^{-1/6}\, \varkappa^{-1/2}$$

stimmt mit dieser Formel überein mit einem Faktor C von der Größenordnung Eins. Der Zahlenfaktor, der sich aus meiner früheren Betrachtung ergibt, ist in befriedigender Übereinstimmung mit der Erfahrung.[1]) So berechnet man für Kupfer nach meiner Formel aus der Kompressibilität

$$\nu = 5{,}7\cdot 10^{12},$$

während sich mit Hilfe der im § 2 besprochenen Nernstschen Formel aus der spezifischen Wärme

$$\nu = 6{,}6\cdot 10^{12}$$

ergibt. Dieser Wert von ν ist aber nicht als „wahre Eigenfrequenz" aufzufassen. Von letzterer wissen wir nur, daß sie zwischen Nernsts ν und der Hälfte dieses Wertes liegt. Es liegt am nächsten, in Ermangelung einer genauen Theorie $\frac{\nu + \nu/2}{2}$ als „wahre Eigenfrequenz" aufzufassen, für welche Größe man nach Nernst für Kupfer den Wert

$$\nu = 5{,}0\cdot 10^{12}$$

erhält, in naher Übereinstimmung mit dem aus der Kompressibilität berechneten Wert.

---

1) Bezüglich der Annäherung, mit der die Formel gilt, vgl. den letzten Absatz dieses Paragraphen.

Wir wenden uns zu Lindemanns Formel.¹) Wir nehmen wieder an, daß zunächst die Masse eines Atoms und der Abstand d zweier Nachbaratome auf die Eigenfrequenz von Einfluß sind. Außerdem nehmen wir an, es gebe mit einer hier genügenden Annäherung ein Gesetz der übereinstimmenden Zustände für den festen Zustand. Dann muß durch Hinzufügung einer weiteren charakteristischen Größe der Substanz, welche durch die vorgenannten noch nicht bestimmt ist, das Verhalten der Substanz, also auch die Eigenfrequenz, vollkommen bestimmt sein. Als diese dritte Größe nehmen wir die Schmelztemperatur T_s. Diese ist natürlich für Dimensionalbetrachtungen nicht ohne weiteres verwendbar, da sie nicht im C.G.S.-System unmittelbar gemessen werden kann. Wir wählen deshalb statt T_s die Energiegröße $\tau = RT_s/N$ als Temperaturmaß. τ ist ein Drittel der Energie, welche ein Atom beim Schmelzpunkt nach der kinetischen Theorie der Wärme besitzt (R = Gaskonstante, N = Zahl der Atome im Grammatom). Die Dimensionalbetrachtung liefert unmittelbar

$$\nu = C \cdot \sqrt{\frac{\tau}{m\,d^2}} = C \cdot R^{1/2} N^{1/3} \sqrt{\frac{T_s}{M\,v^{2/3}}} = C \cdot 0{,}77 \cdot 10^{12} \sqrt{\frac{T_s}{M\,v^{2/3}}}.$$

Die Lindemannsche Formel lautet:

$$\nu = 2{,}12 \cdot 10^{12} \sqrt{\frac{T_s}{M\,v^{2/3}}}.$$

Auch hier ist also die dimensionslose Konstante C von der Größenordnung Eins.

Die Untersuchungen Nernsts und seiner Schüler²) zeigen, daß diese Formel, trotzdem sie auf einer sehr gewagten Annahme ruht, überraschend gute Übereinstimmung mit den aus der spezifischen Wärme bestimmten ν-Werten liefert. Es scheint daraus hervorzugehen, daß das Gesetz der übereinstimmenden Zustände für einfache Körper im festen und flüssigen Zustande mit bemerkenswerter Annäherung gilt. Die Lindemannsche Formel scheint sogar viel besser zu stimmen als meine auf weniger gewagter Grundlage ruhende Formel.

---

1) F. Lindemann, Physik. Zeitschr. **11**. p. 609. 1910.
2) Vgl. insbesondere W. Nernst, Sitzungsber. d. preuß. Akad. d. Wiss. **13**. p. 311. 1911.

Dies ist um so merkwürdiger, als meine Formel natürlich auch aus dem Gesetz der übereinstimmenden Zustände gefolgert werden kann. Sollte sowohl meine wie Lindemanns Formel zutreffen, so müßte, wie durch Division beider Formeln folgt, $M/\varrho T_s \varkappa$ von der Natur des Stoffes unabhängig sein, eine Beziehung, die übrigens auch direkt aus dem Gesetz der übereinstimmenden Zustände gefolgert werden kann. Unter Zugrundelegung der Grüneisenschen[1]) Werte für die Kompressibilität der Metalle erhält man für diese Größe indessen Werte, die etwa zwischen $6 \cdot 10^{-15}$ und $15 \cdot 10^{-15}$ schwanken! Dies ist in Verbindung mit der Tatsache, daß sich das Gesetz der übereinstimmenden Zustände im Falle der Lindemannschen Formel so befriedigend bewährt, recht sonderbar. Wäre es nicht vielleicht möglich, daß in allen Bestimmungen der kubischen Kompressibilität der Metalle noch systematische Fehler stecken? Die Kompression unter allseitig gleichem Druck ist noch nicht zur Messung verwendet worden, wohl wegen der bedeutenden experimentellen Schwierigkeiten. Vielleicht würden derartige Messungen bei Deformation ohne Winkeldeformation zu beträchtlich anderen Werten von $\varkappa$ führen als die bisherigen Messungen. Vom theoretischen Standpunkt aus liegt dieser Verdacht wenigstens nahe.

§ 4. Bemerkungen über das thermische Leitvermögen von Isolatoren.

Das in § 1 gefundene Resultat läßt einen Versuch gerechtfertigt erscheinen, das thermische Leitvermögen fester, nicht metallisch leitender Substanzen angenähert zu berechnen. Es sei nämlich ε die mittlere kinetische Energie eines Atoms,

Fig. 2.

dann gibt nach § 1 das Atom in der Zeit einer halben Schwingung im Mittel eine Energie von der Größe $\alpha \cdot \varepsilon$ an die umgebenden Atome ab, wobei α ein Koeffizient von der Größenordnung Eins, aber kleiner als Eins ist. Denken wir uns die Atome in einem Gitter gelagert und betrachten wir ein Atom A, welches unmittelbar neben einer gedachten Ebene

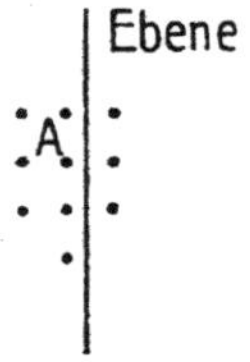

---

1) E. Grüneisen, Ann. d. Phys. **25**. p. 848. 1900.

liegt, die kein Molekül schneidet, so wird im Mittel etwa die Energie

$$\alpha \cdot \varepsilon \frac{9}{26}$$

vom Molekül A während der Zeit einer halben Schwingung durch die Ebene hindurchgesandt werden, in der Zeiteinheit also die Energie

$$\alpha \, \varepsilon \cdot \frac{9}{26} \cdot 2\nu \, .$$

Ist d der kleinste Abstand benachbarter Atome, so liegen pro Flächeneinheit $(1/d)^2$ Atome auf einer Seite an der Ebene an, die zusammen die Energie

$$\alpha \cdot \frac{9}{13} \nu \cdot \frac{1}{d^2} \varepsilon$$

pro Flächeneinheit in der einen Richtung (Richtung der wachsenden x) durch die Flächeneinheit der Ebene senden. Da die Moleküle auf der anderen Seite der Schicht in der Zeiteinheit die Energiemenge

$$-\alpha \frac{9}{13} \nu \frac{1}{d^2} \left(\varepsilon + \frac{d\varepsilon}{dx} \cdot d \right)$$

in der Richtung der negativen x durch die Flächeneinheit senden, so ist die ganze Energieströmung

$$-\alpha \cdot \frac{9}{13} \nu \cdot \frac{1}{d} \frac{d\varepsilon}{dx} \, .$$

Benutzen wir, daß $d = (v/N)^{1/3}$ und bezeichnen wir mit W den Wärmeinhalt des Grammatoms bei der Temperatur T, so erhalten wir den Ausdruck

$$-\alpha \frac{9}{13} \nu \, v^{-1/3} N^{-2/3} \frac{dW}{dT} \frac{dT}{dx} \, ,$$

also für den Wärmeleitungskoeffizienten k

$$k = \alpha \cdot \frac{9}{13} \nu \, v^{-1/3} N^{-2/3} \frac{dW}{dT} \, .$$

Wird W in Kalorien gemessen, so erhält man k im üblichen Maß (cal/cm sec grad). Erfüllt der Stoff in dem in Betracht kommenden Temperaturbereich das Gesetz von Dulong-Petit, so kann man, weil

$$\frac{dW}{dT} = \frac{3\,R}{\text{Wärmeäquivalent}} = \frac{3 \cdot 8{,}3 \cdot 10^7}{4{,}2 \cdot 10^7} \sim 6 \, ,$$

hierfür etwa setzen

$$k = \alpha \cdot 4 N^{-2/3} \nu v^{-1/3}.$$

Diese Formel wenden wir zunächst auf KCl an, welches sich nach Nernst bezüglich seiner spezifischen Wärme ähnlich wie ein Stoff mit lauter gleichen Atomen verhält, und erhalten, indem wir für ν den von Nernst aus dem Verlaufe der spezifischen Wärme ermittelten Wert $3{,}5 \cdot 10^{12}$ nehmen,

$$k = \alpha \cdot 4 \cdot (6{,}3 \cdot 10^{23})^{-2/3} \cdot 3{,}5 \cdot 10^{12} \cdot \left(\frac{74{,}4}{2 \cdot 2}\right)^{-1/3} = \alpha \cdot 0{,}0007,$$

während die Erfahrung bei gewöhnlicher Temperatur etwa

$$k = 0{,}016$$

ergibt.[1]) Die Wärmeleitung ist also viel größer als nach unserer Betrachtung zu erwarten wäre. Aber nicht nur dies. Nach unserer Formel[2]) sollte innerhalb der Gültigkeit des Dulong-Petitschen Gesetzes k von der Temperatur unabhängig sein. Nach Euckens Resultaten ist aber das tatsächliche Verhalten kristallinischer Nichtleiter ein ganz anderes; $\varkappa$ ändert sich annähernd wie $1/T$. Wir müssen daraus schließen, daß die Mechanik nicht imstande ist, die thermische Leitfähigkeit der Nichtleiter zu erklären.[3]) Es ist hinzuzufügen, daß auch die Annahme von einer quantenhaften Verteilung der Energie zur Erklärung von Euckens Resultaten nichts beiträgt.

Man kann auf Euckens wichtiges Resultat, daß die Wärmeleitungsfähigkeit kristallinischer Isolatoren nahezu proportional $1/T$ ist, eine sehr interessante Dimensionalbetrachtung gründen. Wir definieren die „Wärmeleitfähigkeit in natürlichem Maße" k_{nat} durch die Gleichung:

Wärmefluß pro Flächeneinheit und Sekunde $= -k_{\text{nat}} \dfrac{d\tau}{dx},$

wobei der Wärmefluß in absoluten Einheiten ausgedrückt zu denken ist und $\tau = RT/N$ gesetzt ist. k_{nat} ist eine im C.G.S.-System zu messende Größe von der Dimension $[l^{-1} t^{-1}]$.

---

1) Vgl. A. Eucken, Ann. d. Phys. **34**. p. 217. 1911.
2) bzw. nach einer auf der Hand liegenden Ähnlichkeitsbetrachtung.
3) Es muß bemerkt werden, daß hierdurch auch die Betrachtungen der §§ 1 und 2 unsicher werden.

Diese Größe kann bei einem einatomigen festen Isolator abhängen von den Größen:

d (Abstand benachbarter Atome; Dimension l),
m (Masse eines Atoms; Dimension m),
ν (Frequenz des Atoms; Dimension t^{-1}),
τ (Temperaturmaß; Dimension $m^1 l^2 t^{-2}$).

Nehmen wir eine Abhängigkeit von weiteren Größen nicht an, so zeigt die Dimensionalbetrachtung, daß k_{nat} sich durch eine Gleichung von der Form

$$k_{\text{nat}} = C \cdot d^{-1} \nu^1 \varphi \left(\frac{m^1 d^2 \nu^2}{\tau^1} \right)$$

ausdrücken lassen muß, wobei C wieder eine Konstante von der Größenordnung Eins und φ eine a priori willkürliche Funktion bedeutet, die aber nach dem mechanischen Bilde bei Annahme quasielastischer Kräfte zwischen den Atomen gleich einer Konstanten sein müßte. Nach Euckens Resultaten haben wir aber annähernd φ dem Argument proportional zu setzen, damit k_{nat} dem absoluten Temperaturmaß τ umgekehrt proportional werde. Wir erhalten also

$$k_{\text{nat}} = C m^1 d^1 \nu^3 \tau^{-1},$$

wobei C eine andere Konstante von der Größenordnung Eins bedeutet. Führen wir statt k_{nat} wieder k ein, indem wir zur Messung des Wärmestromes die Kalorie und zur Messung des Temperaturgefälles den Celsiusgrad verwenden, und ersetzen wir m, d, τ durch ihre Ausdrücke in M, v, T, so erhalten wir

$$k = \frac{1}{4{,}2 \cdot 10^7} \cdot \frac{R}{N} \cdot C \cdot \frac{M}{N} \cdot \left(\frac{v}{N} \right)^{1/3} \cdot \nu^3 \cdot \frac{N}{RT} = C \frac{N^{-4/3}}{4{,}2 \cdot 10^7} \frac{M v^{1/3} \nu^3}{T}.$$

Diese Gleichung spricht eine Beziehung zwischen der Wärmeleitfähigkeit, dem Atomgewicht, dem Atomvolumen und der Eigenfrequenz aus. Für KCl bekommen wir aus dieser Formel

$$k_{273} = C \cdot 0{,}007.$$

Die Erfahrung ergibt $k_{273} = 0{,}0166$, so daß C in der Tat von der Größenordnung Eins wird. Wir müssen dies als eine Bestätigung der unserer Dimensionalbetrachtung zugrunde liegenden Annahmen ansehen. Ob C einigermaßen unabhängig ist von der Natur der Substanz, wird die Erfahrung entscheiden müssen; Aufgabe der Theorie wird es sein, die Molekularmechanik so zu modifizieren, daß sie sowohl das Gesetz der spezifischen Wärme als auch das dem Anscheine nach so einfache Gesetz der thermischen Leitfähigkeit liefert.

Prag, Mai 1911.

(Eingegangen 4. Mai 1911.)

Nachtrag zur Korrektur.

Zur Verdeutlichung der letzten Absätze von § 2 sei folgendes bemerkt. Bezeichnet man mit $\varphi(\nu/\nu_0)$ eine als zeitliche Häufigkeit der momentanen Frequenz ν aufzufassende Funktion, mit $\Phi(\nu_0/T)$ die spezifische Wärme des monochromatischen Gebildes von der Frequenz ν_0, so kann man die spezifische Wärme des nicht monochromatischen Gebildes durch die Formel ausdrücken

$$\tau = \int_{x=0}^{x=\infty} \Phi\left(\frac{\nu_0 x}{T}\right) \varphi(x)\, dx.$$

Zu Nernsts Formel kommt man, wenn man der Funktion $\varphi(x)$ nur für die Argumente 1 und $1/2$ von Null verschiedene Werte gibt.

898

4. *Über den Einfluß der Schwerkraft auf die Ausbreitung des Lichtes;* von *A. Einstein.*

Die Frage, ob die Ausbreitung des Lichtes durch die Schwere beinflußt wird, habe ich schon an einer vor 3 Jahren erschienenen Abhandlung zu beantworten gesucht.[1]) Ich komme auf dies Thema wieder zurück, weil mich meine damalige Darstellung des Gegenstandes nicht befriedigt, noch mehr aber, weil ich nun nachträglich einsehe, daß eine der wichtigsten Konsequenzen jener Betrachtung der experimentellen Prüfung zugänglich ist. Es ergibt sich nämlich, daß Lichtstrahlen, die in der Nähe der Sonne vorbeigehen, durch das Gravitationsfeld derselben nach der vorzubringenden Theorie eine Ablenkung erfahren, so daß eine scheinbare Vergrößerung des Winkelabstandes eines nahe an der Sonne erscheinenden Fixsternes von dieser im Betrage von fast einer Bogensekunde eintritt.

Es haben sich bei der Durchführung der Überlegungen auch noch weitere Resultate ergeben, die sich auf die Gravitation beziehen. Da aber die Darlegung der ganzen Betrachtung ziemlich unübersichtlich würde, sollen im folgenden nur einige ganz elementare Überlegungen gegeben werden, aus denen man sich bequem über die Voraussetzungen und den Gedankengang der Theorie orientieren kann. Die hier abgeleiteten Beziehungen sind, auch wenn die theoretische Grundlage zutrifft, nur in erster Näherung gültig.

§ 1. Hypothese über die physikalische Natur des Gravitationsfeldes.

In einem homogenen Schwerefeld (Schwerebeschleunigung γ) befinde sich ein ruhendes Koordinatensystem K, das so orientiert sei, daß die Kraftlinien des Schwerefeldes in Richtung

---

1) A. Einstein, Jahrb. f. Radioakt. u. Elektronik IV. 4.

der negativen z-Achse verlaufen. In einem von Gravitationsfeldern freien Raume befinde sich ein zweites Koordinatensystem K', das in Richtung seiner positiven z-Achse eine gleichförmig beschleunigte Bewegung (Beschleunigung γ) ausführe. Um die Betrachtung nicht unnütz zu komplizieren, sehen wir dabei von der Relativitätstheorie vorläufig ab, betrachten also beide Systeme nach der gewohnten Kinematik und in denselben stattfindende Bewegungen nach der gewöhnlichen Mechanik.

Relativ zu K, sowie relativ zu K', bewegen sich materielle Punkte, die der Einwirkung anderer materieller Punkte nicht unterliegen, nach den Gleichungen:

$$\frac{d^2 x_\nu}{dt^2} = 0, \quad \frac{d^2 y_\nu}{dt^2} = 0, \quad \frac{d^2 z_\nu}{dt^2} = -\gamma.$$

Dies folgt für das beschleunigte System K' direkt aus dem Galileischen Prinzip, für das in einem homogenen Gravitationsfeld ruhende System K aber aus der Erfahrung, daß in einem solchen Felde alle Körper gleich stark und gleichmäßig beschleunigt werden. Diese Erfahrung vom gleichen Fallen aller Körper im Gravitationsfelde ist eine der allgemeinsten, welche die Naturbeobachtung uns geliefert hat; trotzdem hat dieses Gesetz in den Fundamenten unseres physikalischen Weltbildes keinen Platz erhalten.

Wir gelangen aber zu einer sehr befriedigenden Interpretation des Erfahrungssatzes, wenn wir annehmen, daß die Systeme K und K' physikalisch genau gleichwertig sind, d. h. wenn wir annehmen, man könne das System K ebenfalls als in einem von einem Schwerefeld freien Raume befindlich annehmen; dafür müssen wir K dann aber als gleichförmig beschleunigt betrachten. Man kann bei dieser Auffassung ebensowenig von der *absoluten Beschleunigung* des Bezugssystems sprechen, wie man nach der gewöhnlichen Relativitätstheorie von der *absoluten Geschwindigkeit* eines Systems reden kann.[1]

---

[1] Natürlich kann man ein *beliebiges* Schwerefeld nicht durch einen Bewegungszustand des Systems ohne Gravitationsfeld ersetzen, ebensowenig, als man durch eine Relativitätstransformation alle Punkte eines beliebig bewegten Mediums auf Ruhe transformieren kann.

Bei dieser Auffassung ist das gleiche Fallen aller Körper in einem Gravitationsfelde selbstverständlich.

Solange wir uns auf rein mechanische Vorgänge aus dem Gültigkeitsbereich von Newtons Mechanik beschränken, sind wir der Gleichwertigkeit der Systeme K und K' sicher. Unsere Auffassung wird jedoch nur dann tiefere Bedeutung haben, wenn die Systeme K und K' in bezug auf alle physikalischen Vorgänge gleichwertig sind, d. h. wenn die Naturgesetze in bezug auf K mit denen in bezug auf K' vollkommen übereinstimmen. Indem wir dies annehmen, erhalten wir ein Prinzip, das, falls es wirklich zutrifft, eine große heuristische Bedeutung besitzt. Denn wir erhalten durch die theoretische Betrachtung der Vorgänge, die sich relativ zu einem gleichförmig beschleunigten Bezugssystem abspielen, Aufschluß über den Verlauf der Vorgänge in einem homogenen Gravitationsfelde.[1]) Im folgenden soll zunächst gezeigt werden, inwiefern unserer Hypothese vom Standpunkte der gewöhnlichen Relativitätstheorie aus eine beträchtliche Wahrscheinlichkeit zukommt.

§ 2. Über die Schwere der Energie.

Die Relativitätstheorie hat ergeben, daß die träge Masse eines Körpers mit dem Energieinhalt desselben wächst; beträgt der Energiezuwachs E, so ist der Zuwachs an träger Masse gleich E/c^2, wenn c die Lichtgeschwindigkeit bedeutet. Entspricht nun aber diesem Zuwachs an träger Masse auch ein Zuwachs an gravitierender Masse? Wenn nicht, so fiele ein Körper in demselben Schwerefelde mit verschiedener Beschleunigung je nach dem Energieinhalte des Körpers. Das so befriedigende Resultat der Relativitätstheorie, nach welchem der Satz von der Erhaltung der Masse in dem Satze von der Erhaltung der Energie aufgeht, wäre nicht aufrecht zu erhalten; denn so wäre der Satz von der Erhaltung der Masse zwar für die *träge* Masse in der alten Fassung aufzugeben, für die gravitierende Masse aber aufrecht zu erhalten.

---

1) In einer späteren Abhandlung wird gezeigt werden, daß das hier in Betracht kommende Gravitationsfeld nur in erster Annäherung homogen ist.

Dies muß als sehr unwahrscheinlich betrachtet werden. Andererseits liefert uns die gewöhnliche Relativitätstheorie kein Argument, aus dem wir folgern könnten, daß das Gewicht eines Körpers von dessen Energieinhalt abhängt. Wir werden aber zeigen, daß unsere Hypothese von der Äquivalenz der Systeme K und K' die Schwere der Energie als notwendige Konsequenz liefert.

Es mögen sich die beiden mit Meßinstrumenten versehenen körperlichen Systeme S_1 und S_2 in der Entfernung h voneinander auf der z-Achse von K befinden[1]), derart, daß das Gravitationspotential in S_2 um $\gamma \cdot h$ größer ist, als das in S_1. Es wurde von S_2 gegen S_1 eine bestimmte Energiemenge E in Form von Strahlung gesendet. Die Energiemengen mögen dabei in S_1 und S_2 mit Vorrichtungen gemessen werden, die — an *einen* Ort des Systems z gebracht und dort miteinander verglichen — vollkommen gleich seien. Über den Vorgang dieser Energieübertragung durch Strahlung läßt sich a priori nichts aussagen, weil wir den Einfluß des Schwerefeldes auf die Strahlung und die Meßinstrumente in S_1 und S_2 nicht kennen.

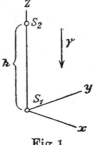

Fig 1.

Nach unserer Voraussetzung von der Äquivalenz von K und K' können wir aber an Stelle des im homogenen Schwerefelde befindlichen Systems K das schwerefreie, im Sinne der positiven z gleichförmig beschleunigt bewegte System K' setzen, mit dessen z-Achse die körperlichen Systeme S_1 und S_2 fest verbunden sind.

Den Vorgang der Energieübertragung durch Strahlung von S_2 auf S_1 beurteilen wir von einem System K_0 aus, das beschleunigungsfrei sei. In bezug auf K_0 besitze K' in dem Augenblick die Geschwindigkeit Null, in welchem die Strahlungsenergie E_2 von S_2 gegen S_1 abgesendet wird. Die Strahlung wird in S_1 ankommen, wenn die Zeit h/c verstrichen ist (in erster Annäherung). In diesem Momente besitzt aber S_1 in bezug auf K_0 die Geschwindigkeit $\gamma \cdot h/c = v$. Deshalb besitzt nach der gewöhnlichen Relativitätstheorie die in S_1

---

[1]) S_1 und S_2 werden als gegenüber h unendlich klein betrachtet.

ankommende Strahlung nicht die Energie E_2, sondern eine größere Energie E_1, welche mit E_2 in erster Annäherung durch die Gleichung verknüpft ist[1]):

(1) $$E_1 = E_2\left(1 + \frac{v}{c}\right) = E_2\left(1 + \frac{\gamma h}{c^2}\right).$$

Nach unserer Annahme gilt genau die gleiche Beziehung, falls derselbe Vorgang in dem nicht beschleunigten, aber mit Gravitationsfeld versehenen System K stattfindet. In diesem Falle können wir γh ersetzen durch das Potential Φ des Gravitationsvektors in S_2, wenn die willkürliche Konstante von Φ in S_1 gleich Null gesetzt wird. Es gilt also die Gleichung:

(1a) $$E_1 = E_2 + \frac{E_2}{c^2}\Phi.$$

Diese Gleichung spricht den Energiesatz für den ins Auge gefaßten Vorgang aus. Die in S_1 ankommende Energie E_1 ist größer als die mit gleichen Mitteln gemessene Energie E_2, welche in S_2 emittiert wurde, und zwar um die potentielle Energie der Masse E_2/c^2 im Schwerefelde. Es zeigt sich also, daß man, damit das Energieprinzip erfüllt sei, der Energie E vor ihrer Aussendung in S_2 eine potentielle Energie der Schwere zuschreiben muß, die der (schweren) Masse E/c^2 entspricht. Unsere Annahme der Äquivalenz von K und K' hebt also die am Anfang dieses Paragraphen dargelegte Schwierigkeit, welche die gewöhnliche Relativitätstheorie übrig läßt.

Besonders deutlich zeigt sich der Sinn dieses Resultates bei Betrachtung des folgenden Kreisprozesses:

1. Man sendet die Energie E (in S_2 gemessen) in Form von Strahlung in S_2 ab nach S_1, wo nach dem soeben erlangten Resultat die Energie $E(1 + \gamma h/c^2)$ aufgenommen wird (in S_1 gemessen).

2. Man senkt einen Körper W von der Masse M von S_2 nach S_1, wobei die Arbeit $M\gamma h$ nach außen abgegeben wird.

3. Man überträgt die Energie E von S_1 auf den Körper W, während sich W in S_1 befindet. Dadurch ändere sich die schwere Masse M, so daß sie den Wert M' erhält.

---

[1] A. Einstein, Ann. d. Phys. **17**. p. 913 u. 914. 1905.

Einfluß der Schwerkraft auf die Ausbreitung des Lichtes.

4. Man hebe W wieder nach S_2, wobei die Arbeit $M'\gamma h$ aufzuwenden ist.

5. Man übertrage E von W wieder auf S_2.

Der Effekt dieses Kreisprozesses besteht einzig darin, daß S_1 den Energiezuwachs $E(\gamma h/c^2)$ erlitten hat, und daß dem System die Energiemenge

$$M'\gamma h - M\gamma h$$

in Form von mechanischer Arbeit zugeführt wurde. Nach dem Energieprinzip muß also

$$E\frac{\gamma h}{c^2} = M'\gamma h - M\gamma h$$

oder

(1b) $$M' - M = \frac{E}{c^2}$$

sein. Der Zuwachs an *schwerer* Masse ist also gleich E/c^2, also gleich dem aus der Relativitätstheorie sich ergebenden Zuwachs an *träger* Masse.

Noch unmittelbarer ergibt sich das Resultat aus der Äquivalenz der Systeme K und K', nach welcher die *schwere* Masse in bezug auf K der *trägen* Masse in bezug auf K' vollkommen gleich ist; es muß deshalb die Energie eine *schwere* Masse besitzen, die ihrer *trägen* Masse gleich ist. Hängt man im System K' eine Masse M_0 an einer Federwaage auf, so wird letztere wegen der Trägheit von M_0 das scheinbare Gewicht $M_0\gamma$ anzeigen. Überträgt man die Energiemenge E auf M_0, so wird die Federwaage nach dem Satz von der Trägheit der Energie $\left(M_0 + \frac{E}{c^2}\right)\gamma$ anzeigen. Nach unserer Grundannahme muß ganz dasselbe eintreten bei Wiederholung des Versuches im System K, d. h. im Gravitationsfelde.

§ 3. Zeit und Lichtgeschwindigkeit im Schwerefelde.

Wenn die im gleichförmig beschleunigten System K' in S_2 gegen S_1 emittierte Strahlung mit Bezug auf die in S_2 befindliche Uhr die Frequenz ν_2 besaß, so besitzt sie in bezug auf S_1 bei ihrer Ankunft in S_1 in bezug auf die in S_1 befindliche gleich beschaffene Uhr nicht mehr die Frequenz ν_2 sondern eine größere Frequenz ν_1, derart, daß in erster Annäherung

(2) $$\nu_1 = \nu_2\left(1 + \frac{\gamma h}{c^2}\right).$$

Führt man nämlich wieder das beschleunigungsfreie Bezugssystem K_0 ein, relativ zu welchem K' zur Zeit der Lichtaussendung keine Geschwindigkeit besitzt, so hat S_1 in bezug auf K_0 zur Zeit der Ankunft der Strahlung in S_1 die Geschwindigkeit $\gamma(h/c)$, woraus sich die angegebene Beziehung vermöge des Dopplerschen Prinzipes unmittelbar ergibt.

Nach unserer Voraussetzung von der Äquivalenz der Systeme K' und K gilt diese Gleichung auch für das ruhende, mit einem gleichförmigen Schwerefeld versehene Koordinatensystem K, falls in diesem die geschilderte Strahlungsübertragung stattfindet. Es ergibt sich also, daß ein bei bestimmtem Schwerepotential in S_2 emittierter Lichtstrahl, der bei seiner Emission — mit einer in S_2 befindlichen Uhr verglichen — die Frequenz ν_2 besitzt, bei seiner Ankunft in S_1 eine andere Frequenz ν_1 besitzt, falls letztere mittels einer in S_1 befindlichen gleich beschaffenen Uhr gemessen wird. Wir ersetzen γh durch das Schwerepotential Φ von S_2 in bezug auf S_1 als Nullpunkt und nehmen an, daß unsere für das *homogene* Gravitationsfeld abgeleitete Beziehung auch für anders gestaltete Felder gelte; es ist dann

$$(2\,\mathrm{a}) \qquad \nu_1 = \nu_2 \left(1 + \frac{\Phi}{c^2}\right).$$

Dies (nach unserer Ableitung in erster Näherung gültige) Resultat gestattet zunächst folgende Anwendung. Es sei ν_0 die Schwingungszahl eines elementaren Lichterzeugers, gemessen mit einer an demselben Orte gemessenen Uhr U. Diese Schwingungszahl ist dann unabhängig davon, wo der Lichterzeuger samt der Uhr aufgestellt wird. Wir wollen uns beide etwa an der Sonnenoberfläche angeordnet denken (dort befindet sich unser S_2). Von dem dort emittierten Lichte gelangt ein Teil zur Erde (S_1), wo wir mit einer Uhr U von genau gleicher Beschaffenheit als der soeben genannten die Frequenz ν des ankommenden Lichtes messen Dann ist nach (2a)

$$\nu = \nu_0 \left(1 + \frac{\Phi}{c^2}\right),$$

wobei Φ die (negative) Gravitationspotentialdifferenz zwischen Sonnenoberfläche und Erde bedeutet. Nach unserer Auffassung müssen also die Spektrallinien des Sonnenlichtes gegenüber den entsprechenden Spektrallinien irdischer Lichtquellen etwas nach dem Rot verschoben sein, und zwar um den relativen Betrag

$$\frac{\nu_0 - \nu}{\nu_0} = \frac{-\Phi}{c^2} = 2 \cdot 10^{-6}.$$

Wenn die Bedingungen, unter welchen die Sonnenlinien entstehen, genau bekannt wären, wäre diese Verschiebung noch der Messung zugänglich. Da aber anderweitige Einflüsse (Druck, Temperatur) die Lage des Schwerpunktes der Spektrallinien beeinflussen, ist es schwer zu konstatieren, ob der hier abgeleitete Einfluß des Gravitationspotentials wirklich existiert.[1]

Bei oberflächlicher Betrachtung scheint Gleichung (2) bzw. (2a) eine Absurdität auszusagen. Wie kann bei beständiger Lichtübertragung von S_2 nach S_1 in S_1 eine andere Anzahl von Perioden pro Sekunde ankommen, als in S_2 emittiert wird? Die Antwort ist aber einfach. Wir können ν_2 bzw. ν_1 nicht als Frequenzen schlechthin (als Anzahl Perioden pro Sekunde) ansehen, da wir eine Zeit im System K noch nicht festgelegt haben. ν_2 bedeutet die Anzahl Perioden, bezogen auf die Zeiteinheit der Uhr U in S_2, ν_1 die Anzahl Perioden, bezogen auf die Zeiteinheit der gleich beschaffenen Uhr U in S_1. Nichts zwingt uns zu der Annahme, daß die in verschiedenen Gravitationspotentialen befindlichen Uhren U als gleich rasch gehend aufgefaßt werden müssen. Dagegen müssen wir die Zeit in K sicher so definieren, daß die Anzahl der Wellenberge und Wellentäler, die sich zwischen S_2 und S_1 befinden, von dem Absolutwerte der Zeit unabhängig ist; denn der ins Auge gefaßte Prozeß ist seiner Natur nach ein stationärer. Würden wir diese Bedingung nicht erfüllen, so kämen wir zu einer Zeitdefinition, bei deren Anwendung die Zeit explizite in die Naturgesetze einginge, was sicher unnatürlich und unzweckmäßig wäre. Die Uhren in S_1 und S_2 geben also

---

[1] L. F. Jewell (Journ. de phys. **6**. p. 84. 1897) und insbesondere Ch. Fabry u. H. Boisson (Compt. rend. **148**. p. 688—690. 1909) haben derartige Verschiebungen feiner Spektrallinien nach dem roten Ende des Spektrums von der hier berechneten Größenordnung tatsächlich konstatiert, aber einer Wirkung des Druckes in der absorbierenden Schicht zugeschrieben.

nicht beide die „Zeit" richtig an. Messen wir die Zeit in S_1 mit der Uhr U, *so müssen wir die Zeit in S_2 mit einer Uhr messen, die $1 + \Phi/c^2$ mal langsamer läuft als die Uhr U, falls sie mit der Uhr U an derselben Stelle verglichen wird.* Denn mit einer solchen Uhr gemessen ist die Frequenz des oben betrachteten Lichtstrahles bei seiner Aussendung in S_2

$$\nu_2 \left(1 + \frac{\Phi}{c^2}\right),$$

also nach (2a) gleich der Frequenz ν_1 desselben Lichtstrahles bei dessen Ankunft in S_1.

Hieraus ergibt sich eine Konsequenz von für diese Theorie fundamentaler Bedeutung. Mißt man nämlich in dem beschleunigten, gravitationsfeldfreien System K' an verschiedenen Orten die Lichtgeschwindigkeit unter Benutzung gleich beschaffener Uhren U, so erhält man überall dieselbe Größe. Dasselbe gilt nach unserer Grundannahme auch für das System K. Nach dem soeben Gesagten müssen wir aber an Stellen verschiedenen Gravitationspotentials uns verschieden beschaffener Uhren zur Zeitmessung bedienen. Wir müssen zur Zeitmessung an einem Orte, der relativ zum Koordinatenursprung das Gravitationspotential Φ besitzt, eine Uhr verwenden, die — an den Koordinatenursprung versetzt — $(1 + \Phi/c^2)$ mal langsamer läuft als jene Uhr, mit welcher am Koordinatenursprung die Zeit gemessen wird. Nennen wir c_0 die Lichtgeschwindigkeit im Koordinatenanfangspunkt, so wird daher die Lichtgeschwindigkeit c in einem Orte vom Gravitationspotential Φ durch die Beziehung

$$(3) \qquad c = c_0 \left(1 + \frac{\Phi}{c^2}\right)$$

gegeben sein. Das Prinzip von der Konstanz der Lichtgeschwindigkeit gilt nach dieser Theorie nicht in derjenigen Fassung, wie es der gewöhnlichen Relativitätstheorie zugrunde gelegt zu werden pflegt.

§ 4. Krümmung der Lichtstrahlen im Gravitationsfeld.

Aus dem soeben bewiesenen Satze, daß die Lichtgeschwindigkeit im Schwerefelde eine Funktion des Ortes ist, läßt sich leicht mittels des Huygensschen Prinzipes schließen, daß quer

Einfluß der Schwerkraft auf die Ausbreitung des Lichtes. 907

zu einem Schwerefeld sich fortpflanzende Lichtstrahlen eine Krümmung erfahren müssen. Sei nämlich ε eine Ebene gleicher Phase einer ebenen Lichtwelle zur Zeit t, P_1 und P_2 zwei Punkte in ihr, welche den Abstand 1 besitzen. P_1 und P_2 liegen in der Papierebene, die so gewählt ist, daß der in der Richtung ihrer Normale genommene Differentialquotient von Φ also auch von c verschwindet. Die entsprechende Ebene gleicher Phase bzw. deren Schnitt mit der Papierebene, zur Zeit $t + dt$ erhalten wir, indem wir um die Punkte P_1 und P_2 mit den Radien $c_1 dt$ bzw. $c_2 dt$ Kreise und an diese die Tangente legen, wobei c_1 bzw. c_2 die Lichtgeschwindigkeit in den Punkten P_1 bzw. P_2 bedeutet. Der Krümmungswinkel des Lichtstrahles auf dem Wege $c\, dt$ ist also

$$\frac{(c_1 - c_2)\, dt}{1} = -\frac{\partial c}{\partial n'} dt,$$

falls wir den Krümmungswinkel positiv rechnen, wenn der Lichtstrahl nach der Seite der wachsenden n' hin gekrümmt

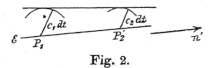

Fig. 2.

wird. Der Krümmungswinkel pro Wegeinheit des Lichtstrahles ist also

$$-\frac{1}{c}\frac{\partial c}{\partial n'}$$

oder nach (3) gleich

$$-\frac{1}{c^2}\frac{\partial \Phi}{\partial n'}.$$

Endlich erhalten wir für die Ablenkung α, welche ein Lichtstrahl auf einem beliebigen Wege (s) nach der Seite n' erleidet, den Ausdruck

$$(4) \qquad \alpha = -\frac{1}{c^2} \int \frac{\partial \Phi}{\partial n'}\, ds.$$

Dasselbe Resultat hätten wir erhalten können durch unmittelbare Betrachtung der Fortpflanzung eines Lichtstrahles in dem gleichförmig beschleunigten System K' und Übertragung des Resultates auf das System K und von hier auf den Fall, daß das Gravitationsfeld beliebig gestaltet ist.

Nach Gleichung (4) erleidet ein an einem Himmelskörper vorbeigehender Lichtstrahl eine Ablenkung nach der Seite sinkenden Gravitationspotentials, also nach der dem Himmelskörper zugewandten Seite von der Größe

$$\alpha = \frac{1}{c^2} \int_{\vartheta=-\frac{\pi}{2}}^{\vartheta=+\frac{\pi}{2}} \frac{kM}{r^2} \cos\vartheta \cdot ds = \frac{2kM}{c^2 \Delta},$$

wobei k die Gravitationskonstante, M die Masse des Himmelskörpers, Δ den Abstand des Lichtstrahles vom Mittelpunkt des Himmelskörpers bedeutet. *Ein an der Sonne vorbeigehender Lichtstrahl erlitte demnach eine Ablenkung vom Betrage* $4\cdot 10^{-6}$ $= 0{,}83$ *Bogensekunden.* Um diesen Betrag erscheint die Winkeldistanz des Sternes vom Sonnenmittelpunkt durch die Krümmung des Strahles vergrößert. Da die Fixsterne der der Sonne zugewandten Himmelspartien bei totalen Sonnenfinsternissen sichtbar werden, ist diese Konsequenz der Theorie mit der Erfahrung vergleichbar. Beim Planeten Jupiter erreicht die zu erwartende Verschiebung etwa $1/100$ des angegebenen Betrages. Es wäre dringend zu wünschen, daß sich Astronomen der hier aufgerollten Frage annähmen, auch wenn die im vorigen gegebenen Überlegungen ungenügend fundiert oder gar abenteuerlich erscheinen sollten. Denn abgesehen von jeder Theorie muß man sich fragen, ob mit den heutigen Mitteln ein Einfluß der Gravitationsfelder auf die Ausbreitung des Lichtes sich konstatieren läßt.

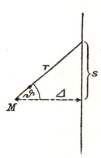

Fig. 3.

Prag, Juni 1911.

(Eingegangen 21. Juni 1911.)

ANNALEN DER PHYSIK.

BEGRÜNDET UND FORTGEFÜHRT DURCH

F. A. C. GREN, L. W. GILBERT, J. C. POGGENDORFF, G. u. E. WIEDEMANN, P. DRUDE.

VIERTE FOLGE.

BAND 37.

DER GANZEN REIHE 342. BAND.

KURATORIUM:

M. PLANCK, G. QUINCKE,
W. C. RÖNTGEN, W. VOIGT, E. WARBURG.

UNTER MITWIRKUNG
DER DEUTSCHEN PHYSIKALISCHEN GESELLSCHAFT

HERAUSGEGEBEN VON

W. WIEN UND M. PLANCK.

MIT FÜNF FIGURENTAFELN.

LEIPZIG, 1912.
VERLAG VON JOHANN AMBROSIUS BARTH.

11. *Thermodynamische Begründung des photochemischen Äquivalentgesetzes;* von *A. Einstein.*

Im folgenden wird auf wesentlich thermodynamischem Wege gleichzeitig das Wiensche Strahlungsgesetz und das photochemische Äquivalentgesetz abgeleitet. Unter dem letzteren verstehe ich den Satz, daß es zur Zersetzung eines Grammäquivalentes durch einen photochemischen Vorgang der absorbierten Strahlungsenergie $Nh\nu$ bedarf, falls man mit N die Zahl der Moleküle im Gramm-Mol, mit h die bekannte Konstante in Plancks Strahlungsformel, mit ν die Frequenz der wirksamen Strahlung bezeichnet.[1]) Das Gesetz erscheint im wesentlichen als eine Konsequenz der Voraussetzung, daß die Zahl der pro Zeiteinheit zersetzten Moleküle der Dichte der wirksamen Strahlung proportional ist; doch ist hervorzuheben, daß die thermodynamischen Zusammenhänge und das Strahlungsgesetz es nicht gestatten, diese Annahme durch eine beliebige andere zu ersetzen, wie am Schlusse der Arbeit kurz gezeigt wird.

Aus dem Folgenden geht ferner klar hervor, daß das Äquivalentgesetz bzw. die zu demselben führenden Annahmen nur so lange gelten, als die wirksame Strahlung dem Gültigkeitsbereiche des Wienschen Gesetzes angehört. Für solche Strahlung aber ist nun an der Gültigkeit des Gesetzes kaum mehr zu zweifeln.

§ 1. Über das thermodynamische Gleichgewicht zwischen Strahlung und einem teilweise dissoziierten Gase vom Standpunkt des Massenwirkungsgesetzes.

Es sei in einem Volumen V eine Mischung dreier chemisch verschiedener Gase mit den Molekulargewichten m_1, m_2, m_3

---

[1]) Vgl. A. Einstein, Ann. d. Phys. 4. (17). p. 132.

vorhanden. n_1 sei die Anzahl g-Mole des ersten, n_2 die des zweiten, n_3 die des dritten Gases.[1]) Zwischen diesen drei Molekülarten sei eine Reaktion möglich, darin bestehend, daß ein Molekül erster Art zerfällt in ein Molekül zweiter und ein Molekül dritter Art. Bei thermodynamischem Gleichgewichte besteht gleiche Häufigkeit der Reaktionen

$$m_1 \rightarrow m_2 + m_3$$

und

$$m_2 + m_3 \rightarrow m_1.$$

Wir wollen den Fall ins Auge fassen, daß der Zerfall von Molekülen m_1 ausschließlich durch die Wirkung der Wärmestrahlung erfolge, und zwar unter der Wirkung eines Teiles der Wärmestrahlung, dessen Frequenz sich wenig von einer gewissen Frequenz ν_0 unterscheidet. Die bei einem derartigen Zerfall im Mittel absorbierte Strahlungsenergie sei ε. In diesem Falle muß umgekehrt bei dem Prozeß der Vereinigung von m_2 und m_3 zu m_1 Strahlung von der Frequenzgegend ν_0 emittiert werden, und zwar ausschließlich Strahlung von der Frequenzgegend ν_0, und es muß die bei einem Wiedervereinigungsprozeß emittierte Strahlungsenergie im Mittel ebenfalls gleich ε sein, da sonst das Strahlungsgleichgewicht durch die Existenz des Gases gestört würde; denn die Zahl der Zerfallprozesse ist gleich der Zahl der Vereinigungsprozesse.

Besitzt das Gasgemisch die Temperatur T, so wird thermodynamisches Gleichgewicht des Systems jedenfalls bestehen können, wenn die im Raum befindliche Strahlung in der Umgebung der Frequenz ν_0 diejenige (monochromatische) Dichte ϱ besitzt, welche zur Wärmestrahlung der Temperatur T gehört. Wir analysieren nun die beiden einander gerade aufhebenden Reaktionen genauer, indem wir über den Mechanismus derselben gewisse Annahmen machen.

Der Zerfall eines Moleküls erster Art geschehe so, wie wenn die übrigen Moleküle nicht da wären (Annahme I). Daraus folgt, daß wir die Zahl der pro Zeiteinheit zerfallenden Moleküle erster Art deren Anzahl (n_1) unter sonst gleichen Umständen proportional, und daß wir die Zahl der pro Zeiteinheit

---

[1]) Natürlich kann eines der Gase mit den Indizes 2 und 3 aus Elektronen bestehen.

zerfallenden Moleküle als von den drei Gasdichten unabhängig anzusetzen haben. Außerdem nehmen wir an, daß die Wahrscheinlichkeit dafür, daß ein Molekül erster Art in einem Zeitteilchen zerfalle, der monochromatischen Strahlungsdichte ϱ proportional sei (Annahme II).

Hauptsächlich von der zweiten dieser Annahmen muß hervorgehoben werden, daß ihre Richtigkeit durchaus nicht selbstverständlich ist. Sie enthält die Aussage, daß die chemische Wirkung einer auf einen Körper fallenden Strahlung nur von der Gesamtmenge der wirkenden Strahlung abhänge, aber nicht von der Bestrahlungsintensität; die Existenz einer unteren Wirksamkeitsschwelle der Strahlung wird durch diese Annahme vollkommen ausgeschlossen. Wir setzen uns durch letztere in Widerspruch mit den Ergebnissen zweier Arbeiten von E. Warburg[1]), durch die ich die Anregung für die vorliegende Arbeit empfing.

Aus den beiden Annahmen folgt, daß die Zahl Z der pro Zeiteinheit zerfallenden Moleküle erster Art gegeben ist durch den Ausdruck

(1) $$Z = A \varrho\, n_1.$$

Der Proportionalitätsfaktor A kann nach dem Gesagten nur von der Gastemperatur T abhängen. Nach dem Vorangehenden gilt die Gleichung auch in dem Falle, daß die Strahlungsdichte ϱ (bei der Frequenz ν_0) eine andere ist, als zur Temperatur T des Gases gehört.

Von dem Wiedervereinigungsprozeß nehmen wir an, daß es ein gewöhnlicher Vorgang zweiter Ordnung im Sinne des Massenwirkungsgesetzes sei, daß also die Zahl der pro Volumeinheit und Zeiteinheit sich bildenden Moleküle erster Art dem Produkt der Konzentrationen n_2/V und n_3/V proportional sei, wobei der Proportionalitätskoeffizient nur von der Gastemperatur, aber nicht von der Dichte der vorhandenen Strahlung abhänge (Annahme III). Die Zahl Z' der sich in der Zeiteinheit bildenden Moleküle erster Art ist also

(2) $$Z' = A' \cdot V \cdot \frac{n_2}{V} \cdot \frac{n_3}{V}.$$

---

1) E. Warburg, Verh. d. Deutsch. Physik. Ges. **9.** p. 24. 1908 und **9.** p. 21. 1909.

Das von uns betrachtete, aus Strahlung und Gasgemisch bestehende System befindet sich stets im thermodynamischen Gleichgewicht, wenn die Zahl Z der Zerfallsprozesse gleich ist der Zahl Z' der Vereinigungsprozesse; denn es bleibt in diesem Falle nicht nur die Menge einer jeden Gasart, sondern auch die Menge der vorhandenen Strahlung ungeändert.[1]) Diese Bedingung lautet

$$(3) \qquad \frac{\frac{n_2}{V} \frac{n_3}{V}}{\frac{n_1}{V}} = \frac{\eta_2 \eta_3}{\eta_1} = \frac{A}{A'} \varrho,$$

wobei A und A' nur von der Temperatur der Gasmischung abhängen. Eine eigentümliche Konsequenz dieser Betrachtung ist die, daß bei gegebener Gastemperatur und beliebig gegebener Strahlungsdichte (d. h. auch Strahlungstemperatur) ein thermodynamisches Gleichgewicht möglich sein soll. Es liegt aber hierin kein Verstoß gegen den zweiten Hauptsatz, was damit zusammenhängt, daß mit einem Wärmeübergang von der Strahlung zum Gase ein bestimmter chemischer Prozeß zwangläufig verbunden ist; man kann mit Hilfe des von uns betrachteten Systems kein Perpetuum mobile zweiter Art konstruieren.

§ 2. Thermodynamische Gleichgewichtsbedingung für das im § 1 betrachtete System.

Ist S_s die Entropie der im Volumen V enthaltenen Strahlung, S_g diejenige des Gasgemisches, so muß für jeden der im vorigen Paragraph gefundenen Gleichgewichtszustände die Bedingung bestehen, daß für jede unendlich kleine virtuelle Änderung der Zustände von Strahlung und Gas die Änderung der Gesamtentropie verschwindet. Die zu betrachtende virtuelle Änderung besteht darin, daß die Energiemenge $N\varepsilon$ (aus der Umgebung von v_0) der Strahlung in Energie des Gasgemisches übergeht unter gleichzeitigem Zerfall eines Gasmoleküls (g-Mol) erster Art. Bei einer solchen virtuellen Änderung würde sich die Temperatur des Gemisches um einen nicht zu vernachlässigenden

---

[1]) Beim Lesen der Korrektur bemerke ich, daß dieser für das Folgende wesentliche Schluß nur unter der Voraussetzung gilt, daß bei gegebener Gastemperatur ε von ϱ unabhängig ist.

Betrag verändern. Um dies zu vermeiden, denken wir uns in bekannter Weise das Gasgemisch mit einem unendlich großen Wärmereservoir von derselben Temperatur T in dauernder wärmeleitender Verbindung. Bei der virtuellen Änderung ändert sich dann die Temperatur des Gasgemisches nicht; dagegen ist zu berücksichtigen, daß das Wärmereservoir die Energie $-(\delta E_s + \delta E_g)$ bei der virtuellen Änderung in Form von Wärme aufnimmt, falls man mit E_s die Energie der Strahlung, mit E_g diejenige des Gases bezeichnet. Die Gleichgewichtsbedingung lautet deshalb

$$\text{(4)} \qquad \delta S_s + \delta S_g - \frac{\delta E_s + \delta E_g}{T} = 0.$$

Wir haben nun die einzelnen Glieder dieser Gleichung zu berechnen. Es ist zunächst für die von uns betrachtete virtuelle Änderung

$$\delta E_s = -N\varepsilon,$$
$$\delta S_s = -\frac{N\varepsilon}{T_s},$$

wenn man mit T_s die zur Strahlungsdichte ϱ gehörige Temperatur bezeichnet. Die auf das Gas bezüglichen Variationen berechnen wir nach in der Thermodynamik geläufigen Methoden, wobei wir — was für das Folgende nicht wesentlich ist — die spezifischen Wärmen als von der Temperatur unabhängig behandeln. Man erhält zunächst

$$E_g = \sum n_1 \{c_{\nu_1} T + b_1\},$$
$$S_g = \sum n_1 \left\{c_{\nu_1} \lg T + c_1 - R \lg \frac{n_1}{V}\right\}.$$

Dabei bedeutet

c_{ν_1} die Wärmekapazität pro g-Mol bei konstantem Volum,
b_1 die Energie pro g-Mol der 1. Gasart bei $T = 0$,
c_1 eine Integrationskonstante der Entropie der ersten Gasart.

Aus diesen Gleichungen folgen unmittelbar die folgenden

$$\delta E_g = \sum \delta n_1 \{c_{\nu_1} T + b_1\},$$
$$\delta S_g = \sum \delta n_1 \left\{c_{\nu_1} \lg T + c_1 - R - R \lg \frac{n_1}{V}\right\},$$

wobei
$$\text{(4a')} \qquad \delta n_1 = -1, \qquad \delta n_2 = +1, \qquad \delta n_3 = +1$$

zu setzen ist. Gleichung (4) nimmt vermöge dieser Gleichungen für die Variationen und der Gleichung (3) die Form an

$$(4\mathrm{a}) \qquad -\frac{N\varepsilon}{RT_s} + \lg \alpha - \lg\left(\frac{A}{A'}\varrho\right) = 0,$$

wobei zur Abkürzung gesetzt ist

$$(4\mathrm{a}'') \qquad \lg \alpha = \frac{N\varepsilon}{RT} + \frac{1}{R}\sum \delta n_1 \left\{c_{\nu_1}\lg T + c_1 - R - c_{\nu_1} - \frac{b_1}{T}\right\}.$$

Die mit α bezeichnete Größe ist von T_s unabhängig.

§ 3. Schlußfolgerungen aus der Gleichgewichtsbedingung.

Wir schreiben nun (4a) in der Form

$$(4\mathrm{b}) \qquad \varrho = \frac{A'\alpha}{A} e^{-\frac{N\varepsilon}{RT_s}}.$$

Da die Beziehung zwischen T_s und ϱ unabhängig sein muß von T, müssen die Größen $A'\alpha/A$ und ε von T unabhängig sein. Da diese Größen auch von T_s unabhängig sind, so sind wir damit zu demjenigen Zusammenhang zwischen ϱ und T_s gelangt, der der Wienschen Strahlungsformel entspricht. Wir schließen hieraus:

Die im § 1 zugrunde gelegten Annahmen über den Verlauf photochemischer Vorgänge sind mit dem empirisch bekannten Gesetze der Wärmestrahlung nur vereinbar, insofern die wirkende Strahlung in den Gültigkeitsbereich des Wienschen Strahlungsgesetzes fällt; in diesem Falle aber ist Wiens Gesetz eine Konsequenz unserer Annahmen.

Schreiben wir Wiens Strahlungsformel unter Einführung der Planckschen Konstanten in der Form

$$\varrho = \frac{8\pi h \nu^3}{c^3} e^{-\frac{h\nu}{\varkappa T_s}},$$

so sehen wir durch Vergleichung mit (4b), daß die Gleichungen

$$(5) \qquad \varepsilon = h\nu_0,$$

$$(6) \qquad \frac{A'\alpha}{A} = \frac{8\pi h \nu_0^3}{c^3}$$

erfüllt sein müssen. Als wichtigste Konsequenz folgt also (5), *daß ein Gasmolekül, welches unter Absorption von Strahlung von* der Frequenz ν_0 zerfällt, bei seinem Zerfall (im Mittel) die Strahlungsenergie $h\nu_0$ absorbiert. Wir haben die einfachste Art der Reaktion vorausgesetzt, hätten aber Gleichung (5) auf demselben Wege wie hier auch für andere unter Lichtabsorption vor sich gehende Gasreaktionen ableiten können. Ebenso liegt es auf der Hand, daß die Beziehung in ähnlicher Weise für verdünnte Lösungen bewiesen werden kann. Sie dürfte wohl allgemein gültig sein.

Wir ersetzen ferner mit Hilfe von (6) in (4a″) die Größe α, so erhalten wir mit Berücksichtigung von (3), indem wir zur Abkürzung $\eta_2\eta_3/\eta_1 = \varkappa$ setzen und Wiens Strahlungsgesetz anwenden

$$\lg \varkappa = \frac{Nh\nu_0}{RT} - \frac{Nh\nu_0}{RT_s} + \frac{1}{R}\sum \delta n_1 \left\{ c_{\nu_1}\lg T + c_1 - (c_{\nu_1} + R) - \frac{b_1}{T} \right\}.$$

Diese Gleichung geht für $T = T_s$ in die bekannte Gleichung für das Dissoziationsgleichgewicht für Gase über, ein Beweis dafür, daß die vorstehende Theorie mit der thermodynamischen Theorie der Dissoziation nicht in Widerspruch gerät.

Prag, Januar 1912.

(Eingegangen 18. Januar 1912.)

ANNALEN DER PHYSIK.

BEGRÜNDET UND FORTGEFÜHRT DURCH

F. A. C. GREN, L. W. GILBERT, J. C. POGGENDORFF, G. u. E. WIEDEMANN, P. DRUDE.

VIERTE FOLGE.

BAND 38.

DER GANZEN REIHE 343. BAND.

KURATORIUM:

M. PLANCK, G. QUINCKE,
W. C. RÖNTGEN, W. VOIGT, E. WARBURG.

UNTER MITWIRKUNG

DER DEUTSCHEN PHYSIKALISCHEN GESELLSCHAFT

HERAUSGEGEBEN VON

W. WIEN UND M. PLANCK.

MIT ACHT FIGURENTAFELN.

LEIPZIG, 1912.
VERLAG VON JOHANN AMBROSIUS BARTH.

3. *Lichtgeschwindigkeit und Statik des Gravitationsfeldes;*
von A. Einstein.

In einer letztes Jahr erschienenen Arbeit[1]) habe ich aus der Hypothese, daß Schwerefeld und Beschleunigungszustand des Koordinatensystems physikalisch gleichwertig seien, einige Folgerungen gezogen, welche sich den Ergebnissen der Relativitätstheorie (Theorie der Relativität der gleichförmigen Bewegung) sehr gut angliedern. Es zeigte sich dabei aber, daß die Gültigkeit des einen Grundsatzes jener Theorie, nämlich des Satzes von der Konstanz der Lichtgeschwindigkeit, nur für Raum-Zeitgebiete konstanten Gravitationspotentials Gültigkeit beanspruchen kann. Trotzdem dies Resultat die allgemeine Anwendbarkeit der Lorentztransformation ausschließt, darf es uns nicht von der weiteren Verfolgung des eingeschlagenen Weges abschrecken; wenigstens hat meiner Meinung nach die Hypothese, daß das „Beschleunigungsfeld" ein Spezialfall des Gravitationsfeldes sei, eine so große Wahrscheinlichkeit, insbesondere mit Rücksicht auf die bereits in der ersten Arbeit gezogenen Folgerungen betreffend die schwere Masse des Energieinhaltes, das eine genauere Durchführung der Folgerungen jener Äquivalenzhypothese geboten erscheint.

Seitdem hat Abraham eine Theorie der Gravitation aufgestellt[2]), welche die in meiner ersten Arbeit gezogenen Folgerungen als Spezialfälle enthält. Wir werden aber im folgenden sehen, daß sich das Gleichungssystem Abrahams mit der Äquivalenzhypothese nicht in Einklang bringen läßt, und daß dessen Auffassung von Zeit und Raum sich schon vom rein mathematisch formalen Standpunkte aus nicht aufrecht erhalten läßt.

---

1) A. Einstein, Ann. d. Phys. **4**. p. 35. 1911.
2) M. Abraham, Physik. Zeitschr. **13**. Nr. 1. 1912.

§ 1. Raum und Zeit im Beschleunigungsfeld.

Das Bezugssystem K (Koordinaten x, y, z) befinde sich im Zustande gleichförmiger Beschleunigung in Richtung seiner x-Koordinate. Diese Beschleunigung sei eine gleichförmige im Bornschen Sinne; d. h. die Beschleunigung seines Anfangspunktes, bezogen auf ein beschleunigungsfreies System, in bezug auf welches die Punkte von K gerade keine, bzw. eine unendlich kleine Geschwindigkeit besitzen, sei eine konstante Größe. Ein solches System K ist nach der Äquivalenzhypothese streng gleichwertig einem ruhenden System, in welchem ein massenfreies statisches Gravitationsfeld[1]) bestimmter Art sich befindet. Die räumliche Ausmessung von K geschieht durch Maßstäbe, welche — im Ruhezustande an der nämlichen Stelle von K miteinander verglichen — die gleiche Länge besitzen; es sollen die Sätze der Geometrie gelten für so gemessene Längen, also auch für die Beziehungen zwischen den Koordinaten x, y, z und anderen Längen. Diese Festsetzung ist nicht selbstverständlich erlaubt, sondern enthält physikalische Annahmen, die sich eventuell als unrichtig erweisen könnten; sie gelten z. B. höchst wahrscheinlich nicht in einem gleichförmig rotierenden Systeme, in welchem wegen der Lorentzkontraktion das Verhältnis des Kreisumfanges zum Durchmesser bei Anwendung unserer Definition für die Längen von π verschieden sein müßte. Der Maßstab sowie die Koordinatenachsen sind als starre Körper aufzufassen. Dies ist erlaubt, trotzdem der starre Körper nach der Relativitätstheorie keine reale Existenz besitzen kann. Denn man kann den starren Meßkörper durch eine große Anzahl kleiner nicht starrer Körper ersetzt denken, die so aneinander gereiht werden, daß sie aufeinander keine Druckkräfte ausüben, indem jeder besonders gehalten wird. Die Zeit t im System K denken wir durch Uhren gemessen von solcher Beschaffenheit und solcher fester Anordnung in den Raumpunkten des Systems K, daß die Zeitspanne, welche — mit ihnen gemessen — ein Lichtstrahl braucht, um von einem Punkt A nach einem Punkte B des Systems K zu gelangen, nicht von dem Zeitpunkt der Aussendung des Licht-

---

1) Die Massen, welche dies Feld hervorbringen, hat man sich im Unendlichen zu denken.

strahles in A abhängig ist. Es wird sich ferner zeigen, daß widerspruchsfrei die Gleichzeitigkeit dadurch definiert werden kann, daß bezüglich des *Richtens* der Uhren die Fortsetzung getroffen wird, daß alle Lichtstrahlen, welche einen Punkt A von K passieren, in A dieselbe, von der Richtung unabhängige Fortpflanzungsgeschwindigkeit besitzen.

Wir denken uns nun das Bezugssystem $K(x, y, z, t)$ von einem beschleunigungsfreien Bezugssystem (von konstantem Gravitationspotential) $\Sigma(\xi, \eta, \zeta, \tau)$ aus betrachtet. Wir setzen voraus, daß die x-Achse dauernd in die ξ-Achse falle und die y-Achse dauernd der η-Achse, die z-Achse dauernd der ζ-Achse parallel sei. Diese Festsetzung ist möglich unter der Annahme, daß der Zustand der *Beschleunigung* auf die Gestalt von K in bezug auf Σ nicht von Einfluß sei. Diese physikalische Annahme legen wir zugrunde. Aus ihr folgt, daß für beliebige τ

(1) $$\begin{cases} \eta = y, \\ \zeta = z \end{cases}$$

sein muß, so daß wir nur noch die Beziehung aufzusuchen haben, welche zwischen ξ und τ einerseits, x und t andererseits, besteht. Zur Zeit $\tau = 0$ mögen beide Bezugssysteme zusammenfallen; dann müssen die gesuchten Substitutionsgleichungen jedenfalls von der Form sein

(2) $$\begin{cases} \xi = \lambda + \alpha t^2 + \ldots \\ \tau = \beta + \gamma t + \delta t^2 + \ldots \end{cases}$$

Die Koeffizienten dieser für genügend kleine positive und negative Werte von t gültigen Reihen sind als vorläufig unbekannte Funktionen von x anzusehen. Indem wir uns auf die angeschriebenen Glieder beschränken, erhalten wir durch Differenziation

(3) $$\begin{cases} d\xi = (\lambda' + \alpha' t^2)dx + 2\alpha t\, dt, \\ d\tau = (\beta' + \gamma' t + \delta' t^2)dx + (\gamma + 2\delta t)dt. \end{cases}$$

Im System Σ denken wir uns die Zeit derart gemessen, daß die Lichtgeschwindigkeit gleich 1 wird. Wir können dann die Gleichung einer Schale, die sich mit Lichtgeschwindigkeit von einem beliebigen Raum-Zeitpunkt ausbreitet, indem wir uns auf die unendlich kleine Umgebung des Raum-Zeitpunktes beschränken, in der Form schreiben

$$d\xi^2 + d\eta^2 + d\zeta^2 - d\tau^2 = 0.$$

Dieselbe Schale muß im System K die Gleichung haben

$$dx^2 + dy^2 + dz^2 - c^2 dt^2 = 0.$$

Die Substitutionsgleichungen (2) müssen derart sein, daß diese beiden Gleichungen äquivalent sind. Dies verlangt wegen (1) die Identität

(4) $$d\xi^2 - d\tau^2 = dx^2 - c^2 dt^2.$$

Setzt man in die linke Seite dieser Gleichung die Ausdrücke in dx und dt vermittelst (3) ein und setzt links und rechts die Koeffizienten von dx^2, dt^2 und $dx\,dt$ einander gleich, so erhält man die Gleichungen

$$1 = (\lambda' + \alpha' t^2)^2 - (\beta' + \gamma' t + \delta' t^2)^2,$$
$$-c^2 = 4\alpha^2 t^2 - (\gamma + 2\delta t)^2,$$
$$0 = (\lambda' + \alpha' t^2)\cdot 2\alpha t - (\beta' + \gamma' t + \delta' t^2)(\gamma + 2\delta t).$$

Diese Gleichungen gelten in t identisch bis zu so hohen Potenzen von t, daß die in (2) weggelassenen Terme noch keinen Einfluß haben, also die erste Gleichung bis zur zweiten, die zweite und dritte bis zur ersten Potenz von t. Hieraus fließen die Gleichungen

$$1 = \lambda'^2 \beta'^2, \quad 0 = \beta'\gamma', \quad 2\lambda\alpha' - \gamma'^2 - 2\beta'\delta' = 0,$$
$$-c^2 = -\gamma^2, \quad 0 = \gamma\delta,$$
$$0 = \beta'\gamma, \quad 0 = 2\alpha\lambda' - 2\beta'\delta - \gamma\gamma'.$$

Da γ nicht verschwinden kann, folgt aus der ersten Gleichung der dritten Zeile $\beta' = 0$. β ist also eine Konstante, die wir bei passender Wahl der Anfangspunkte der Zeit gleich Null setzen dürfen. Der Koeffizient γ muß ferner positiv sein; es ist also nach der ersten Gleichung der zweiten Zeile

$$\gamma = c.$$

Nach der zweiten Gleichung der zweiten Zeile ist

$$\delta = 0.$$

Weil β' verschwindet, und wir x mit ξ wachsend annehmen können, so folgt aus der ersten Gleichung der ersten Zeile
$$\lambda' = 1,$$
also, wenn für $t = 0$ und $\xi = 0$, $x = 0$ sein soll,
$$\lambda = x.$$
Endlich folgen aus der dritten Gleichung der ersten und der zweiten Gleichung der dritten Zeile unter Benutzung der schon gefundenen Relationen die Differentialgleichungen
$$2\alpha' - c'^2 = 0,$$
$$2\alpha - c c' = 0.$$
Aus ihnen folgt, wenn wir mit c_0 und a Integrationskonstante bezeichnen
$$c = c_0 + ax,$$
$$2\alpha = a(c_0 + ax) = ac.$$

Damit ist die gesuchte Substitution für genügend kleine Werte von t ermittelt. Es gelten bei Vernachlässigung der dritten und höheren Potenzen von t die Gleichungen

(4)
$$\begin{cases} \xi = x + \dfrac{ac}{2} t^2, \\ \eta = y, \\ \zeta = z, \\ \tau = ct, \end{cases}$$

wobei die Lichtgeschwindigkeit c im System K, welche nur von x, aber nicht von t abhängen kann, durch die soeben abgeleitete Beziehung

(5) $$c = c_0 + ax$$

gegeben ist. Die Konstante c_0 hängt davon ab, mit einer wie rasch laufenden Uhr wir die Zeit im Anfangspunkte von K messen. Die Bedeutung der Konstante a ergibt sich in folgender Weise. Die erste und vierte der Gleichungen (4) liefert für den Anfangspunkt ($x = 0$) von K mit Rücksicht auf (5) die Bewegungsgleichung
$$\xi = \frac{a}{2 c_0} \tau^2.$$

a/c_0 ist also die Beschleunigung des Anfangspunktes von K in bezug auf Σ, gemessen in dem Zeitmaße, in welchem die Lichtgeschwindigkeit gleich 1 ist.

§ 2. Differentialgleichung des statischen Gravitationsfeldes, Bewegungsgleichung eines materiellen Punktes im statischen Gravitationsfelde.

Aus der früheren Arbeit geht schon hervor, daß im statischen Gravitationsfeld eine Beziehung zwischen c und dem Gravitationspotential existiert, oder mit anderen Worten, daß das Feld durch c bestimmt ist. In demjenigen Gravitationsfelde, welches dem im § 1 betrachteten Beschleunigungsfelde entspricht, ist nach (5) und dem Äquivalenzprinzip die Gleichung.

$$(5\,\mathrm{a}) \qquad \Delta c = \frac{\partial^2 c}{\partial x^2} + \frac{\partial^2 c}{\partial y^2} + \frac{\partial^2 c}{\partial z^2} = 0\,.$$

erfüllt, und es liegt die Annahme nahe, daß wir diese Gleichung als in jedem massenfreien statischen Gravitationsfelde gültig anzusehen haben.[1]) Jedenfalls ist diese Gleichung die einfachste mit (5) vereinbare.

Es ist leicht, diejenige vermutlich gültige Gleichung aufzustellen, welche derjenigen von Poisson entspricht. Es folgt nämlich aus der Bedeutung von c unmittelbar, daß c nur bis auf einen konstanten Faktor bestimmt ist, der davon abhängt, mit einer wie beschaffenen Uhr man t im Anfangspunkte von K mißt. Die der Poissonschen Gleichung entsprechende muß also in c homogen sein. Die einfachste Gleichung dieser Art ist die lineare Gleichung

$$(5\,\mathrm{b}) \qquad \Delta c = k c \varrho\,,$$

wenn unter k die (universelle) Gravitationskonstante, unter ϱ die Dichte der Materie verstanden wird. Letztere muß so definiert sein, daß sie durch die Massenverteilung bereits gegeben, d. h. bei gegebener Materie im Raumelement von c unabhängig ist. Dies erzielen wir, indem wir die Masse eines Kubikzentimeter Wasser gleich 1 setzen, in was für einem Gravitationspotential er sich auch befinden möge; ϱ ist dann das Verhältnis der im Kubikzentimeter enthaltenen Masse zu dieser Einheit.

---

[1]) In einer in kurzem nachfolgender Arbeit wird gezeigt werden, daß die Gleichung (5a) und (5b) noch nicht exakt richtig sein können. In dieser Arbeit sollen sie vorläufig benutzt werden.

Wir suchen nun das Bewegungsgesetz eines materiellen Punktes im statischen Schwerefeld zu ermitteln. Zu diesem Zwecke suchen wir das Bewegungsgesetz eines kräftefrei bewegten materiellen Punktes in dem im § 1 betrachteten Beschleunigungsfelde. Im System Σ ist dies Bewegungsgesetz

$$\xi = A_1 \tau + B_1,$$
$$\eta = A_2 \tau + B_2,$$
$$\zeta = A_3 \tau + B_3,$$

wobei die A und B Konstante sind. Diese Gleichungen gehen vermöge (4) in die für genügend kleine t gültigen Gleichungen über:

$$x = A_1 c t + B_1 - \frac{a c}{2} t^2,$$
$$y = A_2 c t + B_2,$$
$$z = A_3 c t + B_3.$$

Durch einmaliges und nochmaliges Differenzieren erhält man aus der ersten Gleichung, indem man in dieselben $t = 0$ einsetzt, die beiden Gleichungen[1])

$$\dot{x} = A_1 c,$$
$$\ddot{x} = 2 A_1 \dot{c} - a c.$$

Aus diesen beiden Gleichungen folgt durch Eliminieren von A_1

$$c \ddot{x} - 2 \dot{c} \dot{x} = - a c^2,$$

oder die Gleichung

$$\frac{d}{dt}\left(\frac{\dot{x}}{c^2}\right) = -\frac{a}{c^2}.$$

Auf analoge Weise resultieren für die beiden anderen Komponenten die Gleichungen

$$\frac{d}{dt}\left(\frac{\dot{y}}{c^2}\right) = 0,$$
$$\frac{d}{dt}\left(\frac{\dot{z}}{c^2}\right) = 0.$$

Diese drei Gleichungen gelten zunächst im Augenblick $t = 0$. Sie gelten aber allgemein, weil dieser Zeitpunkt durch nichts

---

1) Die in (2) weggelassenen Glieder machen sich bei dieser zweimaligen Differenziation und nachherigem Nullsetzen von t im Resultat nicht bemerkbar.

von den übrigen ausgezeichnet ist als dadurch, daß wir ihn zum Anfangspunkt unserer Reihenentwickelung gemacht haben. Die so gefundenen Gleichungen sind die gesuchten Bewegungsgleichungen des kräftefrei bewegten Punktes im konstanten Beschleunigungsfelde. Berücksichtigen wir, daß $a = \partial c / \partial x$, und daß $(\partial c/\partial y) = (\partial c/\partial z) = 0$ ist, so können wir diese Gleichungen auch in der Form schreiben:

$$(6) \quad \begin{cases} \dfrac{d}{dt}\left(\dfrac{\dot{x}}{c^2}\right) = -\dfrac{1}{c}\dfrac{\partial c}{\partial x}, \\ \dfrac{d}{dt}\left(\dfrac{\dot{y}}{c^2}\right) = -\dfrac{1}{c}\dfrac{\partial c}{\partial y}, \\ \dfrac{d}{dt}\left(\dfrac{\dot{z}}{c^2}\right) = -\dfrac{1}{c}\dfrac{\partial c}{\partial z}. \end{cases}$$

In dieser Form der Gleichungen ist die x-Richtung nicht mehr ausgezeichnet; beide Seiten haben Vektorcharakter. Wir haben diese Gleichungen deshalb wohl auch als die Bewegungsgleichungen eines materiellen Punktes im statischen Gravitationsfelde aufzufassen, falls der Punkt nur der Einwirkung der Schwere unterliegt.

Aus (6) folgt zunächst, in welcher Beziehung die in (5b) auftretende Konstante k zu der Gravitationskonstante K im gewöhnlichen Sinne steht. Im Falle gegen c kleiner Geschwindigkeiten ist nämlich nach (6)

$$\ddot{x} = -c\frac{\partial c}{\partial x} = -\frac{\partial \Phi}{\partial x},$$

so daß (5b) bei Vernachlässigung gewisser Glieder in

$$\Delta \Phi = k c^2 \varrho$$

übergeht. Es ist also

$$K = k c^2.$$

Die Gravitationskonstante K ist also keine universelle Konstante, sondern nur der Quotient K/c^2.

Multiplizieren wir die Gleichungen (6) der Reihe nach mit $\dot{x}/c^2$, $\dot{y}/c^2$, $\dot{z}/c^2$, und addieren wir, so ergibt sich, wenn

$$q^2 = \dot{x}^2 + \dot{y}^2 + \dot{z}^2$$

gesetzt wird,

$$\frac{d}{dt}\left(\frac{1}{2}\frac{q^2}{c^4}\right) = -\frac{\dot{c}}{c^3} = \frac{d}{dt}\left(\frac{1}{2c^2}\right),$$

oder
$$\frac{d}{dt}\left[\frac{1}{c^2}\left(1-\frac{q^2}{c^2}\right)\right] = 0,$$

oder

(7) $$\frac{c}{\sqrt{1-\frac{q^2}{c^2}}} = \text{konst.}$$

Diese Gleichung enthält das Energieprinzip für den im stationären Gravitationsfeld bewegten materiellen Punkt. Die linke Seite dieser Gleichung hängt von q genau in derselben Weise ab, wie die Energie des materiellen Punktes nach der gewöhnlichen Relativitätstheorie von q abhängt. Wir haben daher die linke Seite der Gleichung bis auf einen (nur vom Massenpunkt selbst abhängigen) Faktor als die Energie E des Punktes anzusehen. Dieser Faktor ist offenbar gleich der Masse m im obigen festgesetzten Sinne zu setzen, weil jene Definition die Masse unabhängig vom Gravitationspotential festlegt. Es ist also

(8) $$E = \frac{mc}{\sqrt{1-\frac{q^2}{c^2}}},$$

oder angenähert

(8a) $$E = mc + \frac{m}{2c}q^2.$$

Aus dem zweiten Gliede dieser Entwickelung geht zunächst hervor, daß die von uns als Energie bezeichnete Größe eine von der gewohnten abweichende Dimension besitzt. Dementsprechend wird auch die Maßzahl der einzelnen Energiegröße eine andere, nämlich eine c mal kleinere als in dem uns geläufigen System. Es hängt ferner die „kinetische Energie", welche allerdings nach (8) genau genommen von der Gravitationsenergie nicht getrennt werden kann, nicht nur von m und q, sondern auch von c, d. h. vom Gravitationspotential ab. Aus (8) folgt ferner das wichtige Resultat, daß die Energie des im Schwerefeld ruhenden Punktes mc ist. Wenn wir somit an der Beziehung

Kraft · Weg = zugeführte Energie

festhalten wollen, so ist die auf den ruhenden materiellen Punkt im Schwerefelde ausgeübte Kraft $\mathfrak{K}$

$$\mathfrak{K} = - m \operatorname{grad} c.$$

Wir wollen nun die Bewegungsgleichungen des materiellen Punktes in einem beliebigen statischen Schwerefelde für den Fall ableiten, daß außer der Schwere noch andere Kräfte auf den Punkt wirken. Wir bemerken, daß die Gleichungen (6) den in der Relativitätsmechanik geltenden Bewegungsgleichungen nicht ähnlich sind. Multiplizieren wir sie aber mit der linken Seite von (7), so erhalten wir die den Gleichungen (6) äquivalenten Gleichungen:

$$(6\,\mathrm{a}) \qquad \frac{d}{dt}\left\{ \frac{\dfrac{\dot{x}}{c}}{\sqrt{1-\dfrac{q^2}{c^2}}} \right\} = - \frac{\dfrac{\partial c}{\partial x}}{\sqrt{1-\dfrac{q^2}{c^2}}} \quad \text{usw.}$$

Die linke Seite hat, abgesehen von dem in der gewöhnlichen Relativitätstheorie belanglosen, im Zähler auftretenden Faktor $1/c$ genau dieselbe Form wie in der gewöhnlichen Relativitätstheorie. Wir werden deshalb die Klammergröße als x-Komponente der Bewegungsgröße zu bezeichnen haben (für einen Punkt der Masse 1). Wir haben ferner soeben gezeigt, daß $-\partial c/\partial x$ als x-Komponente der vom Gravitationsfeld auf einen unbewegten Massenpunkt ausgeübten Kraft aufzufassen ist. Die auf einen beliebig bewegten Massenpunkt von der Masse 1 vom Schwerefeld ausgeübte Kraft kann sich hiervon nur durch einen mit q verschwindenden Faktor unterscheiden. Die soeben aufgestellte Gleichung führt dazu, diese Kraft $\mathfrak{K}_g$ gleich $-\dfrac{\partial c/\partial x}{\sqrt{1-q^2/c^2}}$ zu setzen. Die rechte Seite der aufgestellten Gleichung wird dann $\mathfrak{K}_g$. Es ist also die zeitliche Ableitung des Impulses gleich der wirkenden Kraft. Wirkt auf den Punkt noch eine andere Kraft $\mathfrak{K}$, so werden wir auf der rechten Seite der Gleichung noch ein Glied $\mathfrak{K}/m$ zu addieren haben, so daß die Bewegungsgleichung eines Punkts von der Masse m die Form annimmt:

$$(6\,\mathrm{b}) \qquad \frac{d}{dt}\left\{ \frac{m\dfrac{\dot{x}}{c}}{\sqrt{1-\dfrac{q^2}{c^2}}} \right\} = - \frac{m\dfrac{\partial c}{\partial x}}{\sqrt{1-\dfrac{q^2}{c^2}}} + \mathfrak{K}_x \quad \text{usw.}$$

Diese Gleichung ist aber nur dann zulässig, wenn das Energieprinzip in der Form

$$\mathfrak{K}\,q = \dot{E}$$

erfüllt ist. Dies läßt sich in folgender Weise dartun.

Schreibt man (6b) in der Form

$$\frac{d}{dt}\left\{\frac{\dot{x}}{c^2}E\right\} + \frac{1}{c}\frac{\partial c}{\partial x}E = \mathfrak{K}_x \text{ usw.}$$

und multipliziert man diese Gleichungen der Reihe nach mit $\dot{x}/c^2$ usw., und addiert dieselben, so findet man

$$\frac{1}{2}\frac{q^2}{c^4}\dot{E} + \frac{1}{2}E\frac{d}{dt}\left(\frac{q^2}{c^4}\right) + E\frac{\dot{c}}{c^3} = \frac{\mathfrak{K}\,q}{c^2}\,.$$

Hieraus ergibt sich die gesuchte Relation, wenn man berücksichtigt, daß wegen (8)

$$\frac{q^2}{c^4} = \frac{1}{c^2} - \frac{m^2}{E^2}$$

und

$$\frac{d}{dt}\left(\frac{q^2}{c^4}\right) = -\frac{\dot{c}}{c^3} + \frac{m^2 E}{E^3}$$

ist. Die Beziehungen der Kraft zum Impuls- und Energiesatz bleiben also erhalten.

§ 3. Bemerkungen über die physikalische Bedeutung des statischen Schwerepotentials.

Messen wir in einem Raume von nahezu konstantem Schwerepotential die Lichtgeschwindigkeit, indem wir mittels einer bestimmten Uhr die Zeit messen, welche das Licht zum Durchlaufen eines geschlossenen Weges von bestimmter Länge braucht, so erhalten wir für die Lichtgeschwindigkeit immer dieselbe Zahl, ganz unabhängig davon, in einem Raume von wie großem Schwerepotential wir diese Messung ausführen.[1]) Es folgt dies unmittelbar aus dem Äquivalenzprinzip. Wenn wir sagen, daß die Lichtgeschwindigkeit in einem Punkte P c/c_0 mal größer sei als in einem Punkte P_0, so bedeutet dies

---

1) Die zur Zeitmessung benutzte Uhr ist dabei immer die nämliche; sie wird immer an die Stelle gebracht, für die c ermittelt werden soll.

also, daß wir uns in P zur Zeitmessung[1]) einer Uhr bedienen müssen, welche c/c_0 mal langsamer läuft als die zur Zeitmessung in P_0 zu benutzende Uhr, falls der Gang beider Uhren an demselben Orte miteinander verglichen wird. Anders ausgedrückt: eine Uhr läuft desto schneller, an eine Stelle von je größerem c wir sie bringen. Diese Abhängigkeit der Raschheit des zeitlichen Ablaufes vom Gravitationspotential (c) gilt für den zeitlichen Ablauf beliebiger Vorgänge. Dies wurde bereits in der früheren Arbeit dargelegt.

Ebenso hängt die Spannkraft einer in bestimmter Weise gespannten Feder, überhaupt die Kraft bzw. die Energie eines beliebigen Systems stets davon ab, an einem Orte von wie großem c sich das System befindet. Dies geht leicht aus folgender elementaren Überlegung hervor. Wenn wir nacheinander in mehreren kleinen Raumteilen von verschiedenem c experimentieren und uns stets derselben Uhr, derselben Maßstäbe usw. bedienen, so finden wir überall — abgesehen von etwaigen Verschiedenheiten der Intensität des Schwerefeldes — dieselben Gesetzmäßigkeiten mit denselben Konstanten. Dies folgt aus dem Äquivalenzprinzip. Als Uhr können wir uns dabei etwa zweier Spiegel von der Distanz 1 cm bedienen, indem wir die Zahl der Hin- und Hergänge eines Lichtsignals zählen; wir operieren dann mit einer Art Lokalzeit, welche Abraham mit l bezeichnet. Diese steht dann mit der universellen Zeit in der Relation

$$dl = c\,dt.$$

Messen wir die Zeit durch l, so wird man mittels der Deformationsenergie einer bestimmten, in einer bestimmten Weise gespannten Feder einer Masse m eine bestimmte Geschwindigkeit dx/dl erteilen, unabhängig davon, an einem Orte von wie großem c dieser Prozeß vor sich geht. Es ist

$$\frac{dx}{dl} = \frac{dx}{c\,dt} = a,$$

wobei a von c unabhängig ist. Nach (8) kann aber die dieser Bewegung entsprechende kinetische Energie gleich

---

[1]) Nämlich zur Messung der in den Gleichungen mit „t" bezeichneten Zeit.

Lichtgeschwindigkeit und Statik des Gravitationsfeldes. 367

$$\frac{m}{2c} q^2 = \frac{m}{2c} \left(\frac{dx}{dt}\right)^2 = \frac{m}{2c} a^2 c^2 = \frac{m a^2}{2} \cdot c$$

gesetzt werden. Die Energie der Feder ist also c proportional, und es gilt ein Gleiches für Energie und Kräfte irgend eines Systems.

Diese Abhängigkeit hat eine unmittelbare physikalische Bedeutung. Denke ich mir z. B. einen masselosen Faden zwischen zwei Punkten P_1 und P_2 verschiedenen Gravitationspotentials gespannt. Eine von zwei vollkommen gleich beschaffenen Federn ziehe in P_1, die zweite in P_2 an dem Faden, derart, daß Gleichgewicht besteht. Die Verlängerungen l_1 und l_2, welche die beiden Federn dabei erfahren, werden aber nicht gleich sein, sondern die Gleichgewichtsbedingung wird lauten[1]

$$l_1 c_1 = l_2 c_2.$$

Schließlich sei noch erwähnt, daß mit diesem allgemeinen Ergebnis auch die Gleichung (5b) in Übereinstimmung ist. Aus dieser Gleichung und aus dem Umstande, daß die auf eine Masse m wirkende Gravitationskraft gleich $-m \operatorname{grad} c$ ist, folgt nämlich, daß die Kraft $\mathfrak{K}$, mit der sich zwei im Potential c in der Entfernung r befindliche Massen anziehen, in erster Annäherung gegeben ist durch

$$\mathfrak{K} = c k \frac{m m'}{4 \pi r^2}.$$

Es ist also auch diese Kraft c proportional. Denken wir uns ferner eine „Gravitationsuhr" bestehend aus einer Masse m, die um eine festgehaltene Masse m' bei konstantem Abstand R unter alleiniger Wirkung der Gravitationskraft von m' umläuft, so geschieht dies nach (6b) in erster Näherung nach den Gleichungen

$$m \ddot{x} = c \mathfrak{K}_x \text{ usw.}$$

Hieraus folgt

$$m \omega^2 R = c^2 k \frac{m m'}{4 \pi R^2}.$$

Die Ganggeschwindigkeit ω der Gravitationsuhr ist also c proportional, wie dies für Uhren jeder Art der Fall sein soll.

---

[1] Hierbei ist allerdings vorausgesetzt, daß auf den gespannten masselosen Faden im Gravitationsfeld keine Kraft wirkt. Dies wird in einer bald folgenden Arbeit begründet werden.

§ 4. Allgemeine Bemerkungen über Raum und Zeit.

In was für einem Verhältnis steht nun die vorstehende Theorie zu der alten Relativitätstheorie (d. h. zu der Theorie des universellen c)? Nach Abrahams Meinung sollen die Transformationsgleichungen von Lorentz nach wie vor im unendlich Kleinen gelten, d. h. es soll eine x-t-Transformation geben, so daß

$$dx' = \frac{dx - v\,dt}{\sqrt{1 - \frac{v^2}{c^2}}},$$

$$dt' = \frac{-\frac{v}{c^2}dx + dt}{\sqrt{1 - \frac{v^2}{c^2}}}$$

gelten. dx' und dt' müssen vollständige Differentiale sein. Es sollen also die Gleichungen gelten

$$\frac{\partial}{\partial t}\left\{\frac{1}{\sqrt{1-\frac{v^2}{c^2}}}\right\} = \frac{\partial}{\partial x}\left\{\frac{-v}{\sqrt{1-\frac{v^2}{c^2}}}\right\},$$

$$\frac{\partial}{\partial t}\left\{\frac{-\frac{v}{c^2}}{\sqrt{1-\frac{v^2}{c^2}}}\right\} = \frac{\partial}{\partial x}\left\{\frac{1}{\sqrt{1-\frac{v^2}{c^2}}}\right\}.$$

Es sei nun im ungestrichenen System das Gravitationsfeld ein statisches. Dann ist c eine beliebig gegebene Funktion von x, von t aber unabhängig. Soll das gestrichene System ein „gleichförmig" bewegtes sein, so muß v bei festgehaltenem x jedenfalls von t unabhängig sein. Es müssen daher die linken Seiten der Gleichungen, somit auch die rechten Seiten verschwinden. Letzteres ist aber unmöglich, da bei beliebig in Funktionen von x gegebenem c nicht beide rechten Seiten zum Verschwinden gebracht werden können, indem man v in Funktion von x passend wählt. Damit ist also erwiesen, daß man auch für unendlich kleine Raum–Zeitgebiete nicht an der Lorentztransformation festhalten kann, sobald man die universelle Konstanz von c aufgibt.

Mir scheint das Raum–Zeitproblem wie folgt zu liegen. Beschränkt man sich auf ein Gebiet von konstantem Gravitationspotential, so werden die Naturgesetze von ausgezeichnet einfacher und invarianter Form, wenn man sie auf ein Raum-Zeitsystem derjenigen Mannigfaltigkeit bezieht, welche durch die Lorentztransformationen mit konstantem c miteinander verknüpft sind. Beschränkt man sich nicht auf Gebiete von konstantem c, so wird die Mannigfaltigkeit der äquivalenten Systeme, sowie die Mannigfaltigkeit der die Naturgesetze ungeändert lassenden Transformationen eine größere werden, aber es werden dafür die Gesetze komplizierter werden.

Prag, Februar 1912.

(Eingegangen 26. Februar 1912.)

8. *Zur Theorie des statischen Gravitationsfeldes;* von *A. Einstein.*

In einer jüngst erschienenen Arbeit habe ich aus einer Hypothese, die ich als Äquivalenzprinzip bezeichnet habe, die Bewegungsgleichungen eines in einem solchen Felde bewegten materiellen Punktes abgeleitet. Im folgenden soll exakt abgeleitet werden, welchen Einfluß ein statisches Schwerefeld auf die elektromagnetischen und thermischen Vorgänge nach dem Äquivalenzprinzip hat. Die erste dieser beiden Fragen habe ich schon früher in erster Näherung behandelt. Zuletzt wird die Differentialgleichung für das statische Gravitationsfeld selbst abgeleitet.

§ 1. Ableitung der elektromagnetischen Gleichungen unter Berücksichtigung des (statischen) Gravitationsfeldes.

Der Weg, den wir hier einschlagen, ist genau derselbe, welcher uns in der früheren Arbeit die Bewegungsgleichungen des materiellen Punktes geliefert hat. Wir suchen nämlich die elektromagnetischen Gleichungen, welche relativ zu einem (im Bornschen Sinne) gleichförmig beschleunigten System $K(x, y, z, t)$ gelten, und nehmen nach der Äquivalenzhypothese an, daß diese Gleichungen auch im statischen Schwerefeld gelten. Um die in bezug auf K gültigen Gleichungen zu finden, gehen wir aus von den bekannten Gleichungen, welche in bezug auf ein unbeschleunigtes System $\sum(\xi, \eta, \zeta, \tau)$ gelten. Wählen wir in letzterem die Zeiteinheit so, daß die Lichtgeschwindigkeit gleich 1 wird, so haben diese Gleichungen für das Vakuum die bekannte Form:

$$(1) \quad \begin{cases} \mathfrak{v}' \varrho' + \dfrac{\partial \mathfrak{E}'}{\partial \tau} = \operatorname{rot}' \mathfrak{H}', \\ 0 = \operatorname{div}' \mathfrak{H}', \\ \dfrac{\partial \mathfrak{H}'}{\partial \tau} = -\operatorname{rot}' \mathfrak{E}', \\ \varrho' = \operatorname{div}' \mathfrak{E}'. \end{cases}$$

Die Zeichen für die in diesen Gleichungen auftretenden Skalare, Vektoren und Operatoren sind gestrichelt, um ihre Zugehörigkeit zum System Σ anzudeuten. Diese Gleichungen sind auf das gleichförmig beschleunigte System K zu transformieren nach Gleichungen, die für genügend kleine t und bei geeigneter Wahl der Koordinatenachsen und Anfangspunkte für die Zeiten sich in der Form schreiben lassen:

(2)
$$\begin{cases} \xi = x + \frac{ac}{2}t^2, \\ \eta = y, \\ \zeta = z, \\ \tau = ct, \end{cases}$$

wobei
$$c = c_0 + ax.$$

Auch die Feldvektoren $\mathfrak{E}'$ und $\mathfrak{H}'$ wollen wir aufs beschleunigte System K transformieren. Dies tun wir auf Grund der Festsetzung, daß die auf K bezogenen Feldvektoren $\mathfrak{E}$, $\mathfrak{H}$ identisch sein sollen mit den Feldvektoren $\mathfrak{E}'$, $\mathfrak{H}'$ desjenigen unbeschleunigten Systems Σ, in bezug auf welches das System K gerade die Geschwindigkeit Null hat. Für $t = \tau = 0$ aus dieser Festsetzung unmittelbar:
$$\mathfrak{E} = \mathfrak{E}',$$
$$\mathfrak{H} = \mathfrak{H}'.$$

Analoges setzen wir für die elektrische Dichte fest, so daß für $t = \tau = 0$
$$\varrho = \varrho'$$
ist. Nun bemerken wir, daß es genügt, wenn wir die den Gleichungen (1) entsprechenden transformierten Gleichungen für $t = \tau = 0$ aufstellen, da ja diese Gleichungen für jedes t die nämlichen sein müssen. Für $t = \tau = 0$ gilt nach (2)
$$\frac{\partial}{\partial \xi} = \frac{\partial}{\partial x}, \quad \frac{\partial}{\partial \eta} = \frac{\partial}{\partial y}, \quad \frac{\partial}{\partial \zeta} = \frac{\partial}{\partial z}.$$

Aus dem bisher Gesagten folgt schon, daß die rechten Seiten von (1) durch Weglassung der Striche ungeändert bleiben, ebenso die linken Seiten der zweiten und vierten der Gleichungen (1). Einiges Nachdenken erfordert nur die Umformung der linken Seiten der ersten und dritten der Gleichungen (1).

Zunächst folgt aus (2), daß für einen bewegten Punkt zur Zeit $t = 0$ gilt:

$$(2\,\mathrm{a}) \quad \begin{cases} dx = d\xi, \\ dy = d\eta, \\ dz = d\zeta, \\ dt = \dfrac{1}{c} d\tau, \end{cases}$$

woraus unmittelbar folgt, daß

$$\mathfrak{v} = c\,\mathfrak{v}' \quad \text{oder} \quad \mathfrak{v}' = \frac{1}{c}\mathfrak{v}, \quad \text{wenn} \quad \mathfrak{v}_x = \frac{dx}{dt} \text{ usw.}$$

gesetzt wird. Wir bezeichnen ferner mit $d\mathfrak{E}$ die Änderung, welche $\mathfrak{E}$ in einer unendlich kurzen Zeit in einem Systempunkt von K erfährt, mit $d'\mathfrak{E}'$ die entsprechende Änderung, welche $\mathfrak{E}'$ in dem momentan koinzidierenden Punkte von $\sum$ in der entsprechenden Zeit erfährt. Im Anfang der unendlich kleinen Zeitstrecke dt bzw. $d\tau$ sei $t = \tau = 0$; zu dieser Zeit ist $\mathfrak{E} = \mathfrak{E}'$. Diese letztere Gleichung gilt aber am Ende von dt bzw. $d\tau$ aus zwei Gründen nicht mehr genau. Erstens fällt nämlich am Ende von $d\tau$ der Systempunkt von K nicht mehr mit dem von $\sum$ zusammen; hiervon kann jedoch Abstand genommen werden, da diese Verrückung unendlich klein zweiter Ordnung ist. Zweitens aber erlangt während der betrachteten unendlich kleinen Zeit der Systempunkt von K eine Geschwindigkeit $\mathfrak{g}\,d\tau$ in Richtung der ξ-Achse; man hat also, um $\mathfrak{E}$ am Ende von $d\tau$ zu erhalten, das elektromagnetische Feld auf ein beschleunigungsfreies System zu beziehen, welches gegenüber $\sum$ im Sinne der positiven ξ-Achse mit der Geschwindigkeit $\mathfrak{g}\,d\tau$ bewegt ist. Dabei transformiert sich das elektromagnetische Feld in bekannter Weise. Mit Rücksicht auf die angedeuteten Überlegungen erhält man:

$$d\mathfrak{E} = d'\mathfrak{E}' + [\mathfrak{g}\,\mathfrak{H}']\,dt,$$

oder mit Rücksicht auf die letzte der Gleichungen (2a):

$$\frac{\partial \mathfrak{E}'}{\partial \tau} = \frac{1}{c}\frac{\partial \mathfrak{E}}{\partial t} - \frac{1}{c}[\mathfrak{g}\,\mathfrak{H}].$$

Nun erhält man aber aus den Gleichungen (2)

$$|\mathfrak{g}| = \frac{a}{c} = \frac{1}{c}\frac{dc}{dx},$$

also, weil c von y und z unabhängig ist,
$$\mathfrak{g} = \frac{1}{c}\,\mathrm{grad}\,c\,.$$
Man erhält also endlich
$$\frac{\partial\,\mathfrak{E}'}{\partial\,\tau} = \frac{1}{c}\frac{\partial\,\mathfrak{E}}{\partial\,t} = \frac{1}{c}\,[\mathrm{grad}\,c, \mathfrak{H}]$$
und auf ganz analoge Weise
$$\frac{\partial\,\mathfrak{H}'}{\partial\,\tau} = \frac{1}{c}\frac{\partial\,\mathfrak{H}}{\partial\,t} + \frac{1}{c}\,[\mathrm{grad}\,c, \mathfrak{E}]\,.$$
Berücksichtigt man nun noch, daß nach den Regeln der Vektorrechnung
$$c\,\mathrm{rot}\,\mathfrak{H} + [\mathrm{grad}\,c, \mathfrak{H}] = \mathrm{rot}\,(c\,\mathfrak{H})$$
ist, und daß die analoge Gleichung für $\mathrm{rot}\,(c\,\mathfrak{E})$ besteht, so erhält man mit Rücksicht auf die Resultate der bereits angegebenen Überlegungen aus den Gleichungen (1) die folgenden auf das System K bezüglichen:

(1a)
$$\begin{cases} \mathfrak{v}\varrho + \dfrac{\partial\,\mathfrak{E}}{\partial\,t} = \mathrm{rot}\,(c\,\mathfrak{H}), \\ \quad\quad 0 = \mathrm{div}\,\mathfrak{H}, \\ \dfrac{\partial\,\mathfrak{H}}{\partial\,t} = -\,\mathrm{rot}\,(c\,\mathfrak{E}), \\ \quad\quad \varrho = \mathrm{div}\,\mathfrak{E}\,. \end{cases}$$

Die physikalische Bedeutung der in diesen Gleichungen auftretenden Größen ist dabei eine vollkommen bestimmte. x, y, z werden durch am starren System K angelegte Maßstäbe gemessen. t ist die Zeit im System K, welche durch verschieden beschaffene, in den Systempunkten von K ruhend angeordnete Uhren gemessen wird; t ist durch die Festsetzungen definiert, daß die Lichtgeschwindigkeit in K nicht von der Zeit und nicht von der Richtung abhängen soll. $\mathfrak{v}$ ist die mit der Zeit t gemessene Geschwindigkeit der Elektrizität. ϱ ist die Dichte der Elektrizität, gemessen in Einheiten folgender Art: In einem nicht beschleunigten System sollen zwei solche Einheiten im Abstand 1 cm aufeinander die Kraft 1 aufeinander ausüben, wobei die Kraft 1 diejenige ist, welche einem Gramm die Beschleunigung 1 erteilt, falls man als Zeiteinheit die Zeit wählt, welche das Licht braucht, um 1 cm zu durchlaufen (Lichtzeit). Der Feldvektor $\mathfrak{E}$ hat folgende Bedeutung. Hat man eine

Federwage so graduiert, daß sie in dem nicht mitbeschleunigten[1] System $\sum$ die Kraft unter Zugrundelegung der Licht–Zeiteinheit mißt, und befestigt man am Angriffspunkt dieser Federwage die Einheit der Elektrizität, so mißt diese Federwage direkt die Feldintensität $|\mathfrak{E}|$. Analog gestaltet sich die Definition von $\mathfrak{H}$. —

Nach dem Äquivalenzprinzip hat man die Gleichungen (1a) als die elektromagnetischen Grundgleichungen in einem statischen Schwerefelde anzusehen. Sie sind insofern als exakt anzusehen, als sie mit gleicher Annäherung gelten sollen, wie sehr auch das Gravitationspotential mit dem Orte variieren möge. Hingegen könnten sie aus dem Grunde unexakt sein, weil das elektromagnetische Feld das Gravitationsfeld derart beeinflussen könnte, daß letzteres kein statisches Feld mehr ist. Sie erlauben ferner, auch in den Fällen, in denen sie genau gelten, nicht, den Einfluß zu berechnen, welchen das elektromagnetische Feld auf das statische Gravitationsfeld (c) ausübt.

§ 2. Bemerkungen über den Inhalt der abgeleiteten Gleichungen.

Ich will die im letzten Paragraph bei der anschaulichen Interpretation der Feldvektoren eingeführte Federwage nach einem mündlichen Vorschlag P. Ehrenfests als „Taschen"-Federwage bezeichnen. Es sollen überhaupt mit der Bezeichnung „Taschen" solche physikalische Einrichtungen bezeichnet werden, welche an Orte verschiedenen Gravitationspotentials gebracht gedacht werden, und deren Angaben stets benutzt werden, an einem Orte von wie großem c sie sich auch gerade befinden mögen.[2] So kann man die Uhr, welche die „Lichtzeit" angibt, als „Taschenuhr"- bezeichnen, die mit der Elektrizitätseinheit im Angriffspunkte versehene Federwage als „Taschenfeldmesser" usw.

Aus der früheren Arbeit geht nun hervor, daß die Angabe einer „Taschenfederwage" nicht direkt die von ihr ausgeübte

---

1) Natürlich ist dasjenige System $\sum$ gemeint, welches in dem betreffenden Augenblick keine Relativgeschwindigkeit in bezug auf K hat.
2) Mit der Bezeichnung „Taschen"- soll angedeutet werden, daß die Dinge transportiert werden können, nicht nur an einem Orte benutzt werden.

Kraft mißt. Letztere ist vielmehr der mit c multiplizierten Angabe der Taschenfederwage gleichzusetzen. Hieraus ergibt sich unmittelbar, daß die auf die in K ruhende Elektrizitätseinheit ausgeübte ponderomotorische Kraft nicht gleich $\mathfrak{E}$, sondern gleich $c \cdot \mathfrak{E}$ zu setzen ist. Entsprechendes gilt für den Feldvektor $\mathfrak{H}$.

Da nach der dritten der Gleichungen (1a) in einem statischen elektrischen Felde $\operatorname{rot}(c\,\mathfrak{E}) = 0$ ist, das Linienintegral des Vektors $c\,\mathfrak{E}$ über eine geschlossene Kurve also verschwindet, sieht man, daß es unmöglich ist, durch Führen einer Elektrizitätsmenge über eine geschlossene Bahn unbegrenzt Arbeit zu erhalten.

Wir stellen nun Coulombs Gesetz für einen Raum von konstantem c auf. Aus der letzten der Gleichungen (1a) folgt, daß das Feld einer Punktladung ε durch $|\mathfrak{E}| = \dfrac{\varepsilon}{4\pi r^2}$ gegeben ist, falls man mit v den Abstand von der Punktladung bezeichnet. Befindet sich in diesem Falle eine zweite elektrische Masse ε', so ist die auf sie ausgeübte Kraft gleich $c\,\varepsilon'\,|\mathfrak{E}|$ oder gleich $c\,\dfrac{\varepsilon\,\varepsilon'}{4\pi r^2}$, also wie nach der früheren Arbeit jede Kraft eines beliebigen „Taschensystems" in bestimmtem Zustande — proportional c. Mit diesem Resultat hängt das Folgende eng zusammen. Wir bringen von zwei genau gleichen Kondensatoren C und C' mit den Belegungen a, b bzw. $a'\,b'$ den einen an einen Ort vom Gravitationspotential c, den anderen an einen Ort vom Gravitationspotential c'. a sei mit a', b mit b' leitend verbunden. Laden wir die Kondensatoren, so ist wegen $\operatorname{rot}(c\,\mathfrak{E}) = 0$ die Ladung beider Kondensatoren nicht dieselbe; es ist vielmehr $c\,\mathfrak{E} = c'\,\mathfrak{E}'$ und wegen $\varrho = \operatorname{div}\mathfrak{E}$ auch $c\,\varepsilon = c'\,\varepsilon'$, wenn man mit ε bzw. ε' die Ladungen der beiden Kondensatoren bezeichnet.

Aus dem für das Coulombsche Gesetz gefundenen Ausdruck geht hervor, daß wir nicht $\tfrac{1}{2}(\mathfrak{E}^2 + \mathfrak{H}^2)$, sondern den Ausdruck $c/2\,(\mathfrak{E}^2 + \mathfrak{H}^2)$ der Dichte der elektromagnetischen Energie gleichzusetzen haben. Wir werden also die dem Energieprinzip entsprechende Gleichung dadurch erhalten, daß wir die erste der Gleichungen (1a) skalar mit $c\,\mathfrak{E}$, die dritte skalar mit $c\,\mathfrak{H}$ multiplizieren und beide addieren, und hierauf über einen beliebigen geschlossenen Raum integrieren. Es ergibt sich so in bekannter Weise:

$$(3) \quad \int \mathfrak{v} c \mathfrak{E} \varrho \, d\tau + \frac{d}{dt}\left\{\int \frac{c}{2}(\mathfrak{E}^2 + \mathfrak{H}^2) \, d\tau\right\} = \int [c\mathfrak{E}, c\mathfrak{H}] \mathfrak{n} \, d\sigma,$$

falls man mit $d\tau$ das Raumelement, mit $d\sigma$ das Element der Begrenzungsfläche, mit $\mathfrak{n}$ deren nach innen gerichtete Normale bezeichnet. Das Energieprinzip ist also erfüllt, wobei der Vektor $c^2[\mathfrak{E}, \mathfrak{H}]$ dem Energiestrom gleich ist.

Wir leiten nun den Impulssatz ab, indem wir die erste der Gleichungen (1a) vektoriell mit $\mathfrak{H}$, die dritte derselben mit $-\mathfrak{E}$ multiplizieren und addieren. Setzen wir als Ausdruck der Maxwellschen Spannungen

$$X_x = c(\mathfrak{E}_x^2 + \mathfrak{H}_x^2 - \tfrac{1}{2}\mathfrak{E}^2 - \tfrac{1}{2}\mathfrak{H}^2), \quad X_y = c(\mathfrak{E}_x \mathfrak{E}_y + \mathfrak{H}_x \mathfrak{H}_y),$$
$$X_z = c(\mathfrak{E}_x \mathfrak{E}_z + \mathfrak{H}_x \mathfrak{H}_z)$$

usw., so erhalten wir:

$$(4) \quad \begin{cases} \varrho(c\mathfrak{E}_x + [\mathfrak{v}, \mathfrak{H}]_x) + \dfrac{d}{dt}[\mathfrak{E}, \mathfrak{H}]_x \\ \qquad = \left(\dfrac{\partial X_x}{\partial x} + \dfrac{\partial X_y}{\partial y} + \dfrac{\partial X_z}{\partial z}\right) - \tfrac{1}{2}(\mathfrak{E}^2 + \mathfrak{H}^2)\dfrac{\partial c}{\partial x}, \end{cases}$$

sowie die hieraus durch zyklische Vertauschung entstehenden Gleichungen. In dieser Gleichung drückt das erste Glied die X-Komponente der Impulsgröße aus, welche durch die elektrischen Massen pro Zeiteinheit und Volumeinheit an die ponderabeln Massen des Systems abgegeben wird. Der Ausdruck der ponderomotorischen Kraft ist also bis auf den Faktor c der von H. A. Lorentz angegebene. Das zweite Glied der linken Seite drückt den Zuwachs der Volumeinheit an elektromagnetischem Impuls aus. Verschwinden die räumlichen Differentialquotienten von c, d. h. ist kein Schwerefeld vorhanden, so wird die der linken Seite entsprechende Zunahme des Impulses der Volumeinheit durch die elektromagnetischen Spannungen bewirkt, wie in der Elektrodynamik ohne Berücksichtigung des Schwerefeldes. Für den Fall aber, daß ein Gravitationsfeld vorhanden ist, ergibt sich aus dem letzten Gliede der rechten Seite, daß dieses für das elektromagnetische Feld als Impulsquelle anzusehen ist. Die elektromagnetische Feldenergie empfängt aus dem Schwerefeld einen Impuls, genau wie eine ponderable ruhende Masse; denn in der früheren Arbeit ergab es sich, daß das Gravitationsfeld auf die ruhende Masse m pro Zeiteinheit den Impuls $-m\,\mathrm{grad}\,c$ überträgt. Es ergibt sich also z. B., daß die Hohlraumstrahlung eine ihrer trägen Masse genau entsprechende schwere Masse besitzt; dies Resultat ist in den Gleichungen (1a) und dem Ausdruck für die auf die Elektrizitätsmengen wirkenden ponderomotorischen Kräfte bereits enthalten, da die zuletzt angeschriebene Impulsgleichung eine Folge der Gleichungen (1a) ist. Zu bemerken ist, daß die Größe $\frac{1}{2}(\mathfrak{E}^2+\mathfrak{H}^2)$, nicht die eigentliche Energiedichte $c/2\,(\mathfrak{E}^2+\mathfrak{H}^2)$, für die Schwere des elektromagnetischen Feldes maßgebend, d. h. einer räumlichen Dichte unbewegter träger Masse äquivalent ist. Dies ist auch zu erwarten; denn der Ausdruck $\frac{1}{2}(\mathfrak{E}^2+\mathfrak{H}^2)$ ist die Energiedichte, wie sie von einem mit „Tascheninstrumenten" messenden Beobachter erscheint. Diese Größe ist es also, welche der trägen Masse nach der von uns benutzten Definition für letztere analog ist.

Es geht aus diesen Überlegungen hervor, daß das elektromagnetische Feld auch umgekehrt eine Rückwirkung auf das Gravitationsfeld besitzt, dessen Ausdruck für den statischen Fall sich nach den angegebenen Überlegungen ohne weiteres ergibt, da die Raumfunktion $\frac{1}{2}(\mathfrak{E}^2+\mathfrak{H}^2)$ einer gleich großen Dichte unbewegter ponderabler Masse äquivalent ist. Hierauf soll aber an dieser Stelle nicht näher eingegangen werden. Ebensowenig will ich mich hier mit dem in den Gleichungen (1a) enthaltenen Gesetze der Krümmung der Lichtstrahlen im Schwerefelde befassen, weil dieses in erster Annäherung bereits in der voriges Jahr über den Gegenstand erschienenen Abhandlung angegeben ist.

§ 3. Thermische Größen und Gravitationsfeld.

An zwei voneinander entfernten Orten mit den Lichtgeschwindigkeiten c_1 bzw. c_2 seien zwei Wärmebehälter W_1 bzw. W_2 angeordnet. Dieselben sollen insofern gleiche Temperaturen besitzen, als ein und dasselbe Thermometer („Taschenthermometer"), mit ihnen nacheinander in Berührung gebracht, in beiden Fällen die nämliche Temperatur („Taschenthermometer"-Temperatur) T^* haben sollen. Unter „Temperatur" (T) schlechtweg sei jene Temperatur verstanden, wie sie durch

Carnotsche Kreisprozesse definiert wird. Wir fragen nach der Beziehung, die zwischen den Temperaturen der Wärmebehälter W_1 und W_2 besteht.

Wir denken uns folgenden Kreisprozeß. Mit einem Körper von der Taschentemperatur T^* werde dem Behälter W_1 die Taschenwärmemenge Q^* entzogen, der Körper hierauf zum Behälter W_2 bewegt. Dann wird vom Körper dieselbe Taschenwärmemenge Q^* auf den Wärmebehälter W_2 bei der Taschentemperatur T^* übertragen und endlich der Körper wieder zum Behälter W_1 zurückbewegt.

Nach den Ergebnissen der früheren Arbeit ist dabei die den Behältern in Wahrheit entzogene bzw. zugeführte Wärme

$$Q_1 = Q^* c_1,$$
$$Q_2 = Q^* c_2.$$

Die bekannte Relation

$$\frac{Q_1}{T_1} = \frac{Q_2}{T_2}$$

liefert also sofort

$$\frac{c_1}{c_2} = \frac{T_1}{T_2}.$$

Haben also zwei Wärmebehälter — mit Taschenthermometern gemessen — gleiche Temperatur T^*, so verhalten sich ihre wahren (thermodynamischen) Temperaturen wie die Lichtgeschwindigkeiten der betreffenden Orte. Man kann dies auch so ausdrücken: Man erhält die wahre Temperatur, indem man die Angabe eines Taschenthermometers mit c multipliziert:

$$T = c T^*.$$

Hieraus folgt andererseits, daß zwei Wärmebehälter, welche sich an Orten verschiedenen Gravitationspotentials befinden und in wärmeleitender Verbindung stehen, nicht dieselben Taschentemperaturen annehmen, sondern daß letztere beim Temperaturgleichgewicht sich umgekehrt verhalten wie die Lichtgeschwindigkeiten.

Dagegen ist die Entropie eines Körpers nur von seinem mit Tascheninstrumenten gemessenen Zustande, nicht aber von dem Gravitationspotential abhängig. Es folgt dies einmal daraus, daß der Körper ohne Änderung seines mit Tascheninstrumenten gemessenen Zustandes ohne Zufuhr von Wärme nach einer Stelle von anderem Gravitationspotential gebracht werden kann, andererseits aus den soeben gefundenen Relationen. Denn es ist für zwei gleichbeschaffene Körper, die an verschiedenen Orten — mit Tascheninstrumenten gemessen — dieselben Änderungen erfahren:

$$\frac{Q_1}{T_1} = \frac{Q^*}{T^*} = \frac{Q_2}{T_2}.$$

§ 4. Differentialgleichung des statischen Gravitationsfeldes.

In der ersten Arbeit wurde aus der letzten der Gleichungen (2)

$$c = c_0 + a x$$

auf dem Wege der Verallgemeinerung für das statische Gravitationsfeld die Gleichung

(3) $$\Delta c = 0$$

für den materiefreien Raum, und die Gleichung

(3a) $$\Delta c = k c \sigma$$

für den mit Materie erfüllten Raum abgeleitet. Es zeigt sich aber, daß die Gleichung (3a) zusammen mit unserem in der früheren Abhandlung gefundenen Ausdruck für die Kraft $\mathfrak{F}$, welche auf die in der Volumeinheit befindliche ponderable Materie σ wirkt, zu einem Widerspruch führt. Ruht die Materie, so soll nämlich gelten

(4) $$\mathfrak{F} = -\sigma \operatorname{grad} c$$

Bilden wir das Integral

$$\int \mathfrak{F} \, d\tau$$

über einen Raum, für welchen im Unendlichen c konstant ist, so verlangt das Prinzip der Gleichheit von actio und reactio, daß dieses Integral verschwinde. Anderenfalls würde sich die Gesamtheit der in dem betrachteten Raume befindlichen Massen, die wir auf einem starren, masselosen Gerüste uns befestigt denken wollen, sich in Bewegung zu setzen streben. Es ist aber nach (4) und (3a)

$$\int \mathfrak{F} \, d\tau = -\int \sigma \operatorname{grad} c \, d\tau = -\frac{1}{k} \int \frac{\Delta c}{c} \operatorname{grad} c \, d\tau,$$

und man beweist von dem letzten dieser Integrale leicht, daß es im allgemeinen nicht verschwindet.

Wir sind also zu einem recht bedenklichen Resultat gelangt, das geeignet ist, Zweifel an der Zulässigkeit der ganzen hier entwickelten Theorie zu erzeugen. Sicherlich deutet dieses Resultat auf eine tief liegende Lücke des Fundamentes unserer beiden Untersuchungen hin; denn es dürfte kaum gelingen, aus dem für c für das gleichförmig beschleunigte System gefundenen Ausdruck $c_0 + ax$ eine andere in Betracht zu ziehende Gleichung als Gleichung (3) zu entnehmen, welche ihrerseits die Gleichung (3a) mit Notwendigkeit nach sich zieht.

Um diese Schwierigkeit zu lösen, wird man sich zunächst mit Rücksicht auf die Resultate der alten Relativitätstheorie bewogen fühlen, dem Spannungen unterworfenen Gerüst eine schwere Masse zuzuschreiben, so daß zu den Kräften, die das Gravitationsfeld auf die Massen von der Dichte σ ausübt, Kräfte hinzu kämen, die es auf Spannungen unterworfene Gerüstteile ausübt. Die folgende Betrachtung führt aber zur Verwerfung einer derartigen Hypothese.

In einem statischen Schwerefeld befinde sich ein Kasten mit spiegelnden Wänden, in den Strahlung eingeschlossen sei, deren mit „Tascheninstrumenten" gemessene Energie E sei; d. h. es sei

$$E = \frac{1}{2} \int (\mathfrak{E}^2 + \mathfrak{H}^2)\, d\tau.$$

Ist die Ausdehnung des Kastens klein genug, so ergibt sich aus Gleichung (4) dieser Arbeit, daß die Summe der Kräfte, welche die Strahlung auf die Kastenwände ausübt, den Wert

$$- E \operatorname{grad} c$$

besitzt. Diese Kräftesumme muß gleich sein der Resultierenden der Kräfte, welche das Schwerefeld auf das ganze System (Kasten samt Strahlung) ausübt, wenn der Kasten masselos ist, und wenn der Umstand, daß die Kastenwände infolge des Strahlungsdruckes Spannungen unterworfen sind, nicht zur Folge hat, daß das Schwerefeld auf die Kastenwände wirkt. Wäre letzteres der Fall, so würde die Resultierende der von dem Schwerefeld auf den Kasten (samt Inhalt) ausgeübten Kräfte von dem Werte $-E \operatorname{grad} c$ verschieden sein, d. h. die schwere Masse des Systems wäre von E verschieden.

Befindet sich andererseits unser Strahlungskasten in einem Raum von konstantem c, so gelten für ihn die Resultate der alten Relativitätstheorie. Speziell folgt dann, daß die *träge* Masse des Systems gleich E ist.

Will man also an der Proportionalität von schwerer und träger Masse solcher Gebilde, welche sich als materielle Punkte auffassen lassen, festhalten, so muß man annehmen, daß die *schwere* Masse unseres Systems ebenfalls gleich E sei. Dies ist aber nach obiger Überlegung nur dann der Fall, wenn wir Kräfte des Gravitationsfeldes auf Spannungen unterworfene, masselose Wände *nicht* annehmen.

Eine ganz analoge Betrachtung läßt sich an die in der früheren Arbeit gefundenen Bewegungsgleichungen materieller Punkte anknüpfen. Man betrachte nämlich einen Kasten, in dem materielle Punkte hin- und herfliegen, die an den Wänden vollkommen elastisch abprallen (Modell eines einatomigen Gases). Ganz wie im Falle des Strahlungskastens findet man, daß die schwere und die träge Masse des ganzen Systems nur in dem Falle gleich sind, wenn vom Schwerefeld auf in Spannungszuständen befindliche masselose Gerüste Kräfte nicht ausgeübt werden.

Die in Gleichungen (3a) und (4) enthaltene Verletzung des Reaktionsprinzips bleibt also bestehen. Der Ausdruck (4) für die im Gravitationsfelde auf ruhende Massen wirkende Kraft geht mit Notwendigkeit aus unseren Bewegungsgleichungen für den materiellen Punkt hervor. Es liegt deshalb nahe, an dem Zutreffen dieser Gleichungen zu zweifeln; daß letztere aber schwerlich abzuändern sein dürften, geht aus folgender Überlegung hervor.

Soll die Bewegungsgröße eines materiellen Punktes — wie es die alte Relativitätstheorie fordert — in einem Raume von konstantem c durch $\dfrac{m\dot{x}}{\sqrt{1-v^2/c^2}} dc$ gegeben sein, so darf sich der Ausdruck der Bewegungsgröße im allgemeinen Falle von diesem nur durch einen Faktor unterscheiden, der Funktion von c allein ist.[1]) Dieser Faktor wird aus Dimensionsgründen eine Potenz von c sein müssen (c^α). Die Bewegungsgleichungen müssen also von der Form sein

---

1) Eigentlich müßte man noch zulassen, daß die Bewegungsgröße auch von den räumlichen Ableitungen von c abhängt. Wir wollen aber annehmen, daß dies nicht der Fall sei.

$$\frac{d}{dt}\left\{\frac{m\dot{x}c^{\alpha}}{\sqrt{1-\frac{q^2}{c^2}}}\right\} = \Re_{xs} + \Re_{xa},$$

falls man mit $\Re_{xs}$ die x-Komponente der vom Schwerefelde auf den Punkt ausgeübten Kraft, mit $\Re_{xa}$ die x-Komponente der Resultierenden der Kräfte anderen Ursprunges bezeichnet. Es frägt sich nun, durch was für einen Ausdruck $\Re_s$ gegeben sein kann. Handelt es sich um einen Punkt, für den gerade $q = 0$ ist, so wird die Kraft dem Vektor $-m\,\mathrm{grad}\,c$ proportional sein müssen, wenn man nur annimmt, daß das statische Schwerefeld durch c charakterisiert ist. Diese Kraft wird sich von $-m\,\mathrm{grad}\,c$ nur durch einen Faktor unterscheiden können, der von c allein abhängt; auch dieser Faktor wird aus Dimensionsgründen eine Potenz von c sein müssen (c^β). In dem Falle, daß $q \neq 0$ ist, würde die Kraft auch noch von q abhängen; und zwar muß die Abhängigkeit eine derartige sein, daß die schwere Masse eines bewegte elastische materielle Punkte enthaltenden Kastens von der Geschwindigkeit der Bewegung der Punkte in gleicher Weise abhängt wie die schwere Masse. Dies dürfte sich mit Rücksicht auf die Resultate der alten Relativitätstheorie nur durch den Ansatz

$$\Re_s = \frac{-m\,\mathrm{grad}\,c\cdot c^\beta}{\sqrt{1-\frac{q^2}{c^2}}}\cdot \mathrm{konst.}$$

erzielen lassen. Setzt man $\Re_{xs}$ demgemäß in die Bewegungsgleichungen ein, so kann man beweisen, daß $\Re_{xa}\dot{x} + \Re_{ya}\dot{y} + \Re_{za}\dot{z}$ sich nur dann als Differentialquotient nach der Zeit darstellen läßt, wenn den Konstanten α und β solche Werte gegeben werden, daß die in der früheren Arbeit angegebenen Bewegungsgleichungen resultieren. Man wird also wohl an diesen und an dem aus ihnen resultierenden Ausdruck (4) für die Kraft festhalten müssen, wenn man nicht die ganze Theorie (Bestimmtheit des statischen Gravitationsfeldes durch c) aufgeben will.

Eine Beseitigung des genannten Widerspruches gegen das **Reaktionsprinzip** scheint also nur dadurch möglich zu sein, daß man die Gleichungen (3) und (3a) durch andere in c homogene Gleichungen ersetzt, für welche das Reaktionsprinzip bei Anwendung des Kraftansatzes (4) erfüllt ist. Zu diesem

Schritt entschließe ich mich deshalb schwer, weil ich mit ihm den Boden des unbedingten Äquivalenzprinzips verlasse. Es scheint, daß sich letzteres nur für unendlich kleine Felder aufrecht erhalten läßt. Unsere Ableitungen der Gleichungen der Bewegung des materiellen Punktes und der elektromagnetischen Gleichungen werden dadurch nicht illusorisch, weil sie die Gleichungen (2) nur für unendlich kleine Räume anwenden. Man kann diese Ableitungen z. B. auch an die allgemeineren Gleichungen

$$\xi = x + \frac{c\frac{dc}{dx}}{2} t^2,$$
$$\eta = y,$$
$$\zeta = z,$$
$$\tau = ct$$

anknüpfen, wobei c eine beliebige Funktion von x ist. —

Durch passende Umformung des über einen beliebigen Raum erstreckten Integrales

$$\int \frac{\Delta c}{c} \operatorname{grad} c \, d\tau$$

überzeugt man sich leicht, daß dem Reaktionsprinzip genügt wird, wenn wir unter Beibehaltung von (4) die Gleichung (3a) durch die Gleichung

(3b) $$c \Delta c - \tfrac{1}{2}(\operatorname{grad} c)^2 = k c^2 \sigma,$$

die sich auch in die Form

(3b') $$\Delta(\sqrt{c}) = \frac{k}{2}\sqrt{c}\,\sigma$$

bringen läßt, wobei σ die Dichte der ponderabeln Materie bzw. die Dichte der ponderabeln Materie vermehrt um die mit Tascheninstrumenten gemessene Energiedichte bedeutet. Aus diesen Gleichungen folgt

(5) $$\begin{cases} \mathfrak{F}_x = -\sigma \frac{\partial c}{\partial x} = \frac{\partial X_x}{\partial x} + \frac{\partial X_y}{\partial y} + \frac{\partial X_z}{\partial z} \text{ etc.,} \\ \text{wobei} \\ c k X_x = \frac{\partial c}{\partial x}\frac{\partial c}{\partial x} - \frac{1}{2}(\operatorname{grad} c)^2, \quad c k X_y = \frac{\partial c}{\partial x}\frac{\partial c}{\partial y}, \\ \qquad\qquad\qquad\qquad\qquad\qquad c k X_z = \frac{\partial c}{\partial x}\frac{\partial c}{\partial z} \end{cases}$$

usw. gesetzt ist. Das Reaktionsprinzip ist also in der Tat erfüllt. Das in Gleichung (3b) zur Befriedigung des Reaktionsprinzipes hinzugesetzte Glied gewinnt unser Vertrauen durch die folgenden Überlegungen.

Wenn jegliche Energiedichte (σc) eine (negative) Divergenz der Kraftlinien der Gravitation erzeugt, so muß dies auch für die Energiedichte der Gravitation selbst gelten. Schreibt man (3 b) in der Form

$$\Delta c = k \left\{ c\sigma + \frac{1}{2k} \frac{\operatorname{grad}^2 c}{c} \right\},$$

so erkennt man also sogleich, daß das zweite Glied der Klammer als die Energiedichte des Gravitationsfeldes aufzufassen ist.[1]) Wir haben nur noch zu zeigen, daß auch nach dem Energieprinzip dieses Glied die Dichte der Energie des Gravitationsfeldes bedeutet.

Zu diesem Zweck denken wir uns eine im endlichen befindliche Raumbelegung ponderabler Massen (Dichte σ), welche durch eine unendlich ferne Fläche eingeschlossen sei; im Unendlichen strebe c, soweit es die Gleichung (3 b) bzw. 3 b') zuläßt, einem konstanten Werte zu. Wir haben dann zu beweisen, daß für eine beliebige unendlich kleine Verschiebung der Massen ($\delta x, \delta y, \delta z$) die dem System zuzuführende Arbeit δA gleich sei der Vermehrung δE des über den ganzen Raum erstreckten Integrales der totalen, in der Klammer der obigen Gleichung angegebenen Energiedichte.

Vermöge (4) erhält man zunächst

$$\delta A = \int \sigma \left(\frac{\partial c}{\partial x} \delta x + \frac{\partial c}{\partial y} \delta y + \frac{\partial c}{\partial z} \delta z \right) d\tau$$
$$= -\int c \left(\frac{\partial (\sigma \delta x)}{\partial x} + \ldots d\tau \right) = \int c \, \delta\sigma \, d\tau.$$

Für die Berechnung von δE schicken wir voraus, daß

$$\delta \left\{ \int \frac{\operatorname{grad}^2 c}{c} d\tau \right\} = \delta \left\{ 4 \int \operatorname{grad}^2 \sqrt{c} \, d\tau \right\} = \delta \left\{ 4 \int \operatorname{grad}^2 u \, d\tau \right\}$$
$$= 8 \int \left[\frac{\partial u}{\partial x} \delta \left(\frac{\partial u}{\partial x} \right) + \ldots \right] d\tau = 8 \left\{ \int \delta u \cdot \frac{\partial u}{\partial n} ds - \int \Delta u \, \delta u \, d\tau \right\}.$$

Von diesen Integralen verschwindet das erste (Flächenintegral über die unendlich ferne Fläche), weil mit wachsendem Radiusvektor R die Größen δu und $\partial u/\partial n$ wie $1/R$ bzw. wie $1/R^2$

---

1) Es sei hervorgehoben, daß diese — wie bei Abraham — einen positiven Wert erhält.

zu null herabsinken. Das zweite Integral aber läßt sich vermöge der Feldgleichung (3b′) umformen, so daß man erhält

$$\delta\left\{\int \frac{\operatorname{grad}^2 c}{c}\, d\tau\right\} = -4k\int u\,\delta u\,\sigma\, d\tau = -2k\int \sigma\,\delta c\, d\tau.$$

Unter Benutzung hiervon erhält man:

$$\delta E = \int (c\,\delta\sigma + \sigma\,\delta c - \sigma\,\delta c)\, d\tau = \delta A.$$

Damit ist also bewiesen, daß $\frac{1}{2k}\frac{\operatorname{grad}^2 c}{c}$ tatsächlich als die Energiedichte des Gravitationsfeldes aufzufassen ist.

(Eingegangen 23. März 1912.)

Nachtrag zur Korrektur.

Es ist bemerkenswert, daß die Bewegungsgleichungen des materiellen Punktes im Schwerefeld

$$\frac{d}{dt}\left\{\frac{\frac{\dot x}{c}}{\sqrt{1-\frac{q^2}{c^2}}}\right\} = -\frac{\frac{\partial c}{\partial x}}{\sqrt{1-\frac{q^2}{c^2}}} + \frac{\mathfrak{K}_x}{m}\ \text{usw.}$$

eine sehr einfache Form annehmen, wenn man ihnen die Form der Gleichungen von Lagrange gibt. Setzt man nämlich

$$H = -m\sqrt{c^2 - q^2},$$

so lauten sie

$$\frac{d}{dt}\left(\frac{\partial H}{\partial \dot x}\right) - \frac{\partial H}{\partial x} = \mathfrak{K}_x\ \text{usw.}$$

Für den im statischen Gravitationsfeld ohne Einwirkung äußerer Kräfte bewegten materiellen Punkt gilt demnach

$$\delta\left\{\int H\, dt\right\} = 0,$$

oder

$$\delta\left\{\int \sqrt{c^2\, dt^2 - dx^2 - dy^2 - dz^2}\right\} = 0.$$

Auch hier zeigt sich — wie dies für die gewöhnliche Relativitätstheorie von Planck dargetan wurde —, daß den Gleichungen der analytischen Mechanik eine über die Newtonsche Mechanik weit hinausreichende Bedeutung zukommt. Die zuletzt hingeschriebene Hamiltonsche Gleichung läßt ahnen, wie die Bewegungsgleichungen des materiellen Punktes im dynamischen Gravitationsfelde gebaut sind.

9. Nachtrag zu meiner Arbeit: „Thermodynamische Begründung des photochemischen Äquivalentgesetzes"; von A. Einstein.

In der genannten Arbeit[1]) wird auf wesentlich thermodynamischem Wege unter Zugrundelegung gewisser durch die Erfahrung nahe gelegter Annahmen gezeigt, daß bei der photochemischen Zersetzung eines Gasmoleküls durch (verdünnte) Strahlung von der Frequenz v_0 die Strahlungsenergie hv_0 (im Mittel) absorbiert wird. Jene Untersuchung bedarf in einem wichtigen Punkte der Ergänzung. Es wurde nämlich bei jener Betrachtung die Annahme zugrunde gelegt, daß nur ein unendlich kleiner Frequenzbereich photochemisch auf das Gas zu wirken vermöge. Man erhält deshalb keine Antwort auf die Frage, ob für die Größe der pro Molekülzerfall absorbierten Energie die Frequenz der absorbierten Strahlung oder die Eigenfrequenz des absorbierenden Moleküls maßgebend sei.

Eine Antwort auf jene Frage läßt sich nur gewinnen, wenn man den Fall ins Auge faßt, daß ein endlicher Frequenzbereich auf das Molekül zersetzend zu wirken vermag. Die Untersuchung dieses Falles wird mir auch durch persönliche Mitteilung des Hrn. Warburg nahe gelegt, der den photochemischen Zerfall von Ozon untersucht; Hr. Warburg teilte mir nämlich mit, daß auf das O_3-Molekül Strahlung eines gegen v_0 durchaus nicht verschwindenden Frequenzbereiches photochemisch wirksam ist.

Wir legen also jetzt der Betrachtung den Fall zugrunde, daß auf das betrachtete Molekül beliebig viele elementare Frequenzbereiche wirken, die zusammen einen kontinuierlichen endlichen Bereich bilden können; $v^{(1)}$, $v^{(2)}$ usw. seien die mittleren

---

1) A. Einstein, Ann. d. Phys. **37**. p. 832. 1912.

Frequenzen dieser Elementarbereiche. Wir fügen den in der ersten Arbeit gemachten Voraussetzungen die hinzu, daß die Anzahl der pro Zeiteinheit zerfallenden Moleküle gleich sei die Summe der Anzahl der pro Zeiteinheit zerfallenden Moleküle, welche die Strahlungen der einzelnen Frequenzbereiche für sich allein liefern würden. Dann erhalten wir für die Zahl der in der Zeiteinheit zerfallenden Moleküle erster Art (vgl. Formel (1) p. 834 der ersten Abhandlung)

(1a) $$Z = n_1 (A^{(1)} \varrho^{(1)} + A^{(2)} \varrho^{(2)} \ldots).$$

Gleichung (2) für die Anzahl Z' der pro Zeiteinheit stattfindenden Wiedervereinigungen bleibt unverändert gültig.

Auch in dem jetzt betrachteten Falle gibt es den Fall des „gewöhnlichen" thermodynamischen Gleichgewichtes, für welchen die Strahlung *schwarze* Strahlung von der nämlichen Temperatur ist wie die Temperatur des Gasgemisches. Ebenso ergeben sich bei gegebener Gastemperatur unendlich viele Konstitutionen der Strahlung, für welche „außergewöhnliches" thermodynamisches Gleichgewicht herrschen muß, falls $\eta_2 \eta_3 / \eta_1$ einen geeigneten Wert hat. Aber es ist in dem jetzt untersuchten Falle $Z = Z'$ nicht mehr eine *hinreichende* Bedingung für das thermodynamische Gleichgewicht. Damit letzteres vorhanden sei, muß nämlich außerdem gefordert werden, daß für jedes wirksame Elementargebiet der Strahlungsfrequenz die pro Zeiteinheit absorbierte gleich der pro Zeiteinheit neu erzeugten Strahlungsenergie sei.

Man kann leicht zeigen, daß Fälle des „außergewöhnlichen" thermodynamischen Gleichgewichtes existieren müssen. Bezeichnen wir nämlich mit

$$\eta_{10}, \quad \eta_{20}, \quad \eta_{30},$$
$$\varrho_0^{(1)}, \quad \varrho_0^{(2)} \ldots$$

die Molekularkonzentrationen, bzw. Strahlungsdichten in einem Falle „gewöhnlichen" thermodynamischen Gleichgewichtes, wobei sowohl das Gasgemisch, als auch die wirksame Strahlung der einzelnen Elementarbereiche die Temperatur T besitzen, so sind

$$\frac{\eta_{10}}{x}, \quad \eta_{20}, \quad \eta_{30},$$
$$x \varrho_0^{(1)}, \quad x \varrho_0^{(2)} \ldots$$

Werte für die Molekülkonzentrationen bzw. für die Strahlungsdichten, bei welchen „außergewöhnliches" thermodynamisches Gleichgewicht bei beliebigem Werte von x besteht, falls nur das Gasgemisch die Temperatur T besitzt. Denn es folgt aus (1a) und (2), daß die Bedingung $Z = Z'$ erfüllt bleibt; es ändert sich ferner nichts an der pro Zeiteinheit erzeugten Strahlungsenergie z. B. des ersten Bereiches, weil η_2 und η_3 ungeändert geblieben sind, und es ändert sich auch nichts an der Zeiteinheit z. B. aus der Strahlung des ersten Elementarbereiches absorbierten Energie, weil das Produkt $\eta_1 \cdot \varrho^{(1)}$ ungeändert geblieben ist.

Diese Zustände außergewöhnlichen thermodynamischen Gleichgewichtes, welche zur Gemischtemperatur T gehören, sind dadurch ausgezeichnet, daß sich die Dichten $\varrho^{(1)}$, $\varrho^{(2)}$ usw. der Elementarbereiche zueinander verhalten wie die entsprechenden Dichten $\varrho_0^{(1)}$, $\varrho_0^{(2)}$ usw., welche diesen Bereichen bei der Gemischtemperatur T beim gewöhnlichen thermodynamischen Gleichgewichte zukommen. Ist diese notwendige Bedingung für das außergewöhnliche thermodynamische Gleichgewicht

$$(5) \qquad \frac{\varrho^{(1)}}{\varrho_0^{(1)}} = \frac{\varrho^{(2)}}{\varrho_0^{(2)}} \text{ usw.}$$

erfüllt, so kann man (1a) in folgender Weise umformen:

$$Z = n_1 \left(A^{(1)} \varrho^{(1)} + A^{(2)} \left(\frac{\varrho_0^{(2)}}{\varrho_0^{(1)}} \varrho^{(1)} + \cdots \right) \right.$$
$$= n_1 \left(A^{(1)} + A^{(2)} \frac{\varrho_0^{(2)}}{\varrho_0^{(1)}} + \cdots \right) \varrho^{(1)},$$

oder endlich in kürzerer Form

$$(1\,\mathrm{b}) \qquad Z = A^{(1)*} \varrho^{(1)} n_1,$$

wobei $A^{(1)*}$ nur von T allein (Gemischtemperatur) abhängt.

Unter Benutzung von (1 b) und (2) der ersten Arbeit erhält man statt Gleichnng (3), p. 835 die entsprechende Gleichung

$$(3\,\mathrm{a}) \qquad \frac{\frac{n_2}{V} \frac{n_3}{V}}{\frac{n_1}{V}} = \frac{\eta_2 \eta_3}{\eta_1} = \frac{A^{(1)*}}{A'} \varrho^{(1)}.$$

Ist diese Gleichung sowie (5) erfüllt, so besteht „außergewöhnliches" thermodynamisches Gleichgewicht.

Haben wir einen Fall außergewöhnlichen thermodynamischen Gleichgewichtes vor uns, so werden wir uns eine virtuelle Änderung des Systems als zulässig zu denken haben, bei welcher ein Grammol der ersten Molekülart des Gemisches zersetzt wird unter Absorption der Energie $N\varepsilon^{(1)}$ aus der Strahlung des ersten Elementarbereiches derart, daß die Energiemengen der übrigen Elementarbereiche der Strahlung ungeändert bleiben. Bei dieser virtuellen Änderung muß die Bedingung $\delta S_{\text{total}} = 0$ erfüllt sein wie in dem zuerst betrachteten Fall, daß nur Strahlung eines einzigen Elementarbereiches photochemisch wirksam sei.[1]

Die rechnerische Durchführung stimmt genau überein mit derjenigen, welche in der Arbeit für den monochromatischen Fall gegeben ist, mit dem einzigen Unterschiede, daß die auf die Strahlung sich beziehenden Größen auf den ersten Elementarbereich zu beziehen sind. Speziell erhalten wir an Stelle von (5) die Gleichung

$$(5\,\mathrm{a}) \qquad \varepsilon^{(1)} = h\,\nu^{(1)}.$$

Es folgt also aus den angedeuteten Überlegungen, daß die pro Molekülzerfall absorbierte Energie nicht von der Eigenfrequenz des absorbierenden Moleküls sondern von der Frequenz der den Zerfall bewirkenden Strahlung abhängt. Sollte dies bei (5a) aber nicht zutreffen, so müßte man meiner Meinung nach daraus schließen, daß Absorption bzw. Emission der verschiedenen wirksamen Frequenzbereiche nicht unabhängig voneinander erfolgen, sondern zwangläufig miteinander verbunden sind. Es wäre dann eben die von uns betrachtete virtuelle Verschiebung als eine mit den Elementargesetzen nicht vereinbare anzusehen.

Prag, Mai 1912.

---

[1] Dieser Modus wäre nur dann unzulässig, wenn die elementaren Gesetze der Absorption und Emission so beschaffen wären, daß mit der Absorption bzw. Emission von Strahlung einer Frequenz Absorption bzw. Emission anderer Frequenzen zwangläufig verbunden wäre.

(Eingegangen 12. Mai 1912.)

11. *Antwort auf eine Bemerkung von J. Stark: „Über eine Anwendung des Planckschen Elementargesetzes..."*; *von A. Einstein.*

J. Stark hat zu einer kürzlich von mir publizierten Arbeit[1]) eine Bemerkung verfaßt zum Zwecke der Verteidigung seines geistigen Eigentums.[2]) Auf die aufgeworfene Frage der Priorität gehe ich nicht ein, weil sie kaum jemanden interessieren dürfte, zumal es sich bei dem photochemischen Äquivalentgesetz um eine ganz selbstverständliche Folgerung der Quantenhypothese handelt.[3]) Ich sehe aber aus Starks Bemerkung, daß ich das Ziel meiner Arbeit nicht genügend klar hervorgehoben habe. Es sollte gezeigt werden, daß man zur Ableitung jenes Äquivalentgesetzes nicht der Quantenhypothese bedarf, sondern daß dasselbe aus gewissen einfachen Annahmen über den photochemischen Prozeß auf *thermodynamischem* Wege gefolgert werden kann.

Prag, 30. Mai 1912.

---

1) A. Einstein, Ann. d. Phys. **37**. p. 832. 1912.
2) J. Stark, Ann. d. Phys. **38**. p. 467. 1912.
3) Für den Fall, daß das photochemisch empfindliche Molekül in Ionen gespalten wird, habe ich übrigens das Gesetz bereits in meiner ersten Arbeit über die Quantenhypothese (Ann. d. Phys. **17**. p. 148. 1905) besonders ausgesprochen.

(Eingegangen 30. Mai 1912.)

1059

12. *Relativität und Gravitation. Erwiderung auf eine Bemerkung von M. Abraham;* von *A. Einstein.*

In einer in diesen Annalen erscheinenden Notiz hat M. Abraham auf einige von mir geäußerte kritische Bedenken zu seinen Untersuchungen über Gravitation geantwortet, sowie seinerseits an meinen Arbeiten über diesen Gegenstand Kritik geübt. Ich will im folgenden auf die von ihm berührten Punkte einzeln eingehen und insbesondere meine Ansichten über den gegenwärtigen Stand der Relativitätstheorie den von ihm geäußerten gegenüberstellen.

Abraham bemerkt, ich hätte durch das Aufgeben des Postulates von der Konstanz der Lichtgeschwindigkeit und durch den damit zusammenhängenden Verzicht auf die Invarianz der Gleichungssysteme gegenüber Lorentztransformationen der Relativitätstheorie den Gnadenstoß gegeben. Um hierauf zu antworten, bedarf es einer Überlegung über die Grundlagen der Relativitätstheorie.

Die gegenwärtig als „Relativitätstheorie" bezeichnete Theorie ruht auf zwei Prinzipen, die voneinander durchaus unabhängig sind, nämlich

1. dem Relativitätsprinzip (bezüglich gleichförmiger Translation),

2. dem Prinzip von der Konstanz der Lichtgeschwindigkeit.

Ich will diese beiden Prinzipe genauer formulieren, nicht in der Meinung, etwas Neues dabei vorzubringen, sondern nur, um mich nachher bequemer ausdrücken zu können. Wir stellen zwei Formulierungen des Relativitätsprinzipes einander gegenüber:

1. Beziehen wir die physikalischen Systeme auf ein solches Koordinatensystem K, daß die Naturgesetze möglichst einfach werden, so gibt es unendlich viele Koordinatensysteme, in bezug auf welche jene Gesetze dieselben sind, nämlich alle diejenigen Koordinatensysteme, die sich in gleichförmiger Translationsbewegung relativ zu K befinden.

2. Es sei Σ ein von allen übrigen physikalischen Systemen (im Sinne der geläufigen Sprache der Physik) isoliertes System, und es sei Σ auf ein solches Koordinatensystem K bezogen, daß die Gesetze, welchen die räumlich-zeitlichen Änderungen von Σ gehorchen, möglichst einfache werden; dann gibt es unendlich viele Koordinatensysteme, in bezug auf welche jene Gesetze die gleichen sind, nämlich alle diejenigen Koordinatensysteme, die sich relativ zu K in gleichförmiger Translationsbewegung befinden.

Es ist leicht einzusehen, daß lediglich das Relativitätsprinzip in der Form 2 durch die uns gegebenen Erfahrungen nahe gelegt wird. Es bezeichne nämlich Σ wieder das betrachtete „isolierte" System, U die Gesamtheit aller übrigen Systeme der Welt. Um das Relativitätsprinzip in der Form 1 zu prüfen, müßte man zwei Versuche ausführen, in deren ersten U und Σ relativ zu K in genau denselben Zustand gebracht werden, wie im zweiten Versuche relativ zu K'. Dies ist niemals möglich gewesen und wird nie möglich sein. Um das Prinzip in der Form 2 zu prüfen, hat man dagegen nur Σ allein in verschiedene Zustände zu bringen, ohne sich um U zu kümmern; man hat zwei Versuche auszuführen, in deren ersten Σ allein relativ zu K in denselben Zustand gebracht wird wie in dem zweiten Versuche relativ zu K'.

Die Auseinanderhaltung dieser beiden Formulierungen war bisher überflüssig, da man dem „Restsystem" U keinerlei Einfluß auf die Vorgänge in bezug auf Σ einräumte. Aber meine und Abrahams Überlegungen über die Gravitation lassen eine solche Auffassung nicht zu. Nach diesen Überlegungen hängt der Ablauf der Vorgänge in Σ (z. B. die Lichtgeschwindigkeit) vom Zustande von U (z. B. vom mittleren Abstand der U konstituierenden Einzelsysteme von Σ) ab. Es muß aber daran festgehalten werden, daß das Relativitätsprinzip in der Form 2 durch den Charakter unserer gesamten physikalischen Erfahrung und insbesondere durch den Versuch von Michelson und Morley derart gestützt wird, daß es mächtiger Argumente bedürfte, um einen Zweifel in jenem Prinzip zu begründen. Man kann das Relativitätspostulat in der durch die Erfahrung gestützten Form 2 abgekürzt, aber weniger präzis auch so aussprechen:

Relativität und Gravitation.

„Die Relativgeschwindigkeit des Bezugssystems K gegen das Restsystem U geht in die physikalischen Gesetze nicht ein."

Die im vorigen angedeuteten Überlegungen bringen es nach meiner Ansicht mit sich, daß jede Theorie abzulehnen ist, welche *ein* Bezugssystem gegenüber den relativ zu ihm in gleichförmiger Translation befindlichen Bezugssystemen auszeichnet. Abraham macht sogar den Versuch, ein derartiges ausgezeichnetes Bezugssystem festzulegen mit den Worten: „Wenn unter allen Bezugssystemen dasjenige ausgezeichnet ist, in welchem das Schwerefeld statisch oder quasi-statisch ist, so ist es erlaubt, eine auf dieses System bezogene Bewegung „absolut" zu nennen usw." Dies scheint mir selbst dann nicht richtig zu sein, wenn man jedes Element eines dynamischen Schwerefeldes durch eine Geschwindigkeitstransformation auf ein statisches transformieren könnte. Denn daß eine derartige Transformation gleichzeitig *alle* Elemente eines dynamischen Gravitationsfeldes in dieser Weise transformieren würde, ist ausgeschlossen; es kann also durch eine derartige Festsetzung kein Bezugssystem gegenüber allen relativ zu ihm gleichförmig bewegten ausgezeichnet werden.

Es ist allgemein bekannt, daß auf das Relativitätsprinzip allein eine Theorie der Transformationsgesetze von Raum und Zeit nicht gegründet werden kann. Es hängt dies bekanntlich mit der Relativität der Begriffe „Gleichzeitigkeit" und „Gestalt bewegter Körper" zusammen. Um diese Lücke auszufüllen, führte ich das der H. A. Lorentzschen Theorie des ruhenden Lichtäthers entlehnte Prinzip von der Konstanz der Lichtgeschwindigkeit ein, das ebenso wie das Relativitätsprinzip eine physikalische Voraussetzung enthält, die nur durch die einschlägigen Erfahrungen gerechtfertigt erschien (Versuche von Fizeau, Rowland usw.). Dies Prinzip besagt:

Es existiert ein Bezugssystem K, in dem sich jeder Lichtstrahl im Vakuum mit der universellen Geschwindigkeit c fortpflanzt, unabhängig davon, ob der lichtaussendende Körper relativ zu K ruht oder bewegt ist.

Aus diesen beiden Prinzipien heraus läßt sich diejenige Theorie entwickeln, welche gegenwärtig unter dem Namen „Relativitätstheorie" bekannt ist. Diese Theorie ist in dem Umfange richtig, als die beiden ihr zugrunde gelegten Prinzipe zutreffen. Da diese in weitem Umfange zuzutreffen scheinen, so scheint auch die Relativitätstheorie in ihrer jetzigen Form einen wichtigen Fortschritt zu bedeuten; ich glaube nicht, daß sie die Fortentwickelung der theoretischen Physik gehemmt hat!

Wie steht es nun aber mit der Grenze der Gültigkeit der beiden Prinzipe? An der allgemeinen Gültigkeit des Relativitätsprinzips zu zweifeln, haben wir — wie schon hervorgehoben — nicht den geringsten Grund. Dagegen bin ich der Ansicht, daß das Prinzip der Konstanz der Lichtgeschwindigkeit sich nur insoweit aufrecht erhalten läßt, als man sich auf raum-zeitliche Gebiete von konstantem Gravitationspotential beschränkt. Hier liegt nach meiner Meinung die Grenze der Gültigkeit zwar nicht des Relativitätsprinzips wohl aber des Prinzips der Konstanz der Lichtgeschwindigkeit und damit unserer heutigen Relativitätstheorie. Zu dieser Meinung führen mich die im folgenden angedeuteten Überlegungen.

Eines der wichtigsten Resultate der Relativitätstheorie ist die Erkenntnis, daß jegliche Energie E eine ihr proportionale Trägheit (E/c^2) besitzt. Da nun jede träge Masse zugleich eine schwere Masse ist, soweit unsere Erfahrung reicht, können wir nicht umhin, einer jeden Energie E auch eine schwere Masse E/c^2 zuzuschreiben.[1]) Hieraus folgt sofort, daß die Schwere auf einen bewegten Körper stärker wirkt, als auf denselben Körper, falls dieser ruht.

Wenn sich das Schwerefeld im Sinne unserer heutigen Relativitätstheorie deuten läßt, so kann dies wohl nur auf zwei Arten geschehen. Man kann den Gravitationsvektor entweder als Vierervektor oder als Sechservektor auffassen. Für jeden dieser beiden Fälle ergeben sich Transformationsformeln für den Übergang zu einem gleichförmig bewegten Bezugssystem. Mittels dieser Transformationsformeln und der Transformationsformeln für die ponderomotorischen Kräfte gelingt es dann, für beide Fälle die auf in einem statischen Schwere-

---

1) Hr. Langevin machte mich mündlich darauf aufmerksam, daß man zu einem Widerspruch mit der Erfahrung kommt, wenn man diese Annahme nicht macht. Da nämlich beim radioaktiven Zerfall große Energiemengen abgegeben werden, muß dabei die *träge* Masse der Materie abnehmen. Nähme die schwere Masse nicht proportional ab, so müßte die Schwerebeschleunigung von aus verschiedenen Elementen bestehenden Körpern in demselben Schwerefelde eine nachweisbar verschiedene sein.

feld bewegte materielle Punkte wirkenden Kräfte zu finden. Man kommt hierbei aber zu Ergebnissen, die den genannten Konsequenzen aus dem Satz von der schweren Masse der Energie widerstreiten. Es scheint also, daß der Gravitationsvektor sich in das Schema der heutigen Relativitätstheorie nicht widerspruchsfrei einordnen läßt.

Diese Sachlage bedeutet nach meiner Ansicht aber keineswegs das Scheitern der auf das Relativitätsprinzip gegründeten Methode, ebensowenig als die Entdeckung und richtige Deutung der Brownschen Bewegung dazu führt, die Thermodynamik und Hydromechanik als Irrlehren anzusehen. Die heutige Relativitätstheorie wird nach meiner Ansicht stets ihre Bedeutung behalten als einfachste Theorie für den wichtigen Grenzfall des zeiträumlichen Geschehens bei konstantem Gravitationspotential. Aufgabe der nächsten Zukunft muß es sein, ein relativitätstheoretisches Schema zu schaffen, in welchem die Äquivalenz zwischen träger und schwerer Masse ihren Ausdruck findet. Einen ersten, recht bescheidenen Beitrag zur Erreichung dieses Zieles habe ich in meinen Arbeiten über das statische Gravitationsfeld zu geben gesucht. Dabei ging ich von der nächstliegenden Auffassung aus, daß die Äquivalenz von träger und schwerer Masse dadurch auf einer Wesensgleichheit dieser beiden elementaren Qualitäten der Materie bzw. der Energie zurückzuführen sei, daß das statische Gravitationsfeld als physikalisch wesensgleich mit einer Beschleunigung des Bezugssystems aufgefaßt wird. Es ist zuzugestehen, daß ich diese Auffassung nur für unendlich kleine Räume widerspruchsfrei durchführen konnte, und daß ich hierfür keinen befriedigenden Grund anzugeben weiß. Aber ich sehe hierin keinen Grund, jenes Äquivalenzprinzip auch für das unendlich Kleine abzuweisen; niemand wird leugnen können, daß dies Prinzip eine natürliche Extrapolation einer der allgemeinsten Erfahrungssätze der Physik ist. Andererseits eröffnet uns dies Äquivalenzprinzip die interessante Perspektive, daß die Gleichungen einer auch die Gravitation umfassenden Relativitätstheorie auch bezüglich Beschleunigungs- (und Drehungs-) Transformationen invariant sein dürften. Allerdings scheint der Weg zu diesem Ziele ein recht schwieriger zu sein. Man sieht schon aus dem bisher behandelten, höchst speziellen

Falle der Gravitation ruhender Massen, daß die Raum—Zeit-Koordinaten ihre einfache physikalische Deutung einbüßen werden, und es ist noch nicht abzusehen, welche Form die allgemeinen raumzeitlichen Transformationsgleichungen haben könnten. Ich möchte alle Fachgenossen bitten, sich an diesem wichtigen Problem zu versuchen!

Nun noch einige Bemerkungen zu Abrahams Notiz. In seiner Erwiderung sagt Hr. Abraham über seine Theorie: „Es kann von irgend einer Art von Relativität, d. h. von einer Korrespondenz der beiden Systeme, die sich in Gleichungen zwischen ihren Raum-Zeit-Parametern x, y, z, t und x', y', z', t' ausdrücken würde, keine Rede sein." Ich will mir kein Urteil darüber anmaßen, ob dies Abrahams ursprüngliche Annahme war oder nicht. Jedenfalls verliert beim Aufgeben des Relativitätsprinzips das von Abraham in seiner Theorie als Richtschnur benutzte relativitätstheoretische Schema jegliche überzeugende Kraft. Abraham macht mich ferner darauf aufmerksam, daß er bereits in seiner Arbeit[1]) den Ausdruck

$$\frac{mc}{\sqrt{1-\frac{q^2}{c^2}}}$$

für die Energie des materiellen Punktes im Schwerefeld angegeben hat; ich hatte dies leider übersehen. Allerdings ist dies Resultat mit den Grundgleichungen von Abrahams Theorie im Widerspruch. Es folgt nämlich aus diesem Ausdruck für die Energie, daß die auf einen im Schwerfeld ruhenden materiellen Punkt wirkende Kraft $-m\,\mathrm{grad}\,c$ sei; dem widersprechend folgt aber für dieselbe Größe aus den Gleichungen (2) und (6) von Abrahams Arbeit der Ausdruck $-mc\,\mathrm{grad}\,c$. Abraham behauptet ferner, ich hätte seine Ausdrücke für die Energiedichte und für die Spannungen im Schwerefeld benutzt. Dies trifft nicht zu; nach Abraham ist beispielsweise die Energiedichte im statischen Schwerefeld $\frac{c^2}{\gamma}\mathrm{grad}^2 c$, nach meiner Theorie $\frac{1}{2k}\frac{\mathrm{grad}^2 c}{c}$. Das Eingehen von c ist in beiden Theorien verschieden.

---

1) M. Abraham, Physik. Zeitschr. **13**. Nr. 19. p. 2. 1912.

(Eingegangen 4. Juli 1912.)

ANNALEN DER PHYSIK.

BEGRÜNDET UND FORTGEFÜHRT DURCH

F. A. C. GREN, L. W. GILBERT, J. C. POGGENDORFF, G. u. E. WIEDEMANN, P. DRUDE

VIERTE FOLGE.

BAND 39.

DER GANZEN REIHE 344. BAND.

KURATORIUM:

M. PLANCK, G. QUINCKE,
W. C. RÖNTGEN, W. VOIGT, E. WARBURG.

UNTER MITWIRKUNG

DER DEUTSCHEN PHYSIKALISCHEN GESELLSCHAFT

HERAUSGEGEBEN VON

W. WIEN UND **M. PLANCK**.

MIT ZEHN FIGURENTAFELN.

LEIPZIG, 1912.
VERLAG VON JOHANN AMBROSIUS BARTH.

10. Bemerkung zu Abrahams vorangehender Auseinandersetzung „Nochmals Relativität und Gravitation"; von A. Einstein.

Da jeder von uns beiden seinen Standpunkt mit der nötigen Ausführlichkeit vertreten hat, halte ich es nicht für nötig, auf Abrahams vorliegende Notiz wieder zu antworten. Ich möchte hier einstweilen den Leser nur darum ersuchen, mein Schweigen nicht als Einverständnis zu deuten.

Zürich, August 1912.

(Eingegangen 2. September 1912.)

ANNALEN DER PHYSIK.

BEGRÜNDET UND FORTGEFÜHRT DURCH

F. A. C. GREN, L. W. GILBERT, J. C. POGGENDORFF, G. u. E. WIEDEMANN, P. DRUDE.

VIERTE FOLGE.

BAND 40.

DER GANZEN REIHE 345. BAND.

KURATORIUM:

M. PLANCK, G. QUINCKE,
W. C. RÖNTGEN, W. VOIGT, E. WARBURG.

UNTER MITWIRKUNG

DER DEUTSCHEN PHYSIKALISCHEN GESELLSCHAFT

HERAUSGEGEBEN VON

W. WIEN UND **M. PLANCK.**

MIT EINER FIGURENTAFEL.

LEIPZIG, 1913.

VERLAG VON JOHANN AMBROSIUS BARTH.

8. *Einige Argumente für die Annahme einer molekularen Agitation beim absoluten Nullpunkt;* von *A. Einstein und O. Stern.*

Der Ausdruck für die Energie eines Resonators lautet nach der ersten Planckschen Formel:

$$(1) \qquad E = \frac{h\nu}{e^{\frac{h\nu}{kT}} - 1},$$

nach der zweiten:

$$(2) \qquad E = \frac{h\nu}{e^{\frac{h\nu}{kT}} - 1} + \frac{h\nu}{2}.$$

Der Grenzwert für hohe Temperaturen wird, wenn wir die Entwickelung von $e^{\frac{h\nu}{kT}}$ mit dem quadratischen Gliede abbrechen, für (1):

$$\lim_{T=\infty} E = kT - \frac{h\nu}{2},$$

für (2):

$$\lim_{T=\infty} E = kT.$$

Die Energie als Funktion der Temperatur, wie sie in Fig. 1 dargestellt ist, beginnt also nach Formel (1) für $T = 0$ mit Null, dem von der klassischen Theorie geforderten Werte, bleibt aber bei hohen Temperaturen ständig um das Stück $h\nu/2$ kleiner als dieser. Nach Formel (2) hat der Resonator beim absoluten Nullpunkt die Energie $h\nu/2$, im Widerspruch zur klassischen Theorie, erreicht aber bei hohen Temperaturen asymptotisch die von dieser geforderte Energie. Dagegen ist der Differentialquotient der Energie nach der Temperatur, d. h. die spezifische Wärme, in beiden Fällen gleich.

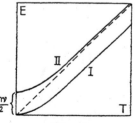

Fig. 1.

Für Gebilde mit unveränderlichem v sind diese Formeln also gleichwertig, während die Theorie solcher Gebilde, deren v für verschiedene Zustände verschiedene Werte hat, durch die Annahme einer Nullpunktsenergie wesentlich beeinflußt wird. Der ideale Fall wäre der eines aus monochromatischen Gebilden bestehenden Systems, dessen v-Wert unabhängig von der Temperatur willkürlich geändert werden kann. Die Abhängigkeit der Energie von der Frequenz bei konstanter Temperatur würde wesentlich von der Existenz einer Nullpunktsenergie abhängen. Leider liegen Erfahrungen über ein derartiges Gebilde nicht vor. Wohl aber kennen wir in den rotierenden Gasmolekülen Gebilde, deren thermische Bewegungen mit denen monochromatischer Gebilde eine weitgehende Ähnlichkeit aufweisen[1]), und bei welchen die mittlere Frequenz mit der Temperatur veränderlich ist. An diesen Gebilden ist also die Berechtigung der Annahme einer Nullpunktsenergie in erster Linie zu prüfen. Im folgenden soll zunächst untersucht werden, inwiefern wir aus der Planckschen Formel auf das theoretische Verhalten solcher Gebilde Rückschlüsse ziehen können.

Die spezifische Wärme des Wasserstoffs bei tiefen Temperaturen.

Es handelt sich um die Frage, wie die Energie der Rotation eines zweiatomigen Moleküls von der Temperatur abhängt. Analog wie bei der Theorie der spezifischen Wärme fester Stoffe sind wir zu der Annahme berechtigt, daß die mittlere kinetische Energie der Rotation davon unabhängig ist, ob das Molekül in Richtung seiner Symmetrieachse ein elektrisches Moment besitzt oder nicht. Im Falle, daß das Molekül ein solches Moment besitzt, darf es das thermodynamische Gleichgewicht zwischen Gasmolekülen und Strahlung nicht stören. Hieraus kann man schließen, daß das Molekül unter der Einwirkung der Strahlung allein dieselbe kinetische Energie der Rotation annehmen muß, die es durch die Zusammenstöße mit anderen Molekülen erhalten würde. Die Frage ist also, bei welchem

---

[1]) Hierauf hat zuerst Nernst aufmerksam gemacht, vgl. Zeitschr. f. Elektroch. **17**. p. 270 u. 825. 1911.

Mittelwerte der Rotationsenergie sich ein träger, starrer Dipol mit Strahlung von bestimmter Temperatur im Gleichgewicht befindet. Wie die Gesetze der Ausstrahlung auch sein mögen, so wird doch wohl daran festzuhalten sein, daß ein rotierender Dipol doppelt so viel Energie pro Zeiteinheit ausstrahlt als ein eindimensionaler Resonator, bei dem die Amplitude des elektrischen und mechanischen Moments gleich dem elektrischen und mechanischen Moment des Dipols ist. Analoges wird auch von dem Mittelwert der absorbierten Energie gelten. Machen wir nun noch die vereinfachende Näherungsannahme, daß bei gegebener Temperatur alle Dipole unseres Gases gleich rasch rotieren, so werden wir zu dem Schluß geführt, daß im Gleichgewicht die kinetische Energie eines Dipols doppelt so groß sein muß, wie die eines eindimensionalen Resonators von gleicher Frequenz. Bei den gemachten Annahmen können wir die Ausdrücke (1) bzw. (2) direkt zur Berechnung der kinetischen Energie eines mit zwei Freiheitsgraden rotierenden Gasmoleküls anwenden, wobei bei jeder Temperatur zwischen E und ν die Gleichung

$$E = \frac{J}{2}(2\pi\nu)^2$$

besteht (J Trägheitsmoment des Moleküls).

So ergibt sich für die Energie der Rotation pro Mol:

(3) $$E = N_0 \cdot \frac{J}{2}(2\pi\nu)^2 = N_0 \frac{h\nu}{e^{\frac{h\nu}{kT}} - 1}$$

bzw.

(4) $$E = N_0 \cdot \frac{J}{2}(2\pi\nu)^2 = N_0 \left(\frac{h\nu}{e^{\frac{h\nu}{kT}} - 1} + \frac{h\nu}{2} \right).$$

Da nun ν und T durch eine transzendente Gleichung verknüpft sind, ist es nicht möglich, dE/dT als explizite Funktion von T auszudrücken, sondern man erhält, falls man zur Abkürzung $2\pi^2 J = p$ setzt, als Formel für die spezifische Wärme der Rotation:

(5) $$c_r = \frac{dE}{dT} = \frac{dE}{d\nu} \cdot \frac{d\nu}{dT} = N_0 \, 2 p \nu \frac{\nu}{T\left(1 + \frac{kT}{p\nu^2 + h\nu}\right)}$$

bzw.

(6) $$c_r = \frac{dE}{dT} = \frac{dE}{d\nu} \cdot \frac{d\nu}{dT} = N_0\, 2\, p\, \nu \cdot \frac{\nu}{T\left(1 + \dfrac{kT}{p\nu^2 - \dfrac{h^2}{4p}}\right)},$$

wobei ν und T durch die Gleichung:

(5a) $$T = \frac{h}{k} \frac{\nu}{\ln\left(\dfrac{h}{p\nu} + 1\right)}$$

bzw.

(6a) $$T = \frac{h}{k} \frac{\nu}{\ln\left(\dfrac{h}{p\nu - \dfrac{h}{2}} + 1\right)}$$

verbunden sind. In Fig. 2 stellt die Kurve I die auf Grund von (6) und (6a) berechnete spezifische Wärme dar, wobei p

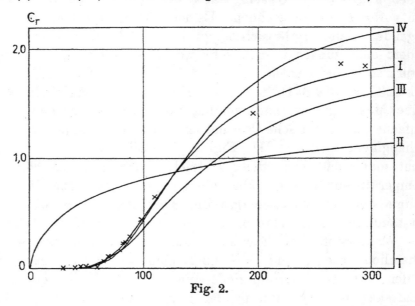

Fig. 2.

den Wert $2{,}90 \cdot 10^{-40}$ hat;[1]) Kurve II ist aus (5) und (5a) mit Hilfe von $p = 2 \cdot 10^{-40}$ berechnet. Die Kreuzchen bezeichnen die von Eucken[2]) gemessenen Werte. Wie man sieht, zeigt die Kurve II einen Verlauf, der mit den Versuchen in völligem

---

1) Berechnet man den zu diesem Trägheitsmoment gehörigen Moleküldurchmesser, so ergibt er sich zu $9 \cdot 10^{-9}$, etwa halb so groß, als der gastheoretisch ermittelte Wert.

2) Eucken, Sitzungsber. d. preuß. Akad. p. 141. 1912.

Widerspruch steht, während Kurve I, die auf der Annahme einer Nullpunktsenergie basiert, die Resultate der Messungen in vorzüglicher Weise widergibt. Um festzustellen, welchen Wert nach Formel (4) ν für die Grenze $T = 0$ annimmt, schreiben wir (4) in folgender Form:

$$e^{\frac{h\nu}{kT}} = \frac{h}{p\nu - \frac{h}{2}} - 1 = \frac{p\nu + \frac{h}{2}}{p\nu - \frac{h}{2}}.$$

Dann sieht man, daß für $T = 0$ ν nicht gleich Null werden kann, da die rechte Seite dann gegen -1 konvergieren würde, während auf der linken eine Potenz von e steht. Es muß also für $\lim T = 0$ ν endlich bleiben, und zwar muß die rechte Seite ebenso wie die linke gegen ∞ konvergieren, es muß daher $p\nu_0 - h/2 = 0$ sein, falls wir mit ν_0 den Grenzwert von ν für $T = 0$ bezeichnen. Es ist also $\nu_0 = h/2p$. Im vorliegenden Falle ergibt sich ν_0 zu $11,3 \cdot 10^{12}$. Der Wert von ν ändert sich zunächst auch sehr wenig mit steigender Temperatur; so ist bei 102^0 abs. $\nu = 11,4 \cdot 10^{12}$, bei 189^0 $\nu = 12,3 \cdot 10^{12}$, bei 323^0 $\nu = 14,3 \cdot 10^{12}$. Dies erklärt nun, weshalb Eucken seine Messungen verhältnismäßig noch am besten durch die einfache Einsteinsche Formel mit von der Temperatur unabhängigem ν (Kurve III, Fig. 2) darstellen konnte. Jedoch sieht man, daß auch diese Formel, namentlich bei höheren Temperaturen, versagt, abgesehen davon, daß ohne die Annahme der Nullpunktsenergie die Konstanz von ν völlig unverständlich bleibt. Man sieht also, daß die spezifische Wärme des Wasserstoffs die Existenz einer Nullpunktsenergie wahrscheinlich macht, und es handelt sich nur noch darum, zu prüfen, wie weit der spezielle Wert von $h\nu/2$ als gesichert anzusehen ist. Da nun in der folgenden Untersuchung über das Strahlungsgesetz der Betrag der Nullpunktsenergie zu $h\nu$ angenommen werden muß, haben wir die spezifische Wärme des Wasserstoffs auch für diese Annahme berechnet ($p = 5,60 \cdot 10^{-40}$, Kurve IV, Fig. 2). Es ist ersichtlich, daß die Kurve bei höheren Temperaturen zu steil und zu hoch ist. Andererseits ist zu bemerken, daß bei Berücksichtigung der Geschwindigkeitsverteilung unter den Molekülen die Kurve jedenfalls etwas flacher ausfallen dürfte. Es ist demnach zwar unwahrscheinlich, aber nicht mit Sicherheit auszuschließen, daß die Nullpunktsenergie den Wert $h\nu$ besitzt.[1])

Die Ableitung des Strahlungsgesetzes.

Im folgenden soll gezeigt werden, wie sich auf Grund der Annahme einer Nullpunktsenergie die Plancksche Strahlungsformel in ungezwungener, wenn auch nicht ganz strenger Weise ableiten läßt, und zwar ohne jede Annahme über irgendwelche Diskontinuitäten. Der Weg, den wir hierzu einschlagen, ist im wesentlichen derselbe, den Einstein und Hopf[2]) in einer vor 2 Jahren erschienenen Abhandlung benutzten. Wir betrachten die fortschreitende Bewegung eines freibeweglichen Resonators, der etwa an einem Gasmolekül festsitzt, unter dem Einflusse eines ungeordneten Strahlungsfeldes. Im thermischen Gleichgewicht muß dann die mittlere kinetische Energie, die das Gasmolekül durch die Strahlung erhält, gleich derjenigen sein, die es durch Zusammenstöße mit anderen Molekülen bekommen würde. Man erhält so den Zusammenhang zwischen der Dichte der schwarzen Strahlung und der mittleren kinetischen Energie einer Gasmolekel, d. h. der Temperatur. Einstein und Hopf finden auf diese Weise das Rayleigh-Jeanssche Gesetz. Wir wollen nun dieselbe Be-

---

[1]) Nimmt man die Entropie rotierender Gebilde gleich der fester Stoffe nach dem Nernstschen Theorem für $T = 0$ zu Null an, so ergibt sich der gesamte von der Rotation der zweiatomigen Moleküle herrührende Anteil der Entropie eines Mols zu

$$S_r = \int_0^T \frac{c_r}{T} dT = \int_{\nu_0}^{\nu} \ln \frac{\nu + \nu_0}{\nu - \nu_0} d\nu = \frac{2 p \nu^2}{T} + k \ln\left[\left(\frac{p\nu}{h}\right)^2 - 1\right].$$

Für hohe Temperaturen wird:

$$S_r = R \ln T + 2R + R \ln \frac{2\pi^2 J k}{h^2}.$$

Nach Sackur (Nernst-Festschrift p. 414. 1912) ist die Entropiekonstante der Rotation:

$$R + R \ln \frac{16 \pi^3 J k}{h^2},$$

in der Hauptsache, nämlich dem Ausdruck Jk/h^2, mit dem obigen Ausdruck übereinstimmend. Dasselbe Resultat erhält man übrigens, wenn man für c_r nicht Formel (5), sondern Formel (6) einsetzt.

[2]) A. Einstein u. L. Hopf, Ann. d. Phys. **33**. p. 1105—1115. 1910.

trachtung unter der Annahme einer Nullpunktsenergie durchführen. Der Einfluß, den die Strahlung ausübt, läßt sich nach Einstein und Hopf in zwei verschiedene Wirkungen zerlegen. Erstens einmal erleidet die geradlinig fortschreitende Bewegung des Resonatormoleküls eine Art Reibung, veranlaßt durch den Strahlungsdruck auf den bewegten Oszillator. Diese Kraft K ist proportional der Geschwindigkeit v, also $K = -Pv$, wenigstens falls v klein gegen die Lichtgeschwindigkeit ist. Der Impuls, den das Resonatormolekül in der kleinen Zeit τ, während deren sich v nicht merklich ändern soll, erhält, ist also $-Pv\tau$. Zweitens erteilt die Strahlung dem Resonatormolekül Impulsschwankungen $\varDelta$, die von der Bewegung des Moleküls in erster Annäherung unabhängig und für alle Richtungen gleich sind, so daß nur ihr quadratischer Mittelwert $\overline{\varDelta^2}$ während der Zeit τ für die kinetische Energie maßgebend ist. Soll nun diese den von der statistischen Mechanik geforderten Wert $k(T/2)$ besitzen (der Oszillator soll der Einfachheit halber nur in der x-Richtung beweglich sein und nur in der z-Richtung schwingen), so muß nach Einstein und Hopf (l. c. p. 1107) folgende Gleichung gelten:

$$\overline{\varDelta^2} = 2kTP\tau.$$

Was nun die Berechnung von P anlangt, so können wir annehmen, daß hierfür nur die von der Strahlung selbst angeregten Schwingungen in Betracht kommen, und daß man diese so berechnen kann, als ob die Nullpunktsenergie nicht vorhanden wäre. Wir können also den von Einstein und Hopf berechneten Wert (l. c. p. 1111):

$$P = \frac{3c\sigma}{10\pi\nu}\left(\varrho - \frac{\nu}{3}\frac{d\varrho}{d\nu}\right)$$

benutzen.

Um nun $\overline{\varDelta^2}$ zu berechnen, setzen wir (l. c. p. 1111) den Impuls, welchen der Oszillator während der Zeit τ in der x-Richtung erfährt:

$$J = \int_0^\tau k_x\, dt = \int_0^\tau \frac{\partial E_z}{\partial x} f\, dt,$$

wobei f das Moment des Oszillators ist. Wir wollen zunächst nur den Fall betrachten, daß die Energie der durch die Strahlung angeregten Schwingung zu vernachlässigen ist gegen die Nullpunktsenergie des Resonators, was bei genügend tiefen Temperaturen sicher erlaubt ist. Bezeichnen wir mit f_0 das maximale Moment des Resonators, so ist:

$$f = f_0 \cos \frac{2 \pi n_0 t}{T},$$

wobei T eine große Zeit und $n_0/T = \nu_0$ die Frequenz des Resonators ist. $\partial \mathfrak{E}_z / \partial x$ setzen wir als Fouriersche Reihe an:

$$\frac{\partial \mathfrak{E}_z}{\partial x} = \sum C_n \cos\left(2 \pi n \frac{t}{T} - \vartheta_n\right).$$

Dann wird:

$$J = \int_0^\tau \sum C_n \cos\left(2 \pi n \frac{t}{T} - \vartheta_n\right) f_0 \cos\left(2 \pi n_0 \frac{t}{T}\right) dt$$

$$= f_0 \sum C_n \frac{T}{2 \pi (n_0 - n)} \sin\left(\pi \frac{n_0 - n}{T} \tau\right) \cdot \cos\left(\pi \frac{n_0 - n}{T} \tau - \vartheta_n\right),$$

da das mit $1/n_0 + n$ behaftete Glied wegfällt, weil $n_0 + n$ eine sehr große Zahl ist. Setzt man nun $n/T = \nu$ und quadriert, so wird:

$$\overline{J^2} = \overline{\varDelta^2} = f_0^2 \overline{C_n^2} \frac{T}{8} \int_{-\infty}^{+\infty} \frac{\sin^2 \pi (\nu_0 - \nu) \tau}{[\pi (\nu_0 - \nu)]^2} d\nu,$$

oder:

$$\overline{\varDelta^2} = \frac{1}{8} f_0^2 \cdot \overline{C_n^2} T \cdot \tau.$$

Nun ist (l. c. p. 1114):

$$\overline{C_n^2} T = \frac{64}{15} \frac{\pi^3 \nu^2}{c^2} \varrho.$$

Also ist:

$$\overline{\varDelta^2} = \frac{8}{15} \frac{\pi^3 \nu^2}{c^2} \varrho \tau \cdot f_0^2.$$

Besitzt nun der Resonator die Nullpunktsenergie $h \nu$[1]), so ist:

$$\tfrac{1}{2} K f_0^2 = h \nu \text{[2])} \quad \text{oder} \quad f_0^2 = \frac{2 h \nu}{K} = \frac{3}{8} \frac{h \sigma c^3}{\pi^4 \nu^2}. \text{[2])}$$

---

1) Es hat sich gezeigt, daß bei der hier skizzierten Rechnungsweise die Nullpunktsenergie gleich $h \nu$ gesetzt werden muß, um zur Planckschen Strahlungsformel zu gelangen. Spätere Untersuchungen müssen zeigen, ob die Diskrepanz zwischen dieser Annahme und der bei der Untersuchung über den Wasserstoff zugrunde gelegten Annahme bei strengerer Rechnung verschwindet.

2) M. Planck, Wärmestrahlung 6. Aufl. p. 112 (Gleichung (168)).

Mithin ist:
$$\overline{\Delta^2} = \frac{1}{5\pi} h c \sigma \varrho \tau.$$

Setzt man dies in die Gleichung
$$\overline{\Delta^2} = 2 k T P \tau$$

ein, so gelangt man zum Wienschen Strahlungsgesetz. Wir wollen hier jedoch gleich die Voraussetzung, daß die durch die Strahlung angeregte Schwingung des Resonators zu vernachlässigen sei, aufgeben. Nehmen wir nun an, daß die Energie der dem Resonator von der Strahlung erteilten Schwingungen Impulsschwankungen liefert, die von den der Nullpunktsenergie entsprechenden Schwankungen unabhängig sind, so können wir den quadratischen Mittelwert beider Impulsschwankungen addieren.[1]) Wir haben also zu dem oben berechneten Wert für $\overline{\Delta^2}$ noch den von Einstein und Hopf (l. c. p. 1114, Gleichung (15)) hinzuzufügen und erhalten:

$$\overline{\Delta^2} = \frac{1}{5\pi} h c \sigma \varrho \tau + \frac{c^4 \sigma \tau}{40 \pi^2 \nu^3} \varrho^2.$$

Andererseits ist:
$$\overline{\Delta^2} = 2 k T P \tau = 2 k T \tau \cdot \frac{3 c \sigma}{10 \pi \nu} \left(\varrho - \frac{\nu}{3} \frac{d\varrho}{d\nu} \right).$$

Es ergibt sich demnach als Differentialgleichung für ϱ:
$$h \varrho + \frac{c^3}{8 \pi \nu^3} \varrho^2 = 3 k T \left(\varrho - \frac{\nu}{3} \frac{d\varrho}{d\nu} \right).$$

Die Auflösung dieser Gleichung liefert:
$$\varrho = \frac{8 \pi \nu^2}{c^3} \frac{h\nu}{e^{\frac{h\nu}{kT}} - 1},$$

das Plancksche Strahlungsgesetz, und die Energie des Resonators ergibt sich zu:
$$E = \frac{h\nu}{e^{\frac{h\nu}{kT}} - 1} + h\nu.$$

---

1) Es braucht kaum betont zu werden, daß diese Art des Vorgehens sich nur durch unsere Unkenntnis der tatsächlichen Resonatorgesetze rechtfertigen läßt.

Zusammenfassung.

1. Die Euckensche Resultate über die spezifische Wärme des Wasserstoffs machen die Existenz einer Nullpunktsenergie vom Betrage $h\nu/2$ wahrscheinlich.

2. Die Annahme der Nullpunktsenergie eröffnet einen Weg, die Plancksche Strahlungsformel ohne Zuhilfenahme irgendwelcher Diskontinuitäten abzuleiten. Es erscheint jedoch zweifelhaft, ob auch die anderen Schwierigkeiten sich ohne die Annahme von Quanten werden bewältigen lassen.

Zürich, Dezember 1912.

(Eingegangen 5. Januar 1913.)

Anmerkung bei der Korrektur:

Hr. Prof. Weiß machte uns darauf aufmerksam, daß auch die Curieschen Messungen über den Paramagnetismus des gasförmigen Sauerstoffs darauf hinweisen, daß dessen Rotationsenergie bei hohen Temperaturen den von der klassischen Theorie geforderten Wert und nicht einen um $h\nu/2$ kleineren besitzt, wie dies ohne die Annahme einer Nullpunktsenergie zu erwarten sein würde. Es läßt sich leicht zeigen, daß in letzterem Falle bei der Genauigkeit der Curieschen Messungen sich Abweichungen vom Curieschen Gesetz hätten zeigen müssen.

ANNALEN
DER
PHYSIK.

BEGRÜNDET UND FORTGEFÜHRT DURCH

F. A. C. GREN, L. W. GILBERT, J. C. POGGENDORFF, G. u. E. WIEDEMANN, P. DRUDE.

VIERTE FOLGE.

BAND 44.

DER GANZEN REIHE 349. BAND.

KURATORIUM:

M. PLANCK, G. QUINCKE,
W. C. RÖNTGEN, W. VOIGT, E. WARBURG.

UNTER MITWIRKUNG

DER DEUTSCHEN PHYSIKALISCHEN GESELLSCHAFT

HERAUSGEGEBEN VON

W. WIEN UND M. PLANCK.

MIT ZWEIUNDZWANZIG FIGURENTAFELN.

LEIPZIG, 1914.

VERLAG VON JOHANN AMBROSIUS BARTH.

8. *Die Nordströmsche Gravitationstheorie vom Standpunkt des absoluten Differentialkalküls;* von *A. Einstein und A. D. Fokker.*

Bei allen bisherigen Darstellungen der Nordströmschen Theorie der Gravitation[1]) wurde als invarianten-theoretisches Hilfsmittel lediglich die Minkowskische Kovariantentheorie benutzt, d. h. es wurde von den Gleichungen der Theorie lediglich verlangt, daß sie linearen orthogonalen Raum-Zeittransformationen gegenüber kovariant sein sollten. Diese den Gleichungen a priori auferlegte Bedingung schränkt aber die theoretischen Möglichkeiten nicht in dem Maße ein, daß man ohne Zuhilfenahme spezieller physikalischer Voraussetzungen. zwanglos zu den Grundgleichungen der Theorie gelangen kann. Im folgenden soll dargetan werden, daß man zu einer in formaler Hinsicht vollkommen geschlossenen und befriedigenden Darstellung der Theorie gelangen kann, wenn man, wie dies bei der Einstein-Großmannschen Theorie bereits geschehen ist, das invarianten-theoretische Hilfsmittel benutzt, welches uns in dem absoluten Differentialkalkül gegeben ist. Da in der Natur Bezugssysteme, auf die wir die Dinge beziehen können, sich uns nicht darbieten, beziehen wir die vierdimensionale Mannigfaltigkeit zunächst auf ganz beliebige Koordinaten (entsprechend den Gaussschen Koordinaten in der Flächentheorie), und beschränken die Wahl des Bezugssystems erst dann, wenn uns das behandelte Problem selbst Veranlassung hierzu bietet.

Es erweist sich hierbei, daß man zur Nordströmschen Theorie statt zur Einstein-Großmannschen gelangt, wenn man die einzige Annahme macht, es sei eine Wahl bevorzugter Bezugssysteme in solcher Weise möglich, daß das Prinzip von der Konstanz der Lichtgeschwindigkeit gewahrt ist.

---

1) Vgl. G. Nordström, Ann. d. Phys. **42**. p. 533. 1913; A. Einstein, Phys. Zeitschr. **14.** p. 1251. 1913.

§ 1. *Charakteristik des Gravitationsfeldes. Einfluß des Gravitationsfeldes auf physikalische Vorgänge.*

Wir nehmen an[1]), daß für einen sich in einem Gravitationsfelde bewegenden Punkt ein Bewegungsgesetz gelte, das in Hamiltonscher Form lautet:

$$\delta \int ds = 0, \tag{1}$$

worin

$$ds^2 = \sum_{\mu\nu} g_{\mu\nu}\, dx_\mu\, dx_\nu. \tag{2}$$

Das Gravitationsfeld wird dann charakterisiert durch die zehn Raum-Zeitfunktionen $g_{\mu\nu}$. ds ist eine Invariante bezüglich beliebiger Substitutionen, welche in der auf dem absoluten Differentialkalkül begründeten allgemeinen Relativitätstheorie dieselbe Rolle spielt wie das Euklidische Linienelement in der Minkowskischen Invariantentheorie. Als der einzige Skalar, der sich auf zwei benachbarte Raum-Zeitpunkte bezieht, hat ds die Bedeutung des „natürlich gemessenen" Abstandes dieser zwei Raum-Zeitpunkte.

Da jeder vektoranalytischen Größe, bzw. jeder vektoranalytischen Operation in der Euklidischen Mannigfaltigkeit eine allgemeinere vektoranalytische Größe bzw. Operation in der durch ein beliebiges Linienelement gegebenen Mannigfaltigkeit entspricht, lassen sich den Gesetzen der ursprünglichen Relativitätstheorie für die physikalischen Erscheinungen entsprechende Gesetze der verallgemeinerten Relativitätstheorie zuordnen. Die so erhaltenen Gesetze, welche allgemein kovariant sind, enthalten den Einfluß des Gravitationsfeldes auf die physikalischen Vorgänge.

Von allen jenen die physikalischen Vorgänge beschreibenden Gesetzen geben wir hier nur ein einziges an, von allgemeinster Bedeutung: nämlich dasjenige, das dem Erhaltungssatz des Impulses und der Energie in der ursprünglichen Theorie der Relativität entspricht. In jener Theorie wurden die energetischen Eigenschaften der Vorgänge ausgedrückt durch einen Spannungs-Energietensor $(T_{\mu\nu})$. Diesen Größen $T_{\mu\nu}$ entsprechen in der verallgemeinerten Theorie Größen $\mathfrak{T}_{\sigma\nu}$, welche die mit $\sqrt{-g}$

---

[1]) Vgl. A. Einstein, Entwurf einer verallgemeinerten Relativitätstheorie und einer Theorie der Gravitation, Zeitschr. f. Math. u. Phys. **62**. p. 6. 1913.

multiplizierten Komponenten eines gemischten Tensors bilden, der aus einen symmetrischen kontravarianten Tensor $(\Theta_{\mu\nu})$ durch die gemischte Multiplikation

$$\frac{1}{\sqrt{-g}}\mathfrak{T}_{\sigma\nu} = \sum_{\mu} g_{\sigma\mu}\Theta_{\mu\nu}$$

hervorgeht (g bedeutet die Determinante aus den Größen $g_{\mu\nu}$).

Besteht z. B. das physikalische System in einer bewegten kontinuierlichen Massenverteilung von der Ruhedichte ϱ_0, so ist

$$\Theta_{\mu\nu} = \varrho_0 \frac{dx_\mu}{ds}\frac{dx_\nu}{ds},$$

und die physikalische Bedeutung der $\mathfrak{T}_{\sigma\nu}$ geht aus folgender Tabelle hervor[1]):

$\mathfrak{T}_{11}$	$\mathfrak{T}_{12}$	$\mathfrak{T}_{13}$	$\mathfrak{T}_{14}$		$-X_x$	$-X_y$	$-X_z$	$-i_x$
$\mathfrak{T}_{21}$	$\mathfrak{T}_{22}$	$\mathfrak{T}_{23}$	$\mathfrak{T}_{24}$	=	$-Y_x$	$-Y_y$	$-Y_z$	$-i_y$
$\mathfrak{T}_{31}$	$\mathfrak{T}_{32}$	$\mathfrak{T}_{33}$	$\mathfrak{T}_{34}$		$-Z_x$	$-Z_y$	$-Z_z$	$-i_z$
$\mathfrak{T}_{41}$	$\mathfrak{T}_{42}$	$\mathfrak{T}_{43}$	$\mathfrak{T}_{44}$		f_x	f_y	f_z	η.

X_x usw. bezeichnen die Komponenten des Flächendrucks, i_x usw. die Komponenten der Impulsdichte, f_x usw. die Komponenten der Strömungsdichte der Energie, und η die Energiedichte.

Die erwähnten Erhaltungssätze haben in der allgemeinen Theorie die allgemein-kovariante Form:

(3) $$\sum_\nu \frac{\partial \mathfrak{T}_{\sigma\nu}}{\partial x_\nu} = \frac{1}{2}\sum_{\mu\nu\tau}\frac{\partial g_{\mu\nu}}{\partial x_\sigma}\gamma_{\mu\tau}\mathfrak{T}_{\tau\nu}.$$

Die rechte Seite dieser Gleichung drückt aus, daß der betrachtete Vorgang für sich allein die Erhaltungssätze nicht erfüllt, da von dem Gravitationsfeld Impuls und Energie an das materielle System abgegeben wird.

Allgemein beziehen sich die Komponenten $\mathfrak{T}_{\sigma\nu}$ auf alle physikalischen Vorgänge im Raume, mit Ausschluß der das Gravitationsfeld selbst betreffenden.

Wir wissen aus der ursprünglichen Relativitätstheorie, daß der Energietensor allein maßgebend ist für die Trägheitseigenschaften eines Systems. Aus der rechten Seite von (3) geht hervor, daß auch die Einwirkung eines Gravitationsfeldes

---

[1]) In der Tabelle, so wie sie in der Phys. Zeitschr. XIV, p. 1257 gegeben wurde, findet sich ein Vorzeichenfehler.

nur durch die Komponenten des Energietensors bestimmt wird. Es entspricht dies durchaus den Erfahrungsgesetzen von der Gleichheit der trägen und der schweren Masse. Wir werden im folgenden annehmen, daß auch für die Erzeugung eines Gravitationsfeldes durch ein materielles System der Energietensor allein maßgebend ist.

§ 2. *Differentialgleichung für das Gravitationsfeld im Falle der Nordströmschen Theorie.*

Das bisher Gesagte gilt ebenso für die Nordströmsche wie für die Einstein-Großmannsche Theorie; der Unterschied beider Theorien aber besteht im folgenden:

Das Gravitationsfeld wird von zehn Größen $g_{\mu\nu}$ bestimmt. Gemäß der Einstein-Großmannschen Theorie werden für diese zehn Größen zehn formal gleichwertige Gleichungen angegeben. Der Nordströmschen Theorie aber liegt die Annahme zugrunde, daß es möglich sei, durch passende Wahl des Bezugssystems dem Prinzip von der Konstanz der Lichtgeschwindigkeit zu genügen. Wir wollen sogleich zeigen, daß dies auf die Annahme herauskommt, daß sich die zehn Größen $g_{\mu\nu}$ bei passender Wahl des Bezugssystems auf eine einzige Größe Φ^2 reduzieren lassen.

Damit nämlich das Prinzip von der Konstanz der Lichtgeschwindigkeit erfüllt sei, muß die für die Lichtausbreitung maßgebende Gleichung

$$\sum_{\mu\nu} g_{\mu\nu} dx_\mu dx_\nu = 0$$

in die Gleichung

$$dx^2 + dy^2 + dz^2 - c^2 dt^2 = 0$$

übergehen. Daraus folgt, daß bei einer solchen Wahl des Bezugssystems sein muß:

$$\sum_{\mu\nu} g_{\mu\nu} dx_\mu dx_\nu = \Phi^2 dx_1^2 + \Phi^2 dx_2^2 + \Phi^2 dx_3^2 - \Phi^2 dx_4^2$$

wobei jetzt $x_1 = x$, $x_2 = y$, $x_3 = z$ und $x_4 = ct$ gesetzt ist.

Das System der $g_{\mu\nu}$ degeneriert also in

$$(4) \quad \begin{matrix} \Phi^2 & 0 & 0 & 0 \\ 0 & \Phi^2 & 0 & 0 \\ 0 & 0 & \Phi^2 & 0 \\ 0 & 0 & 0 & -\Phi^2 \end{matrix}$$

Zur Bestimmung der einen Größe Φ^2 brauchen wir eine einzige Differentialgleichung, die wie die Poissonsche Gleichung skalaren Charakter haben wird. Diese Gleichung wollen wir ebenso wie die früheren in allgemein kovarianter Form aufstellen, d. h. ohne zunächst die durch das Prinzip von der Konstanz der Lichtgeschwindigkeit nahegelegte Spezialisierung des Bezugssystems auszuführen. Die gesuchte Gleichung ist vollständig bestimmt durch die Annahme, daß sie von der zweiten Ordnung ist, wenn man noch berücksichtigt, daß sie eine Verallgemeinerung der Poissonschen Gleichung sein muß. Offenbar wird sie von der Form sein

$$(5) \qquad \Gamma = \varkappa \mathfrak{T},$$

wobei Γ ein Skalar ist, der aus den Größen $g_{\mu\nu}$ und deren ersten und zweiten Ableitungen gebildet ist, und $\mathfrak{T}$ ein Skalar, der durch den materiellen Vorgang, nach dem Gesagten also durch die $\mathfrak{T}_{\sigma\nu}$, bestimmt ist. $\varkappa$ bedeutet eine Konstante.

Aus den Untersuchungen der Mathematiker über die Differentialtensoren einer mehrdimensionalen Mannigfaltigkeit geht hervor, daß der einzige Ausdruck, der für Γ in Betracht kommt, eine Funktion ist von

$$\sum_{iklm} \gamma_{im} \gamma_{kl} (ik, lm).$$

Dabei bedeutet (ik, lm) den bekannten Riemann-Christoffelschen Tensor vierten Ranges, der mit dem Krümmungsmaße der Flächentheorie zusammenhängt, und durch die Gleichung

$$(ik, lm) = \tfrac{1}{2} \left(\frac{\partial^2 g_{im}}{\partial x_k \partial x_l} + \frac{\partial^2 g_{kl}}{\partial x_i \partial x_m} - \frac{\partial^2 g_{il}}{\partial x_k \partial x_m} - \frac{\partial^2 g_{mk}}{\partial x_i \partial x_l} \right)$$
$$+ \sum_{\varrho\sigma} \gamma_{\varrho\sigma} \left(\begin{bmatrix} im \\ \varrho \end{bmatrix} \begin{bmatrix} kl \\ \sigma \end{bmatrix} - \begin{bmatrix} il \\ \varrho \end{bmatrix} \begin{bmatrix} km \\ \sigma \end{bmatrix} \right)$$

definiert ist, wobei $\begin{bmatrix} im \\ \varrho \end{bmatrix}$ bedeutet $\tfrac{1}{2} \left(\frac{\partial g_{i\varrho}}{\partial x_m} + \frac{\partial g_{m\varrho}}{\partial x_i} - \frac{\partial g_{im}}{\partial x_\varrho} \right)$.

Ferner ist aus der allgemeinen Kovariantentheorie klar, daß zu den $\mathfrak{T}_{\sigma\nu}$ nur der Skalar $\dfrac{1}{\sqrt{-g}} \sum_\tau \mathfrak{T}_{\tau\tau}$ gehört (bzw. eine Funktion dieser Größe).

Hieraus geht hervor, daß die gesuchte Gleichung die Form

$$(5\,\mathrm{a}) \qquad \sum_{iklm} \gamma_{im} \gamma_{kl} (ik, lm) = \varkappa \frac{1}{\sqrt{-g}} \sum_\tau \mathfrak{T}_{\tau\tau}$$

erhalten muß. Dabei ist allerdings *vorausgesetzt*, daß in der gesuchten Gleichung die zweiten Ableitungen der $g_{\mu\nu}$ und die $\mathfrak{T}_{\sigma\nu}$ *linear* eingehen.

Die Gleichung (5a), die wir jetzt aufgestellt haben, und die Gleichungen (3) enthalten die Nordströmsche Theorie der Gravitation vollständig mit Bezug auf beliebige Raum-Zeitkoordinaten, wenn man die *Bedingungen hinzunimmt*, welche die $g_{\mu\nu}$ erfüllen müssen, damit das Prinzip der Konstanz der Lichtgeschwindigkeit für ein passend gewähltes Bezugssystem erfüllt sei.

§ 3. *Die Grundgleichungen der Nordströmschen Theorie mit Bezug auf die dem Prinzip der Konstanz der Lichtgeschwindigkeit angepaßten Bezugssysteme.*

Wir denken uns jetzt diejenigen Bezugssysteme bevorzugt in bezug auf welche das Prinzip der Konstanz der Lichtgeschwindigkeit erfüllt ist. Die Komponenten $g_{\mu\nu}$ des Fundamentaltensors sind dann durch die in (4) geschriebenen Werte gegeben. Die zugehörigen $g_{\mu\nu}$ findet man in der Tabelle

$$(4\,\mathrm{a})\quad \begin{matrix} +\dfrac{1}{\Phi^2} & 0 & 0 & 0 \\ 0 & +\dfrac{1}{\Phi^2} & 0 & 0 \\ 0 & 0 & +\dfrac{1}{\Phi^2} & 0 \\ 0 & 0 & 0 & -\dfrac{1}{\Phi^2} \end{matrix}$$

In diesem Falle erhält man $ds = \Phi\sqrt{dx_1^2 + dx_2^2 + dx_3^2 - dx_4^2}$. Wie schon erwähnt, ist ds der „natürlich gemessene" Abstand zweier benachbarter Raum-Zeitpunkte. Jetzt kann man die Fälle unterscheiden, wo der Verbindungsvektor raumartig oder zeitartig ist. Im ersten Falle kann durch passende Wahl des Bezugssystems der Vektor zu einem rein räumlichen gemacht werden; man erhält dann als Zusammenhang der „natürlich" und der im Koordinatenmaß gemessenen Längen

$$ds = \Phi\sqrt{dx^2 + dy^2 + dz^2},$$

d. h. ein Maßstab von der natürlichen Länge ds hat die Koordinatenlänge ds/Φ.

Für einen zeitartigen Verbindungsvektor verschwinden bei passender Wahl des Bezugssystems die räumlichen Komponenten, und man erhält

$$ds = \Phi \sqrt{-dx_4{}^2}, \quad \text{oder} \quad \frac{ds}{i} = \Phi\, dx_4.$$

ds/i ist nichts anderes als die mit einer Uhr von bestimmter Beschaffenheit gemessene Zeitdauer. $ds/\Phi i$ ist also die Zeitdifferenz im Koordinatenmaßstab.

$1/\Phi$ ist also der Faktor, mit dem die natürlich gemessenen Zeiten und Längen multipliziert werden müssen, um Koordinatenzeiten bzw. Koordinatenlängen zu ergeben.

Aus der Form des Linienelementes

$$ds^2 = \Phi^2(dx^2 + dy^2 + dz^2 - c^2 dt^2)$$

folgt, daß die Gleichungen der Nordströmschen Theorie nicht nur bezüglich den Lorentz-Transformationen, sondern auch bezüglich Ähnlichkeitstransformationen kovariant sind.

Die Impuls- und Energiegleichungen (3) für die Materie nehmen die Form an

(3a) $$\sum_\nu \frac{\partial \mathfrak{T}_{\sigma\nu}}{\partial x_\nu} = \frac{\partial \log \Phi}{\partial x_\sigma} \sum_\tau \mathfrak{T}_{\tau\tau}.$$

Es ist bemerkenswert, daß für den Einfluß des Gravitationsfeldes auf ein System gemäß dieser Gleichung nur der Skalar $(1/\sqrt{-g}) \sum_\tau \mathfrak{T}_{\tau\tau}$ maßgebend ist. Es ist dies im Einklang mit der Erwägung, die wir bei der Ableitung der Gleichung (5a) gegeben haben.

Die Differentialgleichung des Gravitationsfeldes (5a) nimmt die Form an

(5b) $$\frac{1}{\Phi^3}\left[\frac{\partial^2 \Phi}{\partial x_1{}^2} + \frac{\partial^2 \Phi}{\partial x_2{}^2} + \frac{\partial^2 \Phi}{\partial x_3{}^2} - \frac{\partial^2 \Phi}{\partial x_4{}^2}\right] = \frac{k}{\Phi^4}\sum_\tau \mathfrak{T}_{\tau\tau}$$

(wobei k eine neue Konstante bedeutet), oder

$$\Phi \,\square\, \Phi = k \sum_\tau \mathfrak{T}_{\tau\tau}.$$

Da das Verhältnis der natürlichen und der Koordinatenlängen an *einem* Orte beliebig gewählt werden kann, kann über die Wahl der Konstante k noch beliebig verfügt werden. Man kann z. B. nach dem Vorgange von Nordström $k=1$ setzen.

Man sieht, daß die abgeleiteten Gleichungen mit den von Nordström gegebenen vollkommen übereinstimmen.

§ 4. *Schlußbemerkungen.*

Im vorstehenden konnte gezeigt werden, daß man bei Zugrundelegung des Prinzips von der Konstanz der Lichtgeschwindigkeit durch rein formale Erwägungen, d. h. ohne Zuhilfenahme weiterer physikalischen Hypothesen zur Nordströmschen Theorie gelangen kann. Es scheint uns deshalb, daß diese Theorie allen anderen Gravitationstheorien gegenüber, die an diesem Prinzip festhalten, den Vorzug verdient. Vom physikalischen Standpunkt ist dies um so mehr der Fall, als diese Theorie dem Satz von der Äquivalenz der trägen und schweren Masse strenge Genüge leistet.

Wir bemerken, daß nur die Verwendung der Invariantentheorie des absoluten Differentialkalküls uns eine klare Einsicht in den formalen Inhalt der Nordströmschen Theorie zu geben vermag. Ferner setzt uns diese Methode in den Stand, die Beeinflussung beliebiger physikalischer Vorgänge durch das Gravitationsfeld, so wie sie nach der Nordströmschen Theorie zu erwarten ist, ohne Hinzuziehung neuer Hypothesen anzugeben. Auch tritt die Beziehung der Nordströmschen Theorie zur Einstein-Großmannschen mit voller Deutlichkeit hervor.

Endlich legt die Rolle, welche bei der vorliegenden Untersuchung der Riemann-Christoffelsche Differentialtensor spielt, den Gedanken nahe, daß er auch für eine von physikalischen Annahmen unabhängige Ableitung der Einstein-Großmannschen Gravitationsgleichungen einen Weg öffnen würde. Der Beweis der Existenz oder Nichtexistenz eines derartigen Zusammenhanges würde einen wichtigen theoretischen Fortschritt bedeuten.[1])

---

1) Die in § 4, p. 36, des „Entwurfs einer verallgemeinerten Relativitätstheorie" angegebene Begründung für die Nichtexistenz eines derartigen Zusammenhanges hält einer genaueren Überlegung nicht stand.

(Eingegangen 19. Februar 1914.)

ANNALEN DER PHYSIK.

BEGRÜNDET UND FORTGEFÜHRT DURCH

F. A. C. GREN, L. W. GILBERT, J. C. POGGENDORFF, G. u. E. WIEDEMANN, P. DRUDE.

VIERTE FOLGE.

BAND 47.

DER GANZEN REIHE 352. BAND.

KURATORIUM:

M. PLANCK, G. QUINCKE,
W. C. RÖNTGEN, W. VOIGT, E. WARBURG.

UNTER MITWIRKUNG

DER DEUTSCHEN PHYSIKALISCHEN GESELLSCHAFT

HERAUSGEGEBEN VON

W. WIEN UND M. PLANCK.

MIT ZEHN FIGURENTAFELN.

LEIPZIG, 1915.
VERLAG VON JOHANN AMBROSIUS BARTH.

2. Antwort auf eine Abhandlung M. v. Laues „Ein Satz der Wahrscheinlichkeitsrechnung und seine Anwendung auf die Strahlungstheorie"; von A. Einstein.

In der zitierten Arbeit bringt Laue die mathematische Grundlage der Statistik der Strahlung in eine Form, die an Prägnanz und Schönheit nichts zu wünschen übrigläßt. Was aber die Anwendung jener Grundlage auf die Strahlungstheorie anbelangt, so scheint er mir einem bedenklichen Irrtume zum Opfer gefallen zu sein, der dringend Berichtigung fordert. Wenn Laues Behauptung, daß die Koeffizienten der Fourierentwicklung der bei natürlicher Strahlung auftretenden örtlichen Schwingung nicht voneinander statistisch unabhängig zu sein brauchten, berechtigt wäre, böte sich wirklich ein höchst aussichtsreicher Weg zur Überwindung der Schwierigkeiten dar, welche in der theoretischen Unverdaulichkeit aller Gesetze besteht, in denen das Plancksche „h" eine Rolle spielt. Dies war eben der Grund, der mich vor fünf Jahren veranlaßte, in einer mit L. Hopf zusammen publizierten Arbeit diese Frage näher zu prüfen.

Das Resultat jener in ihrer Durchführung nicht ganz einwandfreien Arbeit, wird von Laue als richtige Konsequenz der zugrunde gelegten Voraussetzungen anerkannt. Aber Laue bestreitet die Zulässigkeit der Grundvoraussetzung, die sich so formulieren läßt:

Wenn ich dadurch eine vollkommen ungeordnete Strahlung (statistisch unabhängige Fourierkoeffizienten) erhalte daß ich unendlich viele vollkommen gegebene, ganz miteinander übereinstimmende Strahlungen derart superponiere, daß bei dieser Superposition die Gesamtphasen dieser superponierten Strahlungen zufällig gewählt werden, so muß die natürliche Strahlung erst recht statistisch ungeordnet sein.

Diese Grundvoraussetzung schien mir damals evident. Der Umstand aber, daß sie von einem so erfahrenen Fachmann, wie Laue, nicht geteilt wird, beweist das Gegenteil. Ich will deshalb im folgenden einen Beweis geben, der von einer derartigen Voraussetzung frei ist und — wie ich hoffe — unwiderleglich dartut, daß unsere Undulationstheorie die statistische Unabhängigkeit der Fourierkoeffizienten unbedingt fordert. Bevor ich diesen Beweis beginne, will ich aber zeigen, warum die in den Teilen II und III der Laueschen Abhandlung gegebene Betrachtung nach meiner Ansicht nicht beweisend ist.

Laue betrachtet eine Strahlung, die durch eine große Anzahl unregelmäßig über eine Schicht von der Dicke $c\tau$ verteilter Resonatoren senkrecht zu dieser Schicht emittiert wird. Im Teile II seiner Abhandlung nimmt er an, daß alle diese Resonatoren gleichzeitig und nach demselben Gesetze schwingen; im Teile III, daß die Schwingungen aller Resonatoren durch dasselbe, als gegeben zu denkende statistische Gesetz beherrscht seien. In beiden Fällen ergibt sich nicht die statistische Unabhängigkeit der Fourierkoeffizienten der Entwicklung für die resultierende Strahlung. Hieraus darf aber nach meiner Meinung keineswegs die Zulässigkeit der Hypothese gefolgert werden, daß auch *bei der natürlichen Strahlung* jene Unabhängigkeit nicht vorhanden sei. Denn es ist doch gar nicht gesagt, daß der Grad von Unordnung, welchen jene ungeordnete Verteilung der Resonatoren über die Schicht von der Dicke $c\tau$ mit sich bringt, derselbe sei wie bei der natürlichen Strahlung.

Dieser Verdacht erhebt sich um so dringender, als nach Laues rechnerischen Ergebnissen der Grad der statistischen Abhängigkeit zweier durch die Indizes p und p' charakterisierten Glieder der Entwicklung für die resultierende Strahlung wesentlich durch den Wert

$$\frac{\pi(p-p')\tau}{T}$$

bedingt werde, d. h. *durch eine von der Schichtdicke abhängige Größe*, während doch eine derartige statistische Abhängigkeit bei der natürlichen Strahlung — falls eine solche vorhanden wäre — nichts zu tun haben dürfte mit der besonderen Erzeugungsart der betrachteten Strahlung.

Nach meiner Ansicht ist daher keiner der von Laue betrachteten Fälle der natürlichen Strahlung bezüglich

Unordnung äquivalent, so daß aus seinen Ergebnissen über die natürliche Strahlung nichts gefolgert werden kann. Ich halte vielmehr meine frühere Behauptung aufrecht und suche dieselbe im folgenden durch einen neuen Beweis zu stützen, indem ich mich der von Laue in seiner Arbeit dargelegten Sätze aus der Wahrscheinlichkeitsrechnung bediene.

§ 1. Statistische Eigenschaften einer Strahlung, die durch Superposition unendlich vieler, voneinander unabhängig erzeugter Strahlungen entstanden ist.

Jede der betrachteten Teilstrahlungen sei durch eine Fouriersche Entwicklung von der Form

$$(1) \qquad \sum_n a_n^{(\nu)} \cos 2\pi n \frac{t}{T} + b_n^{(\nu)} \sin 2\pi n \frac{t}{T}$$

für das Zeitintervall 0 bis T dargestellt, wobei die Koeffizienten dem Wahrscheinlichkeitsgesetz

$$(2) \qquad dW = f^{(\nu)}(a_1^{(\nu)} \ldots a_z^{(\nu)} \ldots, b_1^{(\nu)} \ldots b_z^{(\nu)}) \, da_1^{(\nu)} \ldots db_z^{(\nu)} \ldots$$

genügen sollen, welches Gesetz für jedes (ν), d. h. für jede der betrachteten Teilstrahlungen ein besonderes sein kann. Das Gesetz sei ferner ein solches, daß

$$(3) \qquad \begin{cases} \overline{a_n^\nu} = \int a_n^\nu f^{(\nu)} \, da_1^{(\nu)} \ldots db_z^{(\nu)} = 0 \\ \overline{b_n^\nu} = \int b_n^\nu f^{(\nu)} \, da_1^{(\nu)} \ldots db_z^{(\nu)} = 0. \end{cases}$$

Die resultierende Strahlung ist für das Zeitintervall 0 bis T durch die Entwicklung

$$(4) \qquad \begin{cases} \sum_n A_n \cos 2\pi n \frac{t}{T} + B_n \sin 2\pi n \frac{t}{T} \\ = \sum_\nu \sum_n \left(a_n^{(\nu)} \cos 2\pi n \frac{t}{T} + b_n^{(\nu)} \sin 2\pi n \frac{t}{T} \right) \end{cases}$$

gegeben, woraus die Gültigkeit der Beziehungen

$$(5) \qquad \begin{cases} A_n = \sum_\nu a_n^{(\nu)} \\ B_n = \sum_\nu b_n^{(\nu)} \end{cases}$$

hervorgeht. Welches statistische Gesetz folgt für die Fourierkoeffizienten $A_1 \ldots B_z$?

Aus einer Betrachtung, die der im Teile I der Laueschen Arbeit durchgeführten ganz analog ist, findet man, daß das gesuchte statistische Gesetz das folgende ist:

(6) $$dW = \text{konst.}\, e^{-\sum_{mn}(\alpha_{mn} A_m A_n + \beta_{mn} B_m B_n + 2\gamma_{mn} A_m B_n)} dA_1 \ldots dB_z.$$

Hieraus ersieht man, daß durch Superposition unendlich vieler Teilstrahlungen die statistische Unabhängigkeit der Fourier-Koeffizienten noch keineswegs garantiert wird. Wohl aber gestattet das Gesetz (6) die Frage nach der statistischen Unabhängigkeit der Fourierkoeffizienten auf eine einfachere Frage zu reduzieren. Jene statistische Unabhängigkeit wird nämlich dann und nur dann erfüllt sein, wenn im Exponenten der Exponentialfunktion nur die Quadrate der A_m und B_m, aber keine Produkte dieser Größen auftreten; d. h. es muß sein:

(7) $$\begin{cases} \alpha_{mn} = \beta_{mn} = 0 \quad \text{für } m \neq n \\ \gamma_{mn} = 0. \end{cases}$$

Es ist ferner wegen (3) und (5) klar, daß im Falle statistischer Unabhängigkeit die Beziehungen

(7a) $$\begin{cases} \overline{A_m A_n} = \overline{B_m B_n} = 0 \quad \text{für } m \neq n \\ \overline{A_m B_n} = 0 \end{cases}$$

bestehen müssen. Da die Zahl der Bedingungen (7a) gleich ist der Zahl der Bedingungen (7), und alle Bedingungen (7a) voneinander unabhängig sind, so folgt, daß im Falle der Gültigkeit von (6) die Bedingungen (7a) *hinreichend* sind für die statistische Unabhängigkeit der Fourierkoeffizienten.

Wir gelangen daher zu folgendem vorläufigen Ergebnis: Da wir von der natürlichen Strahlung annehmen müssen, daß ihre statistischen Eigenschaften durch Superposition von inkohärenten Teilstrahlungen nicht geändert werden, so sind die Gleichungen (7a) bei der natürlichen Strahlung hinreichende Bedingungen für die statistische Unabhängigkeit der Fourierkoeffizienten.

§ 2. Nachweis der statistischen Unabhängigkeit der Fourierkoeffizienten bei der natürlichen Strahlung.

Es sei $F(t)$ eine Komponente des Strahlungsvektors stationärer natürlicher Strahlung, gegeben für unendlich lange Zeit. T sei eine gegen die Schwingungsdauer der langwelligsten in der Strahlung auftretenden Lichtart große Zeitdauer. Zwischen den Zeiten t_0 und $t_0 + T$ sei $F(t)$ dargestellt durch die Fourierreihe

$$(4\,\mathrm{a}) \quad \sum_n \left(A_n \cos 2\pi n \frac{t-t_0}{T} + B_n \sin 2\pi n \frac{t-t_0}{T} \right).$$

Es ist klar, daß die zu $F(t)$ gehörigen Fourierkoeffizienten A_n, B_n von der Wahl der Epoche t_0 abhängen werden. Indem wir die Entwicklung für sehr viele, zufällig gewählte t_0 ausgeführt denken, erlangen wir ein statistisches Material zur Ableitung statistischer Eigenschaften der Koeffizienten A_n, B_n, welche wir bei der natürlichen Strahlung notwendig fordern müssen.

Um diese Eigenschaften abzuleiten, entwickeln wir $F(t)$ in eine Fourierreihe zwischen den Zeiten 0 und ϑ, wobei ϑ eine gegenüber T sehr große Zeitdauer sei. Für dies Zeitintervall sei

$$(8) \quad F(t) = \sum_\nu \alpha_\nu \cos\left(2\pi\nu \frac{t}{\vartheta} + \varphi_\nu\right).$$

Wählen wir t_0 zwischen $t = 0$ und $t = \vartheta - T$, so können die Koeffizienten A_n und B_n durch t_0 und die Koeffizienten α_ν und φ_ν der Entwicklung (8) ausgedrückt werden; man erhält zunächst

$$(9) \quad \begin{cases} A_n = \dfrac{2}{T} \sum_\nu \left\{ \displaystyle\int_{t_0}^{t_0+T} \alpha_\nu \cos\left(2\pi\nu \dfrac{t}{\vartheta} + \varphi_\nu\right) \cos\left(2\pi n \dfrac{t-t_0}{T}\right) dt \right\} \\[1em] B_n = \dfrac{2}{T} \sum_\nu \left\{ \displaystyle\int_{t_0}^{t_0+T} \alpha_\nu \cos\left(2\pi\nu \dfrac{t}{\vartheta} + \varphi_\nu\right) \sin\left(2\pi n \dfrac{t-t_0}{T}\right) dt \right\}. \end{cases}$$

Führt man die Integration aus, so erhält man, indem man in bekannter Weise Glieder mit dem Faktor $\dfrac{1}{\pi(\nu/\vartheta + n/T)}$ gegen solche mit dem Faktor $\dfrac{1}{\pi(\nu/\vartheta - n/T)}$ vernachlässigt:

$$(10) \quad \begin{cases} A_n = \displaystyle\sum_\nu \alpha_\nu \dfrac{\sin\pi\left(\nu \dfrac{T}{\vartheta} - n\right) \cos\left(\chi_{\nu n} + 2\pi\nu \dfrac{t_0}{\vartheta}\right)}{\pi\left(\nu \dfrac{T}{\vartheta} - n\right)} \\[2em] B_n = -\displaystyle\sum_\nu \alpha_\nu \dfrac{\sin\pi\left(\nu \dfrac{T}{\vartheta} - n\right) \sin\left(\chi_{\nu n} + 2\pi\nu \dfrac{t_0}{\vartheta}\right)}{\pi\left(\nu \dfrac{T}{\vartheta} - n\right)}, \end{cases}$$

wobei
$$\chi_{\nu n} = \pi \left(\nu \frac{T}{\vartheta} + n \right) + \varphi_\nu$$
gesetzt ist. Die Formeln (10) gelten nur für Werte von t_0 zwischen $t_0 = 0$ und $t_0 = \vartheta - T$, weil die Entwicklung gemäß (8) nur für das Zeitintervall $0 - \vartheta$ gilt. Wir erlauben uns jedoch, die Formel (8) für das Intervall $0 - (\vartheta + T)$ anzuwenden. Damit ersetzen wir zwischen den Zeitwerten ϑ und $\vartheta + T$ die Funktion $F(t)$ durch die Werte von $F(t)$ zwischen den Zeiten 0 und T. Durch dieses Vorgehen werden im folgenden unsere Mittelwertbetrachtungen gefälscht, aber nur relativ unendlich wenig, weil das Zeitintervall T gegen ϑ unendlich klein ist. Von dieser Erwägung ausgehend, werden wir die Gleichungen (10) so anwenden, wie wenn sie im ganzen Intervall $0 < t_0 < \vartheta$ gelten würden.

Wir bilden nun mit Hilfe von (10) den Mittelwert $\overline{A_m A_n}$, d. h. die Größe
$$\overline{A_m A_n} = \frac{1}{\vartheta} \int_0^\vartheta A_m A_n \, dt_0 \, .$$
Dabei tritt das Integral
$$\int_0^\vartheta \cos\left(\chi_{\mu m} + 2\pi\mu \frac{t_0}{\vartheta}\right) \cos\left(\chi_{\nu n} + 2\pi\nu \frac{t_0}{\vartheta}\right) dt_0$$
auf. Dieses verschwindet wegen der Ganzzahligkeit von μ und ν, wenn $\mu \neq \nu$, und hat für $\mu = \nu$ den Wert $\frac{\vartheta}{2}(-1)^{m-n}$. Mit Rücksicht darauf ergibt die erste der Gleichungen (10)

$$(11) \quad \overline{A_m A_n} = \frac{(-1)^{m-n}}{2} \sum_\nu a_\nu^2 \frac{\sin\pi\left(\nu\frac{T}{\vartheta} - m\right) \sin\pi\left(\nu\frac{T}{\vartheta} - n\right)}{\pi^2 \left(\nu\frac{T}{\vartheta} - m\right)\left(\nu\frac{T}{\vartheta} - n\right)}$$
$$= \frac{1}{2} \sum a_\nu^2 \frac{\sin^2 \pi\nu\frac{T}{\vartheta}}{\pi^2 \left(\nu\frac{T}{\vartheta} - m\right)\left(\nu\frac{T}{\vartheta} - n\right)} \, .$$

A priori ist klar, daß eine statistische Abhängigkeit nur zwischen Strahlungskomponenten von sehr nahe gleicher Frequenz zu erwarten ist. m und n gehören also demselben engen Spektralbereich an, ebenso jene Werte von ν, welche zu unserer Summe merklich beitragen.

In (11) ist der Bruch auf der rechten Seite eine wegen der Kleinheit von T/ϑ mit ν langsam veränderliche Größe. Deshalb kann bezüglich der Größe α_ν^2 über viele aufeinander folgende Glieder ohne merkbaren Fehler gemittelt werden, und es wird jener Mittelwert $\overline{\alpha_\nu^2}$ als Konstante aus der Summe herausgesetzt werden können, da die Summation überhaupt nur über einen engen Spektralbereich zu erstrecken ist. Die über den Bruch erstreckte Summe kann dann noch in ein Integral verwandelt werden, so daß man erhält:

$$(12) \qquad \overline{A_m A_n} = \frac{1}{2} \overline{\alpha_\nu^2} \frac{\vartheta}{\pi T} \int \frac{\sin^2 x}{(x - m\pi)(x - n\pi)} \, dx.$$

Das Integral kann ohne merklichen Fehler zwischen $-\infty$ und $+\infty$ genommen werden, statt zwischen der durch den vorerwähnten Spektralbereich bestimmten Grenzen.

Dieses Integral hat für $m = n$ den Wert π, verschwindet aber stets [1]), wenn $m \neq n$ (m und n sind ganze Zahlen). Damit ist zunächst das Verschwinden von $\overline{A_m A_n}$ (für $m \neq n$) bewiesen; der Beweis für das Verschwinden von $\overline{B_m B_n}$ (für $m \neq n$) und $\overline{A_m B_n}$ ist analog zu führen. Aus dem Verschwinden dieser Mittelwerte folgt nach § 1 die behauptete statistische Unabhängigkeit der Fourierkoeffizienten.

---

[1]) Das Integral ist nämlich gleich

$$\frac{1}{(m-n)\pi} \left\{ \int_{-\infty}^{+\infty} \frac{\sin^2 x}{x - m\pi} \, dx - \int_{-\infty}^{+\infty} \frac{\sin^2 x}{x - n\pi} \, dx \right\}.$$

Jedes der letzteren Integrale ist gleich

$$\int_{-\infty}^{+\infty} \frac{\sin^2 y}{y} \, dy = 0.$$

Bemerkung zur Korrektur: Statt bei der Auswertung von (11) über viele aufeinanderfolgende Summenglieder zu mitteln, kann man auch unendlich viele, voneinander unabhängige Entwicklungen (8) zugrunde legen und über diese mitteln. Nimmt man an (11) jene Mittelwertbildung vor, so tritt der dementsprechend verstandene Mittelwert $\overline{\alpha_\nu^2}$ vor das Summenzeichen. Das Endresultat bleibt natürlich dasselbe.

(Eingegangen 24. Juni 1915.)

ANNALEN DER PHYSIK.

BEGRÜNDET UND FORTGEFÜHRT DURCH

F. A. C. GREN, L. W. GILBERT, J. C. POGGENDORFF, G. U. E. WIEDEMANN, P. DRUDE.

VIERTE FOLGE.

BAND 49.

DER GANZEN REIHE 354. BAND.

KURATORIUM:

M. PLANCK, G. QUINCKE,
W. C. RÖNTGEN, W. VOIGT, E. WARBURG.

UNTER MITWIRKUNG

DER DEUTSCHEN PHYSIKALISCHEN GESELLSCHAFT

HERAUSGEGEBEN VON

W. WIEN UND M. PLANCK.

MIT EINEM PORTRÄT UND ZEHN FIGURENTAFELN.

LEIPZIG, 1916.
VERLAG VON JOHANN AMBROSIUS BARTH.

1. *Die Grundlage der allgemeinen Relativitätstheorie;* *von A. Einstein.*

Die im nachfolgenden dargelegte Theorie bildet die denkbar weitgehendste Verallgemeinerung der heute allgemein als „Relativitätstheorie" bezeichneten Theorie; die letztere nenne ich im folgenden zur Unterscheidung von der ersteren „spezielle Relativitätstheorie" und setze sie als bekannt voraus. Die Verallgemeinerung der Relativitätstheorie wurde sehr erleichtert durch die Gestalt, welche der speziellen Relativitätstheorie durch Minkowski gegeben wurde, welcher Mathematiker zuerst die formale Gleichwertigkeit der räumlichen Koordinaten und der Zeitkoordinate klar erkannte und für den Aufbau der Theorie nutzbar machte. Die für die allgemeine Relativitätstheorie nötigen mathematischen Hilfsmittel lagen fertig bereit in dem „absoluten Differentialkalkül", welcher auf den Forschungen von Gauss, Riemann und Christoffel über nichteuklidische Mannigfaltigkeiten ruht und von Ricci und Levi-Civita in ein System gebracht und bereits auf Probleme der theoretischen Physik angewendet wurde. Ich habe im Abschnitt B der vorliegenden Abhandlung alle für uns nötigen, bei dem Physiker nicht als bekannt vorauszusetzenden mathematischen Hilfsmittel in möglichst einfacher und durchsichtiger Weise entwickelt, so daß ein Studium mathematischer Literatur für das Verständnis der vorliegenden Abhandlung nicht erforderlich ist. Endlich sei an dieser Stelle dankbar meines Freundes, des Mathematikers Grossmann, gedacht, der mir durch seine Hilfe nicht nur das Studium der einschlägigen mathematischen Literatur ersparte, sondern mich auch beim Suchen nach den Feldgleichungen der Gravitation unterstützte.

A. Prinzipielle Erwägungen zum Postulat der Relativität.
§ 1. Bemerkungen zu der speziellen Relativitätstheorie.

Der speziellen Relativitätstheorie liegt folgendes Postulat zugrunde, welchem auch durch die Galilei-Newtonsche Mechanik Genüge geleistet wird: Wird ein Koordinatensystem K so gewählt, daß in bezug auf dasselbe die physikalischen Gesetze in ihrer einfachsten Form gelten, so gelten *dieselben* Gesetze auch in bezug auf jedes andere Koordinatensystem K', das relativ zu K in gleichförmiger Translationsbewegung begriffen ist. Dieses Postulat nennen wir „spezielles Relativitätsprinzip". Durch das Wort „speziell" soll angedeutet werden, daß das Prinzip auf den Fall beschränkt ist, daß K' eine *gleichförmige Translationsbewegung* gegen K ausführt, daß sich aber die Gleichwertigkeit von K' und K nicht auf den Fall *ungleichförmiger* Bewegung von K' gegen K erstreckt.

Die spezielle Relativitätstheorie weicht also von der klassischen Mechanik nicht durch das Relativitätspostulat ab, sondern allein durch das Postulat von der Konstanz der Vakuum-Lichtgeschwindigkeit, aus welchem im Verein mit dem speziellen Relativitätsprinzip die Relativität der Gleichzeitigkeit sowie die Lorentztransformation und die mit dieser verknüpften Gesetze über das Verhalten bewegter starrer Körper und Uhren in bekannter Weise folgen.

Die Modifikation, welche die Theorie von Raum und Zeit durch die spezielle Relativitätstheorie erfahren hat, ist zwar eine tiefgehende; aber *ein* wichtiger Punkt blieb unangetastet. Auch gemäß der speziellen Relativitätstheorie sind nämlich die Sätze der Geometrie unmittelbar als die Gesetze über die möglichen relativen Lagen (ruhender) fester Körper zu deuten, allgemeiner die Sätze der Kinematik als Sätze, welche das Verhalten von Meßkörpern und Uhren beschreiben. Zwei hervorgehobenen materiellen Punkten eines ruhenden (starren) Körpers entspricht hierbei stets eine Strecke von ganz bestimmter Länge, unabhängig von Ort und Orientierung des Körpers sowie von der Zeit; zwei hervorgehobenen Zeigerstellungen einer relativ zum (berechtigten) Bezugssystem ruhenden Uhr entspricht stets eine Zeitstrecke von bestimmter Länge, unabhängig von Ort und Zeit. Es wird sich bald zeigen, daß die allgemeine Relativitätstheorie an dieser einfachen physikalischen Deutung von Raum und Zeit nicht festhalten kann.

§ 2. Über die Gründe, welche eine Erweiterung des Relativitätspostulates nahelegen.

Der klassischen Mechanik und nicht minder der speziellen Relativitätstheorie haftet ein erkenntnistheoretischer Mangel an, der vielleicht zum ersten Male von E. Mach klar hervorgehoben wurde. Wir erläutern ihn am folgenden Beispiel. Zwei flüssige Körper von gleicher Größe und Art schweben frei im Raume in so großer Entfernung voneinander (und von allen übrigen Massen), daß nur diejenigen Gravitationskräfte berücksichtigt werden müssen, welche die Teile *eines* dieser Körper aufeinander ausüben. Die Entfernung der Körper voneinander sei unveränderlich. Relative Bewegungen der Teile eines der Körper gegeneinander sollen nicht auftreten. Aber jede Masse soll — von einem relativ zu der anderen Masse ruhenden Beobachter aus beurteilt — um die Verbindungslinie der Massen mit konstanter Winkelgeschwindigkeit rotieren (es ist dies eine konstatierbare Relativbewegung beider Massen). Nun denken wir uns die Oberflächen beider Körper (S_1 und S_2) mit Hilfe (relativ ruhender) Maßstäbe ausgemessen; es ergebe sich, daß die Oberfläche von S_1 eine Kugel, die von S_2 ein Rotationsellipsoid sei.

Wir fragen nun: Aus welchem Grunde verhalten sich die Körper S_1 und S_2 verschieden? Eine Antwort auf diese Frage kann nur dann als erkenntnistheoretisch befriedigend[1]) anerkannt werden, wenn die als Grund angegebene Sache eine *beobachtbare Erfahrungstatsache* ist; denn das Kausalitätsgesetz hat nur dann den Sinn einer Aussage über die Erfahrungswelt, wenn als Ursachen und Wirkungen letzten Endes nur *beobachtbare Tatsachen* auftreten.

Die Newtonsche Mechanik gibt auf diese Frage keine befriedigende Antwort. Sie sagt nämlich folgendes. Die Gesetze der Mechanik gelten wohl für einen Raum R_1, gegen welchen der Körper S_1 in Ruhe ist, nicht aber gegenüber einem Raume R_2, gegen welchen S_2 in Ruhe ist. Der berechtigte Galileische Raum R_1', der hierbei eingeführt wird, ist aber eine *bloß fingierte* Ursache, keine beobachtbare Sache. Es ist also klar, daß die Newtonsche Mechanik der Forderung

---

1) Eine derartige erkenntnistheoretisch befriedigende Antwort kann natürlich immer noch *physikalisch* unzutreffend sein, falls sie mit anderen Erfahrungen im Widerspruch ist.

der Kausalität in dem betrachteten Falle nicht wirklich, sondern nur scheinbar Genüge leistet, indem sie die bloß fingierte Ursache R_1 für das beobachtbare verschiedene Verhalten der Körper S_1 und S_2 verantwortlich macht.

Eine befriedigende Antwort auf die oben aufgeworfene Frage kann nur so lauten: Das aus S_1 und S_2 bestehende physikalische System zeigt für sich allein keine denkbare Ursache, auf welche das verschiedene Verhalten von S_1 und S_2 zurückgeführt werden könnte. Die Ursache muß also *außerhalb* dieses Systems liegen. Man gelangt zu der Auffassung, daß die allgemeinen Bewegungsgesetze, welche im speziellen die Gestalten von S_1 und S_2 bestimmen, derart sein müssen, daß das mechanische Verhalten von S_1 und S_2 ganz wesentlich durch ferne Massen mitbedingt werden muß, welche wir nicht zu dem betrachteten System gerechnet hatten. Diese fernen Massen (und ihre Relativbewegungen gegen die betrachteten Körper) sind dann als Träger prinzipiell beobachtbarer Ursachen für das verschiedene Verhalten unserer betrachteten Körper anzusehen; sie übernehmen die Rolle der fingierten Ursache R_1. Von allen denkbaren, relativ zueinander beliebig bewegten Räumen R_1, R_2 usw. darf a priori keiner als bevorzugt angesehen werden, wenn nicht der dargelegte erkenntnistheoretische Einwand wieder aufleben soll. *Die Gesetze der Physik müssen so beschaffen sein, daß sie in bezug auf beliebig bewegte Bezugssysteme gelten.* Wir gelangen also auf diesem Wege zu einer Erweiterung des Relativitätspostulates.

Außer diesem schwerwiegenden erkenntnistheoretischen Argument spricht aber auch eine wohlbekannte physikalische Tatsache für eine Erweiterung der Relativitätstheorie. Es sei K ein Galileisches Bezugssystem, d. h. ein solches, relativ zu welchem (mindestens in dem betrachteten vierdimensionalen Gebiete) eine von anderen hinlänglich entfernte Masse sich geradlinig und gleichförmig bewegt. Es sei K' ein zweites Koordinatensystem, welches relativ zu K in *gleichförmig beschleunigter* Translationsbewegung sei. Relativ zu K' führte dann eine von anderen hinreichend getrennte Masse eine beschleunigte Bewegung aus, derart, daß deren Beschleunigung und Beschleunigungsrichtung von ihrer stofflichen Zusammensetzung und ihrem physikalischen Zustande unabhängig ist.

Kann ein relativ zu K' ruhender Beobachter hieraus

den Schluß ziehen, daß er sich auf einem „wirklich" beschleunigten Bezugssystem befindet? Diese Frage ist zu verneinen; denn das vorhin genannte Verhalten frei beweglicher Massen relativ zu K' kann ebensogut auf folgende Weise gedeutet werden. Das Bezugssystem K' ist unbeschleunigt; in dem betrachteten zeiträumlichen Gebiete herrscht aber ein Gravitationsfeld, welches die beschleunigte Bewegung der Körper relativ zu K' erzeugt.

Diese Auffassung wird dadurch ermöglicht, daß uns die Erfahrung die Existenz eines Kraftfeldes (nämlich des Gravitationsfeldes) gelehrt hat, welches die merkwürdige Eigenschaft hat, allen Körpern dieselbe Beschleunigung zu erteilen.[1]) Das mechanische Verhalten der Körper relativ zu K' ist dasselbe, wie es gegenüber Systemen sich der Erfahrung darbietet, die wir als „ruhende" bzw. als „berechtigte" Systeme anzusehen gewohnt sind; deshalb liegt es auch vom physikalischen Standpunkt nahe, anzunehmen, daß die Systeme K und K' beide mit demselben Recht als „ruhend" angesehen werden können, bzw. daß sie als Bezugssysteme für die physikalische Beschreibung der Vorgänge gleichberechtigt seien.

Aus diesen Erwägungen sieht man, daß die Durchführung der allgemeinen Relativitätstheorie zugleich zu einer Theorie der Gravitation führen muß; denn man kann ein Gravitationsfeld durch bloße Änderung des Koordinatensystems „erzeugen". Ebenso sieht man unmittelbar, daß das Prinzip von der Konstanz der Vakuum-Lichtgeschwindigkeit eine Modifikation erfahren muß. Denn man erkennt leicht, daß die Bahn eines Lichtstrahles in bezug auf K' im allgemeinen eine krumme sein muß, wenn sich das Licht in bezug auf K geradlinig und mit bestimmter, konstanter Geschwindigkeit fortpflanzt.

§ 3. Das Raum-Zeit-Kontinuum. Forderung der allgemeinen Kovarianz für die die allgemeinen Naturgesetze ausdrückenden Gleichungen.

In der klassischen Mechanik sowie in der speziellen Relativitätstheorie haben die Koordinaten des Raumes und der Zeit eine unmittelbare physikalische Bedeutung. Ein Punktereignis hat die X_1-Koordinate x_1, bedeutet: Die nach den

---

[1]) Daß das Gravitationsfeld diese Eigenschaft mit großer Genauigkeit besitzt, hat Eötvös experimentell bewiesen.

Regeln der Euklidischen Geometrie mittels starrer Stäbe ermittelte Projektion des Punktereignisses auf die X_1-Achse wird erhalten, indem man einen bestimmten Stab, den Einheitsmaßstab, x_1 mal vom Anfangspunkt des Koordinatenkörpers auf der (positiven) X_1-Achse abträgt. Ein Punkt hat die X_4-Koordinate $x_4 = t$, bedeutet: Eine relativ zum Koordinatensystem ruhend angeordnete, mit dem Punktereignis räumlich (praktisch) zusammenfallende Einheitsuhr, welche nach bestimmten Vorschriften gerichtet ist, hat $x_4 = t$ Perioden zurückgelegt beim Eintreten des Punktereignisses.[1])

Diese Auffassung von Raum und Zeit schwebte den Physikern stets, wenn auch meist unbewußt, vor, wie aus der Rolle klar erkennbar ist, welche diese Begriffe in der messenden Physik spielen; diese Auffassung mußte der Leser auch der zweiten Betrachtung des letzten Paragraphen zugrunde legen, um mit diesen Ausführungen einen Sinn verbinden zu können. Aber wir wollen nun zeigen, daß man sie fallen lassen und durch eine allgemeinere ersetzen muß, um das Postulat der allgemeinen Relativität durchführen zu können, falls die spezielle Relativitätstheorie für den Grenzfall des Fehlens eines Gravitationsfeldes zutrifft.

Wir führen in einem Raume, der frei sei von Gravitationsfeldern, ein Galileisches Bezugssystem $K(x, y, z, t)$ ein, und außerdem ein relativ zu K gleichförmig rotierendes Koordinatensystem $K'(x', y', z', t')$. Die Anfangspunkte beider Systeme sowie deren Z-Achsen mögen dauernd zusammenfallen. Wir wollen zeigen, daß für eine Raum—Zeitmessung im System K' die obige Festsetzung für die physikalische Bedeutung von Längen und Zeiten nicht aufrecht erhalten werden kann. Aus Symmetriegründen ist klar, daß ein Kreis um den Anfangspunkt in der X-Y-Ebene von K zugleich als Kreis in der X'-Y'-Ebene von K' aufgefaßt werden kann. Wir denken uns nun Umfang und Durchmesser dieses Kreises mit einem (relativ zum Radius unendlich kleinen) Einheitsmaßstabe ausgemessen und den Quotienten beider Meßresultate gebildet. Würde man dieses Experiment mit einem relativ zum Galileischen System

---

1) Die Konstatierbarkeit der „Gleichzeitigkeit" für räumlich unmittelbar benachbarte Ereignisse, oder — präziser gesagt — für das raumzeitliche unmittelbare Benachbartsein (Koinzidenz) nehmen wir an, ohne für diesen fundamentalen Begriff eine Definition zu geben.

K ruhenden Maßstabe ausführen, so würde man als Quotienten die Zahl π erhalten. Das Resultat der mit einem relativ zu K' ruhenden Maßstabe ausgeführten Bestimmung würde eine Zahl sein, die größer ist als π. Man erkennt dies leicht, wenn man den ganzen Meßprozeß vom „ruhenden" System K aus beurteilt und berücksichtigt, daß der peripherisch angelegte Maßstab eine Lorentzverkürzung erleidet, der radial angelegte Maßstab aber nicht. Es gilt daher in bezug auf K' nicht die Euklidische Geometrie; der oben festgelegte Koordinatenbegriff, welcher die Gültigkeit der Euklidischen Geometrie voraussetzt, versagt also mit Bezug auf das System K'. Ebensowenig kann man in K' eine den physikalischen Bedürfnissen entsprechende Zeit einführen, welche durch relativ zu K' ruhende, gleich beschaffene Uhren angezeigt wird. Um dies einzusehen, denke man sich im Koordinatenursprung und an der Peripherie des Kreises je eine von zwei gleich beschaffenen Uhren angeordnet und vom „ruhenden" System K aus betrachtet. Nach einem bekannten Resultat der speziellen Relativitätstheorie geht — von K aus beurteilt — die auf der Kreisperipherie angeordnete Uhr langsamer als die im Anfangspunkt angeordnete Uhr, weil erstere Uhr bewegt ist, letztere aber nicht. Ein im gemeinsamen Koordinatenursprung befindlicher Beobachter, welcher auch die an der Peripherie befindliche Uhr mittels des Lichtes zu beobachten fähig wäre, würde also die an der Peripherie angeordnete Uhr langsamer gehen sehen als die neben ihm angeordnete Uhr. Da er sich nicht dazu entschließen wird, die Lichtgeschwindigkeit auf dem in Betracht kommenden Wege explizite von der Zeit abhängen zu lassen, wird er seine Beobachtung dahin interpretieren, daß die Uhr an der Peripherie „wirklich" langsamer gehe als die im Ursprung angeordnete. Er wird also nicht umhin können, die Zeit so zu definieren, daß die Ganggeschwindigkeit einer Uhr vom Orte abhängt.

Wir gelangen also zu dem Ergebnis: In der allgemeinen Relativitätstheorie können Raum- und Zeitgrößen nicht so definiert werden, daß räumliche Koordinatendifferenzen unmittelbar mit dem Einheitsmaßstab, zeitliche mit einer Normaluhr gemessen werden könnten.

Das bisherige Mittel, in das zeiträumliche Kontinuum in bestimmter Weise Koordinaten zu legen, versagt also, und es scheint sich auch kein *anderer* Weg darzubieten, der gestatten würde, der vierdimensionalen Welt Koordinatensysteme so anzupassen, daß bei ihrer Verwendung eine besonders einfache Formulierung der Naturgesetze zu erwarten wäre. Es bleibt daher nichts anderes übrig, als alle denkbaren[1]) Koordinatensysteme als für die Naturbeschreibung prinzipiell gleichberechtigt anzusehen. Dies kommt auf die Forderung hinaus:

Die allgemeinen Naturgesetze sind durch Gleichungen auszudrücken, die für alle Koordinatensysteme gelten, d. h. die beliebigen Substitutionen gegenüber kovariant (allgemein kovariant) sind.

Es ist klar, daß eine Physik, welche diesem Postulat genügt, dem allgemeinen Relativitätspostulat gerecht wird. Denn in *allen* Substitutionen sind jedenfalls auch diejenigen enthalten, welche allen Relativbewegungen der (dreidimensionalen) Koordinatensysteme entsprechen. Daß diese Forderung der allgemeinen Kovarianz, welche dem Raum und der Zeit den letzten Rest physikalischer Gegenständlichkeit nehmen, eine natürliche Forderung ist, geht aus folgender Überlegung hervor. Alle unsere zeiträumlichen Konstatierungen laufen stets auf die Bestimmung zeiträumlicher Koinzidenzen hinaus. Bestände beispielsweise das Geschehen nur in der Bewegung materieller Punkte, so wäre letzten Endes nichts beobachtbar als die Begegnungen zweier oder mehrerer dieser Punkte. Auch die Ergebnisse unserer Messungen sind nichts anderes als die Konstatierung derartiger Begegnungen materieller Punkte unserer Maßstäbe mit anderen materiellen Punkten bzw. Koinzidenzen zwischen Uhrzeigern, Zifferblattpunkten und ins Auge gefaßten, am gleichen Orte und zur gleichen Zeit stattfindenden Punktereignissen.

Die Einführung eines Bezugssystems dient zu nichts anderem als zur leichteren Beschreibung der Gesamtheit solcher Koinzidenzen. Man ordnet der Welt vier zeiträumliche Variable x_1, x_2, x_3, x_4 zu, derart, daß jedem Punktereignis ein Wertesystem der Variablen $x_1 \ldots x_4$ entspricht. Zwei koinzidierenden Punktereignissen entspricht dasselbe

---

[1]) Von gewissen Beschränkungen, welche der Forderung der eindeutigen Zuordnung und derjenigen der Stetigkeit entsprechen, wollen wir hier nicht sprechen.

Wertesystem der Variablen $x_1 \ldots x_4$; d. h. die Koinzidenz ist durch die Übereinstimmung der Koordinaten charakterisiert. Führt man statt der Variablen $x_1 \ldots x_4$ beliebige Funktionen derselben, x_1', x_2', x_3', x_4' als neues Koordinatensystem ein, so daß die Wertesysteme einander eindeutig zugeordnet sind, so ist die Gleichheit aller vier Koordinaten auch im neuen System der Ausdruck für die raumzeitliche Koinzidenz zweier Punktereignisse. Da sich alle unsere physikalischen Erfahrungen letzten Endes auf solche Koinzidenzen zurückführen lassen, ist zunächst kein Grund vorhanden, gewisse Koordinatensysteme vor anderen zu bevorzugen, d. h. wir gelangen zu der Forderung der allgemeinen Kovarianz.

§ 4. Beziehung der vier Koordinaten zu räumlichen und zeitlichen Meßergebnissen. Analytischer Ausdruck für das Gravitationsfeld.

Es kommt mir in dieser Abhandlung nicht darauf an, die allgemeine Relativitätstheorie als ein möglichst einfaches logisches System mit einem Minimum von Axiomen darzustellen. Sondern es ist mein Hauptziel, diese Theorie so zu entwickeln, daß der Leser die psychologische Natürlichkeit des eingeschlagenen Weges empfindet und daß die zugrunde gelegten Voraussetzungen durch die Erfahrung möglichst gesichert erscheinen. In diesem Sinne sei nun die Voraussetzung eingeführt:

Für unendlich kleine vierdimensionale Gebiete ist die Relativitätstheorie im engeren Sinne bei passender Koordinatenwahl zutreffend.

Der Beschleunigungszustand des unendlich kleinen („örtlichen") Koordinatensystems ist hierbei so zu wählen, daß ein Gravitationsfeld nicht auftritt; dies ist für ein unendlich kleines Gebiet möglich. X_1, X_2, X_3 seien die räumlichen Koordinaten; X_4 die zugehörige, in geeignetem Maßstabe gemessene[1]) Zeitkoordinate. Diese Koordinaten haben, wenn ein starres Stäbchen als Einheitsmaßstab gegeben gedacht wird, bei gegebener Orientierung des Koordinatensystems eine unmittelbare physikalische Bedeutung im Sinne der speziellen Relativitätstheorie. Der Ausdruck

(1) $$ds^2 = -dX_1^2 - dX_2^2 - dX_3^2 + dX_4^2$$

---

1) Die Zeiteinheit ist so zu wählen, daß die Vakuum-Lichtgeschwindigkeit — in dem „lokalen" Koordinatensystem gemessen — gleich 1 wird.

hat dann nach der speziellen Relativitätstheorie einen von der Orientierung des lokalen Koordinatensystems unabhängigen, durch Raum—Zeitmessung ermittelbaren Wert. Wir nennen ds die Größe des zu den unendlich benachbarten Punkten des vierdimensionalen Raumes gehörigen Linienelementes. Ist das zu dem Element $(dX_1 \ldots dX_4)$ gehörige ds^2 positiv, so nennen wir mit Minkowski ersteres zeitartig, im entgegengesetzten Falle raumartig.

Zu dem betrachteten „Linienelement" bzw. zu den beiden unendlich benachbarten Punktereignissen gehören auch bestimmte Differentiale $dx_1 \ldots dx_4$ der vierdimensionalen Koordinaten des gewählten Bezugssystems. Ist dieses sowie ein „lokales" System obiger Art für die betrachtete Stelle gegeben, so werden sich hier die dX_ν durch bestimmte lineare homogene Ausdrücke der dx_σ darstellen lassen:

(2) $$dX_\nu = \sum_\sigma \alpha_{\nu\sigma} dx_\sigma.$$

Setzt man diese Ausdrücke in (1) ein, so erhält man

(3) $$ds^2 = \sum_{\sigma\tau} g_{\sigma\tau} dx_\sigma dx_\tau,$$

wobei die $g_{\sigma\tau}$ Funktionen der x_σ sein werden, die nicht mehr von der Orientierung und dem Bewegungszustand des „lokalen" Koordinatensystems abhängen können; denn ds^2 ist eine durch Maßstab-Uhrenmessung ermittelbare, zu den betrachteten, zeiträumlich unendlich benachbarten Punktereignissen gehörige, unabhängig von jeder besonderen Koordinatenwahl definierte Größe. Die $g_{\sigma\tau}$ sind hierbei so zu wählen, daß $g_{\sigma\tau} = g_{\tau\sigma}$ ist; die Summation ist über alle Werte von σ und τ zu erstrecken, so daß die Summe aus 4×4 Summanden besteht, von denen 12 paarweise gleich sind.

Der Fall der gewöhnlichen Relativitätstheorie geht aus dem hier Betrachteten hervor, falls es, vermöge des besonderen Verhaltens der $g_{\sigma\tau}$ in einem endlichen Gebiete, möglich ist, in diesem das Bezugssystem so zu wählen, daß die $g_{\sigma\tau}$ die konstanten Werte

(4) $$\left\{ \begin{array}{cccc} -1 & 0 & 0 & 0 \\ 0 & -1 & 0 & 0 \\ 0 & 0 & -1 & 0 \\ 0 & 0 & 0 & +1 \end{array} \right.$$

annehmen. Wir werden später sehen, daß die Wahl solcher Koordinaten für endliche Gebiete im allgemeinen nicht möglich ist.

Aus den Betrachtungen der §§ 2 und 3 geht hervor, daß die Größen $g_{\sigma\tau}$ vom physikalischen Standpunkte aus als diejenigen Größen anzusehen sind, welche das Gravitationsfeld in bezug auf das gewählte Bezugssystem beschreiben. Nehmen wir nämlich zunächst an, es sei für ein gewisses betrachtetes vierdimensionales Gebiet bei geeigneter Wahl der Koordinaten die spezielle Relativitätstheorie gültig. Die $g_{\sigma\tau}$ haben dann die in (4) angegebenen Werte. Ein freier materieller Punkt bewegt sich dann bezüglich dieses Systems geradlinig gleichförmig. Führt man nun durch eine beliebige Substitution neue Raum—Zeitkoordinaten $x_1 \ldots x_4$ ein, so werden in diesem neuen System die $g_{\mu\nu}$ nicht mehr Konstante, sondern Raum—Zeitfunktionen sein. Gleichzeitig wird sich die Bewegung des freien Massenpunktes in den neuen Koordinaten als eine krummlinige, nicht gleichförmige, darstellen, wobei dies Bewegungsgesetz unabhängig sein wird von der Natur des bewegten Massenpunktes. Wir werden also diese Bewegung als eine solche unter dem Einfluß eines Gravitationsfeldes deuten. Wir sehen das Auftreten eines Gravitationsfeldes geknüpft an eine raumzeitliche Veränderlichkeit der $g_{\sigma\tau}$. Auch in dem allgemeinen Falle, daß wir nicht in einem endlichen Gebiete bei passender Koordinatenwahl die Gültigkeit der speziellen Relativitätstheorie herbeiführen können, werden wir an der Auffassung festzuhalten haben, daß die $g_{\sigma\tau}$ das Gravitationsfeld beschreiben.

Die Gravitation spielt also gemäß der allgemeinen Relativitätstheorie eine Ausnahmerolle gegenüber den übrigen, insbesondere den elektromagnetischen Kräften, indem die das Gravitationsfeld darstellenden 10 Funktionen $g_{\sigma\tau}$ zugleich die metrischen Eigenschaften des vierdimensionalen Meßraumes bestimmen.

B. Mathematische Hilfsmittel für die Aufstellung allgemein kovarianter Gleichungen.

Nachdem wir im vorigen gesehen haben, daß das allgemeine Relativitätspostulat zu der Forderung führt, daß die Gleichungssysteme der Physik beliebigen Substitutionen der Koordinaten $x_1 \ldots x_4$ gegenüber kovariant sein müssen, haben wir zu überlegen, wie derartige allgemein kovariante Gleichungen gewonnen werden können. Dieser rein mathematischen Aufgabe wenden wir uns jetzt zu; es wird sich dabei zeigen, daß bei deren Lösung die in Gleichung (3) angegebene Invariante ds eine fundamentale Rolle spielt, welche wir in Anlehnung an die Gausssche Flächentheorie als „Linienelement" bezeichnet haben.

Der Grundgedanke dieser allgemeinen Kovariantentheorie ist folgender. Es seien gewisse Dinge („Tensoren") mit Bezug auf jedes Koordinatensystem definiert durch eine Anzahl Raumfunktionen, welche die „Komponenten" des Tensors genannt werden. Es gibt dann gewisse Regeln, nach welchen diese Komponenten für ein neues Koordinatensystem berechnet werden, wenn sie für das ursprüngliche System bekannt sind, und wenn die beide Systeme verknüpfende Transformation bekannt ist. Die nachher als Tensoren bezeichneten Dinge sind ferner dadurch gekennzeichnet, daß die Transformationsgleichungen für ihre Komponenten linear und homogen sind. Demnach verschwinden sämtliche Komponenten im neuen System, wenn sie im ursprünglichen System sämtlich verschwinden. Wird also ein Naturgesetz durch das Nullsetzen aller Komponenten eines Tensors formuliert, so ist es allgemein kovariant; indem wir die Bildungsgesetze der Tensoren untersuchen, erlangen wir die Mittel zur Aufstellung allgemein kovarianter Gesetze.

§ 5. Kontravarianter und kovarianter Vierervektor.

Kontravarianter Vierervektor. Das Linienelement ist definiert durch die vier „Komponenten" dx_ν, deren Transformationsgesetz durch die Gleichung

$$(5) \qquad dx_\sigma' = \sum_\nu \frac{\partial x_\sigma'}{\partial x_\nu} dx_\nu$$

ausgedrückt wird. Die dx_σ' drücken sich linear und homogen durch die dx_ν aus; wir können diese Koordinatendifferentiale dx_ν daher als die Komponenten eines „Tensors" ansehen, den wir speziell als kontravarianten Vierervektor bezeichnen. Jedes Ding, was bezüglich des Koordinatensystems durch vier Größen A^ν definiert ist, die sich nach demselben Gesetz

$$(5\,\text{a}) \qquad A^{\sigma'} = \sum_\nu \frac{\partial x_\sigma'}{\partial x_\nu} A^\nu$$

transformieren, bezeichnen wir ebenfalls als kontravarianten Vierervektor. Aus (5a) folgt sogleich, daß die Summen $(A^\sigma \pm B^\sigma)$ ebenfalls Komponenten eines Vierervektors sind, wenn A^σ und B^σ es sind. Entsprechendes gilt für alle später als „Tensoren" einzuführenden Systeme (Regel von der Addition und Subtraktion der Tensoren).

Kovarianter Vierervektor. Vier Größen A_ν nennen wir die Komponenten eines kovarianten Vierervektors, wenn für jede beliebige Wahl des kontravarianten Vierervektors B^ν

$$(6) \qquad \sum_\nu A_\nu B^\nu = \text{Invariante.}$$

Aus dieser Definition folgt das Transformationsgesetz des kovarianten Vierervektors. Ersetzt man nämlich auf der rechten Seite der Gleichung

$$\sum_\sigma A_\sigma' B^{\sigma'} = \sum_\nu A_\nu B^\nu$$

B^ν durch den aus der Umkehrung der Gleichung (5a) folgenden Ausdruck

$$\sum_\sigma \frac{\partial x_\nu}{\partial x_\sigma'} B^{\sigma'},$$

so erhält man

$$\sum_\sigma B^{\sigma'} \sum_\nu \frac{\partial x_\nu}{\partial x_\sigma'} A_\nu = \sum_\sigma B^{\sigma'} A_\sigma'.$$

Hieraus folgt aber, weil in dieser Gleichung die $B^{\sigma'}$ unabhängig voneinander frei wählbar sind, das Transformationsgesetz

$$(7) \qquad A_\sigma' = \sum \frac{\partial x_\nu}{\partial x_\sigma'} A_\nu.$$

Bemerkung zur Vereinfachung der Schreibweise der Ausdrücke.

Ein Blick auf die Gleichungen dieses Paragraphen zeigt, daß über Indizes, die zweimal unter einem Summenzeichen auftreten [z. B. der Index ν in (5)], stets summiert wird, und zwar *nur* über zweimal auftretende Indizes. Es ist deshalb möglich, ohne die Klarheit zu beeinträchtigen, die Summenzeichen wegzulassen. Dafür führen wir die Vorschrift ein: Tritt ein Index in einem Term eines Ausdruckes zweimal auf, so ist über ihn stets zu summieren, wenn nicht ausdrücklich das Gegenteil bemerkt ist.

Der Unterschied zwischen dem kovarianten und kontravarianten Vierervektor liegt in dem Transformationsgesetz

[(7) bzw. (5)]. Beide Gebilde sind Tensoren im Sinne der obigen allgemeinen Bemerkung; hierin liegt ihre Bedeutung. Im Anschluß an Ricci und Levi-Civita wird der kontravariante Charakter durch oberen, der kovariante durch unteren Index bezeichnet.

§ 6. Tensoren zweiten und höheren Ranges.

Kontravarianter Tensor. Bilden wir sämtliche 16 Produkte $A^{\mu\nu}$ der Komponenten A^μ und B^ν zweier kontravarianten Vierervektoren

(8) $$A^{\mu\nu} = A^\mu B^\nu,$$

so erfüllt $A^{\mu\nu}$ gemäß (8) und (5a) das Transformationsgesetz

(9) $$A^{\sigma\tau\prime} = \frac{\partial x_\sigma'}{\partial x_\mu} \frac{\partial x_\tau'}{\partial x_\nu} A^{\mu\nu}.$$

Wir nennen ein Ding, das bezüglich eines jeden Bezugssystems durch 16 Größen (Funktionen) beschrieben wird, die das Transformationsgesetz (9) erfüllen, einen kontravarianten Tensor zweiten Ranges. Nicht jeder solcher Tensor läßt sich gemäß (8) aus zwei Vierervektoren bilden. Aber es ist leicht zu beweisen, daß sich 16 beliebig gegebene $A^{\mu\nu}$ darstellen lassen als die Summe der $A^\mu B^\nu$ von vier geeignet gewählten Paaren von Vierervektoren. Deshalb kann man beinahe alle Sätze, die für den durch (9) definierten Tensor zweiten Ranges gelten, am einfachsten dadurch beweisen, daß man sie für spezielle Tensoren vom Typus (8) dartut.

Kontravarianter Tensor beliebigen Ranges. Es ist klar, daß man entsprechend (8) und (9) auch kontravariante Tensoren dritten und höheren Ranges definieren kann mit 4^3 usw. Komponenten. Ebenso erhellt aus (8) und (9), daß man in diesem Sinne den kontravarianten Vierervektor als kontravarianten Tensor ersten Ranges auffassen kann.

Kovarianter Tensor. Bildet man andererseits die 16 Produkte $A_{\mu\nu}$ der Komponenten zweier *kovarianter* Vierervektoren A_μ und B_ν

(10) $$A_{\mu\nu} = A_\mu B_\nu,$$

so gilt für diese das Transformationsgesetz

(11) $$A_{\sigma\tau}' = \frac{\partial x_\mu}{\partial x_\sigma'} \frac{\partial x_\nu}{\partial x_\tau'} A_{\mu\nu}.$$

Durch dieses Transformationsgesetz wird der kovariante Tensor zweiten Ranges definiert. Alle Bemerkungen, welche vorher über die kontravarianten Tensoren gemacht wurden, gelten auch für die kovarianten Tensoren.

Bemerkung. Es ist bequem, den Skalar (Invariante) sowohl als kontravarianten wie als kovarianten Tensor vom Range Null zu behandeln.

Gemischter Tensor. Man kann auch einen Tensor zweiten Ranges vom Typus

$$(12) \qquad A_\mu{}^\nu = A_\mu B^\nu$$

definieren, der bezüglich des Index μ kovariant, bezüglich des Index ν kontravariant ist. Sein Transformationsgesetz ist

$$(13) \qquad A_\sigma{}^{\tau'} = \frac{\partial x_\tau'}{\partial x_\beta} \frac{\partial x_\alpha}{\partial x_\sigma'} A_\alpha{}^\beta.$$

Natürlich gibt es gemischte Tensoren mit beliebig vielen Indizes kovarianten und beliebig vielen Indizes kontravarianten Charakters. Der kovariante und der kontravariante Tensor können als spezielle Fälle des gemischten angesehen werden.

Symmetrische Tensoren. Ein kontravarianter bzw. kovarianter Tensor zweiten oder höheren Ranges heißt *symmetrisch*, wenn zwei Komponenten, die durch Vertauschung irgend zweier Indizes auseinander hervorgehen, gleich sind. Der Tensor $A^{\mu\nu}$ bzw. $A_{\mu\nu}$ ist also symmetrisch, wenn für jede Kombination der Indizes

$$(14) \qquad A^{\mu\nu} = A^{\nu\mu},$$

bzw.

$$(14\text{a}) \qquad A_{\mu\nu} = A_{\nu\mu}$$

ist.

Es muß bewiesen werden, daß die so definierte Symmetrie eine vom Bezugssystem unabhängige Eigenschaft ist. (Aus (9) folgt in der Tat mit Rücksicht auf (14)

$$A^{\sigma'\tau'} = \frac{\partial x_\sigma'}{\partial x_\mu} \frac{\partial x_\tau'}{\partial x_\nu} A^{\mu\nu} = \frac{\partial x_\sigma'}{\partial x_\mu} \frac{\partial x_\tau'}{\partial x_\nu} A^{\nu\mu} = \frac{\partial x_\tau'}{\partial x_\mu} \frac{\partial x_\sigma'}{\partial x_\nu} A^{\mu\nu} = A^{\tau'\sigma'}.$$

Die vorletzte Gleichsetzung beruht auf der Vertauschung der Summationsindizes μ und ν (d. h. auf bloßer Änderung der Bezeichnungsweise).

Antisymmetrische Tensoren. Ein kontravarianter bzw. kovarianter Tenor zweiten, dritten oder vierten Ranges heißt antisymmetrisch, wenn zwei Komponenten, die durch Vertauschung irgend zweier Indizes auseinander hervorgehen, *entgegengesetzt gleich* sind. Der Tensor $A^{\mu\nu}$ bzw. $A_{\mu\nu}$ ist also antisymmetrisch, wenn stets

(15) $$A^{\mu\nu} = - A^{\nu\mu},$$

bzw.

(15a) $$A_{\mu\nu} = - A_{\nu\mu}$$

ist.

Von den 16 Komponenten $A^{\mu\nu}$ verschwinden die vier Komponenten $A^{\mu\mu}$; die übrigen sind paarweise entgegengesetzt gleich, so daß nur 6 numerisch verschiedene Komponenten vorhanden sind (Sechservektor). Ebenso sieht man, daß der antisymmetrische Tensor $A^{\mu\nu\sigma}$ (dritten Ranges) nur vier numerisch verschiedene Komponenten hat, der antisymmetrische Tensor $A^{\mu\nu\sigma\tau}$ nur eine einzige. Symmetrische Tensoren höheren als vierten Ranges gibt es in einem Kontinuum von vier Dimensionen nicht.

§ 7. Multiplikation der Tensoren.

Äußere Multiplikation der Tensoren. Man erhält aus den Komponenten eines Tensors vom Range z und eines solchen vom Range z' die Komponenten eines Tensors vom Range $z + z'$, indem man alle Komponenten des ersten mit allen Komponenten des zweiten paarweise multipliziert. So entstehen beispielsweise die Tensoren T aus den Tensoren A und B verschiedener Art

$$\begin{aligned} T_{\mu\nu\sigma} &= A_{\mu\nu} B_{\sigma}, \\ T^{\alpha\beta\gamma\delta} &= A^{\alpha\beta} B^{\gamma\delta}, \\ T^{\gamma\delta}_{\alpha\beta} &= A_{\alpha\beta} B^{\gamma\delta}. \end{aligned}$$

Der Beweis des Tensorcharakters der T ergibt sich unmittelbar aus den Darstellungen (8), (10), (12) oder aus den Transformationsregeln (9), (11), (13). Die Gleichungen (8), (10), (12) sind selbst Beispiele äußerer Multiplikation (von Tensoren ersten Ranges).

„Verjüngung" eines gemischten Tensors. Aus jedem gemischten Tensor kann ein Tensor von einem um zwei kleineren Range gebildet werden, indem man einen Index kovarianten und einen Index kontravarianten Charakters gleichsetzt und

nach diesem Index summiert („Verjüngung"). Man gewinnt so z. B. aus dem gemischten Tensor vierten Ranges $A^{\gamma\delta}_{\alpha\beta}$ den gemischten Tensor zweiten Ranges

$$A^{\delta}_{\beta} = A^{\alpha\delta}_{\alpha\beta} \left(= \sum_{\alpha} A^{\alpha\delta}_{\alpha\beta} \right)$$

und aus diesem, abermals durch Verjüngung, den Tensor nullten Ranges $A = A^{\beta}_{\beta} = A^{\alpha\beta}_{\alpha\beta}$.

Der Beweis dafür, daß das Ergebnis der Verjüngung wirklich Tensorcharakter besitzt, ergibt sich entweder aus der Tensordarstellung gemäß der Verallgemeinerung von (12) in Verbindung mit (6) oder aus der Verallgemeinerung von (13).

Innere und gemischte Multiplikation der Tensoren. Diese bestehen in der Kombination der äußeren Multiplikation mit der Verjüngung.

Beispiele. — Aus dem kovarianten Tensor zweiten Ranges $A_{\mu\nu}$ und dem kontravarianten Tensor ersten Ranges B^{σ} bilden wir durch äußere Multiplikation den gemischten Tensor

$$D^{\sigma}_{\mu\nu} = A_{\mu\nu} B^{\sigma}.$$

Durch Verjüngung nach den Indizes ν, σ entsteht der kovariante Vierervektor

$$D_{\mu} = D^{\nu}_{\mu\nu} = A_{\mu\nu} B^{\nu}.$$

Diesen bezeichnen wir auch als inneres Produkt der Tensoren $A_{\mu\nu}$ und B^{σ}. Analog bildet man aus den Tensoren $A_{\mu\nu}$ und $B^{\sigma\tau}$ durch äußere Multiplikation und zweimalige Verjüngung das innere Produkt $A_{\mu\nu} B^{\mu\nu}$. Durch äußere Produktbildung und einmalige Verjüngung erhält man aus $A_{\mu\nu}$ und $B^{\sigma\tau}$ den gemischten Tensor zweiten Ranges $D^{\tau}_{\mu} = A_{\mu\nu} B^{\nu\tau}$. Man kann diese Operation passend als eine gemischte bezeichnen; denn sie ist eine äußere bezüglich der Indizes μ und τ, eine innere bezüglich der Indizes ν und σ.

Wir beweisen nun einen Satz, der zum Nachweis des Tensorcharakters oft verwendbar ist. Nach dem soeben Dargelegten ist $A_{\mu\nu} B^{\mu\nu}$ ein Skalar, wenn $A_{\mu\nu}$ und $B^{\sigma\tau}$ Tensoren sind. Wir behaupten aber auch folgendes. Wenn $A_{\mu\nu} B^{\mu\nu}$ für jede Wahl *des Tensors* $B^{\mu\nu}$ eine Invariante ist, so hat $A_{\mu\nu}$ Tensorcharakter.

Beweis. — Es ist nach Voraussetzung für eine beliebige Substitution

$$A'_{\sigma\tau} B^{\sigma\tau\prime} = A_{\mu\nu} B^{\mu\nu}.$$

Nach der Umkehrung von (9) ist aber
$$B^{\mu\nu} = \frac{\partial x_\mu}{\partial x_\sigma'} \frac{\partial x_\nu}{\partial x_\tau'} B^{\sigma\tau'}.$$
Dies, eingesetzt in obige Gleichung, liefert:
$$\left(A_{\sigma\tau}' - \frac{\partial x_\mu}{\partial x_\sigma'} \frac{\partial x_\nu}{\partial x_\tau'} A_{\mu\nu}\right) B^{\sigma\tau'} = 0.$$

Dies kann bei beliebiger Wahl von $B^{\sigma\tau'}$ nur dann erfüllt sein, wenn die Klammer verschwindet, woraus mit Rücksicht auf (11) die Behauptung folgt.

Dieser Satz gilt entsprechend für Tensoren beliebigen Ranges und Charakters; der Beweis ist stets analog zu führen.

Der Satz läßt sich ebenso beweisen in der Form: Sind B^μ und C^ν beliebige Vektoren, und ist bei jeder Wahl derselben das innere Produkt
$$A_{\mu\nu} B^\mu C^\nu$$
ein Skalar, so ist $A_{\mu\nu}$ ein kovarianter Tensor. Dieser letztere Satz gilt auch dann noch, wenn nur die speziellere Aussage zutrifft, daß bei beliebiger Wahl des Vierervektors B^μ das skalare Produkt
$$A_{\mu\nu} B^\mu B^\nu$$
ein Skalar ist, falls man außerdem weiß, daß $A_{\mu\nu}$ der Symmetriebedingung $A_{\mu\nu} = A_{\nu\mu}$ genügt. Denn auf dem vorhin angegebenen Wege beweist man den Tensorcharakter von $(A_{\mu\nu} + A_{\nu\mu})$, woraus dann wegen der Symmetrieeigenschaft der Tensorcharakter von $A_{\mu\nu}$ selbst folgt. Auch dieser Satz läßt sich leicht verallgemeinern auf den Fall kovarianter und kontravarianter Tensoren beliebigen Ranges.

Endlich folgt aus dem Bewiesenen der ebenfalls auf beliebige Tensoren zu verallgemeinernde Satz: Wenn die Größen $A_{\mu\nu} B^\nu$ bei beliebiger Wahl des Vierervektors B^ν einen Tensor ersten Ranges bilden, so ist $A_{\mu\nu}$ ein Tensor zweiten Ranges. Ist nämlich C^μ ein beliebiger Vierervektor, so ist wegen des Tensorcharakters $A_{\mu\nu} B^\nu$ das innere Produkt $A_{\mu\nu} C^\mu B^\nu$ bei beliebiger Wahl der beiden Vierervektoren C^μ und B^ν ein Skalar, woraus die Behauptung folgt.

§ 8. Einiges über den Fundamentaltensor der $g_{\mu\nu}$.

Der kovariante Fundamentaltensor. In dem invarianten Ausdruck des Quadrates des Linienelementes
$$ds^2 = g_{\mu\nu} dx_\mu dx_\nu$$

spielt dx_μ die Rolle eines beliebig wählbaren kontravarianten Vektors. Da ferner $g_{\mu\nu} = g_{\nu\mu}$, so folgt nach den Betrachtungen des letzten Paragraphen hieraus, daß $g_{\mu\nu}$ ein kovarianter Tensor zweiten Ranges ist. Wir nennen ihn „Fundamentaltensor". Im folgenden leiten wir einige Eigenschaften dieses Tensors ab, die zwar jedem Tensor zweiten Ranges eigen sind; aber die besondere Rolle des Fundamentaltensors in unserer Theorie, welche in der Besonderheit der Gravitationswirkungen ihren physikalischen Grund hat, bringt es mit sich, daß die zu entwickelnden Relationen nur bei dem Fundamentaltensor für uns von Bedeutung sind.

Der kontravariante Fundamentaltensor. Bildet man in dem Determinantenschema der $g_{\mu\nu}$ zu jedem $g_{\mu\nu}$ die Unterdeterminante und dividiert diese durch die Determinante $g = |g_{\mu\nu}|$ der $g_{\mu\nu}$, so erhält man gewisse Größen $g^{\mu\nu} (= g^{\nu\mu})$, von denen wir beweisen wollen, daß sie einen kontravarianten Tensor bilden.

Nach einem bekannten Determinantensatze ist

$$(16) \quad g_{\mu\sigma} g^{\nu\sigma} = \delta_\mu^\nu,$$

wobei das Zeichen δ_μ^ν 1 oder 0 bedeutet, je nachdem $\mu = \nu$ oder $\mu \ne \nu$ ist. Statt des obigen Ausdruckes für ds^2 können wir auch

$$g_{\mu\sigma} \delta_\nu^\sigma dx_\mu dx_\nu,$$

oder nach (16) auch

$$g_{\mu\sigma} g_{\nu\tau} g^{\sigma\tau} dx_\mu dx_\nu$$

schreiben. Nun bilden aber nach den Multiplikationsregeln des vorigen Paragraphen die Größen

$$d\xi_\sigma = g_{\mu\sigma} dx_\mu$$

einen kovarianten Vierervektor, und zwar (wegen der willkürlichen Wählbarkeit der dx_μ) einen beliebig wählbaren Vierervektor. Indem wir ihn in unseren Ausdruck einführen, erhalten wir

$$ds^2 = g^{\sigma\tau} d\xi_\sigma d\xi_\tau.$$

Da dies bei beliebiger Wahl des Vektors $d\xi_\sigma$ ein Skalar ist und $g^{\sigma\tau}$ nach seiner Definition in den Indizes σ und τ symmetrisch ist, folgt aus den Ergebnissen des vorigen Paragraphen, daß $g^{\sigma\tau}$ ein kontravarianter Tensor ist. Aus (16) folgt noch, daß auch δ_μ^ν ein Tensor ist, den wir den gemischten Fundamentaltensor nennen können.

Determinante des Fundamentaltensors. Nach dem Multiplikationssatz der Determinanten ist

$$|g_{\mu\alpha} g^{\alpha\nu}| = |g_{\mu\alpha}| |g^{\alpha\nu}|.$$

Andererseits ist

$$|g_{\mu\alpha} g^{\alpha\nu}| = |\delta_\mu^{\;\nu}| = 1.$$

Also folgt

(17) $$|g_{\mu\nu}| |g^{\mu\nu}| = 1.$$

Invariante des Volumens. Wir suchen zuerst das Transformationsgesetz der Determinante $g = |g_{\mu\nu}|$. Gemäß (11) ist

$$g' = \left| \frac{\partial x_\mu}{\partial x_\sigma'} \frac{\partial x_\nu}{\partial x_\tau'} g_{\mu\nu} \right|.$$

Hieraus folgt durch zweimalige Anwendung des Multiplikationssatzes der Determinanten

$$g' = \left| \frac{\partial x_\mu}{\partial x_\sigma'} \right| \left| \frac{\partial x_\nu}{\partial x_\tau'} \right| |g_{\mu\nu}| = \left| \frac{\partial x_\mu}{\partial x_\sigma'} \right|^2 g,$$

oder

$$\sqrt{g'} = \left| \frac{\partial x_\mu}{\partial x_\sigma'} \right| \sqrt{g}.$$

Andererseits ist das Gesetz der Transformation des Volumelementes

$$d\tau' = \int dx_1 \, dx_2 \, dx_3 \, dx_4$$

nach dem bekannten Jakobischen Satze

$$d\tau' = \left| \frac{\partial x_\sigma'}{\partial x_\mu} \right| d\tau.$$

Durch Multiplikation der beiden letzten Gleichungen erhält man

(18) $$\sqrt{g'} \, d\tau' = \sqrt{g} \, d\tau.$$

Statt $\sqrt{g}$ wird im folgenden die Größe $\sqrt{-g}$ eingeführt, welche wegen des hyperbolischen Charakters des zeiträumlichen Kontinuums stets einen reellen Wert hat. Die Invariante $\sqrt{-g} \, d\tau$ ist gleich der Größe des im „örtlichen Bezugssystem" mit starren Maßstäben und Uhren im Sinne der speziellen Relativitätstheorie gemessenen vierdimensionalen Volumelementes.

Bemerkung über den Charakter des raumzeitlichen Kontinuums. Unsere Voraussetzung, daß im unendlich Kleinen stets die spezielle Relativitätstheorie gelte, bringt es mit sich, daß sich ds^2 immer gemäß (1) durch die reellen Größen $dX_1 \ldots dX_4$ ausdrücken läßt. Nennen wir $d\tau_0$ das „natürliche" Volumelement $dX_1\, dX_2\, dX_3\, dX_4$, so ist also

(18a) $$d\tau_0 = \sqrt{-g}\, d\tau.$$

Soll an einer Stelle des vierdimensionalen Kontinuums $\sqrt{-g}$ verschwinden, so bedeutet dies, daß hier einem endlichen Koordinatenvolumen ein unendlich kleines „natürliches" Volumen entspreche. Dies möge nirgends der Fall sein. Dann kann g sein Vorzeichen nicht ändern; wir werden im Sinne der speziellen Relativitätstheorie annehmen, daß g stets einen endlichen negativen Wert habe. Es ist dies eine Hypothese über die physikalische Natur des betrachteten Kontinuums und gleichzeitig eine Festsetzung über die Koordinatenwahl.

Ist aber $-g$ stets positiv und endlich, so liegt es nahe, die Koordinatenwahl a posteriori so zu treffen, daß diese Größe gleich 1 wird. Wir werden später sehen, daß durch eine solche Beschränkung der Koordinatenwahl eine bedeutende Vereinfachung der Naturgesetze erzielt werden kann. An Stelle von (18) tritt dann einfach

$$d\tau' = d\tau,$$

woraus mit Rücksicht auf Jakobis Satz folgt

(19) $$\left| \frac{\partial x_\sigma'}{\partial x_\mu} \right| = 1.$$

Bei dieser Koordinatenwahl sind also nur Substitutionen der Koordinaten von der Determinante 1 zulässig.

Es wäre aber irrtümlich, zu glauben, daß dieser Schritt einen partiellen Verzicht auf das allgemeine Relativitätspostulat bedeute. Wir fragen nicht: „Wie heißen die Naturgesetze, welche gegenüber allen Transformationen von der Determinante 1 kovariant sind?" Sondern wir fragen: „Wie heißen die *allgemein* kovarianten Naturgesetze?" Erst nachdem wir diese aufgestellt haben, vereinfachen wir ihren Ausdruck durch eine besondere Wahl des Bezugssystems.

Bildung neuer Tensoren vermittelst des Fundamentaltensors. Durch innere, äußere und gemischte Multiplikation eines Tensors mit dem Fundamentaltensor entstehen Tensoren anderen Charakters und Ranges.

Beispiele:
$$A^\mu = g^{\mu\sigma} A_\sigma,$$
$$A = g_{\mu\nu} A^{\mu\nu}.$$

Besonders sei auf folgende Bildungen hingewiesen:
$$A^{\mu\nu} = g^{\mu\alpha} g^{\nu\beta} A_{\alpha\beta},$$
$$A_{\mu\nu} = g_{\mu\alpha} g_{\nu\beta} A^{\alpha\beta}$$

("Ergänzung" des kovarianten bzw. kontravarianten Tensors) und
$$B_{\mu\nu} = g_{\mu\nu} g^{\alpha\beta} A_{\alpha\beta}.$$

Wir nennen $B_{\mu\nu}$ den zu $A_{\mu\nu}$ gehörigen reduzierten Tensor. Analog
$$B^{\mu\nu} = g^{\mu\nu} g_{\alpha\beta} A^{\alpha\beta}.$$

Es sei bemerkt, daß $g^{\mu\nu}$ nichts anderes ist als die Ergänzung von $g_{\mu\nu}$. Denn man hat
$$g^{\mu\alpha} g^{\nu\beta} g_{\alpha\beta} = g^{\mu\alpha} \delta_\alpha{}^\nu = g^{\mu\nu}.$$

§ 9. Gleichung der geodätischen Linie (bzw. der Punktbewegung).

Da das "Linienelement" ds eine unabhängig vom Koordinatensystem definierte Größe ist, hat auch die zwischen zwei Punkten P_1 und P_2 des vierdimensionalen Kontinuums gezogene Linie, für welche $\int ds$ ein Extremum ist (geodätische Linie), eine von der Koordinatenwahl unabhängige Bedeutung. Ihre Gleichung ist

(20) $$\delta \left\{ \int_{P_1}^{P_2} ds \right\} = 0.$$

Aus dieser Gleichung findet man in bekannter Weise durch Ausführung der Variation vier totale Differentialgleichungen, welche diese geodätische Linie bestimmen; diese Ableitung soll der Vollständigkeit halber hier Platz finden. Es sei λ eine Funktion der Koordinaten x_ν; diese definiert eine Schar von Flächen, welche die gesuchte geodätische Linie sowie alle ihr unendlich benachbarten, durch die Punkte P_1 und P_2 gezogenen Linien schneiden. Jede solche Kurve kann dann dadurch gegeben gedacht werden, daß ihre Koordinaten x_ν in Funktion von λ ausgedrückt werden. Das Zeichen δ entspreche dem Übergang von einem Punkte der gesuchten geodätischen

Linie zu demjenigen Punkte einer benachbarten Kurve, welcher zu dem nämlichen λ gehört. Dann läßt sich (20) durch

(20a) $$\begin{cases} \int_{\lambda_1}^{\lambda_2} \delta w \, d\lambda = 0 \\ w^2 = g_{\mu\nu} \dfrac{dx_\mu}{d\lambda} \dfrac{dx_\nu}{d\lambda} \end{cases}$$

ersetzen. Da aber

$$\delta w = \frac{1}{w}\left\{\frac{1}{2}\frac{\partial g_{\mu\nu}}{\partial x_\sigma}\frac{dx_\mu}{d\lambda}\frac{dx_\nu}{d\lambda}\delta x_\sigma + g_{\mu\nu}\frac{dx_\mu}{d\lambda}\delta\left(\frac{dx_\nu}{d\lambda}\right)\right\},$$

so erhält man nach Einsetzen von δw in (20a) mit Rücksicht darauf, daß

$$\delta\left(\frac{dx_\nu}{d\lambda}\right) = \frac{d\,\delta x_\nu}{d\lambda},$$

nach partieller Integration

(20b) $$\begin{cases} \int_{\lambda_1}^{\lambda_2} d\lambda\, \varkappa_\sigma\, \delta x_\sigma = 0 \\ \varkappa_\sigma = \dfrac{d}{d\lambda}\left\{\dfrac{g_{\mu\nu}}{w}\dfrac{dx_\mu}{\partial\lambda}\right\} - \dfrac{1}{2w}\dfrac{\partial g_{\mu\nu}}{\partial x_\sigma}\dfrac{dx_\mu}{d\lambda}\dfrac{dx_\nu}{d\lambda}. \end{cases}$$

Hieraus folgt wegen der freien Wählbarkeit der δx_σ das Verschwinden der $\varkappa_\sigma$. Also sind

(20c) $$\varkappa_\sigma = 0$$

die Gleichungen der geodätischen Linie. Ist auf der betrachteten geodätischen Linie nicht $ds = 0$, so können wir als Parameter λ die auf der geodätischen Linie gemessene „Bogenlänge" s wählen. Dann wird $w = 1$, und man erhält an Stelle von (20c)

$$g_{\mu\nu}\frac{d^2 x_\mu}{ds^2} + \frac{\partial g_{\mu\nu}}{\partial x_\sigma}\frac{dx_\sigma}{d\lambda}\frac{dx_\mu}{d\lambda} - \frac{1}{2}\frac{\partial g_{\mu\nu}}{\partial x_\sigma}\frac{dx_\mu}{d\lambda}\frac{dx_\nu}{d\lambda} = 0,$$

oder durch bloße Änderung der Bezeichnungsweise

(20d) $$g_{\alpha\sigma}\frac{d^2 x_\alpha}{ds^2} + \begin{bmatrix}\mu\nu\\\sigma\end{bmatrix}\frac{dx_\mu}{ds}\frac{dx_\nu}{ds} = 0,$$

wobei nach Christoffel gesetzt ist

(21) $$\begin{bmatrix}\mu\nu\\\sigma\end{bmatrix} = \frac{1}{2}\left(\frac{\partial g_{\mu\sigma}}{\partial x_\nu} + \frac{\partial g_{\nu\sigma}}{\partial x_\mu} - \frac{\partial g_{\mu\nu}}{\partial x_\sigma}\right).$$

Multipliziert man endlich (20d) mit $g^{\sigma\tau}$ (äußere Multiplikation bezüglich τ, innere bezüglich σ), so erhält man schließlich als endgültige Form der Gleichung der geodätischen Linie

$$(22) \quad \frac{d^2 x_\tau}{ds^2} + \begin{Bmatrix} \mu\,\nu \\ \tau \end{Bmatrix} \frac{dx_\mu}{ds} \frac{dx_\nu}{ds} = 0.$$

Hierbei ist nach Christoffel gesetzt

$$(23) \quad \begin{Bmatrix} \mu\,\nu \\ \tau \end{Bmatrix} = g^{\tau\alpha} \begin{bmatrix} \mu\,\nu \\ \alpha \end{bmatrix}.$$

§ 10. Die Bildung von Tensoren durch Differentiation.

Gestützt auf die Gleichung der geodätischen Linie können wir nun leicht die Gesetze ableiten, nach welchen durch Differentiation aus Tensoren neue Tensoren gebildet werden können. Dadurch werden wir erst in den Stand gesetzt, allgemein kovariante Differentialgleichungen aufzustellen. Wir erreichen dies Ziel durch wiederholte Anwendung des folgenden einfachen Satzes.

Ist in unserem Kontinuum eine Kurve gegeben, deren Punkte durch die Bogendistanz s von einem Fixpunkt auf der Kurve charakterisiert sind, ist ferner φ eine invariante Raumfunktion, so ist auch $d\varphi/ds$ eine Invariante. Der Beweis liegt darin, daß sowohl $d\varphi$ als auch ds Invariante sind.

Da
$$\frac{d\varphi}{ds} = \frac{\partial \varphi}{\partial x_\mu} \frac{dx_\mu}{ds},$$

so ist auch
$$\psi = \frac{\partial \varphi}{\partial x_\mu} \frac{dx_\mu}{ds}$$

eine Invariante, und zwar für alle Kurven, die von einem Punkte des Kontinuums ausgehen, d. h. für beliebige Wahl des Vektors der dx_μ. Daraus folgt unmittelbar, daß

$$(24) \quad A_\mu = \frac{\partial \varphi}{\partial x_\mu}$$

ein kovarianter Vierervektor ist (Gradient von φ).

Nach unserem Satze ist ebenso der auf einer Kurve genommene Differentialquotient

$$\chi = \frac{d\psi}{ds}$$

eine Invariante. Durch Einsetzen von ψ erhalten wir zunächst

$$\chi = \frac{\partial^2 \varphi}{\partial x_\mu \partial x_\nu} \frac{dx_\mu}{ds} \frac{dx_\nu}{ds} + \frac{\partial \varphi}{\partial x_\mu} \frac{d^2 x_\mu}{ds^2}.$$

Hieraus läßt sich zunächst die Existenz eines Tensors nicht ableiten. Setzen wir nun aber fest, daß die Kurve, auf welcher wir differenziiert haben, eine geodätische Kurve sei, so erhalten wir nach (22) durch Ersetzen von d^2x_ν/ds^2:

$$\chi = \left\{ \frac{\partial^2 \varphi}{\partial x_\mu \partial x_\nu} - \begin{Bmatrix} \mu\,\nu \\ \tau \end{Bmatrix} \frac{\partial \varphi}{\partial x_\tau} \right\} \frac{dx_\mu}{ds} \frac{dx_\nu}{ds}.$$

Aus der Vertauschbarkeit der Differentiationen nach μ und ν und daraus, daß gemäß (23) und (21) die Klammer $\begin{Bmatrix} \mu\,\nu \\ \tau \end{Bmatrix}$ bezüglich μ und ν symmetrisch ist, folgt, daß der Klammerausdruck in μ und ν symmetrisch ist. Da man von einem Punkt des Kontinuums aus in beliebiger Richtung eine geodätische Linie ziehen kann, dx_μ/ds also ein Vierervektor mit frei wählbarem Verhältnis der Komponenten ist, folgt nach den Ergebnissen des § 7, daß

(25) $$A_{\mu\nu} = \frac{\partial^2 \varphi}{\partial x_\mu \partial x_\nu} - \begin{Bmatrix} \mu\,\nu \\ \tau \end{Bmatrix} \frac{\partial \varphi}{\partial x_\tau};$$

ein kovarianter Tensor zweiten Ranges ist. Wir haben also das Ergebnis gewonnen: Aus dem kovarianten Tensor ersten Ranges

$$A_\mu = \frac{\partial \varphi}{\partial x_\mu}$$

können wir durch Differentiation einen kovarianten Tensor zweiten Ranges

(26) $$A_{\mu\nu} = \frac{\partial A_\mu}{\partial x_\nu} - \begin{Bmatrix} \mu\,\nu \\ \tau \end{Bmatrix} A_\tau$$

bilden. Wir nennen den Tensor $A_{\mu\nu}$ die „*Erweiterung*" des Tensors A_μ. Zunächst können wir leicht zeigen, daß diese Bildung auch dann auf einen Tensor führt, wenn der Vektor A_μ nicht als ein Gradient darstellbar ist. Um dies einzusehen, bemerken wir zunächst, daß

$$\psi \frac{\partial \varphi}{\partial x_\mu}$$

ein kovarianter Vierervektor ist, wenn ψ und φ Skalare sind. Dies ist auch der Fall für eine aus vier solchen Gliedern bestehende Summe

$$S_\mu = \psi^{(1)} \frac{\partial \varphi^{(1)}}{\partial x_\mu} + \cdot + \cdot + \psi^{(4)} \frac{\partial \varphi^{(4)}}{\partial x_\mu},$$

falls $\psi^{(1)} \varphi^{(1)} \ldots \psi^{(4)} \varphi^{(4)}$ Skalare sind. Nun ist aber klar, daß sich jeder kovariante Vierervektor in der Form S_μ darstellen läßt. Ist nämlich A_μ ein Vierervektor, dessen Komponenten beliebig gegebene Funktionen der x_ν sind, so hat man nur (bezüglich des gewählten Koordinatensystems) zu setzen

$$\begin{aligned}\psi^{(1)} &= A_1, & \varphi^{(1)} &= x_1, \\ \psi^{(2)} &= A_2, & \varphi^{(2)} &= x_2, \\ \psi^{(3)} &= A_3, & \varphi^{(3)} &= x_3, \\ \psi^{(4)} &= A_4, & \varphi^{(4)} &= x_4,\end{aligned}$$

um zu erreichen, daß S_μ gleich A_μ wird.

Um daher zu beweisen, daß $A_{\mu\nu}$ ein Tensor ist, wenn auf der rechten Seite für A_μ ein beliebiger kovarianter Vierervektor eingesetzt wird, brauchen wir nur zu zeigen, daß dies für den Vierervektor S_μ zutrifft. Für letzteres ist es aber, wie ein Blick auf die rechte Seite von (26) lehrt, hinreichend, den Nachweis für den Fall

$$A_\mu = \psi \frac{\partial \varphi}{\partial x_\mu}$$

zu führen. Es hat nun die mit ψ multiplizierte rechte Seite von (25)

$$\psi \frac{\partial^2 \varphi}{\partial x_\mu \partial x_\nu} - \begin{Bmatrix} \mu\,\nu \\ \tau \end{Bmatrix} \psi \frac{\partial \varphi}{\partial x_\tau}$$

Tensorcharakter. Ebenso ist

$$\frac{\partial \psi}{\partial x_\mu} \frac{\partial \varphi}{\partial x_\nu}$$

ein Tensor (äußeres Produkt zweier Vierervektoren). Durch Addition folgt der Tensorcharakter von

$$\frac{\partial}{\partial x_\nu}\left(\psi \frac{\partial \varphi}{\partial x_\mu}\right) - \begin{Bmatrix} \mu\,\nu \\ \tau \end{Bmatrix}\left(\psi \frac{\partial \varphi}{\partial x_\tau}\right).$$

Damit ist, wie ein Blick auf (26) lehrt, der verlangte Nachweis für den Vierervektor

$$\psi \frac{\partial \varphi}{\partial x_\mu},$$

und daher nach dem vorhin Bewiesenen für jeden beliebigen Vierervektor A_μ geführt. —

Mit Hilfe der Erweiterung des Vierervektors kann man leicht die „Erweiterung" eines kovarianten Tensors beliebigen Ranges definieren; diese Bildung ist eine Verallgemeinerung der Erweiterung des Vierervektors. Wir beschränken uns auf die Aufstellung der Erweiterung des Tensors zweiten Ranges, da dieser das Bildungsgesetz bereits klar übersehen läßt.

Wie bereits bemerkt, läßt sich jeder kovariante Tensor zweiten Ranges darstellen[1]) als eine Summe von Tensoren vom Typus $A_\mu B_\nu$. Es wird deshalb genügen, den Ausdruck der Erweiterung für einen solchen speziellen Tensor abzuleiten. Nach (26) haben die Ausdrücke

$$\frac{\partial A_\mu}{\partial x_\sigma} - \left\{{\sigma\,\mu \atop \tau}\right\} A_\tau,$$

$$\frac{\partial B_\nu}{\partial x_\sigma} - \left\{{\sigma\,\nu \atop \tau}\right\} B_\tau$$

Tensorcharakter. Durch äußere Multiplikation des ersten mit B_ν, des zweiten mir A_μ erhält man je einen Tensor dritten Ranges; deren Addition ergibt den Tensor dritten Ranges

(27) $$A_{\mu\nu\sigma} = \frac{\partial A_{\mu\nu}}{\partial x_\sigma} - \left\{{\sigma\,\mu \atop \tau}\right\} A_{\tau\nu} - \left\{{\sigma\,\nu \atop \tau}\right\} A_{\mu\tau},$$

wobei $A_{\mu\nu} = A_\mu B_\nu$ gesetzt ist. Da die rechte Seite von (27) linear und homogen ist bezüglich der $A_{\mu\nu}$ und deren ersten Ableitungen, führt dieses Bildungsgesetz nicht nur bei einem Tensor vom Typus $A_\mu B_\nu$, sondern auch bei einer Summe solcher Tensoren, d. h. bei einem beliebigen kovarianten Tensor zweiten Ranges, zu einem Tensor. Wir nennen $A_{\mu\nu\sigma}$ die Erweiterung des Tensors $A_{\mu\nu}$.

Es ist klar, daß (26) und (24) nur spezielle Fälle von (27) sind (Erweiterung des Tensors ersten bzw. nullten Ranges). Überhaupt lassen sich alle speziellen Bildungsgesetze von Tensoren auf (27) in Verbindung mit Tensormultiplikationen auffassen.

§ 11. Einige Spezialfälle von besonderer Bedeutung.

Einige den Fundamentaltensor betreffende Hilfssätze. Wir leiten zunächst einige im folgenden viel gebrauchte Hilfs-

---

[1]) Durch äußere Multiplikation der Vektoren mit den (beliebig gegebenen) Komponenten A_{11}, A_{12}, A_{13}, A_{14} bzw. 1, 0, 0, 0 entsteht ein Tensor mit den Komponenten

$$\begin{matrix} A_{11} & A_{12} & A_{13} & A_{14} \\ 0 & 0 & 0 & 0 \\ 0 & 0 & 0 & 0 \\ 0 & 0 & 0 & 0 \end{matrix}$$

Durch Addition von vier Tensoren von diesem Typus erhält man den Tensor $A_{\mu\nu}$ mit beliebig vorgeschriebenen Komponenten.

gleichungen ab. Nach der Regel von der Differentiation der Determinanten ist

(28) $$dg = g^{\mu\nu} g \, dg_{\mu\nu} = -g_{\mu\nu} g \, dg^{\mu\nu}.$$

Die letzte Form rechtfertigt sich durch die vorletzte, wenn man bedenkt, daß $g_{\mu\nu} g^{\mu'\nu} = \delta_\mu^{\mu'}$, daß also $g_{\mu\nu} g^{\mu\nu} = 4$, folglich

$$g_{\mu\nu} dg^{\mu\nu} + g^{\mu\nu} dg_{\mu\nu} = 0.$$

Aus (28) folgt

(29) $$\frac{1}{\sqrt{-g}} \frac{\partial \sqrt{-g}}{\partial x_\sigma} = \frac{1}{2} \frac{\partial \lg(-g)}{\partial x_\sigma} = \frac{1}{2} g^{\mu\nu} \frac{\partial g_{\mu\nu}}{\partial x_\sigma} = -\frac{1}{2} g_{\mu\nu} \frac{\partial g^{\mu\nu}}{\partial x_\sigma}.$$

Aus

$$g_{\mu\sigma} g^{\nu\sigma} = \delta_\mu^\nu$$

folgt ferner durch Differentiation

(30) $$\begin{cases} \quad\text{bzw.}\quad & g_{\mu\sigma} dg^{\nu\sigma} = -g^{\nu\sigma} dg_{\mu\sigma} \\ & g_{\mu\sigma} \frac{\partial g^{\nu\sigma}}{\partial x_\lambda} = -g^{\nu\sigma} \frac{\partial g_{\mu\sigma}}{\partial x_\lambda}. \end{cases}$$

Durch gemischte Multiplikation mit $g^{\sigma\tau}$ bzw. $g_{\nu\lambda}$ erhält man hieraus (bei geänderter Bezeichnungsweise der Indizes)

(31) $$\begin{cases} dg^{\mu\nu} = -g^{\mu\alpha} g^{\nu\beta} dg_{\alpha\beta}, \\ \dfrac{\partial g^{\mu\nu}}{\partial x_\sigma} = -g^{\mu\alpha} g^{\nu\beta} \dfrac{\partial g_{\alpha\beta}}{\partial x_\sigma} \end{cases}$$

bzw.

(32) $$\begin{cases} dg_{\mu\nu} = -g_{\mu\alpha} g_{\nu\beta} dg^{\alpha\beta} \\ \dfrac{\partial g_{\mu\nu}}{\partial x_\sigma} = -g_{\mu\alpha} g_{\nu\beta} \dfrac{\partial g^{\alpha\beta}}{\partial x_\sigma}. \end{cases}$$

Die Beziehung (31) erlaubt eine Umformung, von der wir ebenfalls öfter Gebrauch zu machen haben. Gemäß (21) ist

(33) $$\frac{\partial g_{\alpha\beta}}{\partial x_\sigma} = \begin{bmatrix} \alpha & \sigma \\ \beta & \end{bmatrix} + \begin{bmatrix} \beta & \sigma \\ \alpha & \end{bmatrix}.$$

Setzt man dies in die zweite der Formeln (31) ein, so erhält man mit Rücksicht auf (23)

(34) $$\frac{\partial g^{\mu\nu}}{\partial x_\sigma} = -\left(g^{\mu\tau} \begin{Bmatrix} \tau & \sigma \\ \nu & \end{Bmatrix} + g^{\nu\tau} \begin{Bmatrix} \tau & \sigma \\ \mu & \end{Bmatrix} \right).$$

Durch Substitution der rechten Seite von (34) in (29) ergibt sich

(29a) $$\frac{1}{\sqrt{-g}} \frac{\partial \sqrt{-g}}{\partial x_\sigma} = \begin{Bmatrix} \mu & \sigma \\ \mu & \end{Bmatrix}.$$

Divergenz des kontravarianten Vierervektors. Multipliziert man (26) mit dem kontravarianten Fundamentaltensor $g^{\mu\nu}$ (innere Multiplikation), so nimmt die rechte Seite nach Umformung des ersten Gliedes zunächst die Form an

$$\frac{\partial}{\partial x_\nu}(g^{\mu\nu} A_\mu) - A_\mu \frac{\partial g^{\mu\nu}}{\partial x_\nu} - \frac{1}{2} g^{\tau\alpha}\left(\frac{\partial g_{\mu\alpha}}{\partial x_\nu} + \frac{\partial g_{\nu\alpha}}{\partial x_\mu} - \frac{\partial g_{\mu\nu}}{\partial x_\alpha}\right) g^{\mu\nu} A_\tau.$$

Das letzte Glied dieses Ausdruckes kann gemäß (31) und (29) in die Form

$$\frac{1}{2} \frac{\partial g^{\tau\nu}}{\partial x_\nu} A_\tau + \frac{1}{2} \frac{\partial g^{\tau\mu}}{\partial x_\mu} A_\tau + \frac{1}{\sqrt{-g}} \frac{\partial \sqrt{-g}}{\partial x_\alpha} g^{\mu\nu} A_\tau.$$

gebracht werden. Da es auf die Benennung der Summationsindizes nicht ankommt, heben sich die beiden ersten Glieder dieses Ausdruckes gegen das zweite des obigen weg; das letzte läßt sich mit dem ersten des obigen Ausdruckes vereinigen. Setzt man noch

$$g^{\mu\nu} A_\mu = A^\nu,$$

wobei A^ν ebenso wie A_μ ein frei wählbarer Vektor ist, so erhält man endlich

(35) $$\Phi = \frac{1}{\sqrt{-g}} \frac{\partial}{\partial x_\nu}(\sqrt{-g}\, A^\nu).$$

Dieser Skalar ist die *Divergenz* des kontravarianten Vierervektors A^ν.

"Rotation" des (kovarianten) Vierervektors. Das zweite Glied in (26) ist in den Indizes μ und ν symmetrisch. Es ist deshalb $A_{\mu\nu} - A_{\nu\mu}$ ein besonders einfach gebauter (antisymmetrischer) Tensor. Man erhält

(36) $$B_{\mu\nu} = \frac{\partial A_\mu}{\partial x_\nu} - \frac{\partial A_\nu}{\partial x_\mu}.$$

Antisymmetrische Erweiterung eines Sechservektors. Wendet man (27) auf einen antisymmetrischen Tensor zweiten Ranges $A_{\mu\nu}$ an, bildet hierzu die beiden durch zyklische Vertauschung der Indizes μ, ν, σ entstehenden Gleichungen und addiert diese drei Gleichungen, so erhält man den Tensor dritten Ranges

(37) $$B_{\mu\nu\sigma} = A_{\mu\nu\sigma} + A_{\nu\sigma\mu} + A_{\sigma\mu\nu} = \frac{\partial A_{\mu\nu}}{\partial x_\sigma} + \frac{\partial A_{\nu\sigma}}{\partial x_\mu} + \frac{\partial A_{\sigma\mu}}{\partial x_\nu},$$

von welchem leicht zu beweisen ist, daß er antisymmetrisch ist.

Divergenz des Sechservektors. Multipliziert man (27) mit $g^{\mu\alpha} g^{\nu\beta}$ (gemischte Multiplikation), so erhält man ebenfalls einen Tensor. Das erste Glied der rechten Seite von (27) kann man in der Form

$$\frac{\partial}{\partial x_\sigma}(g^{\mu\alpha}g^{\nu\beta}A_{\mu\nu}) - g^{\mu\alpha}\frac{\partial g^{\nu\beta}}{\partial x_\sigma}A_{\mu\nu} - g^{\nu\beta}\frac{\partial g^{\mu\alpha}}{\partial x_\sigma}A_{\mu\nu}$$

schreiben. Ersetzt man $g^{\mu\alpha}g^{\nu\beta}A_{\mu\nu\sigma}$ durch $A_\sigma^{\alpha\beta}$, $g^{\mu\alpha}g^{\nu\beta}A_{\mu\nu}$ durch $A^{\alpha\beta}$ und ersetzt man in dem umgeformten ersten Gliede

$$\frac{\partial g^{\nu\beta}}{\partial x_\sigma} \quad \text{und} \quad \frac{\partial g^{\mu\alpha}}{\partial x_\sigma}$$

vermittelst (34), so entsteht aus der rechten Seite von (27) ein siebengliedriger Ausdruck, von dem sich vier Glieder wegheben. Es bleibt übrig

$$(38) \qquad A_\sigma^{\alpha\beta} = \frac{\partial A^{\alpha\beta}}{\partial x_\sigma} + \begin{Bmatrix}\sigma\varkappa\\ \alpha\end{Bmatrix} A^{\varkappa\beta} + \begin{Bmatrix}\sigma\varkappa\\ \beta\end{Bmatrix} A^{\alpha\varkappa}.$$

Es ist dies der Ausdruck für die Erweiterung eines kontravarianten Tensors zweiten Ranges, der sich entsprechend auch für kontravariante Tensoren höheren und niedrigeren Ranges bilden läßt.

Wir merken an, daß sich auf analogem Wege auch die Erweiterung eines gemischten Tensors A_μ^α bilden läßt:

$$(39) \qquad A_{\mu\sigma}^\alpha = \frac{\partial A_\mu^\alpha}{\partial x_\sigma} - \begin{Bmatrix}\sigma\mu\\ \tau\end{Bmatrix} A_\tau^\alpha + \begin{Bmatrix}\sigma\tau\\ \alpha\end{Bmatrix} A_\mu^\tau.$$

Durch Verjüngung von (38) bezüglich der Indizes β und σ (innere Multiplikation mit δ_β^σ) erhält man den kontravarianten Vierervektor

$$A^\alpha = \frac{\partial A^{\alpha\beta}}{\partial x_\beta} + \begin{Bmatrix}\beta\varkappa\\ \beta\end{Bmatrix} A^{\alpha\varkappa} + \begin{Bmatrix}\beta\varkappa\\ \alpha\end{Bmatrix} A^{\varkappa\beta}.$$

Wegen der Symmetrie von $\begin{Bmatrix}\beta\varkappa\\ \alpha\end{Bmatrix}$ bezüglich der Indizes β und $\varkappa$ verschwindet das dritte Glied der rechten Seite, falls $A^{\alpha\beta}$ ein antisymmetrischer Tensor ist, was wir annehmen wollen; das zweite Glied läßt sich gemäß (29a) umformen. Man erhält also

$$(40) \qquad A^\alpha = \frac{1}{\sqrt{-g}}\frac{\partial(\sqrt{-g}\,A^{\alpha\beta})}{\partial x_\beta}.$$

Dies ist der Ausdruck der Divergenz eines kontravarianten Sechservektors.

Divergenz des gemischten Tensors zweiten Ranges. Bilden wir die Verjüngung von (39) bezüglich der Indizes α und σ, so erhalten wir mit Rücksicht auf (29a)

$$(41) \qquad \sqrt{-g}\, A_\mu = \frac{\partial (\sqrt{-g}\, A_\mu^\sigma)}{\partial x_\sigma} - \begin{Bmatrix} \sigma\mu \\ \tau \end{Bmatrix} \sqrt{-g}\, A_\tau^\sigma.$$

Führt man im letzten Gliede den kontravarianten Tensor $A^{\varrho\sigma} = g^{\varrho\tau} A_\tau^\sigma$ ein, so nimmt es die Form an

$$- \begin{bmatrix} \sigma\mu \\ \varrho \end{bmatrix} \sqrt{-g}\, A^{\varrho\sigma}.$$

Ist ferner der Tensor $A^{\varrho\sigma}$ ein symmetrischer, so reduziert sich dies auf

$$-\tfrac{1}{2} \sqrt{-g}\, \frac{\partial g_{\varrho\sigma}}{\partial x_\mu} A^{\varrho\sigma}.$$

Hätte man statt $A^{\varrho\sigma}$ den ebenfalls symmetrischen kovarianten Tensor $A_{\varrho\sigma} = g_{\varrho\alpha} g_{\sigma\beta} A^{\alpha\beta}$ eingeführt, so würde das letzte Glied vermöge (31) die Form

$$\tfrac{1}{2} \sqrt{-g}\, \frac{\partial g^{\varrho\sigma}}{\partial x_\mu} A_{\varrho\sigma}.$$

annehmen. In dem betrachteten Symmetriefalle kann also (41) auch durch die beiden Formen

$$(41\,\mathrm{a}) \qquad \sqrt{-g}\, A_\mu = \frac{\partial (\sqrt{-g}\, A_\mu^\sigma)}{\partial x_\sigma} - \tfrac{1}{2} \frac{\partial g_{\varrho\sigma}}{\partial x_\mu} \sqrt{-g}\, A^{\varrho\sigma}$$

und

$$(41\,\mathrm{b}) \qquad \sqrt{-g}\, A_\mu = \frac{\partial (\sqrt{-g}\, A_\mu^\sigma)}{\partial x_\sigma} + \tfrac{1}{2} \frac{\partial g^{\varrho\sigma}}{\partial x_\mu} \sqrt{-g}\, A_{\sigma\varrho}$$

ersetzt werden, von denen wir im folgenden Gebrauch zu machen haben.

§ 12. Der Riemann-Christoffelsche Tensor.

Wir fragen nun nach denjenigen Tensoren, welche aus dem Fundamentaltensor der $g_{\mu\nu}$ allein durch Differentiation gewonnen werden können. Die Antwort scheint zunächst auf der Hand zu liegen. Man setzt in (27) statt des beliebig gegebenen Tensors $A_{\mu\nu}$ den Fundamentaltensor der $g_{\mu\nu}$ ein und erhält dadurch einen neuen Tensor, nämlich die Erweiterung des Fundamentaltensors. Man überzeugt sich jedoch leicht, daß diese letztere identisch verschwindet. Man gelangt jedoch auf folgendem Wege zum Ziel. Man setze in (27)

$$A_{\mu\nu} = \frac{\partial A_\mu}{\partial x_\nu} - \begin{Bmatrix} \mu\nu \\ \varrho \end{Bmatrix} A_\varrho,$$

d. h. die Erweiterung des Vierervektors A_ν ein. Dann erhält man (bei etwas geänderter Benennung der Indizes) den Tensor dritten Ranges

$$A_{\mu\sigma\tau} = \frac{\partial^2 A_\mu}{\partial x_\sigma \partial x_\tau}$$
$$- \begin{Bmatrix} \mu\,\sigma \\ \varrho \end{Bmatrix} \frac{\partial A_\varrho}{\partial x_\tau} - \begin{Bmatrix} \mu\,\tau \\ \varrho \end{Bmatrix} \frac{\partial A_\varrho}{\partial x_\sigma} - \begin{Bmatrix} \sigma\,\tau \\ \varrho \end{Bmatrix} \frac{\partial A_\mu}{\partial x_\varrho}$$
$$+ \left[-\frac{\partial}{\partial x_\tau} \begin{Bmatrix} \mu\,\sigma \\ \varrho \end{Bmatrix} + \begin{Bmatrix} \mu\,\tau \\ \alpha \end{Bmatrix} \begin{Bmatrix} \alpha\,\sigma \\ \varrho \end{Bmatrix} + \begin{Bmatrix} \sigma\,\tau \\ \alpha \end{Bmatrix} \begin{Bmatrix} \alpha\,\mu \\ \varrho \end{Bmatrix} \right] A_\varrho.$$

Dieser Ausdruck ladet zur Bildung des Tensors $A_{\mu\sigma\tau} - A_{\mu\tau\sigma}$ ein. Denn dabei heben sich folgende Terme des Ausdruckes für $A_{\mu\sigma\tau}$ gegen solche von $A_{\mu\tau\sigma}$ weg: das erste Glied, das vierte Glied, sowie das dem letzten Term in der eckigen Klammer entsprechende Glied; denn alle diese sind in σ und τ symmetrisch. Gleiches gilt von der Summe des zweiten und dritten Gliedes. Wir erhalten also

(42) $$A_{\mu\sigma\tau} - A_{\mu\tau\sigma} = B^\varrho_{\mu\sigma\tau} A_\varrho,$$

(43) $$\begin{cases} B^\varrho_{\mu\sigma\tau} = -\frac{\partial}{\partial x_\tau} \begin{Bmatrix} \mu\,\sigma \\ \varrho \end{Bmatrix} + \frac{\partial}{\partial x_\sigma} \begin{Bmatrix} \mu\,\tau \\ \varrho \end{Bmatrix} \\ \qquad - \begin{Bmatrix} \mu\,\sigma \\ \alpha \end{Bmatrix} \begin{Bmatrix} \alpha\,\tau \\ \varrho \end{Bmatrix} + \begin{Bmatrix} \mu\,\tau \\ \alpha \end{Bmatrix} \begin{Bmatrix} \alpha\,\sigma \\ \varrho \end{Bmatrix}. \end{cases}$$

Wesentlich ist an diesem Resultat, daß auf der rechten Seite von (42) nur die A_ϱ, aber nicht mehr ihre Ableitungen auftreten. Aus dem Tensorcharakter von $A_{\mu\sigma\tau} - A_{\mu\tau\sigma}$ in Verbindung damit, daß A_ϱ ein frei wählbarer Vierervektor ist, folgt, vermöge der Resultate des § 7, daß $B^\varrho_{\mu\sigma\tau}$ ein Tensor ist (Riemann-Christoffelscher Tensor).

Die mathematische Bedeutung dieses Tensors liegt im folgenden. Wenn das Kontinuum so beschaffen ist, daß es ein Koordinatensystem gibt, bezüglich dessen die $g_{\mu\nu}$ Konstanten sind, so verschwinden alle $R^\varrho_{\mu\sigma\tau}$. Wählt man statt des ursprünglichen Koordinatensystems ein beliebiges neues, so werden die auf letzteres bezogenen $g_{\mu\nu}$ nicht Konstanten sein. Der Tensorcharakter von $R^\varrho_{\mu\sigma\tau}$ bringt es aber mit sich, daß diese Komponenten auch in dem beliebig gewählten Bezugssystem sämtlich verschwinden. Das Verschwinden des Riemannschen Tensors ist also eine notwendige Bedingung dafür, daß durch geeignete Wahl des Bezugssystems die Konstanz der $g_{\mu\nu}$ herbeigeführt werden kann.[1]) In unserem Problem entspricht dies dem Falle, daß bei passender Wahl des Koordinatensystems in endlichen Gebieten die spezielle Relativitätstheorie gilt.

Durch Verjüngung von (43) bezüglich der Indizes τ und ϱ erhält man den kovarianten Tensor zweiten Ranges

$$(44) \quad \begin{cases} B_{\mu\nu} = R_{\mu\nu} + S_{\mu\nu} \\ R_{\mu\nu} = -\dfrac{\partial}{\partial x_\alpha} \begin{Bmatrix} \mu\,\nu \\ \alpha \end{Bmatrix} + \begin{Bmatrix} \mu\,\alpha \\ \beta \end{Bmatrix} \begin{Bmatrix} \nu\,\beta \\ \alpha \end{Bmatrix} \\ S_{\mu\nu} = \dfrac{\partial \lg \sqrt{-g}}{\partial x_\mu \partial x_\nu} - \begin{Bmatrix} \mu\,\nu \\ \alpha \end{Bmatrix} \dfrac{\partial \lg \sqrt{-g}}{\partial x_\alpha}. \end{cases}$$

Bemerkung über die Koordinatenwahl. Es ist schon in § 8 im Anschluß an Gleichung (18a) bemerkt worden, daß die Koordinatenwahl mit Vorteil so getroffen werden kann, daß $\sqrt{-g} = 1$ wird. Ein Blick auf die in den beiden letzten Paragraphen erlangten Gleichungen zeigt, daß durch eine solche Wahl die Bildungsgesetze der Tensoren eine bedeutende Vereinfachung erfahren. Besonders gilt dies für den soeben entwickelten Tensor $B_{\mu\nu}$, welcher in der darzulegenden Theorie eine fundamentale Rolle spielt. Die ins Auge gefaßte Spezialisierung der Koordinatenwahl bringt nämlich das Verschwinden von $S_{\mu\nu}$ mit sich, so daß sich der Tensor $B_{\mu\nu}$ auf $R_{\mu\nu}$ reduziert.

Ich will deshalb im folgenden alle Beziehungen in der vereinfachten Form angeben, welche die genannte Spezialisierung der Koordinatenwahl mit sich bringt. Es ist dann ein Leichtes, auf die *allgemein* kovarianten Gleichungen zurückzugreifen, falls dies in einem speziellen Falle erwünscht erscheint.

C. Theorie des Gravitationsfeldes.

§ 13. Bewegungsgleichung des materiellen Punktes im Gravitationsfeld. Ausdruck für die Feldkomponenten der Gravitation.

Ein frei beweglicher, äußeren Kräften nicht unterworfener Körper bewegt sich nach der speziellen Relativitätstheorie geradlinig und gleichförmig. Dies gilt auch nach der allgemeinen

---

[1]) Die Mathematiker haben bewiesen, daß diese Bedingung auch eine *hinreichende* ist.

Relativitätstheorie für einen Teil des vierdimensionalen Raumes, in welchem das Koordinatensystem K_0 so wählbar und so gewählt ist, daß die $g_{\mu\nu}$ die in (4) gegebenen speziellen konstanten Werte haben.

Betrachten wir eben diese Bewegung von einem beliebig gewählten Koordinatensystem K_1 aus, so bewegt er sich von K_1 aus, beurteilt nach den Überlegungen des § 2 in einem Gravitationsfelde. Das Bewegungsgesetz mit Bezug auf K_1 ergibt sich leicht aus folgender Überlegung. Mit Bezug auf K_0 ist das Bewegungsgesetz eine vierdimensionale Gerade, also eine geodätische Linie. Da nun die geodätische Linie unabhängig vom Bezugssystem definiert ist, wird ihre Gleichung auch die Bewegungsgleichung des materiellen Punktes in bezug auf K_1 sein. Setzen wir

$$(45) \qquad \Gamma^\tau_{\mu\nu} = - \begin{Bmatrix} \mu\,\nu \\ \tau \end{Bmatrix},$$

so lautet also die Gleichung der Punktbewegung in bezug auf K_1

$$(46) \qquad \frac{d^2 x_\tau}{ds^2} = \Gamma^\tau_{\mu\nu} \frac{dx_\nu}{ds} \frac{dx_\nu}{ds}.$$

Wir machen nun die sehr naheliegende Annahme, daß dieses allgemein kovariante Gleichungssystem die Bewegung des Punktes im Gravitationsfeld auch in dem Falle bestimmt, daß kein Bezugssystem K_0 existiert, bezüglich dessen in endlichen Räumen die spezielle Relativitätstheorie gilt. Zu dieser Annahme sind wir um so berechtigter, als (46) nur *erste* Ableitungen der $g_{\mu\nu}$ enthält, zwischen denen auch im Spezialfalle der Existenz von K_0 keine Beziehungen bestehen.[1])

Verschwinden die $\Gamma^\tau_{\mu\nu}$, so bewegt sich der Punkt geradlinig und gleichförmig; diese Größen bedingen also die Abweichung der Bewegung von der Gleichförmigkeit. Sie sind die Komponenten des Gravitationsfeldes.

§ 14. Die Feldgleichungen der Gravitation bei Abwesenheit von Materie.

Wir unterscheiden im folgenden zwischen „Gravitationsfeld" und „Materie", in dem Sinne, daß alles außer dem Gravitationsfeld als „Materie" bezeichnet wird, also nicht nur

---

1) Erst zwischen den zweiten (und ersten) Ableitungen bestehen gemäß § 12 die Beziehungen $B^\varrho_{\mu\sigma\tau} = 0$.

die „Materie" im üblichen Sinne, sondern auch das elektromagnetische Feld.

Unsere nächste Aufgabe ist es, die Feldgleichungen der Gravitation bei Abwesenheit von Materie aufzusuchen. Dabei verwenden wir wieder dieselbe Methode wie im vorigen Paragraphen bei der Aufstellung der Bewegungsgleichung des materiellen Punktes. Ein Spezialfall, in welchem die gesuchten Feldgleichungen jedenfalls erfüllt sein müssen, ist der der ursprünglichen Relativitätstheorie, in dem die $g_{\mu\nu}$ gewisse konstante Werte haben. Dies sei der Fall in einem gewissen endlichen Gebiete in bezug auf ein bestimmtes Koordinatensystem K_0. In bezug auf dies System verschwinden sämtliche Komponenten $B^{\varrho}_{\mu\sigma\tau}$ des Riemannschen Tensors [Gleichung (43)]. Diese verschwinden dann für das betrachtete Gebiet auch bezüglich jedes anderen Koordinatensystems.

Die gesuchten Gleichungen des materiefreien Gravitationsfeldes müssen also jedenfalls erfüllt sein, wenn alle $B^{\varrho}_{\mu\sigma\tau}$ verschwinden. Aber diese Bedingung ist jedenfalls eine zu weitgehende. Denn es ist klar, daß z. B. das von einem Massenpunkte in seiner Umgebung erzeugte Gravitationsfeld sicherlich durch keine Wahl des Koordinatensystems „wegtransformiert", d. h. auf den Fall konstanter $g_{\mu\nu}$ transformiert werden kann.

Deshalb liegt es nahe, für das materiefreie Gravitationsfeld das Verschwinden des aus dem Tensor $B^{\varrho}_{\mu\sigma\tau}$ abgeleiteten symmetrischen Tensors $B_{\mu\nu}$ zu verlangen. Man erhält so 10 Gleichungen für die 10 Größen $g_{\mu\nu}$, welche im speziellen erfüllt sind, wenn sämtliche $B^{\varrho}_{\mu\sigma\tau}$ verschwinden. Diese Gleichungen lauten mit Rücksicht auf (44) bei der von uns getroffenen Wahl für das Koordinatensystem für das materiefreie Feld

$$(47) \quad \begin{cases} \dfrac{\partial \varGamma^{\alpha}_{\mu\nu}}{\partial x_{\alpha}} + \varGamma^{\alpha}_{\mu\beta}\varGamma^{\beta}_{\nu\alpha} = 0 \\ \sqrt{-g} = 1. \end{cases}$$

Es muß darauf hingewiesen werden, daß der Wahl dieser Gleichungen ein Minimum von Willkür anhaftet. Denn es gibt außer $B_{\mu\nu}$ keinen Tensor zweiten Ranges, der aus den

$g_{\mu\nu}$ und deren Ableitungen gebildet ist, keine höheren als zweite Ableitungen enthält und in letzteren linear ist.[1]

Daß diese aus der Forderung der allgemeinen Relativität auf rein mathematischem Wege fließenden Gleichungen in Verbindung mit den Bewegungsgleichungen (46) in erster Näherung das Newtonsche Attraktionsgesetz, in zweiter Näherung die Erklärung der von Leverrier entdeckten (nach Anbringung der Störungskorrektionen übrigbleibenden) Perihelbewegung des Merkur liefern, muß nach meiner Ansicht von der physikalischen Richtigkeit der Theorie überzeugen.

§ 15. Hamiltonsche Funktion für das Gravitationsfeld, Impulsenergiesatz.

Um zu zeigen, daß die Feldgleichungen dem Impulsenergiesatz entsprechen, ist es am bequemsten, sie in folgender Hamiltonscher Form zu schreiben:

(47a)
$$\begin{cases} \delta \left\{ \int H d\tau \right\} = 0 \\ H = g^{\mu\nu} \Gamma^{\alpha}_{\mu\beta} \Gamma^{\beta}_{\nu\alpha} \\ \sqrt{-g} = 1 . \end{cases}$$

Dabei verschwinden die Variationen an den Grenzen des betrachteten begrenzten vierdimensionalen Integrationsraumes.

Es ist zunächst zu zeigen, daß die Form (47a) den Gleichungen (47) äquivalent ist. Zu diesem Zweck betrachten wir H als Funktion der $g^{\mu\nu}$ und der

$$g^{\mu\nu}_{\sigma}\left(=\frac{\partial g^{\mu\nu}}{\partial x_{\sigma}}\right).$$

Dann ist zunächst

$$\delta H = \Gamma^{\alpha}_{\mu\beta}\Gamma^{\beta}_{\nu\alpha}\delta g^{\mu\nu} + 2 g^{\mu\nu}\Gamma^{\alpha}_{\mu\beta}\delta\Gamma^{\beta}_{\nu\alpha}$$
$$= -\Gamma^{\alpha}_{\mu\beta}\Gamma^{\beta}_{\nu\alpha}\delta g^{\mu\nu} + 2\Gamma^{\alpha}_{\mu\beta}\delta(g^{\mu\nu}\Gamma^{\beta}_{\nu\alpha}).$$

Nun ist aber

$$\delta(g^{\mu\nu}\Gamma^{\beta}_{\nu\alpha}) = -\tfrac{1}{2}\delta\left[g^{\mu\nu}g^{\beta\lambda}\left(\frac{\partial g_{\nu\lambda}}{\partial x_{\alpha}} + \frac{\partial g_{\alpha\lambda}}{\partial x_{\nu}} - \frac{\partial g_{\alpha\nu}}{\partial x_{\lambda}}\right)\right].$$

---

[1] Eigentlich läßt sich dies nur von dem Tensor $B_{\mu\nu} + \lambda g_{\mu\nu}(g^{\alpha\beta}B_{\alpha\beta})$ behaupten, wobei λ eine Konstante ist. Setzt man jedoch diesen $= 0$, so kommt man wieder zu den Gleichungen $B_{\mu\nu} = 0$.

Die aus den beiden letzten Termen der runden Klammer hervorgehenden Terme sind von verschiedenem Vorzeichen und gehen auseinander (da die Benennung der Summationsindizes belanglos ist) durch Vertauschung der Indizes μ und β hervor. Sie heben einander im Ausdruck für δH weg, weil sie mit der bezüglich der Indizes μ und β symmetrischen Größe $\Gamma^{\alpha}_{\mu\beta}$ multipliziert werden. Es bleibt also nur das erste Glied der runden Klammer zu berücksichtigen, so daß man mit Rücksicht auf (31) erhält

$$\delta H = -\Gamma^{\alpha}_{\mu\beta}\Gamma^{\beta}_{\nu\alpha}\delta g^{\mu\nu} - \Gamma^{\alpha}_{\mu\beta}\delta g^{\mu\beta}_{\alpha}.$$

Es ist also

(48)
$$\begin{cases} \dfrac{\partial H}{\partial g^{\mu\nu}} = -\Gamma^{\alpha}_{\mu\beta}\Gamma^{\beta}_{\nu\alpha} \\ \dfrac{\partial H}{\partial g^{\mu\nu}_{\sigma}} = \Gamma^{\sigma}_{\mu\nu}. \end{cases}$$

Die Ausführung der Variation in (47a) ergibt zunächst das Gleichungssystem

(47b) $$\frac{\partial}{\partial x_{\alpha}}\left(\frac{\partial H}{\partial g^{\mu\nu}_{\alpha}}\right) - \frac{\partial H}{\partial g^{\mu\nu}} = 0,$$

welches wegen (48) mit (47) übereinstimmt, was zu beweisen war. — Multipliziert man (47b) mit $g^{\mu\nu}_{\sigma}$, so erhält man, weil

$$\frac{\partial g^{\mu\nu}_{\sigma}}{\partial x_{\alpha}} = \frac{\partial g^{\mu\nu}_{\alpha}}{\partial x_{\sigma}}$$

und folglich

$$g^{\mu\nu}_{\sigma}\frac{\partial}{\partial x_{\alpha}}\left(\frac{\partial H}{\partial g^{\mu\nu}_{\alpha}}\right) = \frac{\partial}{\partial x_{\alpha}}\left(g^{\mu\nu}_{\sigma}\frac{\partial H}{\partial g^{\mu\nu}_{\alpha}}\right) - \frac{\partial H}{\partial g^{\mu\nu}_{\alpha}}\frac{\partial g^{\mu\nu}_{\alpha}}{\partial x_{\sigma}}$$

die Gleichung

$$\frac{\partial}{\partial x_{\alpha}}\left(g^{\mu\nu}_{\sigma}\frac{\partial H}{\partial g^{\mu\nu}_{\alpha}}\right) - \frac{\partial H}{\partial x_{\sigma}} = 0$$

oder[1]

(49) $$\begin{cases} \dfrac{\partial t^{\alpha}_{\sigma}}{\partial x_{\alpha}} = 0 \\ -2\varkappa t^{\alpha}_{\sigma} = g^{\mu\nu}_{\sigma}\dfrac{\partial H}{\partial g^{\mu\nu}_{\alpha}} - \delta^{\alpha}_{\sigma}H. \end{cases}$$

---

[1] Der Grund der Einführung des Faktors $-2\varkappa$ wird später deutlich werden.

oder, wegen (48), der zweiten Gleichung (47) und (34)

(50) $\quad \varkappa t_\sigma^a = \tfrac{1}{2} \delta_\sigma^a g^{\mu\nu} \Gamma_{\mu\beta}^a \Gamma_{\nu\alpha}^\beta - g^{\mu\nu} \Gamma_{\mu\beta}^a \Gamma_{\nu\sigma}^\beta$.

Es ist zu beachten, daß t_σ^a kein Tensor ist; dagegen gilt (49) für alle Koordinatensysteme, für welche $\sqrt{-g} = 1$ ist. Diese Gleichung drückt den Erhaltungssatz des Impulses und der Energie für das Gravitationsfeld aus. In der Tat liefert die Integration dieser Gleichung über ein *dreidimensionales* Volumen V die vier Gleichungen

(49a) $\quad \dfrac{d}{dx_4}\left\{\int t_\sigma^4 \, dV\right\} = \int (t_\sigma^1 \alpha_1 + t_\sigma^2 \alpha_2 + t_\sigma^3 \alpha_3)\, dS$,

wobei α_1, α_2, α_3 der Richtungskosinus der nach innen gerichteten Normale eines Flächenelementes der Begrenzung von der Größe dS (im Sinne der euklidischen Geometrie) bedeuten. Man erkennt hierin den Ausdruck der Erhaltungssätze in üblicher Fassung. Die Größen t_σ^a bezeichnen wir als die „Energiekomponenten" des Gravitationsfeldes.

Ich will nun die Gleichungen (47) noch in einer dritten Form angeben, die einer lebendigen Erfassung unseres Gegenstandes besonders dienlich ist. Durch Multiplikation der Feldgleichungen (47) mit $g^{\nu\sigma}$ ergeben sich diese in der „gemischten" Form. Beachtet man, daß

$$g^{\nu\sigma} \dfrac{\partial \Gamma_{\mu\nu}^a}{\partial x_\alpha} = \dfrac{\partial}{\partial x_\alpha}\left(g^{\nu\sigma} \Gamma_{\mu\nu}^a\right) - \dfrac{\partial g^{\nu\sigma}}{\partial x_\alpha} \Gamma_{\mu\nu}^a,$$

welche Größe wegen (34) gleich

$$\dfrac{\partial}{\partial x_\alpha}\left(g^{\nu\sigma} \Gamma_{\mu\nu}^a\right) - g^{\nu\beta} \Gamma_{\alpha\beta}^\sigma \Gamma_{\mu\nu}^a - g^{\sigma\beta} \Gamma_{\beta\alpha}^\nu \Gamma_{\mu\nu}^a,$$

oder (nach geänderter Benennung der Summationsindizes) gleich

$$\dfrac{\partial}{\partial x_\alpha}\left(g^{\sigma\beta} \Gamma_{\mu\beta}^a\right) - g^{mn} \Gamma_{m\beta}^\sigma \Gamma_{n\mu}^\beta - g^{\nu\sigma} \Gamma_{\mu\beta}^a \Gamma_{\nu\alpha}^\beta.$$

Das dritte Glied dieses Ausdrucks hebt sich weg gegen das aus dem zweiten Glied der Feldgleichungen (47) entstehende; an Stelle des zweiten Gliedes dieses Ausdruckes läßt sich nach Beziehung (50)
$$\varkappa (t_\mu^\sigma - \tfrac{1}{2} \delta_\mu^\sigma t)$$
setzen $(t = t_a^a)$. Man erhält also an Stelle der Gleichungen (47)

(51) $\quad \begin{cases} \dfrac{\partial}{\partial x_\alpha}\left(g^{\sigma\beta} \Gamma_{\mu\beta}^a\right) = -\varkappa(t_\mu^\sigma - \tfrac{1}{2}\delta_\mu^\sigma t) \\ \sqrt{-g} = 1 \,. \end{cases}$

§ 16. **Allgemeine Fassung der Feldgleichungen der Gravitation.**

Die im vorigen Paragraphen aufgestellten Feldgleichungen für materiefreie Räume sind mit der Feldgleichung

$$\Delta \varphi = 0$$

der Newtonschen Theorie zu vergleichen. Wir haben die Gleichungen aufzusuchen, welche der Poissonschen Gleichung

$$\Delta \varphi = 4\pi\varkappa\varrho$$

entspricht, wobei ϱ die Dichte der Materie bedeutet.

Die spezielle Relativitätstheorie hat zu dem Ergebnis geführt, daß die träge Masse nichts anderes ist als Energie, welche ihren vollständigen mathematischen Ausdruck in einem symmetrischen Tensor zweiten Ranges, dem Energietensor, findet. Wir werden daher auch in der allgemeinen Relativitätstheorie einen Energietensor der Materie T_σ^α einzuführen haben, der wie die Energiekomponenten t_σ^α [Gleichungen (49) und (50)] des Gravitationsfeldes gemischten Charakter haben wird, aber zu einem symmetrischen kovarianten Tensor gehören wird [1]).

Wie dieser Energietensor (entsprechend der Dichte ϱ in der Poissonschen Gleichung) in die Feldgleichungen der Gravitation einzuführen ist, lehrt das Gleichungssystem (51). Betrachtet man nämlich ein vollständiges System (z. B. das Sonnensystem), so wird die Gesamtmasse des Systems, also auch seine gesamte gravitierende Wirkung, von der Gesamtenergie des Systems, also von der ponderablen und Gravitationsenergie zusammen, abhängen. Dies wird sich dadurch ausdrücken lassen, daß man in (51) an Stelle der Energiekomponenten t_μ^σ des Gravitationsfeldes allein die Summen $t_\mu^\sigma + T_\mu^\sigma$ der Energiekomponenten von Materie und Gravitationsfeld einführt. Man erhält so statt (51) die Tensorgleichung

$$(52) \quad \begin{cases} \dfrac{\partial}{\partial x_\alpha}\left(g^{\sigma\beta}\Gamma_{\mu\beta}^\alpha\right) = -\varkappa\left[(t_\mu^\sigma + T_\mu^\sigma) - \tfrac{1}{2}\delta_\mu^\sigma(t+T)\right] \\ \sqrt{-g} = 1, \end{cases}$$

wobei $T = T_\mu^\mu$ gesetzt ist (Lauescher Skalar). Dies sind die gesuchten allgemeinen Feldgleichungen der Gravitation in ge-

---

1) $g_{\sigma\tau}T_\sigma^\alpha = T_{\sigma\tau}$ und $g^{\sigma\beta}T_\sigma^\alpha = T^{\alpha\beta}$ sollen symmetrische Tensoren sein.

mischter Form. An Stelle von (47) ergibt sich daraus rückwärts das System

$$(53) \quad \begin{cases} \dfrac{\partial \Gamma^\alpha_{\mu\nu}}{\partial x_\alpha} + \Gamma^\alpha_{\mu\beta} \Gamma^\beta_{\nu\alpha} = -\varkappa(T_{\mu\nu} - \tfrac{1}{2} g_{\mu\nu} T), \\ \sqrt{-g} = 1. \end{cases}$$

Es muß zugegeben werden, daß diese Einführung des Energietensors der Materie durch das Relativitätspostulat allein nicht gerechtfertigt wird; deshalb haben wir sie im vorigen aus der Forderung abgeleitet, daß die Energie des Gravitationsfeldes in gleicher Weise gravitierend wirken soll, wie jegliche Energie anderer Art. Der stärkste Grund für die Wahl der vorstehenden Gleichungen liegt aber darin, daß sie zur Folge haben, daß für die Komponenten der Totalenergie Erhaltungsgleichungen (des Impulses und der Energie) gelten, welche den Gleichungen (49) und (49a) genau entsprechen. Dies soll im folgenden dargetan werden.

§ 17. Die Erhaltungssätze im allgemeinen Falle.

Die Gleichung (52) ist leicht so umzuformen, daß auf der rechten Seite das zweite Glied wegfällt. Man verjünge (52) nach den Indizes μ und σ und subtrahiere die so erhaltene, mit $\tfrac{1}{2}\delta^\sigma_\mu$ multiplizierte Gleichung von (52). Es ergibt sich

$$(52\mathrm{a}) \quad \frac{\partial}{\partial x_\alpha}\left(g^{\sigma\beta}\Gamma^\alpha_{\mu\beta} - \tfrac{1}{2}\delta^\sigma_\mu g^{\lambda\beta}\Gamma^\alpha_{\lambda\beta}\right) = -\varkappa(t^\sigma_\mu + T^\sigma_\mu).$$

An dieser Gleichung bilden wir die Operation $\partial/\partial x_\sigma$. Es ist

$$\frac{\partial^2}{\partial x_\alpha \partial x_\sigma}(g^{\sigma\beta}\Gamma^\alpha_{\mu\beta})$$
$$= -\frac{1}{2}\frac{\partial^2}{\partial x_\alpha \partial x_\sigma}\left[g^{\sigma\beta}g^{\alpha\lambda}\left(\frac{\partial g_{\mu\lambda}}{\partial x_\beta} + \frac{\partial g_{\beta\lambda}}{\partial x_\mu} - \frac{\partial g_{\mu\beta}}{\partial x_\lambda}\right)\right].$$

Das erste und das dritte Glied der runden Klammer liefern Beiträge, die einander wegheben, wie man erkennt, wenn man im Beitrage des dritten Gliedes die Summationsindizes α und σ einerseits, β und λ andererseits vertauscht. Das zweite Glied läßt sich nach (31) umformen, so daß man erhält

$$(54) \quad \frac{\partial^2}{\partial x_\alpha \partial x_\sigma}(g^{\sigma\beta}\Gamma^\alpha_{\mu\beta}) = \frac{1}{2}\frac{\partial^3 g^{\alpha\beta}}{\partial x_\alpha \partial x_\beta \partial x_\mu}.$$

Das zweite Glied der linken Seite von (52a) liefert zunächst

$$-\frac{1}{2}\frac{\partial^2}{\partial x_\alpha \partial x_\mu}(g^{\lambda\beta}\Gamma^\alpha_{\lambda\beta})$$

oder

$$\frac{1}{4} \frac{\partial^2}{\partial x_\alpha \partial x_\mu}\left[g^{\lambda\beta} g^{\alpha\delta}\left(\frac{\partial g_{\delta\lambda}}{\partial x_\beta} + \frac{\partial g_{\delta\beta}}{\partial x_\lambda} - \frac{\partial g_{\lambda\beta}}{\partial x_\delta}\right)\right].$$

Das vom letzten Glied der runden Klammer herrührende Glied verschwindet wegen (29) bei der von uns getroffenen Koordinatenwahl. Die beiden anderen lassen sich zusammenfassen und liefern wegen (31) zusammen,

$$-\frac{1}{2} \frac{\partial^3 g^{\alpha\beta}}{\partial x_\alpha \partial x_\beta \partial x_\mu},$$

so daß mit Rücksicht auf (54) die Identität

$$(55) \qquad \frac{\partial^2}{\partial x_\alpha \partial x_\sigma}\left(g^{\sigma\beta} \Gamma^\alpha_{\mu\beta} - \tfrac{1}{2} \delta_\mu^\sigma g^{\lambda\beta} \Gamma^\alpha_{\lambda\beta}\right) \equiv 0$$

besteht. Aus (55) und (52a) folgt

$$(56) \qquad \frac{\partial (t_\mu^\sigma + T_\mu^\sigma)}{\partial x_\sigma} = 0.$$

Aus unseren Feldgleichungen der Gravitation geht also hervor, daß den Erhaltungssätzen des Impulses und der Energie Genüge geleistet ist. Man sieht dies am einfachsten nach der Betrachtung ein, die zu Gleichung (49a) führt; nur hat man hier an Stelle der Energiekomponenten t_μ^σ des Gravitationsfeldes die Gesamtenergiekomponenten von Materie und Gravitationsfeld einzuführen.

§ 18. Der Impulsenergiesatz für die Materie als Folge der Feldgleichungen.

Multipliziert man (53) mit $\partial g^{\mu\nu}/\partial x_\sigma$, so erhält man auf dem in § 15 eingeschlagenen Wege mit Rücksicht auf das Verschwinden von

$$g_{\mu\nu} \frac{\partial g^{\mu\nu}}{\partial x_\sigma}$$

die Gleichung

$$\frac{\partial t_\sigma^\alpha}{\partial x_\alpha} + \frac{1}{2} \frac{\partial g^{\mu\nu}}{\partial x_\sigma} T_{\mu\nu} = 0,$$

oder mit Rücksicht auf (56)

$$(57) \qquad \frac{\partial T_\sigma^\alpha}{\partial x_\alpha} + \frac{1}{2} \frac{\partial g^{\mu\nu}}{\partial x_\sigma} T_{\mu\nu} = 0.$$

Ein Vergleich mit (41b) zeigt, daß diese Gleichung bei der getroffenen Wahl für das Koordinatensystem nichts anderes aussagt als das Verschwinden der Divergenz des Tensors der Energiekomponenten der Materie. Physikalisch zeigt das Auftreten des zweiten Gliedes der linken Seite, daß für die Materie allein Erhaltungssätze des Impulses und der Energie im eigentlichen Sinne nicht, bzw. nur dann gelten, wenn die $g^{\mu\nu}$ konstant sind, d. h. wenn die Feldstärken der Gravitation verschwinden. Dies zweite Glied ist ein Ausdruck für Impuls bzw. Energie, welche pro Volumen und Zeiteinheit vom Gravitationsfelde auf die Materie übertragen werden. Dies tritt noch klarer hervor, wenn man statt (57) im Sinne von (41) schreibt

(57a) $$\frac{\partial T_\sigma{}^\alpha}{\partial x_\alpha} = -\Gamma_{\sigma\beta}^{\alpha} T_\alpha{}^\beta.$$

Die rechte Seite drückt die energetische Einwirkung des Gravitationsfeldes auf die Materie aus.

Die Feldgleichungen der Gravitation enthalten also gleichzeitig vier Bedingungen, welchen der materielle Vorgang zu genügen hat. Sie liefern die Gleichungen des materiellen Vorganges vollständig, wenn letzterer durch vier voneinander unabhängige Differentialgleichungen charakterisierbar ist.[1])

D. Die „materiellen" Vorgänge.

Die unter B entwickelten mathematischen Hilfsmittel setzen uns ohne weiteres in den Stand, die physikalischen Gesetze der Materie (Hydrodynamik, Maxwellsche Elektrodynamik), wie sie in der speziellen Relativitätstheorie formuliert vorliegen, so zu verallgemeinern, daß sie in die allgemeine Relativitätstheorie hineinpassen. Dabei ergibt das allgemeine Relativitätsprinzip zwar keine weitere Einschränkung der Möglichkeiten; aber es lehrt den Einfluß des Gravitationsfeldes auf alle Prozesse exakt kennen, ohne daß irgendwelche neue Hypothese eingeführt werden müßte.

Diese Sachlage bringt es mit sich, daß über die physikalische Natur der Materie (im engeren Sinne) nicht notwendig bestimmte Voraussetzungen eingeführt werden müssen. Insbesondere kann die Frage offen bleiben, ob die Theorie des elektromagnetischen Feldes und des Gravitationsfeldes zu-

---

1) Vgl. hierüber D. Hilbert, Nachr. d. K. Gesellsch. d. Wiss. zu Göttingen, Math.-phys. Klasse. p. 3. 1915.

sammen eine hinreichende Basis für die Theorie der Materie liefern oder nicht. Das allgemeine Relativitätspostulat kann uns hierüber im Prinzip nichts lehren. Es muß sich bei dem Ausbau der Theorie zeigen, ob Elektromagnetik und Gravitationslehre zusammen leisten können, was ersterer allein nicht gelingen will.

§ 19. Eulersche Gleichungen für reibungslose adiabatische Flüssigkeiten.

Es seien p und ϱ zwei Skalare, von denen wir ersteren als den „Druck", letzteren als die „Dichte" einer Flüssigkeit bezeichnen; zwischen ihnen bestehe eine Gleichung. Der kontravariante symmetrische Tensor

$$(58) \qquad T^{\alpha\beta} = - g^{\alpha\beta} p + \varrho \frac{dx_\alpha}{ds} \frac{dx_\beta}{ds}$$

sei der kontravariante Energietensor der Flüssigkeit. Zu ihm gehört der kovariante Tensor

$$(58\,\mathrm{a}) \qquad T_{\mu\nu} = - g_{\mu\nu} p + g_{\mu\alpha} \frac{dx_\alpha}{ds} g_{\mu\beta} \frac{dx_\beta}{ds} \varrho,$$

sowie der gemischte Tensor[1])

$$(58\,\mathrm{b}) \qquad T_\sigma^\alpha = - \delta_\sigma^\alpha p + g_{\sigma\beta} \frac{dx_\beta}{ds} \frac{dx_\alpha}{ds} \varrho.$$

Setzt man die rechte Seite von (58b) in (57a) ein, so erhält man die Eulerschen hydrodynamischen Gleichungen der allgemeinen Relativitätstheorie. Diese lösen das Bewegungsproblem im Prinzip vollständig; denn die vier Gleichungen (57a) zusammen mit der gegebenen Gleichung zwischen p und ϱ und der Gleichung

$$g_{\alpha\beta} \frac{dx_\alpha}{ds} \frac{dx_\beta}{ds} = 1$$

genügen bei gegebenen $g_{\alpha\beta}$ zur Bestimmung der 6 Unbekannten

$$p, \varrho, \frac{dx_1}{ds}, \frac{dx_2}{ds}, \frac{dx_3}{ds}, \frac{dx_4}{ds}.$$

---

[1]) Für einen mitbewegten Beobachter, der im unendlich Kleinen ein Bezugssystem im Sinne der speziellen Relativitätstheorie benutzt, ist die Energiedichte T_4^4 gleich $\varrho - p$. Hierin liegt die Definition von ϱ. Es ist also ϱ nicht konstant für eine inkompressible Flüssigkeit.

Sind auch die $g_{\mu\nu}$ unbekannt, so kommen hierzu noch die Gleichungen (53). Dies sind 11 Gleichungen zur Bestimmung der 10 Funktionen $g_{\mu\nu}$, so daß diese überbestimmt scheinen. Es ist indessen zu beachten, daß die Gleichungen (57a) in den Gleichungen (53) bereits enthalten sind, so daß letztere nur mehr 7 unabhöngige Gleichungen repräsentieren. Diese Unbestimmtheit hat ihren guten Grund darin, daß die weitgehende Freiheit in der Wahl der Koordinaten es mit sich bringt, daß das Problem mathematisch in solchem Grade unbestimmt bleibt, daß drei der Raumfunktionen beliebig gewählt werden können.[1])

§ 20. Maxwellsche elektromagnetische Feldgleichungen für das Vakuum.

Es seien φ_ν die Komponenten eines kovarianten Vierervektors, des Vierervektors des elektromagnetischen Potentials. Aus ihnen bilden wir gemäß (36) die Komponenten $F_{\varrho\sigma}$ des kovarianten Sechservektors des elektromagnetischen Feldes gemäß dem Gleichungssystem

$$(59) \qquad F_{\varrho\sigma} = \frac{\partial \varphi_\varrho}{\partial x_\sigma} - \frac{\partial \varphi_\sigma}{\partial x_\varrho}.$$

Aus (59) folgt, daß das Gleichungssystem

$$(60) \qquad \frac{\partial F_{\varrho\sigma}}{\partial x_\tau} + \frac{\partial F_{\sigma\tau}}{\partial x_\varrho} + \frac{\partial F_{\tau\varrho}}{\partial x_\sigma} = 0$$

erfüllt ist, dessen linke Seite gemäß (37) ein antisymmetrischer Tensor dritten Ranges ist. Das System (60) enthält also im wesentlichen 4 Gleichungen, die ausgeschrieben wie folgt lauten:

$$(60\,\mathrm{a}) \qquad \begin{cases} \dfrac{\partial F_{23}}{\partial x_4} + \dfrac{\partial F_{34}}{\partial x_2} + \dfrac{\partial F_{42}}{\partial x_3} = 0 \\[4pt] \dfrac{\partial F_{34}}{\partial x_1} + \dfrac{\partial F_{41}}{\partial x_3} + \dfrac{\partial F_{13}}{\partial x_4} = 0 \\[4pt] \dfrac{\partial F_{41}}{\partial x_2} + \dfrac{\partial F_{12}}{\partial x_4} + \dfrac{\partial F_{24}}{\partial x_1} = 0 \\[4pt] \dfrac{\partial F_{12}}{\partial x_3} + \dfrac{\partial F_{23}}{\partial x_1} + \dfrac{\partial F_{31}}{\partial x_2} = 0. \end{cases}$$

---

[1]) Bei Verzicht auf die Koordinatenwahl gemäß $g = -1$ blieben *vier* Raumfunktionen frei wählbar, entsprechend den vier willkürlichen Funktionen, über die man bei der Koordinatenwahl frei verfügen kann.

Dieses Gleichungssystem entspricht dem zweiten Gleichungssystem Maxwells. Man erkennt dies sofort, indem man setzt

(61) $$\begin{cases} F_{23} = \mathfrak{h}_x & F_{14} = \mathfrak{e}_x \\ F_{31} = \mathfrak{h}_y & F_{24} = \mathfrak{e}_y \\ F_{12} = \mathfrak{h}_z & F_{34} = \mathfrak{e}_z \end{cases}$$

Dann kann man statt (60a) in üblicher Schreibweise der dreidimensionalen Vektoranalyse setzen

(60b) $$\begin{cases} \dfrac{\partial \mathfrak{h}}{\partial t} + \operatorname{rot} \mathfrak{e} = 0 \\ \operatorname{div} \mathfrak{h} = 0 \end{cases}$$

Das erste Maxwellsche System erhalten wir durch Verallgemeinerung der von Minkowski angegebenen Form. Wir führen den zu $F_{\alpha\beta}$ gehörigen kontravarianten Sechservektor

(62) $$F^{\mu\nu} = g^{\mu\alpha} g^{\nu\beta} F_{\alpha\beta}$$

ein sowie den kontravarianten Vierervektor J^μ der elektrischen Vakuumstromdichte; dann kann man das mit Rücksicht auf (40) gegenüber beliebigen Substitutionen von der Determinante 1 (gemäß der von uns getroffenen Koordinatenwahl) invariante Gleichungssystem ansetzen:

(63) $$\frac{\partial F^{\mu\nu}}{\partial x_\nu} = J^\mu$$

Setzt man nämlich

(64) $$\begin{cases} F^{23} = \mathfrak{h}_x' & F^{14} = -\mathfrak{e}_x' \\ F^{31} = \mathfrak{h}_y' & F^{24} = -\mathfrak{e}_y' \\ F^{12} = \mathfrak{h}_z' & F^{34} = -\mathfrak{e}_z' \end{cases}$$

welche Größen im Spezialfall der speziellen Relativitätstheorie den Größen $\mathfrak{h}_x \ldots \mathfrak{e}_z$ gleich sind, und außerdem

$$J^1 = \mathfrak{i}_x, \quad J^2 = \mathfrak{i}_y, \quad J^3 = \mathfrak{i}_z, \quad J^4 = \varrho,$$

so erhält man an Stelle von (63)

(63a) $$\begin{cases} \operatorname{rot} \mathfrak{h}' - \dfrac{\partial \mathfrak{e}'}{\partial t} = \mathfrak{i} \\ \operatorname{div} \mathfrak{e}' = \varrho \end{cases}$$

Die Gleichungen (60), (62) und (63) bilden also die Verallgemeinerung der Maxwellschen Feldgleichungen des

814   *A. Einstein.*

Vakuums bei der von uns bezüglich der Koordinatenwahl getroffenen Festsetzung.

Die Energiekomponenten des elektromagnetischen Feldes. Wir bilden das innere Produkt

(65) $$\varkappa_\sigma = F_{\sigma\mu} J^\mu.$$

Seine Komponenten lauten gemäß (61) in dreidimensionaler Schreibweise

(65a) $$\begin{cases} \varkappa_1 = \varrho\, \mathfrak{e}_x + [\mathfrak{i},\, \mathfrak{h}]_x \\ \cdots\cdots\cdots \\ \cdots\cdots\cdots \\ \varkappa_4 = -(\mathfrak{i},\, \mathfrak{e}). \end{cases}$$

Es ist $\varkappa_\sigma$ ein kovarianter Vierervektor, dessen Komponenten gleich sind dem negativen Impuls bzw. der Energie, welche pro Zeit- und Volumeinheit auf das elektromagnetische Feld von den elektrischen Massen übertragen werden. Sind die elektrischen Massen frei, d. h. unter dem alleinigen Einfluß des elektromagnetischen Feldes, so wird der kovariante Vierervektor $\varkappa_\sigma$ verschwinden.

Um die Energiekomponenten $T_\sigma{}^\nu$ des elektromagnetischen Feldes zu erhalten, brauchen wir nur der Gleichung $\varkappa_\sigma = 0$ die Gestalt der Gleichung (57) zu geben. Aus (63) und (65) ergibt sich zunächst

$$\varkappa_\sigma = F_{\sigma\mu} \frac{\partial F^{\mu\nu}}{\partial x_\nu} = \frac{\partial}{\partial x_\nu} (F_{\sigma\mu} F^{\mu\nu}) - F^{\mu\nu} \frac{\partial F_{\sigma\mu}}{\partial x_\nu}.$$

Das zweite Glied der rechten Seite gestattet vermöge (60) die Umformung

$$F^{\mu\nu} \frac{\partial F_{\sigma\mu}}{\partial x_\nu} = -\frac{1}{2} F^{\mu\nu} \frac{\partial F_{\mu\nu}}{\partial x_\sigma} = -\frac{1}{2} g^{\mu\alpha} g^{\nu\beta} F_{\alpha\beta} \frac{\partial F_{\mu\nu}}{\partial x_\sigma},$$

welch letzterer Ausdruck aus Symmetriegründen auch in der Form

$$-\frac{1}{4} \left[g^{\mu\alpha} g^{\nu\beta} F_{\alpha\beta} \frac{\partial F_{\mu\nu}}{\partial x_\sigma} + g^{\mu\alpha} g^{\nu\beta} \frac{\partial F_{\alpha\beta}}{\partial x_\sigma} F_{\mu\nu} \right]$$

geschrieben werden kann. Dafür aber läßt sich setzen

$$-\frac{1}{4} \frac{\partial}{\partial x_\sigma} (g^{\mu\alpha} g^{\nu\beta} F_{\alpha\beta} F_{\mu\nu}) + \frac{1}{4} F_{\alpha\beta} F_{\mu\nu} \frac{\partial}{\partial x_\sigma} (g^{\mu\alpha} g^{\nu\beta}).$$

Das erste dieser Glieder lautet in kürzerer Schreibweise

$$-\frac{1}{4} \frac{\partial}{\partial x_\sigma} (F^{\mu\nu} F_{\mu\nu}),$$

das zweite ergibt nach Ausführung der Differentiation nach einiger Umformung

$$-\frac{1}{2} F^{\mu\tau} F_{\mu\nu} g^{\nu\varrho} \frac{\partial g_{\sigma\tau}}{\partial x_\sigma}.$$

Nimmt man alle drei berechneten Glieder zusammen, so erhält man die Relation

(66) $$\varkappa_\sigma = \frac{\partial T_\sigma^\nu}{\partial x_\nu} - \frac{1}{2} g^{\tau\mu} \frac{\partial g_{\mu\nu}}{\partial x_\sigma} T_\tau^\nu,$$

wobei

(66a) $$T_\sigma^\nu = -F_{\sigma\alpha} F^{\nu\alpha} + \frac{1}{4} \delta_\sigma^\nu F_{\alpha\beta} F^{\alpha\beta}.$$

Die Gleichung (66) ist für verschwindendes $\varkappa_\sigma$ wegen (30) mit (57) bzw. (57a) gleichwertig. Es sind also die T_σ^ν die Energiekomponenten des elektromagnetischen Feldes. Mit Hilfe von (61) und (64) zeigt man leicht, daß diese Energiekomponenten des elektromagnetischen Feldes im Falle der speziellen Relativitätstheorie die wohlbekannten **Maxwell-Pointingschen** Ausdrücke ergeben.

Wir haben nun die allgemeinsten Gesetze abgeleitet, welchen das Gravitationsfeld und die Materie genügen, indem wir uns konsequent eines Koordinatensystems bedienten, für welches $\sqrt{-g} = 1$ wird. Wir erzielten dadurch eine erhebliche Vereinfachung der Formeln und Rechnungen, ohne daß wir auf die Forderung der allgemeinen Kovarianz verzichtet hätten: denn wir fanden unsere Gleichungen durch Spezialisierung des Koordinatensystems aus allgemein kovarianten Gleichungen.

Immerhin ist die Frage nicht ohne formales Interesse, ob bei entsprechend verallgemeinerter Definition der Energiekomponenten des Gravitationsfeldes und der Materie auch ohne Spezialisierung des Koordinatensystems Erhaltungssätze von der Gestalt der Gleichung (56) sowie Feldgleichungen der Gravitation von der Art der Gleichungen (52) bzw. (52a) gelten, derart, daß links eine Divergenz (im gewöhnlichen Sinne), rechts die Summe der Energiekomponenten der Materie und der Gravitation steht. Ich habe gefunden, daß beides in der Tat der Fall ist. Doch glaube ich, daß sich eine Mitteilung meiner ziemlich umfangreichen Betrachtungen über diesen Gegenstand nicht lohnen würde, da doch etwas sachlich Neues dabei nicht herauskommt.

E. § 21. Newtons Theorie als erste Näherung.

Wie schon mehrfach erwähnt, ist die spezielle Relativitätstheorie als Spezialfall der allgemeinen dadurch charakterisiert, daß die $g_{\mu\nu}$ die konstanten Werte (4) haben. Dies bedeutet nach dem Vorherigen eine völlige Vernachlässigung der Gravitationswirkungen. Eine der Wirklichkeit näher liegende Approximation erhalten wir, indem wir den Fall betrachten, daß die $g_{\mu\nu}$ von den Werten (4) nur um (gegen 1) kleine Größen abweichen, wobei wir kleine Größen zweiten und höheren Grades vernachlässigen. (Erster Gesichtspunkt der Approximation.)

Ferner soll angenommen werden, daß in dem betrachteten zeiträumlichen Gebiete die $g_{\mu\nu}$ im räumlich Unendlichen bei passender Wahl der Koordinaten den Werten (4) zustreben; d. h. wir betrachten Gravitationsfelder, welche als ausschließlich durch im Endlichen befindliche Materie erzeugt betrachtet werden können.

Man könnte annehmen, daß diese Vernachlässigungen auf Newtons Theorie hinführen müßten. Indessen bedarf es hierfür noch der approximativen Behandlung der Grundgleichungen nach einem zweiten Gesichtspunkte. Wir fassen die Bewegung eines Massenpunktes gemäß den Gleichungen (46) ins Auge. Im Falle der speziellen Relativitätstheorie können die Komponenten

$$\frac{dx_1}{ds}, \frac{dx_2}{ds}, \frac{dx_3}{ds}$$

beliebige Werte annehmen; dies bedeutet, daß beliebige Geschwindigkeiten

$$v = \sqrt{\frac{dx_1^2}{dx_4} + \frac{dx_2^2}{dx_4} + \frac{dx_3^2}{dx_4}}$$

auftreten können, die kleiner sind als die Vakuumlichtgeschwindigkeit ($v < 1$). Will man sich auf den fast ausschließlich der Erfahrung sich darbietenden Fall beschränken, daß v gegen die Lichtgeschwindigkeit klein ist, so bedeutet dies, daß die Komponenten

$$\frac{dx_1}{ds}, \frac{dx_2}{ds}, \frac{dx_3}{ds}$$

als kleine Größen zu behandeln sind, während dx_4/ds bis auf Größen zweiter Ordnung gleich 1 ist (zweiter Gesichtspunkt der Approximation).

Nun beachten wir, daß nach dem ersten Gesichtspunkte der Approximation die Größen $\Gamma^\tau_{\mu\nu}$ alle kleine Größen mindestens erster Ordnung sind. Ein Blick auf (46) lehrt also, daß in dieser Gleichung nach dem zweiten Gesichtspunkt der Approximation nur Glieder zu berücksichtigen sind, für welche $\mu = \nu = 4$ ist. Bei Beschränkung auf Glieder niedrigster Ordnung erhält man an Stelle von (46) zunächst die Gleichungen

$$\frac{d^2 x_\tau}{dt^2} = \Gamma^\tau_{44},$$

wobei $ds = dx_4 = dt$ gesetzt ist, oder unter Beschränkung auf Glieder, die nach dem ersten Gesichtspunkte der Approximation erster Ordnung sind:

$$\frac{d^2 x_\tau}{dt^2} = \begin{bmatrix} 44 \\ \tau \end{bmatrix} (\tau = 1, 2, 3)$$

$$\frac{d^2 x_4}{dt^2} = -\begin{bmatrix} 44 \\ 4 \end{bmatrix}.$$

Setzt man außerdem voraus, daß das Gravitationsfeld ein quasi statisches sei, indem man sich auf den Fall beschränkt, daß die das Gravitationsfeld erzeugende Materie nur langsam (im Vergleich mit der Fortpflanzungsgeschwindigkeit des Lichtes) bewegt ist, so kann man auf der rechten Seite Ableitungen nach der Zeit neben solchen nach den örtlichen Koordinaten vernachlässigen, so daß man erhält

$$(67) \qquad \frac{d^2 x_\tau}{dt^2} = -\frac{1}{2}\frac{\partial g_{44}}{\partial x_\tau} (\tau = 1, 2, 3).$$

Dies ist die Bewegungsgleichung des materiellen Punktes nach Newtons Theorie, wobei $g_{44}/2$ die Rolle des Gravitationspotentiales spielt. Das Merkwürdige an diesem Resultat ist, daß nur die Komponente g_{44} des Fundamentaltensors allein in erster Näherung die Bewegung des materiellen Punktes bestimmt.

Wir wenden uns nun zu den Feldgleichungen (53). Dabei ist zu berücksichtigen, daß der Energietensor der „Materie" fast ausschließlich durch die Dichte ϱ der Materie im engeren Sinne bestimmt wird, d. h. durch das zweite Glied der rechten Seite von (58) [bzw. (58a) oder (58b)]. Bildet man die uns interessierende Näherung, so verschwinden alle Komponenten bis auf die Komponente

$$T_{44} = \varrho = T.$$

Auf der linken Seite von (53) ist das zweite Glied klein von zweiter Ordnung; das erste liefert in der uns interessierenden Näherung

$$+ \frac{\partial}{\partial x_1}\begin{bmatrix}\mu\,\nu\\1\end{bmatrix} + \frac{\partial}{\partial x_2}\begin{bmatrix}\mu\,\nu\\2\end{bmatrix} + \frac{\partial}{\partial x_3}\begin{bmatrix}\mu\,\nu\\3\end{bmatrix} - \frac{\partial}{\partial x_4}\begin{bmatrix}\mu\,\nu\\4\end{bmatrix}.$$

Dies liefert für $\mu = \nu = 4$ bei Weglassung von nach der Zeit differenzierten Gliedern

$$-\frac{1}{2}\left(\frac{\partial^2 g_{44}}{\partial x_1^2} + \frac{\partial^2 g_{44}}{\partial x_2^2} + \frac{\partial^2 g_{44}}{\partial x_3^2}\right) = -\tfrac{1}{2}\Delta g_{44}.$$

Die letzte der Gleichungen (53) liefert also

(68) $$\Delta g_{44} = \varkappa \varrho.$$

Die Gleichungen (67) und (68) zusammen sind äquivalent dem Newtonschen Gravitationsgesetz.

Für das Gravitationspotential ergibt sich nach (67) und (68) der Ausdruck

(68a) $$-\frac{\varkappa}{8\pi}\int\frac{\varrho\,d\tau}{r},$$

während Newtons Theorie bei der von uns gewählten Zeiteinheit

$$-\frac{K}{c^2}\int\frac{\varrho\,d\tau}{r}$$

ergibt, wobei K die gewöhnlich als Gravitationskonstante bezeichnete Konstante $6{,}7 \cdot 10^{-8}$ bedeutet. Durch Vergleich ergibt sich

(69) $$\varkappa = \frac{8\pi K}{c^2} = 1{,}87 \cdot 10^{-27}.$$

§ 22. Verhalten von Masstäben und Uhren im statischen Gravitationsfelde. Krümmung der Lichtstrahlen. Perihelbewegung der Planetenbahnen.

Um die Newtonsche Theorie als erste Näherung zu erhalten, brauchten wir von den 10 Komponenten des Gravitationspotentials $g_{\mu\nu}$ nur g_{44} zu berechnen, da nur diese Komponente in die erste Näherung (67) der Bewegungsgleichung des materiellen Punktes im Gravitationsfelde eingeht. Man sieht indessen schon daraus, daß noch andere Komponenten der $g_{\mu\nu}$ von den in (4) angegebenen Werten in erster Näherung abweichen müssen, daß letzteres durch die Bedingung $g = -1$ verlangt wird.

Für einen im Anfangspunkt des Koordinatensystems befindlichen felderzeugenden Massenpunkt erhält man in erster Näherung die radialsymmetrische Lösung

$$(70) \begin{cases} g_{\varrho\sigma} = -\delta_{\varrho\sigma} - \alpha \frac{x_\varrho x_\sigma}{r^3} & (\varrho \text{ und } \sigma \text{ zwischen 1 und 3}) \\ g_{\varrho 4} = g_{4\varrho} = 0 & (\varrho \text{ zwischen 1 und 3}) \\ g_{44} = 1 - \frac{\alpha}{r} \end{cases}$$

$\delta_{\varrho\sigma}$ ist dabei 1 bzw. 0, je nachdem $\varrho = \sigma$ oder $\varrho \sigma$, r ist die Größe

$$+ \sqrt{x_1^2 + x_2^2 + x_3^2}.$$

Dabei ist wegen (68a)

$$(70\text{a}) \qquad \alpha = \frac{\varkappa M}{8\pi},$$

wenn mit M die felderzeugende Masse bezeichnet wird. Daß durch diese Lösung die Feldgleichungen (außerhalb der Masse) in erster Näherung erfüllt werden, ist leicht zu verifizieren.

Wir untersuchen nun die Beeinflussung, welche die metrischen Eigenschaften des Raumes durch das Feld der Masse M erfahren. Stets gilt zwischen den „lokal" (§ 4) gemessenen Längen und Zeiten ds einerseits und den Koordinatendifferenzen dx_ν andererseits die Beziehung

$$ds^2 = g_{\mu\nu} dx_\mu dx_\nu.$$

Für einen „parallel" der x-Achse gelegten Einheitsmaßstab wäre beispielsweise zu setzen

$$ds^2 = -1; \quad dx_2 = dx_3 = dx_4 = 0,$$

also

$$-1 = g_{11} dx_1^2.$$

Liegt der Einheitsmaßstab außerdem auf der x-Achse, so ergibt die erste der Gleichungen (70)

$$g_{11} = -\left(1 + \frac{\alpha}{r}\right).$$

Aus beiden Relationen folgt in erster Näherung genau

$$(71) \qquad dx = 1 - \frac{\alpha}{2r}.$$

Der Einheitsmaßstab erscheint also mit Bezug auf das Koordinatensystem in dem gefundenen Betrage durch das Vorhandensein des Gravitationsfeldes verkürzt, wenn er radial angelegt wird.

Analog erhält man seine Koordinatenlänge in tangentialer Richtung, indem man beispielsweise setzt

$$ds^2 = -1; \quad dx_1 = dx_3 = dx_4 = 0; \quad x_1 = r, \, x_2 = x_3 = 0.$$

Es ergibt sich

(71a) $$-1 = g_{22} \, dx_2{}^2 = -dx_2{}^2.$$

Bei tangentialer Stellung hat also das Gravitationsfeld des Massenpunktes keinen Einfluß auf die Stablänge.

Es gilt also die Euklidische Geometrie im Gravitationsfelde nicht einmal in erster Näherung, falls man einen und denselben Stab unabhängig von seinem Ort und seiner Orientierung als Realisierung derselben Strecke auffassen will. Allerdings zeigt ein Blick auf (70a) und (69), daß die zu erwartenden Abweichungen viel zu gering sind, um sich bei der Vermessung der Erdoberfläche bemerkbar machen zu können.

Es werde ferner die auf die Zeitkoordinate untersuchte Ganggeschwindigkeit einer Einheitsuhr untersucht, welche in einem statischen Gravitationsfelde ruhend angeordnet ist. Hier gilt für eine Uhrperiode

$$ds = 1; \quad dx_1 = dx_2 = dx_3 = 0.$$

Also ist
$$1 = g_{44} \, dx_4{}^2;$$
$$dx_4 = \frac{1}{\sqrt{g_{44}}} = \frac{1}{\sqrt{1 + (g_{44} - 1)}} = 1 - \frac{g_{44} - 1}{2}$$

oder

(72) $$dx_4 = 1 + \frac{\varkappa}{8\pi} \int \frac{\varrho \, d\tau}{r}.$$

Die Uhr läuft also langsamer, wenn sie in der Nähe ponderabler Massen aufgestellt ist. Es folgt daraus, daß die Spektrallinien von der Oberfläche großer Sterne zu uns gelangenden Lichtes nach dem roten Spektralende verschoben erscheinen müssen.[1]

---

[1] Für das Bestehen eines derartigen Effektes sprechen nach E. Freundlich spektrale Beobachtungen an Fixsternen bestimmter Typen. Eine endgültige Prüfung dieser Konsequenz steht indes noch aus.

Wir untersuchen ferner den Gang der Lichtstrahlen im statischen Gravitationsfeld. Gemäß der speziellen Relativitätstheorie ist die Lichtgeschwindigkeit durch die Gleichung

$$-dx_1^2 - dx_2^2 - dx_3^2 + dx_4^2 = 0$$

gegeben, also gemäß der allgemeinen Relativitätstheorie durch die Gleichung

(73) $$ds^2 = g_{\mu\nu} dx_\mu dx_\nu = 0.$$

Ist die Richtung, d. h. das Verhältnis $dx_1 : dx_2 : dx_3$ gegeben, so liefert die Gleichung (73) die Größen

$$\frac{dx_1}{dx_4}, \quad \frac{dx_2}{dx_4}, \quad \frac{dx_3}{dx_4}$$

und somit die Geschwindigkeit

$$\sqrt{\left(\frac{dx_1}{dx_4}\right)^2 + \left(\frac{dx_2}{dx_4}\right)^2 + \left(\frac{dx_3}{dx_4}\right)^2} = \gamma,$$

im Sinne der Euklidischen Geometrie definiert. Man erkennt leicht, daß die Lichtstrahlen gekrümmt verlaufen müssen mit Bezug auf das Koordinatensystem, falls die $g_{\mu\nu}$ nicht konstant sind. Ist n eine Richtung senkrecht zur Lichtfortpflanzung, so ergibt das Huggenssche Prinzip, daß der Lichtstrahl [in der Ebene (γ, n) betrachtet] die Krümmung $-\partial\gamma/\partial n$ besitzt.

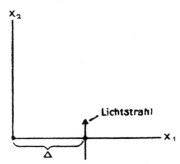

Wir untersuchen die Krümmung, welche ein Lichtstrahl erleidet, der im Abstand Δ an einer Masse M vorbeigeht. Wählt man das Koordinatensystem gemäß der vorstehenden Skizze, so ist die gesamte Biegung B des Lichtstrahles (positiv gerechnet, wenn sie nach dem Ursprung hin konkav ist) in genügender Näherung gegeben durch

$$B = \int_{-\infty}^{+\infty} \frac{\partial \gamma}{\partial x_1} dx_2,$$

während (73) und (70) ergeben

$$\gamma = \sqrt{-\frac{g_{44}}{g_{22}}} = 1 + \frac{\alpha}{2\,r}\left(1 + \frac{x_2^2}{r^2}\right).$$

Die Ausrechnung ergibt

(74) $$B = \frac{2\,\alpha}{\varDelta} = \frac{\varkappa M}{4\,\pi\,\varDelta}.$$

Ein an der Sonne vorbeigehender Lichtstrahl erfährt demnach eine Biegung von 1,7″, ein am Planeten Jupiter vorbeigehender eine solche von etwa 0,02″.

Berechnet man das Gravitationsfeld um eine Größenordnung genauer, und ebenso mit entsprechender Genauigkeit die Bahnbewegung eines materiellen Punktes von relativ unendlich kleiner Masse, so erhält man gegenüber den Kepler-Newtonschen Gesetzen der Planetenbewegung eine Abweichung von folgender Art. Die Bahnellipse eines Planeten erfährt in Richtung der Bahnbewegung eine langsame Drehung vom Betrage

(75) $$\varepsilon = 24\,\pi^3\frac{a^2}{T^2\,c^2(1-e^2)}$$

pro Umlauf. In dieser Formel bedeutet a die große Halbachse, c die Lichtgeschwindigkeit in üblichem Maße, e die Exzentrizität, T die Umlaufszeit in Sekunden.[1])

Die Rechnung ergibt für den Planeten Merkur eine Drehung der Bahn um 43″ pro Jahrhundert, genau entsprechend der Konstatierung der Astronomen (Leverrier); diese fanden nämlich einen durch Störungen der übrigen Planeten nicht erklärbaren Rest der Perihelbewegung dieses Planeten von der angegebenen Größe.

---

1) Bezüglich der Rechnung verweise ich auf die Originalabhandlungen A. Einstein, Sitzungsber. d. Preuß. Akad. d. Wiss. **47**. p. 831. 1915. — K. Schwarzschild, Sitzungsber. d. Preuß. Akad. d. Wiss. **7**. p. 189. 1916.

(Eingegangen 20. März 1916.)

ANNALEN DER PHYSIK.

BEGRÜNDET UND FORTGEFÜHRT DURCH

F. A. C. GREN, L. W. GILBERT, J. C. POGGENDORFF, G. u. E. WIEDEMANN, P. DRUDE.

VIERTE FOLGE.

BAND 51.

DER GANZEN REIHE 356. BAND.

KURATORIUM:

M. PLANCK, G. QUINCKE,
W. C. RÖNTGEN, W. VOIGT, E. WARBURG.

UNTER MITWIRKUNG

DER DEUTSCHEN PHYSIKALISCHEN GESELLSCHAFT

HERAUSGEGEBEN VON

W. WIEN UND M. PLANCK.

MIT VIER FIGURENTAFELN.

LEIPZIG, 1916.
VERLAG VON JOHANN AMBROSIUS BARTH.

5. Über Friedrich Kottlers Abhandlung „Über Einsteins Äquivalenzhypothese und die Gravitation"[1]; von A. Einstein.

Unter den Arbeiten, welche sich kritisch mit der allgemeinen Relativitätstheorie beschäftigen, sind besonders diejenigen Kottlers bemerkenswert, denn dieser Fachgenosse ist wirklich in den Geist der Theorie eingedrungen. Mit der letzten dieser Arbeiten will ich mich hier auseinandersetzen.

Kottler behauptet, ich hätte das von mir aufgestellte „Äquivalenzprinzip", durch welches ich die Begriffe der „trägen Masse" und der „schweren Masse" zu einem einheitlichen Begriffe zu vereinigen strebte, in meinen späteren Arbeiten wieder aufgegeben. Diese Meinung muß darauf beruhen, daß wir beide nicht dasselbe als „Äquivalenzprinzip" bezeichnen; denn nach meiner Auffassung ruht meine Theorie ausschließlich auf diesem Prinzip. Deshalb sei folgendes wiederholt:

1. *Der Grenzfall der speziellen Relativitätstheorie.* Ein raumzeitlich endliches Gebiet sei frei von einem Gravitationsfelde, d. h. es sei möglich, ein Bezugssystem K („Galileisches System") aufzustellen, relativ zu welchem für das genannte Gebiet folgendes gilt. Koordinaten seien in bekannter Weise mit dem Einheitsmaßstab, Zeiten mit der Einheitsuhr unmittelbar meßbar, wie dies in der speziellen Relativitätstheorie vorausgesetzt zu werden pflegt. In bezug auf dieses System bewege sich ein isolierter materieller Punkt geradlinig und gleichförmig, wie es von Galilei vorausgesetzt wurde.

2. *Äquivalenzprinzip.* Ausgehend von diesem Grenzfall der speziellen Relativitätstheorie kann man sich fragen, ob

---

1) Ann. d. Phys. **50**. p. 955. 1916.

ein in dem betrachteten Gebiete relativ zu K gleichförmig beschleunigter Beobachter seinen Zustand als beschleunigt auffassen muß, oder ob ihm nach den (angenähert) bekannten Naturgesetzen eine Auffassung übrig bleibt, vermöge derer er seinen Zustand als „Ruhe" deuten kann. Präziser ausgedrückt: Erlauben uns die in gewisser Annäherung bekannten Naturgesetze ein in bezug auf K gleichförmig beschleunigtes Bezugssystem K' als ruhend zu betrachten? Oder etwas allgemeiner: Läßt sich das Relativitätsprinzip auch auf relativ zueinander (gleichförmig) beschleunigte Bezugssysteme ausdehnen? Die Antwort lautet: Soweit wir die Naturgesetze wirklich kennen, hindert uns nichts daran, das System K' als ruhend zu betrachten, wenn wir relativ zu K' ein (in erster Annäherung homogenes) Schwerefeld als vorhanden annehmen; denn wie in einem homogenen Schwerefeld, so auch in bezug auf unser System K' fallen alle Körper unabhängig von ihrer physikalischen Natur mit derselben Beschleunigung. Die Voraussetzung, daß man in aller Strenge K' als ruhend behandeln dürfe, ohne daß irgendein Naturgesetz in bezug auf K' nicht erfüllt wäre, nenne ich „Äquivalenzprinzip".

3. *Das Schwerefeld nicht nur kinematisch bedingt.* Man kann die vorige Betrachtung auch umkehren. Sei das mit dem oben betrachteten Schwerefelde ausgestaltete System K' das ursprüngliche. Dann kann man ein neues, gegen K' beschleunigtes Bezugssystem K einführen, mit Bezug auf welches sich (isolierte) Massen (in dem betrachteten Gebiete) geradlinig gleichförmig bewegen. Aber man darf nun *nicht* weitergehen und sagen: Ist K' ein mit einem *beliebigen* Gravitationsfeld versehenes Bezugssystem, so ist stets ein Bezugssystem K auffindbar, in bezug auf welches sich isolierte Massen geradlinig gleichförmig bewegen, d. h. in bezug auf welches kein Gravitationsfeld existiert. Die Absurdität einer solchen Voraussetzung liegt auf der Hand. Ist das Gravitationsfeld in bezug auf K' zum Beispiel das eines ruhenden Massenpunktes, so läßt sich dieses Feld für die ganze Umgebung des Massenpunktes gewiß durch kein noch so feines Transformationskunststück hinwegtransformieren. Man darf also keineswegs behaupten, das Gravitationsfeld sei gewissermaßen rein kinematisch zu erklären; eine „kinematische, nicht dynamische Auffassung der Gravitation" ist nicht möglich. Durch bloße Trans-

Über Einsteins Äquivalenzhypothese und die Gravitation. 641

formation aus einem Galileischen System in ein anderes durch Beschleunigungstransformationen lernen wir also nicht *beliebige* Gravitationsfelder kennen, sondern solche ganz spezieller Art, welche aber doch denselben Gesetzen genügen müssen wie alle anderen Gravitationsfelder. Dies ist nur wieder eine andere Formulierung des Äquivalenzpinzips (speziell in seiner Anwendung auf die Gravitation).

Eine Gravitationstheorie verletzt also das Äquivalenzprinzip in dem Sinne, wie ich es verstehe, nur dann, wenn die Gleichungen der Gravitation in *keinem* Bezugssystem K' erfüllt sind, welches relativ zu einem galileischen Bezugssystem ungleichförmig bewegt ist. Daß dieser Vorwurf gegen meine Theorie mit *allgemein* kovarianten Gleichungen nicht erhoben werden kann, ist evident; denn hier sind die Gleichungen bezüglich eines jeden Bezugssystems erfüllt. *Die Forderung der allgemeinen Kovarianz der Gleichungen umfaßt die des Äquivalenzprinzips als ganz speziellen Fall.*

4. *Sind die Kräfte des Gravitationsfeldes „reale" Kräfte?* Kottler rügt es, daß ich in den Bewegungsgleichungen

$$\frac{d^2 x_\nu}{ds^2} + \sum_{\alpha\beta} \begin{Bmatrix} \alpha\beta \\ \nu \end{Bmatrix} \frac{dx_\alpha}{ds} \frac{dx_\beta}{ds} = 0$$

das zweite Glied als den Ausdruck des Einflusses des Schwerefeldes auf den Massenpunkt, das erste Glied gewissermaßen als den Ausdruck der Galileischen Trägheit interpretiere. Dadurch würden „wirkliche Kräfte des Schwerefeldes" eingeführt, was dem Geiste des Äquivalenzprinzipes nicht entspreche. Hierauf antworte ich, daß jene Gleichung als Ganzes allgemein kovariant, also jedenfalls der Äquivalenzhypothese gemäß ist. Die von mir eingeführte Benennung der Teile ist prinzipiell bedeutungslos und einzig dazu bestimmt, unseren physikalischen Denkgewohnheiten entgegenzukommen. Dies gilt auch insbesondere von den Begriffen

$$\Gamma^\nu_{\alpha\beta} = -\begin{Bmatrix} \alpha\beta \\ \nu \end{Bmatrix}$$

(Komponenten des Gravitationsfeldes) und t_σ^ν (Energiekomponenten des Gravitationsfeldes). Die Einführung dieser Benennungen ist prinzipiell unnötig, erscheint mir aber für die Aufrechterhaltung der Kontinuität der Gedanken wenigstens einstweilen nicht wertlos; deshalb habe ich diese Größen eingeführt, trotzdem ihnen kein Tensorcharakter zukommt. Dem Äquivalenzprinzip aber ist stets Genüge geleistet, wenn die Gleichungen kovariant sind.

5. Es ist wahr, daß ich die allgemeine Kovarianz der Gleichungen durch das Aufgeben der gewöhnlichen Zeitmessung und der Euklidischen Raummessung habe erkaufen müssen. Kottler glaubt ohne dies Opfer auskommen zu können. Aber bereits im Falle des von ihm betrachteten im Bornschen Sinne relativ zu einem Galileischen System beschleunigten Systems K' muß man auf die gewöhnliche Zeitmessung verzichten. Da ist vom Standpunkt der Relativitätstheorie der Gedanke schon naheliegend, daß auch die gewöhnliche Raummessung aufgegeben werden müsse. Von dieser Notwendigkeit wird sich Hr. Kottler sicherlich selbst überzeugen, wenn er die ihm vorschwebenden theoretischen Pläne allgemein durchzuführen suchen wird.

Oktober 1916.

(Eingegangen 19. Oktober 1916.)

ANNALEN DER PHYSIK.

BEGRÜNDET UND FORTGEFÜHRT DURCH

F. A. C. GREN, L. W. GILBERT, J. C. POGGENDORFF, G. u. E. WIEDEMANN, P. DRUDE.

VIERTE FOLGE.

BAND 55.

DER GANZEN REIHE 360. BAND.

KURATORIUM:

M. PLANCK, G. QUINCKE,
W. C. RÖNTGEN, W. VOIGT, E. WARBURG.

UNTER MITWIRKUNG

DER DEUTSCHEN PHYSIKALISCHEN GESELLSCHAFT

HERAUSGEGEBEN VON

W. WIEN UND M. PLANCK.

MIT EINER FIGURENTAFEL.

LEIPZIG, 1918.
VERLAG VON JOHANN AMBROSIUS BARTH.

1. *Prinzipielles zur allgemeinen Relativitätstheorie;*
von A. Einstein.

Eine Reihe von Publikationen der letzten Zeit, insbesondere die neulich in diesen Annalen 53. Heft 16 erschienene scharfsinnige Arbeit von Kretschmann, veranlassen mich, nochmals auf die Grundlagen der allgemeinen Relativitätstheorie zurückzukommen. Dabei ist es mein Ziel, lediglich die Grundgedanken herauszuheben, wobei ich die Theorie als bekannt voraussetze.

Die Theorie, wie sie mir heute vorschwebt, beruht auf drei Hauptgesichtspunkten, die allerdings keineswegs voneinander unabhängig sind. Sie seien im folgenden kurz angeführt und charakterisiert und hierauf im nachfolgenden von einigen Seiten beleuchtet:

a) *Relativitätsprinzip:* Die Naturgesetze sind nur Aussagen über zeiträumliche Koinzidenzen; sie finden deshalb ihren einzig natürlichen Ausdruck in allgemein kovarianten Gleichungen.

b) *Äquivalenzprinzip:* Trägheit und Schwere sind wesensgleich. Hieraus und aus den Ergebnissen der speziellen Relativitätstheorie folgt notwendig, daß der symmetrische „Fundamentaltensor" ($g_{\mu\nu}$) die metrischen Eigenschaften des Raumes, das Trägheitsverhalten der Körper in ihm, sowie die Gravitationswirkungen bestimmt. Den durch den Fundamentaltensor beschriebenen Raumzustand wollen wir als „G-Feld" bezeichnen.

c) *Machsches Prinzip*[1]*:* Das G-Feld ist *restlos* durch die Massen der Körper bestimmt. Da Masse und Energie nach

---

[1] Bisher habe ich die Prinzipe a) und c) nicht auseinandergehalten, was aber verwirrend wirkte. Den Namen „Machsches Prinzip" habe ich deshalb gewählt, weil dies Prinzip eine Verallgemeinerung der Machschen Forderung bedeutet, daß die Trägheit auf eine Wechselwirkung der Körper zurückgeführt werden müsse.

den Ergebnissen der speziellen Relativitätstheorie das Gleiche sind und die Energie formal durch den symmetrischen Energietensor ($T_{\mu\nu}$) beschrieben wird, so besagt dies, daß das G-Feld durch den Energietensor der Materie bedingt und bestimmt sei.

Zu a) bemerkt Hr. Kretschmann, das so formulierte Relativitätsprinzip sei keine Aussage über die physikalische Realität, d. h. über den *Inhalt* der Naturgesetze, sondern nur eine Forderung bezüglich der mathematischen *Formulierung*. Da nämlich die gesamte physikalische Erfahrung sich nur auf Koinzidenzen beziehe, müsse es stets möglich sein, Erfahrungen über die gesetzlichen Zusammenhänge dieser Koinzidenzen durch allgemein kovariante Gleichungen darzustellen. Er hält es deshalb für nötig, einen anderen Sinn mit der Relativitätsforderung zu verbinden. Ich halte Hrn. Kretschmanns Argument für richtig, die von ihm vorgeschlagene Neuerung jedoch nicht für empfehlenswert. Wenn es nämlich auch richtig ist, daß man jedes empirische Gesetz in allgemein kovariante Form muß bringen können, so besitzt das Prinzip a) doch eine bedeutende heuristische Kraft, die sich am Gravitationsproblem ja schon glänzend bewährt hat und auf folgendem beruht. Von zwei mit der Erfahrung vereinbarten theoretischen Systemen wird dasjenige zu bevorzugen sein, welches vom Standpunkte des absoluten Differentialkalküls das einfachere und durchsichtigere ist. Man bringe einmal die Newtonsche Gravitationsmechanik in die Form von absolut kovarianten Gleichungen (vierdimensional) und man wird sicherlich überzeugt sein, daß das Prinzip a) diese Theorie zwar nicht theoretisch, aber praktisch ausschließt!

Das Prinzip b) hat den Ausgangspunkt der ganzen Theorie gebildet und erst die Aufstellung des Prinzipes a) mit sich gebracht; es kann sicherlich nicht verlassen werden, solange man am Grundgedanken des theoretischen Systems festhalten will.

Anders ist es mit dem „Machschen Prinzip" c); die Notwendigkeit, an diesem festzuhalten, wird keineswegs von allen Fachgenossen geteilt, ich selbst aber empfinde seine Erfüllung als unbedingt notwendig. Nach c) darf gemäß den Gravitations-Feldgleichungen kein G-Feld möglich sein ohne Materie. Das Postulat c) hängt offenbar aufs engste mit der Frage nach der zeiträumlichen Struktur des Weltganzen zusammen; denn an der Erzeugung des G-Feldes werden alle Massen der Welt teilhaben.

Als allgemein kovariante Feldgleichungen der Gravitation hatte ich zunächst vorgeschlagen

$$(1) \quad G_{\mu\nu} = -\varkappa(T_{\mu\nu} - \tfrac{1}{2} g_{\mu\nu} T),$$

wobei zur Abkürzung

$$G_{\mu\nu} = \sum_{\sigma\tau} g^{\sigma\tau}(\mu\sigma, \tau\nu)$$

gesetzt ist. Diese Feldgleichungen erfüllen aber das Postulat c) nicht; denn sie lassen die Lösung zu

$$g_{\mu\nu} = \text{konst. (für alle } \mu \text{ und } \nu),$$
$$T_{\mu\nu} = 0 \quad \text{(für alle } \mu \text{ und } \nu).$$

Nach den Gleichungen (1) wäre also im Widerspruch mit dem Machschen Postulat ein G-Feld denkbar ohne jede erzeugende Materie.

Das Postulat c) wird aber — soweit meine bisherige Einsicht reicht — erfüllt durch die aus (1) durch Hinzufügung des „λ-Gliedes" gebildeten Feldgleichungen[1])

$$(2) \quad G_{\mu\nu} - \lambda g_{\mu\nu} = -\varkappa(T_{\mu\nu} - \tfrac{1}{2} g_{\mu\nu} T).$$

Ein singularitätenfreies Raum-Zeit-Kontinuum mit überall verschwindendem Energietensor der Materie scheint es nach (2) nicht zu geben. Die einfachste nach (2) denkbare Lösung ist eine statische, in den räumlichen Koordinaten sphärische bzw. elliptische Welt mit gleichmäßig verteilter, ruhender Materie. Man kann sich so aber nicht nur eine Welt *gedanklich konstruieren*, welche dem Machschen Postulat entspricht; man kann sich vielmehr vorstellen, daß unsere wirkliche Welt durch die eben genannte sphärische approximiert wird. In unserer Welt ist zwar die Materie nicht gleichmäßig verteilt, sondern in einzelnen Himmelskörpern konzentriert, nicht ruhend, sondern in (gegen die Lichtgeschwindigkeit langsamer) relativer Bewegung begriffen. Aber es ist sehr wohl möglich, daß die mittlere, („natürlich gemessene") räumliche Dichte der Materie,

---

[1]) Kosmologische Betrachtungen zur allgemeinen Relativitätstheorie. Berl. Ber. 1917, S. 142.

genommen für Räume, die sehr viele Fixsterne umspannen, eine nahezu konstante Größe in der Welt ist. In diesem Falle *müssen* die Gleichungen (1) durch ein Zusatzglied vom Charakter des λ-Gliedes ergänzt werden; es muß dann die Welt in sich geschlossen sein, und ihre Geometrie weicht von der eines sphärischen bzw. elliptischen Raumes nur wenig und nur lokal ab, wie etwa die Gestalt der Erdoberfläche von der eines Ellipsoides abweicht.

(Eingegangen 6. März 1918.)

ANNALEN DER PHYSIK

BEGRÜNDET UND FORTGEFÜHRT DURCH
F. A. C. GREN, L. W. GILBERT, J. C. POGGENDORFF,
G. u. E. WIEDEMANN, P. DRUDE

VIERTE FOLGE

BAND 69

DER GANZEN REIHE 374. BAND

KURATORIUM:
M. PLANCK, G. QUINCKE, W. C. RÖNTGEN, E. WARBURG

UNTER MITWIRKUNG
DER DEUTSCHEN PHYSIKALISCHEN GESELLSCHAFT
HERAUSGEGEBEN VON

W. WIEN UND M. PLANCK

MIT VIER FIGURENTAFELN

1922

LEIPZIG · VERLAG VON JOHANN AMBROSIUS BARTH

2. Bemerkung zu der Franz Seletyschen Arbeit „Beiträge zum kosmologischen System";
(Ann. d. Phys. 68. S. 281. 1922)
von *A. Einstein.*

Es ist zuzugeben, daß die Hypothese vom „molekularhierarchischen" Charakter des Aufbaues der Sternenwelt vom Standpunkt der Newtonschen Theorie manches für sich hat, wenn auch die Hypothese von der Gleichwertigkeit der Spiralnebel mit der Milchstraße durch die letzten Beobachtungen als widerlegt zu betrachten sein dürfte. Diese Hypothese erklärt ungezwungen das Nichtbuchten des Himmelsgrundes und vermeidet den Seeligerschen Konflikt mit dem Newtonschen Gesetz, ohne die Materie als Insel im leeren Raum aufzufassen.

Auch vom Standpunkte der allgemeinen Relativitätstheorie ist die Hypothese vom molekularhierarchischen Bau des Weltalls *möglich*. Aber vom Standpunkt dieser Theorie ist die Hypothese dennoch als unbefriedigend anzusehen. Dies sei im folgenden noch einmal kurz begründet. Wenn die geometrischen und die Inertialeigenschaften des Raumes durch die Materie beeinflußt, bzw. zum Teil bedingt sind, so drängt sich die Ansicht auf, daß diese Bedingtheit eine vollständige sei, wie dies nach der allgemeinen Relativitätstheorie der Fall ist, wenn die mittlere Dichte der Materie endlich und die Welt räumlich geschlossen ist. Ich will dies durch einen einfacheren fingierten Fall — wenn auch unvollkommen — zu illustrieren suchen.

Es sei angenommen, man würde die Gravitation nur durch das genaue Studium der Mechanik solcher Massen kennen, welche uns bei Laboratoriumsversuchen zur Verfügung stehen. Die Kugelgestalt der Erde sei uns unbekannt. Dann könnte man folgende Theorie aufstellen. Es existiert primär ein vertikales „kosmisches" Schwerefeld, welches sich überall ins Unendliche erstreckt. Die Erde erstreckt sich unten ins Unendliche. Ihre Gravitationswirkung sei gegen das kosmische

Schwerefeld zu vernachlässigen.¹) Das kosmische Schwerefeld wird modifiziert durch Gravitationswirkungen von der Erfahrung zugänglichen Massen an der Erdoberfläche.

Obwohl das angenommene kosmische Schwerefeld der Poissonschen Gleichung entspricht ebenso wie die Schwerefelder der dem Experiment zugänglichen Massen an der Erdoberfläche, wäre diese Auffassung deshalb unbefriedigend, weil man das kosmische Feld selbst ohne materielle Ursache angenommen hat. Die Idee, daß das Schwerefeld, welches in der Hauptsache den Fall der Körper an der Erdoberfläche bedingt, nicht selbständig existierend, sondern durch den Erdkörper verursacht sei, würde gewiß als großer Fortschritt empfunden werden.

Daß heute das Bedürfnis einer Zurückführung des metrischen und Inertialfeldes der Welt auf physikalische Ursachen nicht ähnlich intensiv gefordert wird, liegt nur daran, daß dieses letztere Feld als physikalische Realität nicht so deutlich gefühlt wird, wie im obigen Beispiel die physikalische Realität des „kosmischen Schwerefeldes". Einer späteren Generation wird aber diese Genügsamkeit unbegreiflich erscheinen.

Die „molekular-hierarchische Welt" erfüllt ebensowenig wie die „Inselwelt" das Machsche Postulat, nach welchem die Trägheitswirkung des einzelnen Körpers durch die Gesamtheit aller übrigen im gleichen Sinne bedingt sein soll, wie seine Gravitationskraft. Es ist mir schwer verständlich, wieso Hrn. Selety dieser Mangel seines Systems hat entgehen können. Dieser Mangel ist um so schwerwiegender, als man in der allgemeinen Relativitätstheorie auch ohne Betrachtungen kosmologischen Charakters zeigen kann, daß sich die Körper der ersten Näherung so verhalten, wie es nach dem Machschen Gedanken erwartet werden muß. Ich verweise hierüber auf die vierte meiner bei Vieweg erschienenen „Vier Vorlesungen über Relativitätstheorie" (gehalten im Mai 1921 an der Universität Princeton).

Es sei endlich noch ein Punkt erwähnt, der nicht nur in der Seletyschen Abhandlung, sondern vielfach in der ein-

---

1) Daß diese Hypothese zum Newtonschen Gesetz nicht paßt, bitte ich zu entschuldigen.

schlägigen Literatur Verwirrung stiftet. Die Relativitätstheorie sagt: Die Naturgesetze sind unabhängig von jeder besonderen Koordinatenwahl zu formulieren, da dem Koordinatensystem nichts Reales entspricht; die Einfachheit eines hypothetischen Gesetzes ist nur nach seiner allgemein kovarianten Formulierung zu beurteilen. Daraus folgt aber nicht, daß man sich die Beschreibung durch passende Wahl des Bezugssystems nicht erleichtern dürfe, ohne gegen das Relativitätspostulat zu verstoßen. Wenn ich z. B. die wirkliche Welt durch die „Zylinderwelt" mit gleichmäßig verteilter Materie approximiere und dabei die Zeitachse parallel den Erzeugenden des „Zylinders" wähle, so bedeutet dies nicht die Einführung einer „absoluten Zeit". In der Welt gibt es nach wie vor kein Koordinatensystem, welches für die Formulierung der Naturgesetze bevorzugt wäre. Bezüglich der wirklichen Welt ist eine exakte Definition eines derartigen Koordinatensystems übrigens unmöglich, auch dann, wenn sich die wirkliche Welt *durch* jene Zylinderwelt roh approximieren läßt. Das Relativitätsprinzip behauptet nicht, daß die Welt gegenüber allen Koordinatensystemen in gleich einfacher oder gar in gleicher Weise zu beschreiben sei, sondern nur, daß die *allgemeinen Gesetze* der Natur bezüglich aller Systeme die gleichen seien (genauer: daß die hypothetisch möglichen Naturgesetze bezüglich ihrer Einfachheit nur in ihrer allgemein kovarianten Formulierung gegeneinander abzuwägen sind.

September 1922.

(Eingegangen 25. September 1922.)

▶ My Profile ▶ Log In

digital Encyclopedia of Applied Physics

"The 23-volume Encyclopedia of Applied Physics – EAP – is a monumental first in scope, depth, and usability. It demonstrates the synergy between physics and technological applications."

Edited by George L. Trigg

www.physics-encyclopedia.com

This Encyclopedia provides the basic principles and in-depth coverage of all technically pertinent areas of modern-day physics, coupled with technological applications from real life. The print edition filled 25 volumes, containing some 15,500 pages and over 650 articles with a total of 7,800 figures and 980 tables. Now Wiley InterScience brings you the online edition.

Some of the numerous features:

- ▶ new material on recent topics of applied physics
- ▶ annual updates of at least 10% new or revised articles
- ▶ keyword search in the full text of all volumes
- ▶ all references cross-linked

The key subject areas of **The Digital Encyclopedia of Applied Physics** are basics, methods and applications in physics as well as all the neighboring fields of physics and engineering. It thus comprehensively covers optics, laser physics, solid state and semiconductor physics, atom and nuclear physics, and biophysics. Thanks to its clear structure, users can easily access targeted information and quickly research the areas by way of the comprehensive reference section.

Highlights:

- ▶ detailed table of contents for quick access to desired information
- ▶ glossary of unfamiliar terms
- ▶ references providing an introduction to the literature of the field
- ▶ suggestions for further reading
- ▶ uniform terms, abbreviations, symbols and units

www.physics-encyclopedia.com

- ▶ Visit the website above for complete content details, sample articles and information regarding how to order the online *Digital Encyclopedia of Applied Physics*. Click on "How to Order" for licensing details and to determine the price available to your organization.

The Digital Encyclopedia of Applied Physics – Your quick reference in modern physics!

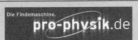  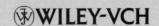